Reviews of *Ancient Boats in North-West Europe:*

"This book is outstanding, of an importance in its field rare in any single publication"

Basil Greenhill, ANTIQUITY

"The result of this accumulated scholarship is truly a manual, indeed an encyclopaedic manual, of every type of craft of N.W. Europe: log rafts, logboats, bark boats, wooden plank boats, bundle rafts and boats, hide boats and float rafts, not only from their varied methods of construction, but also, as the author is a master mariner, of their means of handling. Special sections are devoted to seamanship, pilotage and their employment. The bibliography is particularly rich and the index is a model of perfection. ...this book is, from the very moment of its publication, a classic"

Lucien Basch, THE MARINER'S MIRROR

"De très nombreux tableaux de synthèse, une abondante bibliographie (plus de 700 titres), un index des sites (d'une grande utilité), un index général, viennent remarquablement compléter l'ouvrage de S. McGrail qu'aucun archéologue ne peut désormais ignorer."

Eric Rieth, NEPTUNIA

"This remarkable book is a compendium of information relating to early water transport of all types, up to and including the later Middle Ages: their construction, performance, propulsion and navigation. ...The result is an encyclopedic technical handbook of enormous value"

Andrew Saunders, INTERNATIONAL JOURNAL OF MARITIME HISTORY

"The author of this work has a thoroughly well deserved reputation for exceptional expertise as a marine archaeologist. This book - extremely well researched, finely presented and most intelligently illustrated, and with a bibliography of nearly 700 books and articles – can only add to that reputation."

John de Courcy Ireland, PERITIA

"Es ist wegen der im Text enthaltenen Anmerkungen und des sehr komprimierten Inhalts wegen keine leichte, aber eine sehr lohnende Lektüre und ein wertwolles, wenn auch nicht billiges (sicher viele billige Schriften zum Thema ersetzendes) Nachschlagewerk und Handbuch."

A. Göttlicher, DAS LOGBUCH

"the chapters on seamanship, pilotage and navigation are masterly... an incomparable source for the student"

Margaret Rule, ANTIQUARIES JOURNAL

"There is a substantial literature relating to ships of the Mediterranean, but less attention has been paid until comparatively recently to the ships of the northern world, and there has been no major synthesis of the results obtained in that region. That omission has been filled with the appearance of *Ancient Boats in North-West Europe*. This magisterial survey..."

BRITISH ARCHAEOLOGICAL NEWS

Longman Archaeology Series

Advisory editor: Barry Cunliffe

Already published:
M. Aitken *Science-based dating in archaeology*
Richard Bradley *The social foundations of prehistoric Britain: themes and variations in the archaeology of power*
Seán McGrail *Ancient Boats in North-West Europe: the archaeology of water transport to AD 1500*
D. P. S. Peacock *Pottery in the Roman world: an ethnoarchaeological approach*
D. P. S. Peacock and D. F. Williams *Amphorae and the Roman economy: an introductory guide*
R. F. Tylecote *The early history of metallurgy in Europe*

Seán McGrail
Formerly Chief Archaeologist, National Maritime Museum, Greenwich;
and Professor of Maritime Archaeology, University of Oxford

Ancient Boats in North-West Europe

The archaeology of water transport to AD 1500

Routledge
Taylor & Francis Group
LONDON AND NEW YORK

First published 1987 by Addison Wesley Longman Group Limited

Published 2014 by Routledge
2 Park Square, Milton Park, Abingdon, Oxon OX14 4RN
605 Third Avenue, New York, NY 10017

First issued in hardback 2016

Routledge is an imprint of the Taylor & Francis Group, an informa business

ISBN 13: 978-1-138-17480-1 (hbk)
ISBN 13: 978-0-582-31975-2 (pbk)

British Library Cataloguing in Publication Data

A catalogue record for this book is available from the British Library

Library of Congress Cataloging-in-Publication Data

McGrail, Seán.
Ancient boats in NW Europe.

(Longman archaeology series)
Bibliography: p.
Includes index.
1. Shipbuilding – Europe – History. 2. Shipbuilding – Europe, Northern – History. 3. Ships, Ancient – Europe. 4. Ships, Ancient – Europe, Northern. I. Title. II. Series.
VM55.M36 1986 623.8'21'0936 85–24022
ISBN 0–582–49267–X

Set in Linotron 202 9/11pt Bembo

Contents

List of figures

List of tables

Glossary

Several of the definitions given below are illustrated in the relevant chapter. The reader should find Figures 3.1, 3.3, 4.2, 8.36, 12.13 and 13.1 especially useful.

baulk: a tree trunk which has been roughly squared

beam tie: a transverse strengthening member at the ends of a logboat, may be in the form of a *crook*

bevel: a surface which has been angled to make a fit with another

bilge: region between the sides and the bottom of a boat

blind fastening: one in which the point of the nail does not protrude through the timber

bole: main stem or trunk of a tree

bottom boards: lengths of timber fastened together and laid over the bottom of a boat as flooring

braces: lines to trim yard

brail: rope used to bundle a sail rapidly

bulkhead: a transverse partition which divides the boat into compartments

carvel build: there are several definitions in use: in particular, this term is sometimes taken to be synonymous with skeleton build with flush-laid strakes. The term is confused and ill defined.

caulk: to insert material between two members and thus make the junction watertight. Whether this is done before or after planking is fastened may be an important diagnostic trait.

ceiling: lining of planking fastened over *floor timbers*

clamp: a device for holding elements of a boat together (temporarily)

cleat: a projection to which other fittings may be fastened

clench, rivet: to deform or turn the end of a fastening so that it will not draw out – may be done over a *rove*

clinker built: a form of boatbuilding in which the strakes are placed so that they partly overlap one another - usually upper strake outboard of lower strake, but occasionally the reverse arrangement is found

construction plan: a scale drawing of a boat with a longitudinal section, horizontal plan, and several transverse sections. The position and nature of the scarfs, and other important constructional details and scantlings, may also be given.

crook: a curved piece of wood which has grown into a shape useful for boatbuilding

crossbeam: a timber extending across the vessel

deadrise: angle at which the bottom planking lies to the horizontal

dolly: a metal billet held against the head of a boat nail whilst it is being clenched

double-ended: a boat which is (nearly) symmetrical about the midship transverse plane

draft (draught): 1. the vertical distance between the waterline and the lowest point of the hull. 2. an alternative term for *lines*.

draw-tongue joint: a method of fastening flush-laid planking in which free tenons are fitted into mortices cut in the meeting edges of adjacent planks; after the planking is assembled the tenons are generally pierced by two treenails, one through each plank. Sometimes known as 'mortice and tenon'.

feather-edge: tapering to nothing

fibre saturation point: a theoretical stage in wood-water relations when all the water (free water) has been removed from the cell cavities, but none from the cell walls. For most timbers this is at a *moisture content* of 25 per cent to 30 per cent

flare: the transverse section of a boat increases in breadth towards the *sheer*

floor timber: a transverse member often a *crook* - extending from turn of *bilge* to turn of bilge, and set against the planking (see *frame*)
flush-laid: planking in which adjoining *strakes* are butted and do not overlap
frame: a transverse member made up of more than one piece of timber, and set against the planking (see *rib* and *timber*)
freeboard: height of sides above the waterline

garboard: the *strake* next to the keel
gripe: 1. the tendency of a sailing vessel to come up into the wind 2. the forefoot of a stem
grommet: strand(s) of rope layed up in the form of a ring
guares: a retractable wooden device for combating leeway and for steering

halyard: line to hoist and lower yard and sail

interference fit: said of a *treenail* in a hole or a tenon in a mortice when the wood fibres interlock

joggle: to cut out a notch in a piece of timber so that it will fit close against another member

keel: the main longitudinal strength member, joined to the stems forward and aft
knee: a naturally grown *crook* used as a bracket between two members set at about right-angles to each other

land: that part of a strake which is overlapped by the strake immediately above it - in *clinker* building
layoff: to draw out the lines of a boat full size
lines: the interrelation of sections in different planes which show the shape of a boat's hull. They usually consist of: (a) sheer plan with longitudinal section; (b) half breadth plan with waterlines or horizontal sections; (c) body plan with transverse sections. Diagonal lines, longitudinal section lines on the half-breadth plan, and waterlines on the sheer plan, enable the three plans to be related to each other and checked for fairness. Lines converted to numbers are known as a Table of Offsets.
loom: that part of an oar inboard of the point of pivot. The section of an oar between the loom and the blade is called the shaft.

mast step: fitting used to locate the heel of a mast
metacentric height: distance from the metacentre to the centre of gravity of a loaded boat
moisture content: the weight of water in a specimen of wood expressed as a percentage of the weight of oven drywood. Thus the figure can be greater than 100%
moulds: transverse wooden patterns giving the internal shape of a vessel

painter: light head rope by which a boat is made fast; mooring rope
pay: cover plank seams with a layer of hot pitch, or to coat a ship's bottom with tar or other waterproofing substance
pith: the middle core of a *bole*
plank: a component of a *strake* that is not all in one piece

rabbet, rabet, rebate: a groove or channel worked in a member to accept another, without a lip being formed
radial plane: a longitudinal section of a *bole* through the pith and at right angles to the growth rings
rays: layers of parenchyma cells in horizontal strands running out from the centre of a tree towards the circumference
rib: a simple form of frame. This term may by more appropriate than *frame*, when applied to small open boats.
rocker: fore-and-aft curvature of *keel* or bottom of vessel
rove, roove: a washer-like piece of metal, which is forced over the point of a nail before it is *clenched*
rubbing strake: an extra thick strake (*wale*); or a strake fastened outboard of an existing strake, near the top of a vessel's sides to protect the hull when alongside another vessel or a waterfront

scarf, scraph, scarve: a tapered or wedge-shaped joint between pieces of similar section at the join
seam: juncture of two members required to be watertight
sheer, sheer line: the curve of the upper edge of the hull
sheer strake: the top *strake* of planking
sheet: line used to trim the foot of a sail
shelf (rising): a longitudinal timber or *stringer* fastened inside the planking and/or framing of a vessel to support the ends of *crossbeams*
shore: stout timber used to support (part of) the hull of a vessel, internally or externally, when she takes the ground
shrouds: ropes leading from the masthead to the sides of the boat to support the mast athwartships
spile: to transfer a curved line on to a pattern which, when laid flat, will give the shape to cut the lower edge of a plank

stabilisers: external longitudinal timbers fastened to a boat's sides at the loaded waterline to increase transverse stability

stays: ropes leading from the masthead forward and aft, to support the mast

stocks, set-up: the temporary wooden support on which a boat is built

strake: a single *plank* or combination of planks which stretches from one end of a boat to the other

stretcher, foot timber: an athwartships length of timber against which a rower braces his feet

stringer: a longitudinal strength member along the inside of the planking

tangential plane: a longitudinal section of a *bole* at right angles to the radial plane and tangential to the growth rings

thole: a pin projecting upwards at sheer level to provide a pivot for an oar

thwart: a transverse member (*crossbeam*) used as a seat

timber: used generally referring to any piece of wood used in boatbuilding. One piece *ribs* or *frames*, especially those steamed or bent into place, are frequently called timbers

transition strake: the strake at the transition between bottom and sides of a boat - especially when there is a marked change in the boat's transverse section (*ile*; chine girder)

transom: athwartship bulkhead; in this text it is normally applied to a fitted bulkhead at the stern or the bow

treenail, trunnel: wooden peg or through fastening used to join two members. It may be secured at each or either end by the insertion of a wedge

wale: a strake thicker than the rest

washstrake: an additional strake normally fitted to increase freeboard and to keep out spray and water

Symbols used in the text[1]

A	cross section area(m^2)	normally at midship station
F	freeboard (m)	
T	draft (m)	
D	depth of hull (m)	
L_{OA}	length overall (m)	
L_{WL}	length of waterline (m)	
B_x	maximum beam (m)	
$\overline{KB}$	distance 'keel' to centre of buoyancy (m)	
$\overline{KG}$	distance 'keel' to centre of gravity (m)	
$\overline{BM}$	distance centre of buoyancy to metacentre (m)	
$\overline{KM}_T$	distance 'keel' to transverse metacentre (m)	
$\overline{GM}_T$	transverse metacentric height (m)	
I_T	transverse moment of inertia	

Note: 'keel' is taken to be the lowest point of the hull.

V	speed (kn)
W	deadweight (kg)
Δ	displacement force (kg)
∇	displacement volume (m^3)

Note (a) The following abbreviations appear in the text:

HW/LW high water/low water
fsp fibre saturation point
kn knot
mc moisture content
nm nautical mile
rh relative humidity

(b) The following convention is used: ad, bc date in radiocarbon years
AD, BC date in calendar years

1. The boat geometry symbols are after Rawson and Tupper 1976. See also Figures 3.1 and 3.3.

Acknowledgements

During the 13 years I have been at the National Maritime Museum I have benefited greatly from discussions with and advice from a number of people. I am especially grateful to: Dr. Carl-Olof Cederlund, National Maritime Museum, Stockholm; Dr. Arne-Emil Christensen, Universitetets Oldsaksamling, Oslo; Dr. Ole Crumlin-Pedersen, Viking Ship Museum, Roskilde; Dr. Detlev Ellmers, Deutsches Schiffhartsmuseum, Bremerhaven; Dr. Basil Greenhill, Director of the National Maritime Museum until 1983; the late Harold Kimber, master boatbuilder of Highbridge, Somerset; the late Eric McKee, formerly Caird Research Fellow, National Maritime Museum; Dr. Wolfgang Rudolph, Academie der Wissenshaften der DDR; Christopher Daniel, Alan Stimson and Alan Viner of the National Maritime Museum; Sir David Wilson, Director of the British Museum and my supervisor at University College London 1972–7. Discussions with my colleagues in the Archaeological Research Centre at Greenwich have also been rewarding.

Others have not only helped me in the past but have given specific help with this book by reading and commenting on early drafts of certain chapters: John Coates of Bath, formerly RCNC, on naval architectural matters; Owain Roberts of Angelsey and Eric Kentley of the Archeological Research Centre on boatbuilding; David Waters formerly Deputy Director National Maritime Museum, Greenwich, and Alan Binns, formerly University of Hull, on navigation and pilotage; and Dr. Richard Clarke of the Archaeological Research Centre on wood science. Professor Barry Cunliffe suggested that I should write the book and has been a source of encouragement throughout.

I am grateful to the Trustees of the National Maritime Museum for permission to reproduce illustrations from the Museum's collections, and to Mrs. Marilla Fletcher who typed two complete versions of the book and three drafts of the more difficult parts. Finally, I wish to thank my wife and family for their forebearance during the past three years of working weekends and vacations.

The publishers have attempted to contact the copyright owners of figs 7.4, 8.46, 11.2, 11.3, 12.26 and 12.27 but have been unsuccessful and would appreciate any information that would enable them to do so.

October, 1985

ACKNOWLEDGMENTS

Preface to the paperback edition

I have taken the opportunity afforded by this paperback edition to correct misprints and, within the original pagination, to make minor alterations to the text where the meaning can be clarified. A revised diagram replaces the original Fig. 1.1, Maritime Archaeology, and the drawing of the Hjortspring plank fastenings in Fig. 8.31 has been corrected. A handful of references inadvertently omitted from the first edition have been re-instated, and a small number of topics, previously overlooked, have been indexed. Recent publications have been added to the Index of Sites.

Ancient Boats in North-West Europe is not a conventional narrative history of water transport, as two reviewers of the first edition appear to think it should have been. Such a book cannot yet be written, except as a series of incomplete, non-continuous episodes; and it was not my intention to write such an account. On the other hand, early water transport is clearly the focus of this book, dealing, as it does, with the known finds, and with other simple forms of water transport – known elsewhere and at other times – that we may yet find in the archaeological record of north-west Europe.

The book's prime concern is with the *technical* aspects of these ancient rafts, boats and ships: the raw materials used; 'design' methods; how vessels were built; the nature of the environment in which they were used; methods of propulsion & steering; what sort of speeds they could achieve; how close to the wind they could sail; how much cargo or how many men they could carry; how they were navigated across the trackless sea; and how landing places were chosen. Although plank boats and ships are dealt with in some detail, much emphasis is placed on rafts and non-plank boats, since these are the craft most likely to have been used in earliest times about which we know least. A main aim has been to increase understanding of the ancient boatwright's craft by, for example, presenting the principles behind the choice of hull shape, and the reasoning on which 'design' methods were evidently based, and by identifying the common features that characterise individual parts of rafts and boats. In this search for pattern and regularity of behaviour, by both boatwrights and seamen, it was necessary to seek analogous evidence worldwide, wherever and whenever there was adequate documentation: hence the book contains ethnographic data on boats and their uses from every ocean of the world.

THE PAPERBACK EDITION

During the past decade, eighty & more articles and books have been published dealing with aspects of the subject matter of this book. This is particularly true of logboat (Ch. 6) and plank boats studies (Ch. 8), since prehistoric, Romano-Celtic & medieval finds have recently been excavated. Conversely, little new has been published about log rafts (Ch. 5), bark boats (Ch. 7), bundle rafts, bundle boats (Ch. 9), hide boats and float rafts (Ch. 10). In between these extremes, there have been several informative publications on the performance, propulsion and navigation of boats and ships, and on ancient seafaring (Chs 11 to 14). These matters are discussed below, generally in the order in which they appear in the book.

SOURCES AND TECHNIQUES (CHAPTER 1)

Ethno-archaeology The documentation of rafts and boats of present-day, non-industrial, small-scale societies, has increasingly been recognised as of great value to the study of ancient water transport

(McGrail, 1996: 231–2; Blue, forthcoming A). Such ethno-archaeological studies encourage the boat archaeologist to escape the mental bounds of his own twentieth-century culture and become familiar with technologies beyond his immediate experience. He is then better equipped to interpret excavated material, identify the function of unusual artifacts, and make hypothetical reconstructions of the original form and structure of a boat-find.

Fieldwork on the *masula* and allied sewn plank boats of Sri Lanka and the east coast of India by Dr Eric Kentley (1985; 1987; 1993; 1996) has shed new light on the building and (especially) on the use of the Bronze Age plank boats of N.W. Europe. Other recent ethno-archaeological work in India is leading to a re-appraisal of the evidence for reverse-clinker planking in Medieval Europe (Blue, 1997), and has added a new dimension to research into early frame-first boats and ships (Blue, forthcoming B).

Historical research Gillian Hutchinson (1994) has combined historical and representational evidence with excavated evidence from eleventh- to sixteenth- century Atlantic Europe to compile an account of medieval ships and shipping with special reference to England; whilst Dr Ian Friel (1995) has based his research on financial accounts and inventories concerning English ships and shipbuilding from the early thirteenth to the early sixteenth century.

Dendrological research Dendrological examination of timber from boat finds has just about become a standard procedure: of recent major finds, only the prehistoric sewn plank boat from Dover (Parfitt, 1993) and the Romano–Celtic ship from St Peter Port, Guernsey (Rule, 1993) have not been examined in this way. Dendro-curves, based on tree ring widths measured during such examinations, cannot always be dated because of lack of early regional master curves. Nevertheless, dendrological examination adds tree ring measurements to regional databases, and has the further advantage of revealing much about the growth conditions of the trees used to build the boat, their size and age when felled, and how they were converted into useful timbers. If a boat's dendro-curves can be matched with a master chronology then not only can the boat be dated with greater precision than by radiocarbon assay, but the regional origin of its parent trees is also known.

The Bronze Age sewn plank fragments from Goldcliff in south-east Wales (Bell, 1992) have a dendro-date, but other British prehistoric boat finds from the Humber and Severn regions cannot yet be dated in this way, although site dendro-curves exist.

Dating technological change With an increasing number of rigorously recorded and well-dated boat finds, fine-dating of technological change has become a realistic possibility. The large logboat from Hasholme, north of the Humber estuary, (Millet & McGrail, 1987), with an array of woodworking techniques, has been dendro-dated to 287 – 322 BC. This has led to a revision of earlier work in dating boatbuilding techniques (McGrail, 1987/8). The value of this nautically-derived information would be enhanced if similar data from other wooden artifacts and structures could be blended with it: regretably, there is no sign yet of such research being undertaken by conventional archaeologists.

CLASSIFICATION STUDIES (CHAPTER 2)

The provisional classification scheme published in Chapter 2 of the first edition has not yet been appraised by other archaeologists. Some form of agreed scheme is undoubtedly essential if progress is to be made in comparative studies of excavated and ethnographic boats.

Further study of publications by Cairo (1972) and by Aubaile-Sallenave (1987) has suggested that SE Asian basket boats should be re classified (tables 2.1 and 2.2). The closely woven bamboo structure which forms the main hull of these boats is *not* watertight, as was thought, but has to be payed inside and out with a resin mixture. Conceptually these basket boats are therefore skeleton-built and in Class C12. The basket boats of Arabia (for example, the *tarada*) are made using coiled basketry rather than the woven basketry of the SE Asian boats; they too have to be fully payed, in this case with tar, and are also in class C12.

STABILITY AND SEAWORTHINESS (CHAPTERS 3, 11 & 12)

A revised edition of the naval architectural textbook *Basic Ship Theory* has been published in paperback (Rawson & Tupper, 1983/4).

Ballast has been found with several early wrecks, but its uses and its relevance to stability were insufficiently emphasised in the first edition of this book. Ballast may have been of rubble, rocks, gravel, shingle or sand, or in the form of high cargo density/low stowage factor tradeable goods such as tin ingots, lead pigs, iron bars, granite millstones or marble. Ballast of one or both types would almost certainly have been necessary to maintain the stability of all but the smallest of early cargo-carrying craft. It would have been especially necessary for vessels under sail, not only for stability reasons but also to improve sailing perfomance by adjusting the boat's trim (McGrail, 1989). Although some natural ballast might remain in a vessel throughout its life, set in a dunnage of sand, hopleaves, brushwood, moss and the like, to protect the ship's structure, other (non-tradable) ballast could be discharged into a harbour when a relatively high density cargo was to be carried; or extra ballast could be loaded from the harbour before a low density cargo was brought onboard. This recycling means that an excavated ship need never have been to the region suggested by the provenence of its ballast (McGrail, 1991).

TREES AND ROPES (CHAPTERS 4 & 12)

Data on the parent trees of early boat and ship timbers is now available for the prehistoric period (Millett & McGrail, 1987), the Roman period (Nayling and McGrail, forthcoming) and medieval times (McGrail, 1993).

Prehistoric ropes of yew withies, split willow, lime bast, clematis and nettles are noted in Ch.4. To the data on medieval ropes in Ch.12 may now be added information from a recent publication by Crumlin-Pedersen (1997: 188-190). About three-quarters of the rope samples taken from eleventh– to fourteenth-century Danish ship-finds proved to be of lime bast: the remainder were of oak or elm bast, with a few samples of horse-hair. The earliest find of hemp rope is from the fourteenth-century Vejby cog. Although ships' ropes of seal and whale hide are mentioned in the ninth-century account of Ottar/Ohthere's voyage along the Norwegian coast (Lund, 1984:20), no examples have so far been excavated.

LOGBOAT STUDIES (CHAPTER 6)

The Hasholme logboat of 322 to 277 BC, excavated in late 1984, was briefly noted in the first edition. This was a large (12.78 × 1.4 × 1.25m.) oak logboat of some complexity, with a composite bow, an inserted transom stern with beam-ties, washstrakes fastened to the main hull by wooden keys or 'cotter pins', and dovetail-shaped repairs (Millett & McGrail, 1987). These features, particularly the 'design' and fashioning of the bow, are evidence for a standard of woodworking and boatbuilding hitherto unsuspected in Iron Age Britain.

Studies of the logboats of several European countries have recently been published: central Europe (Arnold, 1995/6); Denmark (Christensen, 1990; Andersen, 1994); and Scotland (Mowat, 1996).

Crumlin-Pedersen (1991) has published about the first to third century AD boat graves at Slusegard, Bornholm; many of these burials were in oak logboats which Crumlin-Pedersen believes were expanded. Some of the nine criteria he used (1991: 254) to identity expansion are qualitative rather than quantitative and are therefore difficult to apply objectively. Furthermore, these are theoretical criteria and it has not yet been demonstrated that oak logboats, unlike those of more malleable species such as aspen, can be sucessfully expanded: the research reported by Gifford (1993) is inconclusive since only small sections of oak were used, whereas the critical task in expansion is to prevent the ends of a hollowed log splitting.

Crumlin-Pedersen also identified expanded logboats from other European countries, including three from Britain, which he considered had been expanded because they had inserted ribs (1991: 261). However these three logboats (Stanley Ferry, Smallburgh & Walton) have other features which show that it is unlikely that they were expanded: all have sides and bottoms which are up to 60mm thick, whereas the criteria require a "thin" hull; the Stanley Ferry boat has flared sides which meet the bottom in a chine, and such an angular section is incompatible with expansion. The support of washstrakes or thwarts, or use as footrests are more likely reasons than expansion for fitting ribs in these three boats.

PLANK BOATS AND SHIPS (CHAPTER 8)

Authors of two recent books deal with plank boats and ships in structural terms: Greenhill (1995) in an enlarged and revised edition of a book first published in 1976; and Steffy (1994) in a more technical work, but one that is more restricted in scope. The first three volumes in Conway's *History of the Ship* (Christensen, 1996; Morrison, 1995; Unger, 1994) contain articles by several authors concerning (mainly European) boats and ships from 'earliest times' to the mid-seventeenth century.

Sewn plank boats There have been further fragmentary finds of Bronze Age sewn plank boats: on the northern foreshore of the Humber estuary at

Ferriby (Wright, 1993, 1994) and at Kilnsea (Middleton, 1996); on the Welsh side of the Severn estuary at Caldicot (McGrail, 1997) and at Goldcliff (Bell, 1992); and west of Southampton, near the mouth of the River Test. A major find was made at Dover in 1992 (Parfitt, 1993: Christensen, 1996: 33–4). These British finds are amongst the earliest plank boats in the world, outside Egypt.

The Brigg 'raft' (McGrail, 1981; 1985) has been re-appraised by Owain Roberts (1992) who concluded that it had been a round-bilged boat capable of short sea crossings rather than a flat-bottomed ferry used on the middle reaches of a Humber tidal creek. This was followed by a paper arguing against this revision (McGrail, 1994), and a further response by Roberts (1995).

Romano-Celtic vessels A Romano-Celtic tradition of boat and ship building was tentatively identified in the first edition, based on earlier work by Marsden (1967) and Ellmers (1969). The recent excavation of an estuary boat at Barland's Farm, Magor, on the Welsh side of the Severn (Nayling, 1994), the publication of the ship from St. Peter Port, Guernsey (Rule & Monaghan, 1993), and the re-assessment of Blackfriars 1 ship (Marsden, 1994) confirmed this identification and emphasised the importance of this tradition, not only in its own right, but also in the development of medieval ship types (McGrail, 1995).

Early frame-first vessels These three Romano-Celtic seagoing and estuary vessels of the second and third centuries AD were built by frame-based methods: first some framing; then some planking; and so on (McGrail, 1995). Their hull shape was determined by the framing and not by planking; their builders were frame-orientated. First to fifth century AD boats from the Rhine region were also of this Celtic tradition (McGrail, 1995: 139; Crumlin-Pedersen, 1997: 187). However, Hockmann (1997: 244) has claimed that the fifth century AD boats from Mainz were 'not built frame-first' but were built 'mould-first': that is, the planking was fastened to active moulds which were later replaced by frames. But this is mere quibbling: the moulds, not the planking, determined the shape of the hull; the builder was as much frame-orientated as any other builder of a Romano-Celtic vessel; the Mainz boat were 'frame based' even if the frame-shaped moulds were subsequently replaced by permanent framing.

Celtic boatwrights and shipwrights are the first known who evidently visualised their vessels in terms of the framing rather than the planking, and built them frame-first. Subsequently this innovative technique becomes archaeologically visible, albeit in a variant form, in the sixth to eleventh-century Mediterranean (McGrail, 1997: fig.1) and from there appears to have spread into Atlantic Europe (McGrail, 1996: 78–80). The 'design' of the framework of such frame-first vessels was first described in the early fifteenth century (Bellabarba, 1993; 1996); known to us as 'Mediterrancan Moulding', this technique enables the shape of the main frames to be derived by simple geometrical means from a master frame (Sarsfield, 1984;1988; Rieth, 1995).

The frames of the European ships which sailed 'all the seas of the world' in the late fifteenth / early sixteenth centuries, appear to have been 'designed' by such lofting methods (Bellabarba, 1993) and similar procedures have recently been found in use in Brazil (Sarsfield, 1988). Comparable methods are still used in Tamil Nadu, on India's eastern coast, where small boats (*vattai, vattal* & *vallam*) and ships of 250 to 600 tonnes cargo capacity (*thoni* & *dinghi*) are built by "traditional" means, even though some have engines installed (Blue, forthcoming B).

Units of measurement De Weerd (1988; 1990; 1994) believes that he has identified a Roman unit of measurement, the *pes monetalis*, in the frame spacing used in several Romano-Celtic vessels. As Arnold (1990) has pointed out, however, precise measurement cannot be derived from the small-scale drawings studied by de Weerd, and the sequence of frame assembly postulated in order to derive a 'standard unit' is hardly practicable. De Weerd's claim to have identified a known unit cannot be sustained: the most that can be deduced is that a unit equivalent to *c.* 0.30m may have been used. Rule (1993: 29) has identified a unit of *c.* 0.56m in the St Peter Port ship; whilst the unit used in the Barland's Farm boat was *c* 0.55m: it seems likely that these measurements were multiples, not of a legal unit, but of one of the natural units of measurements described in Ch. 8 of this book – i.e. approximately two human feet.

Medieval boats and ships Crumlin-Pedersen's hypothesis (1978; 1997: 28) that a certain number of boat types in the N.W. European Iron Age formed the basis for the construction of the three main types of medieval seagoing ship – the Nordic ship, the cog and the hulc – remains unchallenged in general, although more emphasis than heretofore is now placed on the role of the Romano-Celtic tradition in this transformation (McGrail, 1995).

The Nordic tradition – building by eye? How the builders of medieval plank-first vessels, such as those of the Nordic tradition, got the required shape of planked hull is as unclear as how the Romano-Celtic frame-first builders got the shape of their frames. Was the fourth century AD plank-first Nydam boat built entirely "by eye", or were simple building aids used which, perhaps in signs, lengths and angles, encapsulated the shape of the boat?

It is possible to visualise building a boat of Nydam size and relative simplicity without such aids and, from the evidence of plank-first boatbuilding in 20th century West Norway (Christensen, 1972; Crumlin-Pedersen, 1997: 14–5) and other parts of the modern world (Greenhill, 1995:26–72), we know that building a *boat* by eye, with the very minimum guidance on shape, is practicable. Whether 12th and 13th century *ships* could be built this way is questionable; whether such large ships as the early-15th century royal vessel *Grace Dieu* could be built by eye is doubtful. Yet this ship, now a wreck in the River Hamble, (Friel, 1993; Clarke, 1993; McGrail, 1993) was edge-fastened and evidently built 'plank-first' and had other Nordic characteristics on a grand scale with three thicknesses of planking fastened in the clinker manner. For such a large and complex ship as *Grace Dieu*, building without aids would seem to be impracticable: a ship ell, a builder's level, full size moulds, scale models or scale drawings are aids that might have been used, but, to date, there is no archaeological or documentary evidence for them before post-medieval times. Undoubtedly there is much scope for further research.

Parts of boats and ships The investigation of timbers from dismantled medieval boats and ships in Bergen (Christensen, 1985) and in Dublin (McGrail, 1993) has led to the compilation of an attribute list for clinker planking and a standardised scheme for describing structural parts of vessels, and fittings for propulsion and steering, by wood science characteristics & woodworking features & by significant angles, dimensions and ratios (McGrail, 1993: 4–31, 169–71). Use of such a scheme should ensure that timbers are objectively recorded, classification undertaken, and comparisons made between timbers from different sites. In this way, by analogy with timbers in near-complete vessels, the function of loose timbers may be deduced and the size of the parent vessel estimated.

It has been argued (McGrail, 1993: 11) that, within a particular tradition, there is a direct correlation between the size of a vessel (as represented by its overall length) and the size of its individual timbers (as represented by the product (M × S) of each timber's moulded (M) and sided (S) dimensions at a significant cross section): the bigger the vessel, *ceteris paribus*, the more massive her timbers. It also seems possible to infer the approximate size of parent vessel from thickness of planking, plank scarf gradients, and sizes of iron nails, roves and treenails (McGrail, 1993: table 1).

Crumlin-Pedersen (1997: 107–8, 111–2, 116–9) has argued against this method, stating that the (MS) product, or 'McGrail numeral' as he styles it, should 'be abandoned', and that it 'is of little use in determining the size of the parent vessel'. In support of these statements he has published two graphs, plotting the overall length of parent vessel against (i) the (MS) of keels (fig. 5.6) and (ii) the (MS) of frames (fig. 5.13). In fact, these two plots demonstrate clearly that, within the seventh to twelfth century Nordic tradition, there is indeed a direct relationship between each of those two (MS) products and the length of parent vessel. Statistically, there is a sufficiently strong positive correlation in each case to justify using regression equations to predict the probable overall length of parent vessel from the (MS) of loose frames or loose keels. These equations are approximately:

Frames: (M × S) in cm^2 = 22 + 5 (Loa in m)
Keels: (M × S) in cm^2 = 17 + 19 (Loa in m)

A table of the values for (MS) and Loa (rather than the small scale graphs so far published) would enable more accurate equations to be calculated and presented in terms of forecasting Loa from (MS).

In fact, despite his dismissal of the 'McGrail numeral', Crumlin Pedersen uses it: to deduce the size of Hedeby 2 from the size of excavated frames (1997: 98); to distinguish between frames from Nordic ships and those from cogs (1997: 117); to deduce the size of the parent vessel of a frame found at Gretstedbro (1997: 291); and to differentiate between cargo ships and warships (1997: 300).

Cogs and Hulcs Although there are now sufficient examples of excavated cogs for the general characteristics and the main variants of this tradition to be defined, once they are all fully published, the hulc remains virtually invisible archaeologically. Some of the medieval depictions of what are thought to be hulcs appear to show that the planking is reverse-clinker: that is, the upper strake overlaps *in*board of the lower. Apart from some sixteenth-century fragments recorded by Marsden (1996: 136–144), vessels with reverse-clinker planking have not yet been excavated. However, there is no doubt that this method is practicable: boats in the Sylhet and Khulna regions of Bangladesh have reverse-clinker planking (Greenhill, 1971: 101, 104); the *patia* fishing boat of Orissa, on the east coast of India, is built in this way; and there is representational evidence for the use of reverse-clinker planking in this region back to the 18th/19th century and possibly much earlier (Blue, 1997).

ASSESSMENT OF PERFORMANCE (CHAPTER 11)

The basic concepts used in the assessment of performance are dealt with in the third edition of a standard text on naval architecture by Rawson & Tupper (1983/4) and in the second edition of Marchaj's book (1996) on sailing. Computers are being increasingly used to speed the process of estimating performance (Wilson, 1994) and Marsden (1994/1996) has published computer-based assessments of second- to fourteenth-century AD boats and ships excavated from the Thames in London. The practical problems of building full-size experimental reconstructions of excavated vessels, and undertaking sea trials, have been discussed by Crumlin-Pedersen (1996); see also the proceedings of a conference on these matters (Crumlin-Pedersen & Vinner, 1986). The crucial stage in both theoretical and practical assessments is the formulation of a hypothetical reconstruction, as a scale drawing or model, of the structure, propulsion and steering of the ancient vessel (McGrail, 1992; Coates, 1995). Computer-aided design methods are now used in this process

Performance comparison between vessels are done at standard drafts of 60 per cent (cargo ships) and 50 per cent (warships) of the vessel's height of sides amidships (Crumlin-Pedersen, 1997: 103, 202; 308). The stowage factors (the inverse of bulk density) of the goods thought to have been carried need to be considered when assessing the cargo capacities of ancient vessels since, with certain combinations of goods, stability may be inadequate; moreover, if a cargo consists mainly of high stowage factor goods it may not be practicable to use all available cargo space (McGrail, 1989).

SEAFARING AND NAVIGATION (CHAPTERS 13 AND 14)

Early seafaring cannot be understood without reference to the contemporary maritime environment, a principle determinant of which is the sea level (Flemming, 1996). A summary of the evidence for sea level and coastline changes in British waters over the past 5000 years has been published by Tooley (1990). The problems of deducing the region of origin of a wrecked ship and her route on her last voyage have been tackled both in a theoretical way (McGrail, 1991) and in relation to specific wrecks (L'Hour & Veyrat, 1989, 1994; Crumlin-Pedersen, 1997: 94, 98–9, 104).

There have been a number of publications on navigation without instruments: in the South Pacific (Irwin, 1992; Lewis, 1994; Finney, 1994); in the Indian Ocean (Arunachalam, 1996); in early medieval Irish waters (McGrail, 1989); in medieval northern Europe (Schnall, 1996); and in the Classical Mediterranean (McGrail, 1996). There is a striking similarity in the methods used which are all based on perceptive observation of the maritime environment. This tendency to reach similar solutions to universal problems is an aspect of the common inheritance of humanity.

Early Arab seamen in the Indian Ocean appear to have used hand spans and finger breadths when estimating the altitude (angular height) of heavenly bodies (Fatimi, 1996: 283). One of the earliest instruments may have been a wooden tablet or stick with which such altitudes could more accurately be measured. Fatimi (1996: 284) has traced the first documented use of such a tablet, subsequently known as the *kamal*, to the 9th/10th century AD.

At about the same time, the Vikings of northern Europe may also have begun to use instruments for navigation. Vebaek and Thirslund (1992) have interpreted lines inscribed on a wooden disc, excavated from a Viking age site in Greenland, as gnomon curves, and V-shaped notches as compass points. Such a disc, when fitted with a gnomon to cast a shadow, could have been used as a sun compass.

June, 1997.

For Brigid, Dominic, Frances, Joseph,
Luke, Oliver, Robert, Samuel and Tomas

1 Introduction

In 1946 James Hornell, marine biologist by profession but nautical historian by inclination, published the distillation of a lifetime's work in *Water Transport*. In that book, subtitled 'origins and early evolution', Hornell set out 'to marshal in due order the major part of the knowledge within our ken concerning the origins of the many devices upon which men, living in varying stages of culture, launch themselves afloat upon river, lake and sea' (1946: xv). Preoccupation with the idea of tracing the 'evolution' or 'development' of the various types of boat has bedevilled many otherwise objective studies and Hornell's work was similarly biased. Nevertheless, although he may not have written a history of the early use of water transport that would satisfy a prehistoric archaeologist or an ancient historian, Hornell *did* compile, in this book and in his many journal articles, a matchless record of the 'many devices' nineteenth and twentieth century men 'in varying stages of culture' had used, and in some cases were continuing to use; it was thus principally ethnographic in content.

Notwithstanding Hornell's zeal and undoubted ability, it is questionable whether his aim of establishing origins was attainable – the information available and the techniques used at that time were inadequate: indeed such a task would only be slightly less difficult today. I have chosen to include Water Transport in the title of this book, not because I wish to achieve his unattainable aim, nor because I wish to update or fill out his ethnographic reports, but because the two words succinctly identify a whole range of artifacts, each one different from the remainder yet all with sufficient in common to form a group which can be readily recognised by layman and professional alike.

The study of the building and operation of water transport is one of the two foci of the maritime subdiscipline of Archaeology. Until a few years ago I would have written that it was the single focal point of what has been variously called nautical, marine or naval archaeology. But such technological study of rafts and boats, isolated from the study of their use, their geomorphological, climatic and economic environments, and without some consideration of the land-based facilities they require, can easily become overspecialised, uninfluenced by and uninfluencing the rest of the archaeological and historical disciplines. Maritime Archaeology (Fig. 1.1) is more wide-ranging than Nautical Archaeology, but it contains a more natural grouping of topics, internally cohesive and well differentiated at its interface with other research areas, yet with sufficient overlap of interest to ensure some creative conflict and interchange. It may be defined in general terms as

Figure 1.1 The scope of Maritime Archaeology. (Diagram: NMM Greenwich).

the study of man's use of all types of waterways – lakes, rivers and seas (McGrail, 1983).

In a development parallel to the establishment of Nautical Archaeology and its transformation into Maritime Archaeology, urban archaeologists excavating coastal and riverine sites recognised an archaeology of waterfronts, and, as with the artifact *boat*, the study of the structure *waterfront* has been logically extended from mere construction to use. Thus waterfront archaeology has expanded in scope, as may be seen from the titles of papers read at recent waterfront conferences (Milne and Hobley, 1981; Herteig, forthcoming), and has converged on the expanded nautical archaeology. We may now consider these two groups of archaeologists to be studying, from differing viewpoints, the same research field, that of Maritime Archaeology, with two focal points: the boat and the waterfront.

It is too early to write a comprehensive work on Maritime Archaeology, even if restricted to northern and western Europe, and it may be that when the time is ripe such a book could only be a multi-author work by urban and nautical specialists, and possibly an economic historian. A book on water transport is less ambitious, focusing, as this book is intended to do, on the building and use of early rafts, boats and ships with some consideration of maritime structures such as landing places and harbours.

There is indirect evidence for use of water transport from very early times (see Ch. 14), but the question may fairly be posed whether there is sufficient direct evidence to warrant a specialised study. The editor of a recent encyclopedia of archaeology has described his book as 'an attempt to summarise the present state of knowledge over the whole field of archaeological enquiry' (Sherratt, 1980: 9); yet, with the exception of the sections on Australia and Oceania, the maritime dimension of antiquity is scarcely mentioned. Are the distinguished but landlubberly contributors to this encyclopedia ignoring this research area? Or is there no firm evidence to be evaluated? The answer to both these questions must be a qualified 'yes'. The direct evidence for all forms of early water transport is undoubtedly poor and in some cases non-existent, as has been recognised by Clark (1952: 292), Cunliffe (1978: 290) and Hassall (Wilde, 1978: 79) when considering the Bronze Age, the Iron Age and the Roman period in N.W. Europe. On the other hand many archaeologists have failed to evaluate and draw deductions from what evidence there is, and with some notable exceptions – for example Grahame Clark in his *Prehistoric Europe: the economic basis* (1952: 84–90, 282–92) – discussion of the maritime aspects of prehistoric and medieval Europe has been left to non-archaeologists, often with an unsatisfactory outcome.

This book is written from the standpoint of one who was a seaman and is now an archaeologist. I assume that the readership will include seamen and boat enthusiasts on the one hand and archaeologists on the other; not quite mutually exclusive, but certainly disparate, groups who differ in the extent of their knowledge and in their approach to the subject. Archaeologists are generally landsmen, possibly unskilled in the natural and applied sciences, and handicapped by lack of technological knowledge; whereas boat enthusiasts and seamen are often confused about the nature of archaeological evidence (for example its incompleteness and its representativeness or otherwise) and by the constraints on archaeological techniques (for example accuracy of dating methods). A major aim of this book, therefore, is to explain aspects of seamanship to the archaeologist and archaeological matters to the boatman.

The range of this book extends from the late-Palaeolithic to the later Middle Ages in northern and western Europe. Classical ships and shipping are dealt with in the companion volume by Dr Toby Parker and are thus not discussed in detail here. However, as much of the earliest evidence for water transport comes from the Mediterranean region, aspects of Egyptian, Greek and Roman boatbuilding and seafaring are introduced whenever they can throw light on the techniques of N.W. Europe.

Documentary evidence is used when available, but the approach is essentially archaeological and the two sources are not conflated unless it is clear that they are describing the same object or event. Towards the end of the period under discussion increasing documentation on seafaring and shipping becomes available; this requires specialised study closely allied to economic history and such matters are only briefly mentioned here. At about the same time, *c.* AD 1300, vessels which may truly be called ships began to be built in northern and western Europe. Open planked boats, until then the principal form of water transport – albeit of some size and capable of ocean voyages – were generally relegated to local use and replaced for seafaring by vessels developed from the post-Viking tradition and by cogs and hulcs. The study of the complex structure of such ships and their use is outside the scope of this book but aspects are brought into the discussion where they can illuminate some of the features found in the contemporary and earlier boats.

Ethnographic descriptions of material culture in recent, non-industrial, generally illiterate, small-scale societies are of great use in the technological interpretation of excavated material (McGrail, 1983b) as they enable the archaeologist to escape the bounds of his own culture and become aware of other technologies. There are problems in using analogies cross-culturally but the more alike in environmental, technological and economic terms two cultures (one ancient, one modern) can be shown to be, the greater the likelihood that ethnographic studies will be of relevance to the understanding of ancient cultures. Descriptions of recent methods of building and handling simple forms of water transport thus form a significant element in this book; this is particularly important for rafts and boats made principally of hide, bark and reed for which there is little evidence in the excavated record.

Ethnographic analogies must be treated with great caution, however; the fact that hide boats (Hornell, 1946) and reed rafts (Delaney, 1976) were recently used in the west of Ireland does not prove that they were in use there millennia or even centuries ago. Our attention may, however, be drawn to such a possibility, not only in prehistoric Ireland but in similar contexts throughout N.W. Europe. If use of a basic type of water transport can be traced through documentary references and iconographic evidence back to the protohistoric period – as it can be for hide boats (McGrail, 1981a: 20, 26–7) – then a case may be argued for even earlier use. If it can also be shown that appropriate raw materials were available in earlier periods and that techniques analogous to those required to build a simple hide boat were indeed used in those times, albeit in the manufacture of artifacts other than boats and rafts, then the hypothesis may be considered that, despite the non-survival of direct evidence, hide boats may have been built and used in the Bronze Age or even earlier. Such evidence as has survived, both artifactual and environmental, may then be examined and appropriate questions asked to establish whether there is any support for such a hypothesis. It must be borne in mind that, however persuasive arguments based on indirect evidence may be, they cannot constitute proof. As Grahame Clark (1953: 357) said more than thirty years ago 'Comparative ethnography can prompt the right questions, only archaeology in conjunction with the various natural sciences . . . can give the right answers' – with the rider that, in the present state of knowledge no anwers may be possible and any answer will be probabilistic rather than definitive.

In Chapter 2 an attempt is made to identify primary types of water transport, based on structural considerations, so that defined type names can subsequently be used in the book. The second part of the book begins at Chapter 3 with an introduction to the structural aspects of naval architecture, and continues with an investigation into the methods of building those primary types of boat or raft which were used, or it is believed might have been used, in early N.W. Europe. For much of the period under investigation the tree appears to have been the principal source of building material and in Chapter 4 the maritime archaeological aspects of trees and wood are considered in some detail. This is followed by three chapters on the building and use of log rafts, logboats (dugout canoes) and bark boats. Plank boat building, for which there is relatively much more archaeological and documentary evidence and which appears to have dominated the maritime scene in N.W. Europe from an early date, is discussed in some depth in Chapter 8. Bundle rafts and boats are dealt with in Chapter 9, and hide boats and float rafts in Chapter 10.

The third part of the book deals with the operation and use of water transport, principally the planked boat. In Chapters 11 and 12 naval architectural theory and boathandling principles are used to investigate operational performance; and to assess the surviving evidence for propulsion, steering and anchoring in prehistoric and medieval times. The knowledge the early seaman required to handle and operate his boat, the practical problems he faced and the landing places and harbours he used are dealt with in Chapter 13; and the final chapter considers how early mariners found their way by coastal or sea routes, using essentially non-instrumental methods of pilotage and navigation.

Unlike James Hornell's *Water Transport*, this book is not an attempt to seek out origins or evolutionary trends, rather it is an attempt to 'marshal in due order the major part of the knowledge within our ken' of the water transport of early N.W. Europe.

2 Types of water transport

Man has used a wide range of objects as water transport in his exploitation of lake, river and sea – see, for example, those described by James Hornell (1946). To further archaeological and ethnographic studies it is first necessary to summarise and put into order the large body of data available and then to devise a classification scheme which will incorporate most, if not all, individual examples. The aim of establishing such a classification scheme is not to fossilise types, for any scheme must be capable of responding to newly acquired data; nor is the aim to demonstrate any 'evolution' or 'development' of one type from another in a hierarchy of classes (cladogram). On the contrary, a classification scheme is an essential prerequisite to a comprehensive research programme, so that the superficially disparate data can be comprehended, patterns recognised, significant similarities and differences highlighted, and a picture of general groupings emerge. Cultural choices may also be identified and a common frame of reference is provided for international studies.

If primary classes of water transport can be identified which together comprise the full range ever used by man – non-primary types being variants (possibly of increasing complexity) of these primary classes; and if the materials, the techniques and the minimum tool kit needed to build the vessels in each primary class can be recognised; then some light may be shed on the earliest period of man's use of water transport, the Neolithic and earlier for which little, if any, direct evidence has survived. This can be done by indicating, in general terms, the earliest periods or technological states in which each primary class of water transport could have been used, by analogy with the technology, tools and materials known to have been used in the manufacture of other artifacts which have survived (see Chs. 5 to 10).

CLASSIFICATION

Until recently water transport has traditionally been classified, but not systematically, by reference to two features:

1. source and type of buoyancy, e.g. floats, rafts, boats;
2. principal raw material used, e.g. reed, bark, hide, logs, planks, etc.

Thus we have 'reed boats' and 'log rafts'. This binomial scheme has sometimes been supplemented or replaced by regional type-names used outside their region of origin, or by terms which imply a certain shape, e.g.: 'coracle' for a boat of circular form; 'canoe' for an elongated boat. This traditional classification scheme has several disadvantages – for example:

1. One class may include two or more fundamentally different types. Boats of the same principal material are given one class name e.g. plank boat, but they may be built by fundamentally different constructional methods – shell or skeleton sequence, for example.
2. Types which are conceptually similar may be in two or more classes. A variety of materials can be used to build what is essentially the same boat, e.g. boats of tarred flannel or canvas on a wooden framework are analytically the same as hide-on-frame boats ('skin boats'). Boats and rafts built of grass, straw or bark bundles or even bundles of light poles (*ambatch*) are structurally the same as the more-widely known rafts and boats built of reed bundles.
3. The indiscriminate use of names. For example, the S. American term 'balsa' has been applied to hideboats, lografts and reed bundle rafts (Hornell, 1946: 41, 81, 150), thereby causing confusion.

As a first step towards the systematic classification of water transport McGrail (1984c) has proposed an analytical scheme in which structural considerations take precedence over choice of raw materials (an environmentally related attribute) and form (functionally and environmentally determined), and in which the attributes identify, whenever possible, choices which have cultural significance. How a boat is conceived as a three-dimensional object, and how the builder translates idea into artifact (the 'design' of the boat) are both probably culturally determined; as also is the choice of manufacturing techniques used to convert raw materials into boatbuilding elements. These attributes are thus fundamental to an archaeological or ethnographic enquiry and furthermore may be culturally diagnostic.

FLOATS, RAFTS AND BOATS

The scheme analyses water transport by first asking questions about buoyancy and thereby distinguishes floats, rafts and boats. If we may generalise from the iconographic evidence for the use of floats and rafts in the ancient Middle East (Hornell, 1946: 6–7, 26–7, 46–51) and the Mediterranean (Johnstone, 1980: 9–12, 58–60), and from the widespread ethnographic evidence (Hornell, 1942, 1945; 1946: 1–85), these two forms of water transport have made a significant and worldwide contribution to man's use of inland and inshore waters.

In northern and western Europe, however, the use of personal floats has been constrained by water temperature and limited by the endurance of the user. Thus they have had only short-term use on inland waters, although widespread. Rafts, of one form or another, can be used on almost all rivers and the smaller lakes, but their use at sea is restricted to the zones of warmer water, probably between latitudes 40 °N and 40 °S. The lower air and sea temperatures experienced outside this zone, combined with exposure to wind and wetness from rain or sea water, soon make a man numb, thus it is unlikely that seagoing rafts were used off northern and western Europe. Within this environmentally-determined constraint, the choice between boat and raft is mainly a cultural one: whether to invest in a wash-through raft of low-level technology and relatively short life; or in a comparatively dry boat requiring greater investment of material and effort and the use of a higher level technology, but generally with a life measurable in years and possibly decades.

1. *Floats*. The buoyancy of a float is applied direct to the man, that is it is a personal aid to flotation with the man partly immersed in the water.
2. *Rafts*. Rafts may be distinguished from boats by the source of their buoyancy. A raft's is derived from the flotation characteristics of individual elements: the material must have a specific density of less than unity. Such vessels as those from Lobito Bay, W. Africa (Greenhill, 1976: fig. 43) and the ambatch 'canoe' from Kenya (Hornell, 1946: pl. VIIIA) may superficially resemble a boat in shape, but they are nevertheless rafts.
3. *Boats*. On the other hand, boats derive their buoyancy from the flotation characteristics of a hollowed vessel, due to the displacement of water by a continuous watertight outer surface. The specific density of the material is not restricted – although boats of lightweight material will have a correspondingly greater freeboard and thus greater usable capacity.

Certain twentieth-century rafts – none from northern and western Europe – have been described in detail (see, for example, Hornell, 1946: 20–90; Edwards, 1965: Ch. 1, 2, 5 and 6; and Doran, 1971) but archaeological examples are very rare: Ellmers (1972: 106) has listed three from the Baltic lands and two from the R. Rhine at Strasbourg (see Ch. 5). To date, little analytical work has been done, although McKee (1983: 51–3) has made a start in his presentation of theoretical forms of raft structure. Further analytical work based on actual structures and further fieldwork is required before primary classes can be identified. For the present, therefore, it seems best to use the traditional classes based on the principal raw material: bundle rafts; lografts; and rafts with buoyancy from inflated skins, netted gourds or pots (float rafts); and to describe individual structures (Ch. 5, 9 and 10).

On the other hand, more practical and theoretical work has been done on boats – both ethnographic and archaeological – and a strong case has been argued for partitioning plank boats into two main groups depending on the builder's concept of his boat either as a *watertight shell* or as a *waterproofed frame*. In the first of these classes the planking defines the form of the boat; framing may subsequently be inserted – this is the shell sequence of boatbuilding. In the second class the framework or skeleton defines the form, and this is subsequently waterproofed by an enveloping layer moulded to the shape of the framework – this is the *skeleton* sequence of boatbuilding.

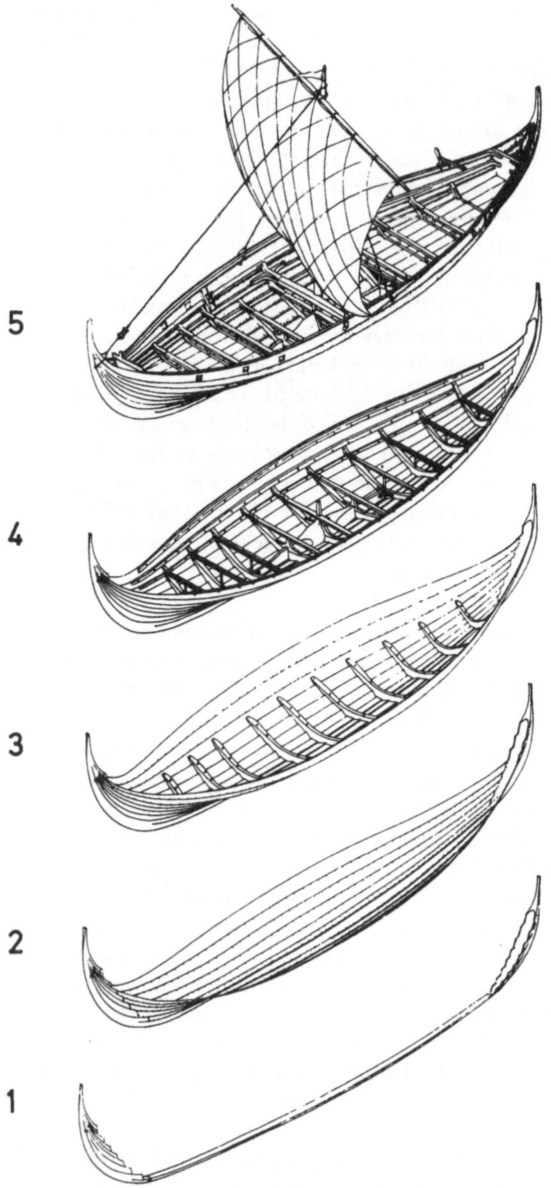

A

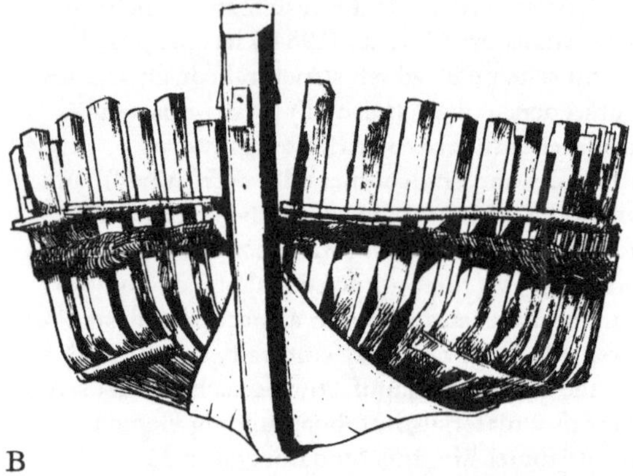

B

Figure 2.1 A Five stages in the shell sequence of building a medieval boat. After Crumlin-Pedersen, 1983b: fig 5.
B The skeleton sequence: planking-up on an already erected framework. (Drawing: NMM Greenwich.)

This fundamental distinction (Fig. 2.1) has long been recognised for plank boats (Hasslöf, 1963; 1972: 27–72; Greenhill, 1976: 60–88), but it may also be recognised in boats built mainly of hide and of bark (McGrail, 1985). Most known examples of hide boat have been thought of, and built, as a wooden or bone framework around which a waterproofing envelope is wrapped (see Hornell, 1946: 111–80), but the S. American *pelota* (Hornell, 1941b; 1946: 150–4) and certain N. American (Adney and Chapelle, 1964: 219–20) and Mongol (Sinor, 1961: 158) boats were built as a watertight shell of hide ('a leather bag') which was sometimes, but not always, reinforced by inserted framework. Bark boats conversely, have generally been built in the shell sequence (Hornell, 1946: 182–6; Greenhill, 1976: 124–8; Birdsell, 1977), but there is some evidence from Sweden, Siberia and British Columbia that boats were built there by sewing or lashing bark to a pre-erected frame (Brindley, 1919; Humbla and von Post, 1937: 11; Eskeröd 1956: 71; Hansen and Madsen, 1981: 4–5; Westerdahl, 1982).

On the other hand, the existence of shell-built bundle boats (for example, of reed) seems unlikely, as an outer waterproofing layer *has* to be applied to the framework of bundles to turn this form of construction into a boat – and this is the skeleton sequence.

These two groups of boats, shell-built and skeleton-built, may each be subdivided according to the techniques the builder uses when converting his material into a boat. If we disregard modern boat-building materials such as metal, GRP, cement and the inflatables, the principal materials from which boats are and have been built are logs, reeds, hide, clay, tar and bone/antler. Three main techniques may be identified, which builders have used to convert these materials into a boat:

1. *Reduction* (or subtractive technique). The raw material is reduced in volume as in the hollowing of a log to make a logboat or in the fashioning of a log to make a keel.

Note: Stone masons have an analogous approach.

2. *Construction* (or additive technique). The junction of several smaller parts (some of which may have been obtained by reduction techniques, e.g. planks or bark). Examples are:
(a) binding reeds to make bundles and linking the bundles together as in the coiled basketry method of building a *quffa* (Hornell, 1946: 101–8).
(b) production of frameworks by weaving, twilling, plaiting or fastening light timbers or split bamboo as in the framework of Arctic skinboats, Vietnamese basket boats, Irish curachs or the nineteenth-century version of the Arabian *tirada* (Hornell, 1946: 57–8, 109–11; Hornell, 1936–38).
(c) the junction of planks by fastenings of sewing, lashing, wood or metal as in edge-fastened plank boats.
Note: Potters who make coil-built pottery and tinkers and tailors use analogous methods.
3. *Transformation* (or moulding technique). Altering the shape of the material without addition or subtraction, as in the expansion of a Satakunta logboat (Johnstone, 1980: figs 5.5 and 5.6) or the inflation of a skin float or the bending of a plank to a new shape.
Note: Potters who throw pots and blacksmiths and founders use analogous methods.

One or more of these techniques may be used to make the form-determining watertight envelope of shell-built boats. The corresponding 'design' element in skeleton-built boats, the inner framework, does not appear to show much variation – all three techniques are used except in reed boats where there is only construction and transformation. Thus classification of skeleton-built boats by the techniques used to make the *outer* waterproofing elements (as for shell-built boats) is considered to be analytically more rewarding at this level of classification.

The analytical scheme resulting from posing these four questions (application of buoyancy? derivation of buoyancy? fundamental concept? principal techniques?) is shown diagrammatically in Fig. 2.2. Fourteen theoretical classes of boat structure (C1 to C14) are distinguished by the states of the 3rd and 4th attributes.

STRUCTURAL CLASSES

C1 *Shell-built by reduction*. Basic logboats hollowed from a single log such as the medieval ones from Llyn Llangorse (McGrail, 1978a: 233–6, figs 29 and 69) and Kew (McGrail, 1978a: 226–9, figs 27 and 68) and the twentieth century boat from the White Nile depicted by Hornell (1946: pl. XXVIIA) are members of this structural class.

C2 *Shell-built by construction*. No examples are known.

C3 *Shell-built by transformation*. Pottery boats such as those Strabo noted in the Nile delta (17.1.4) and the *tigari* from Bengal (Hornell, 1946: fig. 9) are members of this class (but see C6). Shell-built single-hide boats could be members of this class if the hide were not cut to shape (see C5).

C4 *Shell-built by reduction and construction*. Extended logboats such as the medieval Kentmere 1 (McGrail, 1981a: pl. 16) are built by adding washstrakes (planks reduced from logs) to a hollowed log. Basic logboats which have a fitted transom added (e.g. the prehistoric boats from Brigg and Poole), or those with transverse strengthening timbers across the ends (e.g. the medieval boats from Warrington and from Giggleswick) – see McGrail, 1978a & b and McGrail and Switsur, 1979 – are also in this class; as are shell-built plank boats where the planking is *not* bent to shape (transformed). The Brigg 'raft' (McGrail, 1981b) of *c* .650 bc and the three Ferriby boats of *c.* 1500 bc (Wright, 1976) are probably in this class. The so-called 'raft' was a flat-bottomed, box-like boat and thus the planking did not have to be bent to shape; in the Ferriby boats shape was achieved by hollowing (reducing) the planking. Some of the simple twentieth-century plank boats from the S. Baltic rivers described by Rudolph (1974: 5–15), e.g. the Warnemünde *prahm*, the Dievenow *knurrkahn* and the Pomeranian *konger*, are also in this class; as are the tub-boats of Japan and China (Hornell, 1946: fig. 11 and pl. XVIII).

C5 *Shell-built by reduction and transformation*. Expanded logboats are built by reducing a log and then deforming it to the required shape after heat treatment; such boats are known from Finland, Estonia, India, Burma, Siam, British Columbia, Tierra del Fuego, Guianas, Brazil, Amazon Basin (McGrail, 1978a: 38–9). Those that are also extended should be in C7.
The simple bark boats of Australia (Hornell, 1946: 182–6; Birdsell, 1977) and of S. America and Central Africa (Hornell, 1946: 183–6) are made from a one-piece bark shell obtained by reduction from a log or tree. The bark is transformed (sometimes after heat treatment) into the required shape and the ends fastened

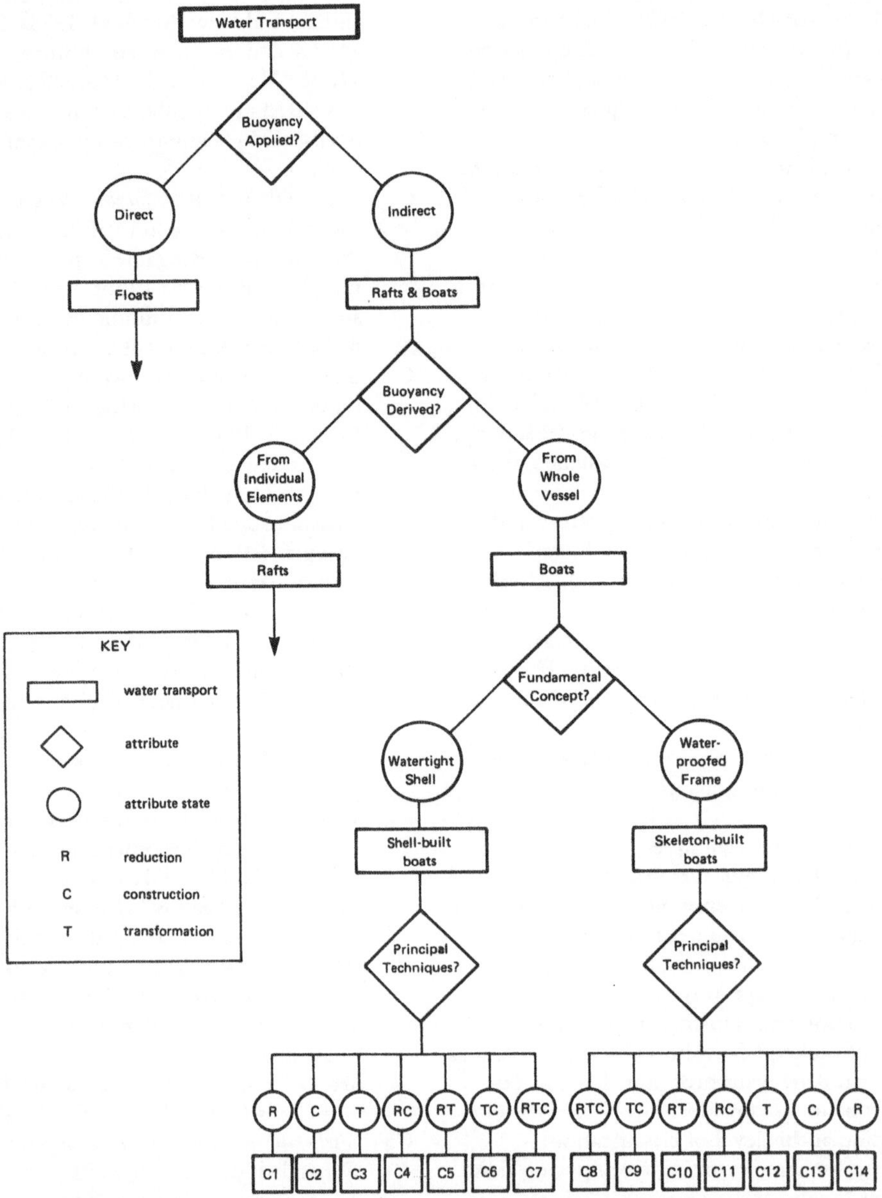

Figure 2.2 Structural classification scheme for water transport – mainly boats. (Diagram: NMM Greenwich.)

(McGrail, 1981a: 65–6). The S. American *pelota* (Hornell, 1941b) and the simple hide boats of the Mongols seen in use by the mid thirteenth-century traveller Iohannes de Plano Carpini (Sinor, 1961: 158; Serruys, 1981: 112) were made from a single hide, cut to shape (reduced) and then moulded to shape by the cargo carried or sometimes by an inserted light framework. The bull boats of the N. American Plains Indians may also have been of this class, but the two principal recorders of these craft (Adney and Chapelle, 1964: 219–20; Hornell, 1946: 148–50) differ in their interpretation of the function of the elementary framework.

C6 *Shell-built by transformation and construction.* If they exist, boats built of coiled pottery, rather than thrown pottery, would be in this class.

C7 *Shell-built by reduction, transformation and construction.* Logboats which have been expanded and then extended (for example, twentieth-century boats from Satakunta, Finland – Johnstone, 1980: fig 5.5 to 5.7) are in this class. Shell-built plank boats in which the planking is stressed and bent (transformed) into the required curved shapes (rather than hewn as in C4) are also in this class; they include the ninth/tenth century AD Graveney boat and all boats built in the Viking tradition (McGrail, 1981a: 31–6). Mediterranean boats of the Classical period (Casson, 1971: 201–13) were generally similar. Many of the plank boats in the Indian sub-continent today are built in this manner (Greenhill, 1976: 50–6).

The more complex Australian bark boats and the majority of those recently built in N. America (Greenhill, 1976: figs 77–8) and S. America (McGrail, 1981a: 77–8, pl. 6) were built from several sheets of bark sewn together and moulded to shape; the inserted framework strengthened this bark shell and held it in shape. The temporary or emergency hide boats of the N. American Malecites were built of two or more skins in a similar manner to the American bark boats (Adney and Chapelle, 1964: 219–20) and thus are also members of this class.

The wooden basketry of the Vietnamese boats described by Nishimura (1931: 36–43), Hornell (1946: 109–111) and Cairo (1972) is built from split bamboo (reduction) which is woven and thus prestressed (addition and transformation). As the resin and dung mixture is generally only applied as a caulking to the interstices of this closely woven basketry (rather than applied overall as a waterproofing envelope) the boat belongs in this class – but see C12.

C8 *Skeleton-built with waterproofing envelope made by reduction, transformation and construction.* The planking of many skeleton-built boats (but see C11) is fashioned by reduction and then forced into shape (transformed) around the framework. In northern and western Europe this technique does not seem to have been used until the fourteenth century (McGrail, 1981a: 42–3) although aspects of it may be seen in some of the boats of the Romano-Celtic tradition and possibly in the cogs of N.W. Europe (McGrail, 1981a: 22–4, 36–8, 42–3). From the fourteenth century the use of this skeleton sequence became increasingly common in Europe, especially for the larger boats and ships (Greenhill, 1976: 60–88, 268–301).

The hide waterproofing envelope of the British and Irish *coracles* and *curachs* (Hornell, 1936–38), the *umiak, kayak, biadara* and *baidarka* of the northern circumpolar lands (Johnstone, 1980: fig. 4.12; Hornell, 1946: pl. XXIVB; McGrail, 1981a: 27) the *parisil* of S. India (Kentley, 1984) and others from Manchuria, Korea and Tibet are made by cutting hides to shape (reduction), fastening them together and moulding them to the pre-erected wooden framework. The boats Herodotus (1.194) saw on the river at Babylon in the fifth century BC were probably built in a similar manner. Waterproofed flannel or canvas has replaced hide on recent Irish and British boats, otherwise their characteristics appear to be retained from earlier times.

Certain bark boats in Sweden (Hansen and Madsen, 1981: 4–5; Westerdahl, 1982) and in E. Siberia and N.W. America (Brindley, 1919: 106) seem to have, or to have had, a bark envelope produced by reduction and construction, and then moulded to a pre-erected wooden or bone framework; they are therefore to be classified here.

C9 *Skeleton-built with waterproofing envelope made by transformation and construction.* None known.

C10 *Skeleton-built with waterproofing envelope made by reduction and transformation.* Single hide and single bark boats built on a frame, if they exist, would be in this class.

C11 *Skeleton-built with waterproofing envelope made by reduction and construction.*

Skeleton-built boats whose planking is not transformed by bending into the required shape would be in this class – some of the second to fourth century AD Romano-Celtic flat-bottomed barge-like boats (McGrail, 1981a: 23–4) of the R. Rhine may have been built in this way – publication of further details is awaited.

C12 *Skeleton-built with waterproofing envelope made by transformation.* The reed bundle, coiled basketry framework of the boats of Sargon (Anderson, 1978: 49), Moses (Exodus 2, 3) and Strabo (16.1.15), the modern *quffa* (McGrail, 1981A: pl. 26), *zaima* (Thesiger, 1978: 128, fig. 45) and *jillabie* (Heyerdahl, 1978: 35) are covered with bitumen which is transformed from a solid mass into a waterproofing envelope. The nineteenth-century *tarada* noted by Layard (1853: 552) are also in this class as they consisted of a woven basket framework completely covered by a waterproofing envelope of tar (Hornell, 1946: 57–8, fig. 7). Possibly some of the Vietnamese basket boats may also be in this class rather than in C7; as also would skeleton-built boats of other materials with a complete waterproofing envelope.

C13 *Skeleton-built with waterproofing envelope made by construction.* None known.

C14 *Skeleton-built with waterproofing envelope made by reduction.* None known.

Five of the seven shell-built classes are known to have members and a sixth may have members. Of the seven skeleton-built classes, two are known to have members; two others may have members but it is impossible to be certain on this point because of inadequate structural information. The classes with certain and with doubtful members are listed in Table 2.1.

Traditionally-named boat types (e.g. bark boats or plank boats) appear in Table 2.1 in several structural classes. This may be more clearly seen in Table 2.2 which is a rearrangement of the data in Table 2.1.

Logboats are all shell construction and may be built using four different combinations of techniques depending upon whether they are to be basic, extended, expanded, or expanded and extended logboats – the complexity and the range of achievable shape increasing generally in that sequence.

Plank boats may be built in shell or skeleton sequence, the class determinants then being whether or not the planking is bent to shape – significantly different approaches to boatbuilding.

Bark and hide boats feature in the same classes, two shell and one skeleton, with a further possible skeleton class. The difference between the two groups of techniques (RT or RTC) is whether the 'design' requires a single hide or bark, or several to be used; this is more a question of size than a significant structural difference. The choice of shell or skeleton for hide and bark boats, while an important conceptual difference, may also be related to size in that the difficulties of visualising the shape of larger boats and indeed stabilising the hull during construction would seem to increase significantly

Table 2.1 Proposed structural classes with their members listed by traditional type name

Class	*Attributes*		*Traditional type name*
	Concept	*Techniques*	
C1	Shell	R	Basic logboats
C3	Shell	T	Pottery boats (some)
C4	Shell	RC	Extended logboats
			Unstressed plank boats (some)
C5	Shell	RT	Expanded logboats
			Basic bark boats (some)
			Basic hide boats (some)
(C6)	Shell	(TC)	Possibly some pottery boats
C7	Shell	RTC	Logboats expanded and extended
			Stressed plank boats (some)
			Complex bark boats (some)
			Complex hide boats (some)
C8	Skeleton	RTC	Stressed plank boats (some)
			Complex hide boats (some)
			Complex bark boats (some)
(C10)	Skeleton	(RT)	Possibly some basic bark boats
			Possibly some basic hide boats
(C11)	Skeleton	(RC)	Possibly some unstressed plank boats
C12	Skeleton	T	Bundle boats
			Basket boats

Notes: 1. R = reduction. T = transformation. C = construction.
2. Unlisted classes have no known members.
3. Items in parentheses are doubtful.
4. Traditional type names are defined in the text by reference to specific, reasonably well-documented examples.

Table 2.2 Traditionally-named boat types classified by two structural attributes

Traditional type name	*Attributes*		*Class*
	Concept	*Techniques*	
Logboats	Shell	R	C1
	Shell	RC	C4
	Shell	RT	C5
	Shell	RTC	C7
Plank boats	Shell	RC	C4
	Shell	RTC	C7
	Skeleton	RTC	C8
	Skeleton	(RC)	(C11)
Bark boats	Shell	RT	C5
	Shell	RTC	C7
	Skeleton	RTC	C8
	Skeleton	(RT)	(C10)
Hide boats	Shell	RT	C5
	Shell	RTC	C7
	Skeleton	RTC	C8
	Skeleton	(RT)	(C10)
Bundle boats	Skeleton	T	C12
Pottery boats	Shell	T	C3
	Shell	(TC)	(C6)
Basket boats	Skeleton	T	C12

Note: Items in parentheses are doubtful.

with size – although this may be merely twentieth-century lack of knowledge and expertise. On the other hand, such considerations, plus the requirement to give large, seagoing ships sufficient strength to withstand the stresses imposed in a sea way (McGrail, 1981a: 42–3) undoubtedly influenced the widespread change from shell sequence to skeleton sequence of plank-built vessels in later Medieval Europe.

Bundle boats by their nature need to have their coiled basketry framework waterproofed, most readily done by the transformation of some mastic-like substance (e.g. tar). Other materials, such as hide, might be used but this seems an overcomplex solution and one which, as far as is known, has not been tried – the *quffa* covered with skins which Layard (1849: 380) mentions, and which Hornell (1946: 104) believes had a reed bundle frame, was in all probability a hide boat with a woven wooden basketry frame. It seems unlikely, therefore, that bundle boats of reed, grass or similar materials could be represented in any other class but C12.

The known *pottery boats* (C3) are said to be made by throwing (Hornell, 1946: 98) but it is conceivable that some could be made by coil techniques and therefore some may be in class C6.

Basket boats, like bundle boats, appear to have a distribution restricted to Indo-China and Arabia. The main hulls of the basket boats of Indo-China are built from a closely-woven split bamboo framework which is made watertight by paying inside and out with a resin-based mixture. These basket boats are in class C12. The Arabian *tarada* has a complete covering of tar and is conceptually the same as a bundle boat and therefore also in class C12.

Individual boat types which are well defined may be identified by extension of the proposed classification scheme (not shown in Fig. 2.2). Thus a group of Irish hide boats generally known as *curachs* and a group of Arctic hide boats generally known as *umiaks* are both in class C8. They may best be distinguished from one another by identifying the states of certain structural attributes; an example would be the distinction proposed by Christensen (1977: 275), and incorporated in Kentley's scheme (1984: fig. 1) as a cross-sectional, shape-determined attribute. Subgroups may be further recognised by other structural considerations, such as fastening methods or other woodworking techniques.

As analytical work on boat structures progresses it may be that significant patterns in human behaviour will be recognised and the use of class names based on structural considerations become appropriate. For the present it seems best to continue to identify boats by traditional names, but supplemented by the structural data described above.

The analytical scheme outlined in Fig. 2.2 and Table 2.2 is used in Chapters 6 to 10 to describe the structure of logboats, bark boats, plank boats, bundle boats and hide boats. Pottery and basket boats are not described as there is no evidence for their use in N.W. Europe. The three principal types of raft, log rafts, bundle rafts and float rafts, are considered in Chapters 5, 9 and 10, but simple floats are not considered further.

3 Stability and seaworthiness

The aim of this chapter is to provide a basic knowledge of naval architecture so that a maritime archaeologist may understand something of the relationship between a boat's shape and structure and the hydrostatic, hydrodynamic and other forces to which the boat is subjected in service. With such knowledge technical papers such as Corlett's (1978) examination of the Graveney boat may be tackled with some confidence, and the reader should be better equipped to examine boat remains and gain an understanding of the techniques used by ancient boatbuilders. This is not to say that the ancient boatbuilder necessarily formulated scientific reasons for his methods but, rather, that by an empirical approach he evolved techniques which enabled him to incorporate desirable qualities in his boat, qualities which we may, from an informed viewpoint, deduce.

Although formulae and diagrams were used to work out hull shapes in fifteenth-century Italy (Anderson, 1925b) the craft approach prevailed until the seventeenth century when the mathematician and master shipwright Phineas Pett made the first known calculations from which displacement and draft could be estimated (Alexander, 1927: 236). Scientific methods for calculating stability, resistance and strength developed only slowly and even today there are some fields in which theoretical predictions are not always fulfilled in practice, which is understandable in the light of the numerous variables confronting the naval architect, not least because boats and ships operate at the interface of two media: air and water. Indeed, testing small-scale models in experimental tanks and wind tunnels is essentially trial and error pursued systematically within a framework of physical laws and established constants.

The twentieth-century naval architect deals mainly with metal ships, mechanically propelled, and he designs them to have certain qualities; the maritime archaeologist on the other hand deals in the main with wooden boats, propelled by oar or sail and he attempts to deduce the qualities inherent in the design. Theoretical and practical studies of sailing and rowing have only recently been undertaken, and structural studies of wooden boats are rare (Coates, 1977, 222). Nevertheless the fundamental elements of naval architectural theory are common to all displacement vessels no matter the size or material. Precise answers to questions about a boat's structure and performance, especially under sail, may not be possible, however, as the problems of a small sailing boat in a seaway are complex; whilst 'some theory helps, the process (of understanding and solving the problem) overall remains much of an art' (Rawson and Tupper, 1976: 622).

The specialised application of naval architecture to logboats, with their size and shape limited in general by the size and outline of the log, has been considered by McGrail (1978a: 94–102) and by Arnold (1980). Very little work has been done anywhere on boats of hide, reed or bark or on any of the rafts. This chapter will therefore concentrate on the problems of open boats built from planks and, in particular, stability and seaworthiness will be considered. The assessment of speed and cargo capacity is dealt with in Chapter 11, and propulsion and steering in Chapter 12.

FLOTATION AND STATIC STABILITY

When a body is partly or totally immersed in a fluid it experiences an upthrust equal to the weight of fluid it displaces: this is Archimedes principle. This upthrust is the resultant force of all the fluid pressure forces acting normally on every element of the immersed area of the body. It is known as *Buoyancy* or *Displacement* force (Δ) and when a body is floating in equilibrium it acts vertically upwards through the centroid (B) of the immersed volume of the body. B

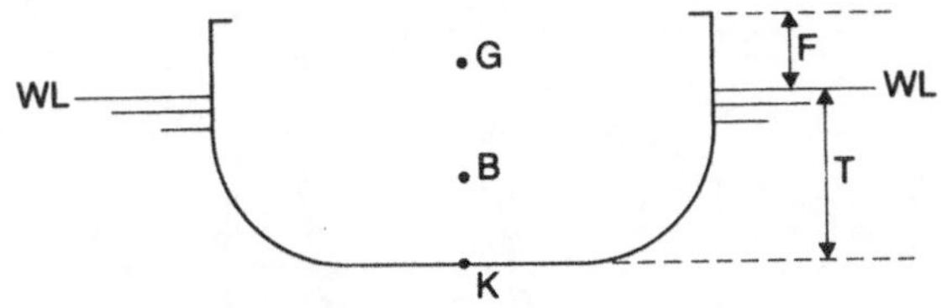

Figure 3.1 Body floating in equilibrium. (Diagram: NMM Greenwich.)
G = centre of mass
B = centre of buoyancy
WL = waterline
K = lowest point on hull
T = draft
F = freeboard

is also known as the *centre of buoyance* of the body (Fig. 3.1). A corollary of this principle, the Law of Flotation, states that the weight of fluid displaced by a freely floating body equals that of the body itself. When the body is in equilibrium its centre of mass (G) will be in a line through B.

The volume of water displaced is known as the *Displacement Volume* (∇): this is equal to the volume of the immersed part of a boat, that part below the waterline (WL). The position of this waterline (and hence ∇) will change depending on the load carried. The vertical distance from the waterline (in any one condition of load) to the bottom of the keel (K) is known as the draft (T); from waterline to the top of the boat's side amidships is the freeboard (F) – see Fig. 3.1.

The weight (W) of a boat with and without loads, and the longitudinal and transverse position of G and B in various load states needs to be known before calculations of upright static stability can be undertaken. The weight of an actual boat without loads (light weight) may be found directly, the most accurate method being to weigh the boat in air and then carry out an inclining experiment to establish the position of G (McKee, 1974: 11–13). For hypothetical reconstruction of a box-shaped boat such as the Brigg 'raft' calculation is relatively simple (Coates, 1981), but for anything more complex, from the basic logboat (McGrail, 1978a) to the relatively complex shape of the Graveney boat (Corlett, 1978), a series of elementary calculations has to be made using data from the drawings. The boat is divided into discrete elements (keel, planks, etc.) whose weight and individual G can be calculated or estimated. Crew and boat's equipment are similarly treated. The moment (weight × orthogonal distance) of each element is then calculated about a vertical transverse plane (say through the bow or the mid-length) and a horizontal plane (say through K). In each case the sum of moments divided by the total weight gives the position of G relative to the bow or mid-length and to K. For a symmetrical boat with an evenly disposed crew and equipment G is in the fore and aft middle plane at the midship section. The centre of buoyancy (B) is established in a similar manner, estimating underwater areas (and thus weights) in all three planes by counting squares on a gridded drawing or by using such numerical methods of integration as Simpson's or Tchebycheff's rules (Rawson and Tupper, 1976: 22–34). The longitudinal ($\overline{KB_L}$) and transverse ($\overline{KB_T}$) positions of B are thus obtained. As with G, B lies in the fore-and-aft middle plane, generally near the midship section.

These calculations are then repeated using several different types of cargo of different density, and possibly different numbers of crew. Such calculations are iterative and readily lead themselves to desk calculator solution. The horizontal and vertical positions and the size of G and B will then be known for the light weight condition (without cargo), and for various loaded states (Appendix 3.1). If valid comparisons between boats are to be made certain parameters must be fixed (McGrail, 1978a: 131–3):

1. Specific density of the wood at the moisture content applicable to operational use, say 25% m.c. (See Ch. 4).
2. Number and normal station of crew, their individual weight (Ch. 11) and position of C of G when standing, sitting and kneeling; and aspects of the rowing geometry (see Ch. 12). Standard crews may be defined for rowing and sailing roles.
3. Bulk density of cargo and the space it is allowed to occupy. For certain finds the type of cargo may be clear, but often there may only be general guidance from the circumstances and period of the find, or from historical documentation. In this latter case it seems best to use standard homogeneous loads of a range of densities, e.g. stone at 2500 kg/m^3; grain at 680 kg/m^3 and turf at 435 kg/m^3.
4. Freeboard must be established; that is, the position of the operational waterplane(s) must be determined. Ethnographic evidence is that, in inland waters, small boats are loaded to very little freeboard (McGrail, 1978a: 91); seagoing data is not readily available, however, some guidance may be available in a medieval Icelandic Law which states that the minimum freeboard (F) of a cargo ship is to be F = 2D/5 (Morken, 1980: 178) where D = depth of hull (near) amidships. Elsewhere I have suggested the use of four

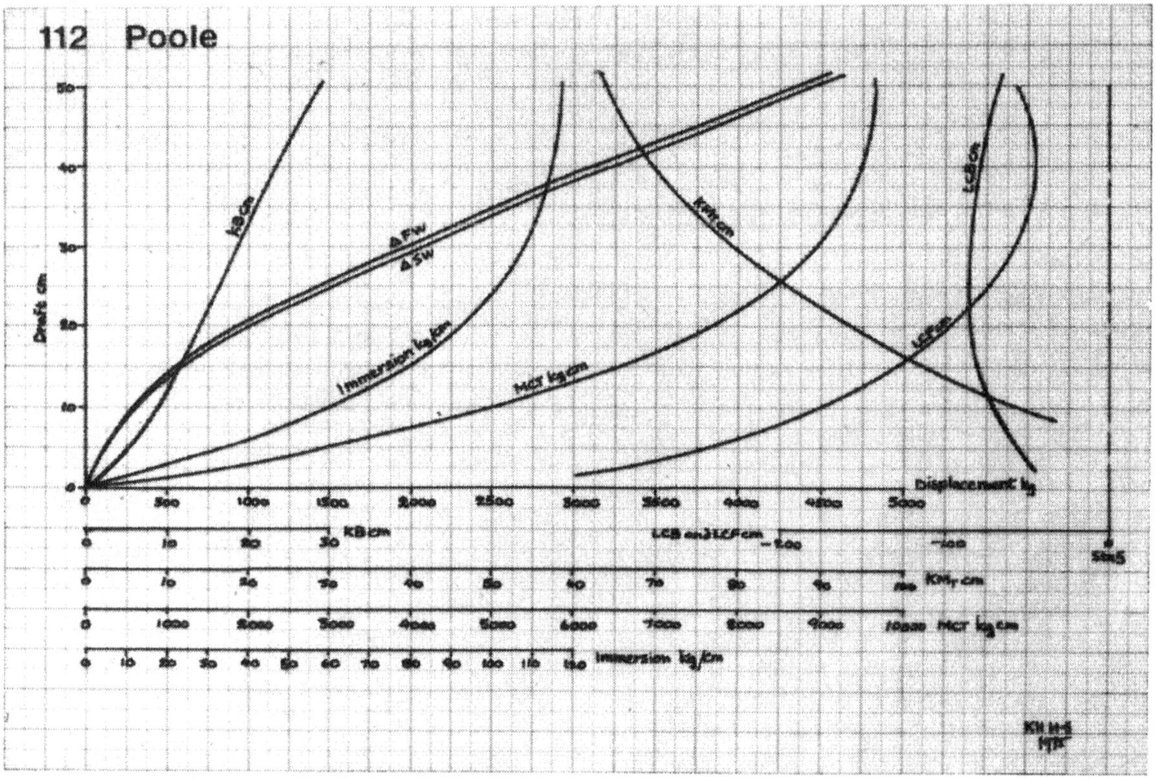

Figure 3.2 Hydrostatic curves for the Poole logboat. Abbreviations are explained in the text. (Diagram: NMM Greenwich.)

standard freeboards (Ch. 11). Other criteria may be chosen: for example, a freeboard which would allow four men to move from one side of a boat to the other without the top edge becoming awash, as used by Coates (1981: 263) in his calculations for the Brigg 'raft'. Whatever freeboards are chosen they should be clearly defined.

HYDROSTATIC CURVES

The weights of displaced water (of the appropriate density) between successive horizontal (water) planes are then calculated and the curve of displacement (Δ) against draft (T) is plotted (Fig. 3.2). Draft (T) and hence freeboard (F) can be established from this curve for any loaded state.

Other hydrostatic curves shown in Fig. 3.2 include TPI, LCB, LCF, MCT, KB_T and KM_T. These curves display the boat's geometry and are independent of the weight of the boat or any added cargo. The shape of the boat and the density of water are all that is required to calculate them. These curves are discussed in the following three sections.

Immersion

It is often useful to know the amount a boat will sink into the water if extra load is added. Tonnes per inch or centimetre immersion (TPI/TPC) or kg/cm may be found by calculating the areas of several waterplanes over the required range of drafts and plotting the curve.

Trim

Trim is the result of an angular rotation about a thwartship horizontal axis and is measured by the difference between draft at bow and stern. Movement of the longitudinal position (LCB) of B as the underwater shape varies with draft may be used to calculate changes of trim as the load is varied. Where there is a change of trim without change of displacement (e.g. man moves from bow to stern) the boat rotates about the centre of flotation (CF) which is at the centroid of the waterplane. The curve of the longitudinal position of CF (LCF) against draft is also plotted (Fig. 3.2). Moments to change trim (MCT) by one inch or one centimetre are calculated for selected waterplanes and plotted as a curve against draft. Thus the effects of adding or subtracting loads at various stations in the boat may be determined.

Initial static stability

That 'mystic but important point' (Gordon, 1978: 347) the *metacentre* (M) of a floating body which has been heeled through a small angle without change in displacement, is the point of intersection of the *new* buoyancy or displacement force (B^1) with the vertical plane through B in the upright condition (Fig. 3.3). The position of M is independent of load and is determined solely by the underwater shape of the boat. The distances KB_T and KM_T are calculated for various drafts and plotted (Fig. 3.2).

From Fig. 3.3 it can be seen that, providing M is higher than G, as soon as a boat is displaced from the upright position the new buoyancy force (B^1) acts through M to correct this displacement, until M is once more vertically above G. An indication of the size of this righting moment is the magnitude of the distance GM_T, the transverse *metacentric height*; this will be different for different loads as the position of G will change depending on the quantity, volume density and position of the cargo (see above). A boat with a large GM_T is said to be *stiff*; an excessively stiff boat may become unbearable to the crew in a resonant beam sea: for example the Graveney boat reconstruction in a certain loaded state has a GM_T of 2 m which is large for a boat of this size and could

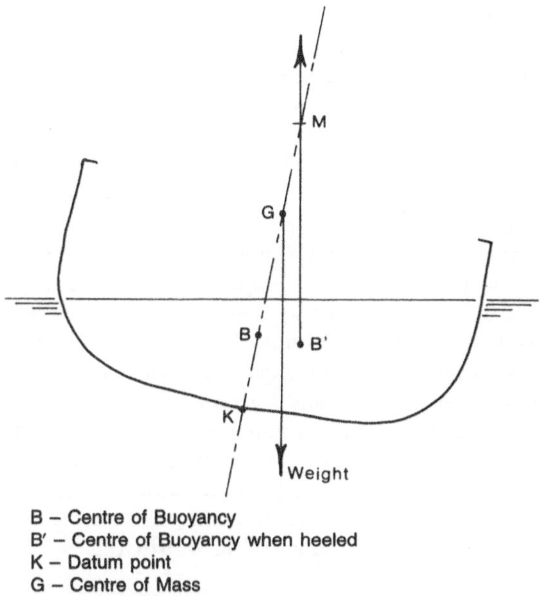

B – Centre of Buoyancy
B′ – Centre of Buoyancy when heeled
K – Datum point
G – Centre of Mass
M – Metacentre
$\overline{GM}$ – Metacentric Height

Figure 3.3 Diagram illustrating transverse stability, (Diagram: NMM Greenwich.)

result in considerable roll being generated (Corlett, 1978, 305). A small GM_T means a weak righting moment; a boat in this condition is said to be *tender* or *crank*, and will be slow to return upright and may become unstable at larger angles of heel or loads.

METHODS OF IMPROVING TRANSVERSE STABILITY

Several interdependent variables influence transverse stability and it is difficult to distinguish the effects of a change in one of them, but general indications may be given (Rawson and Tupper, 1976: 124–7). The metacentric height ($\overline{GM}_T$) may be increased by lowering G or by increasing $\overline{BM}$ and thus raising M (Fig. 3.3). The first method is conceptually simple though it may be difficult in practice. The C of G of the boat herself may be kept low down during the building stage by using appropriate designs with lighter members used in the upper parts of the boat. (See for example the reconstruction model of Ferriby 3 in Fig. 3.5 where the top strake is of leather rather than wood). Cargo and passengers may be loaded low down in the boat and if necessary ballast may be added. If men and animals maintain an upright position by shifting their balance to oppose the heel of the boat their effective C of G is at the level of their feet when standing, or seat when sitting and the problem is less. However, if they move rigidly with the motion of the boat by steadying themselves against the boat structure, their effective C of G should be taken to be at their C of G. In this latter case a kneeling paddler is to be preferred to one standing: in critical cases standing may not be possible without a compensating lowering of G by adding ballast (which also increases draft and hence displacement thereby affecting BM). It may prove impossible to carry the theoretical maximum weight (determined by safe freeboard) of a low density, bulky cargo such as turf, when this would raise G to the critical point where it coincides with M, and $\overline{GM}_T$ becomes zero. On the other hand a boat with this sort of stability may well be able to carry the maximum weight of stone: an example is the Giggleswick Tarn logboat – she is said to be 'volume limited' (McGrail, 1978A: 135–6; McGrail, 1978b).

Variations in beam measurements have a great influence on the magnitude of $\overline{BM}_T$. It can be shown (Rawson and Tupper, 1976: 71) that:

$$\overline{BM}_T = \frac{\text{Inertia of waterplane area}}{\text{Displacement volume}}$$

$$= \frac{I_T}{\nabla} \text{ which varies as } \frac{LB^3}{LBT} \text{ or } \frac{B^2}{T}$$

Thus $\overline{BM}$ is independent of length (L) but varies as the square of the beam (B) and inversely as the draft (T). Greater $\overline{BM}_T$ may therefore be obtained by increasing the beam whilst keeping displacement the same – it is commonly the case that the consequent % increase in the $\overline{GM}_T$ (the metacentric height) is more than twice the % increase in $\overline{BM}_T$ (Rawson and Tupper, 1976: 97). Thus, in general, the broader boat is more stable; some measure of this is given by the Beam/Draft coefficient B/T. A boat with a low B/T (that is relatively deep) will have a relatively small $\overline{BM}_T$ (hence small $\overline{GM}_T$) and be tender: examples of this form are the 'coffin ships' of the early nineteenth century (Coates, 1977: 219). A loaded boat may in fact be less stable than when unloaded (light weight) depending on the relative movements of G and B as the load and hence draft (T) is increased.

STABILITY, DISPLACEMENT AND DRAFT CALCULATIONS

An example of stability/displacement/draft calculations for the Poole logboat is given at the end of this chapter; the boat's operational envelope is determined in terms of her ability to carry loads in four roles with the bare minimum of $\overline{GM}_T$ (+ 0.1 cm). Such a $\overline{GM}_T$ would mean that any displacement from the vertical would be corrected, but very slowly: a rowing eight is in, or very nearly in, this condition in that point of each stroke when the C of G of the crew is highest (Alexander, 1927: 237). Without understanding the concept of metacentric height the state of negative static stability when upright can easily be recognised in a boat which heels over and remains there in a state of loll; this state would have been recognised and avoided in antiquity. It seems likely that underwater shapes of boats and methods of loading were evolved such that metacentric height was greater than the bare minimum and thus the righting moment (which would have been noticeable) would be greater.

The use of a metacentric height ($\overline{GM}_T$) of 0.1 cm in the calculations allows boats to be compared on a common basis. Results for other logboats from southern Britain may be seen in the catalogue entries in McGrail (1978a: 145–306).

An alternative method to near-zero $\overline{GM}_T$ is to define loaded states in precise combinations of men and cargo to be carried (e.g. 50 cattle and 25 men); and to eliminate any role which results in an (defined) inadequate $\overline{GM}_T$ or which results in a (defined) unsafe freeboard. This method is more appropriate for single finds and when the type of cargo is reasonably well known (see Ch. 11).

Taking into consideration the standard of data from boat finds and the approximation inherent in hypothetical reconstructions (see Ch. 11), precise results cannot be expected from these loading and stability calculations; we cannot, for example, claim that the $\overline{GM}_T$ of the Brigg 'raft' under certain conditions was indeed 1.85 m and that she could safely have carried 25 cattle and 10 men. But what may be claimed is that, if the reconstruction is reasonably accurate, the $\overline{GM}_T$ and the safe load must have been of that order. It is thus important to undertake several sets of calculations where more than one reconstruction is postulated (see Ch. 11). The outcome will be performance figures for each reconstruction, which are valid for the parameters chosen and within the assumptions made. By using standard calculations comparisons of relative performance may be made between boats.

There are other stability concepts – large angle stability, and dynamic stability – but initial static stability as represented by the transverse metacentric height ($\overline{GM}_T$) is of fundamental importance. Without such calculations as outlined above any assessment of a reconstruction is incomplete and may be very misleading.

SEAWORTHINESS

A boat in a seaway in rough weather is subject to complex hydrodynamic and aerodynamic forces and those generated in response to them by the boat's mass and stiffness. But boats are also subjected to varying stresses during landing and beaching, and in fact when afloat on the proverbial mill pond. Unless a boat is a rectangular box shape neither her weight nor that of her cargo or crew can readily be distributed along her length in correspondence with the distribution of buoyancy. Thus with a boat of the usual shape, broadest amidships, pointed bow and a rounded stern, although total weight and buoyancy must be equal at equilibrium, there is likely to be an excess of buoyancy amidships, a deficiency of buoyancy at the stern, and a greater deficiency at the bow where there is least immersed volume. This will be the state in still water without load, with the effect that the imposed stress will tend to cause the ends of the boat to droop – a condition known as *hogging* (Fig. 3.4) Unwise loading will accentuate this tendency.

At sea this problem is compounded by the action of waves which have two main effects: (a) they induce roll which causes *racking* or a shearing deformation

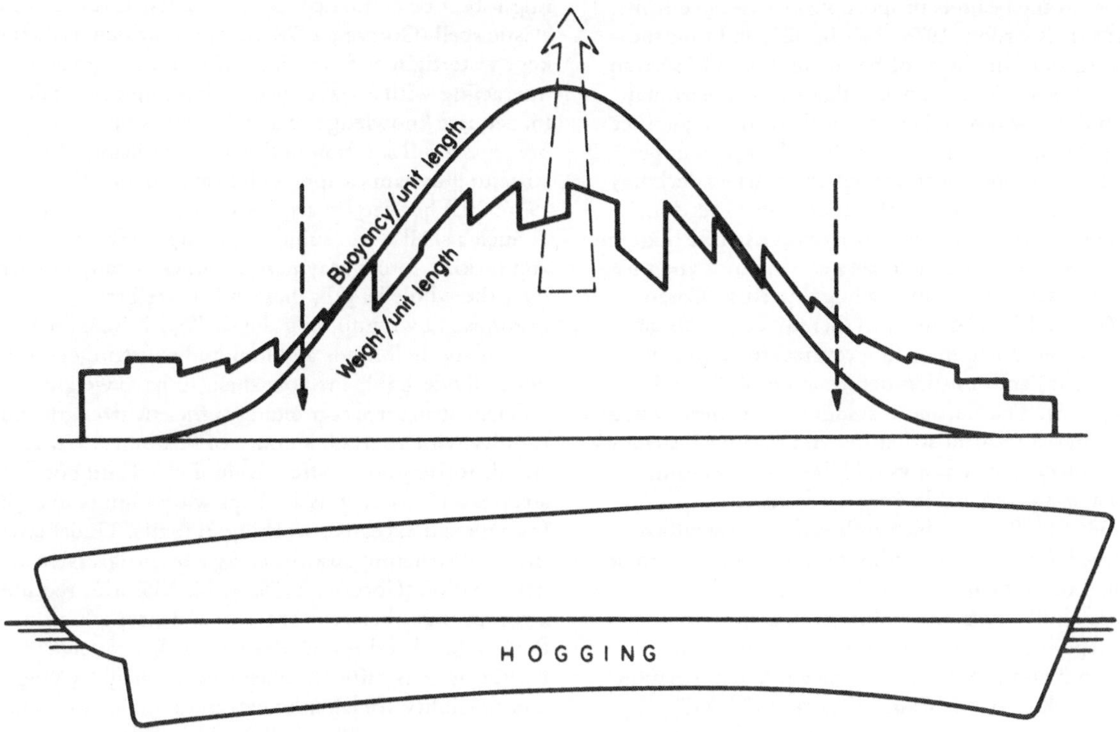

Figure 3.4 Imbalance of weight and buoyancy causes hogging. After Rawson and Tupper, 1976, fig. 6.2.

of the transverse sections of the hull; and (b) they superimpose constantly changing variations in buoyancy along the length of the hull, producing many and various imbalances between weight and buoyancy with consequent stress. When waves are at an oblique angle to the length of the boat they cause the hull to be twisted in torsion. The motion of the boat through the water further adds to these variations. The problem of determining the loadings in these conditions and the boat's reaction to them is 'probably the most formidable and complex of all structural engineering problems' (Rawson and Tupper, 1976: 175).

THE BOAT AS A GIRDER

Analytically, a boat of any appreciable length may be considered to be a girder or floating beam, supported at positions along its length by variable buoyancy forces. The worst condition is taken to be in a sea with a wavelength equal to the length of the boat (Fig. 3.4). Maximum tendency to hog occurs when there is a wave trough at each end with a crest amidships; maximum tendency to *sag* when there is a crest at each end. The boat-beam is deflected under these loadings and bent into a convex or concave shape.

In hogging, the beam is convex up with the upper part of the hull in tension and the lower in compression. In this situation there is a tendency for joints in the upper part of the sides to open, for the sides themselves to be forced inward where curved in plan, and for the plank seams there to be forced open; the seams in the lower part are forced together. The crossbeams are compressed in resisting the inward force from the curved upper planking As the wave train progresses down the boat's sides so that the wave trough is amidships, sagging will occur with a reversal of the tensile and compressive forces described above. These continuing but irregular reversals of stress are imposed on the whole structure, but particularly on the fastenings which may fail or at least loosen; watertight integrity is then impaired. Frames and other transverse members oppose the tendency of the sides to open out and close in in response to hogging and sagging and thus indirectly contribute to the longitudinal strength of the boat.

Between the two regions of tension and compression in a beam there is a horizontal *neutral axis* where the longitudinal stress and strain is zero.

Stress in the beam is proportional to distance from this axis (Gordon, 1978: 242, fig. 26) and thus the most economic forms of beam are those of I section or of box section. A boat with a strong horizontal member low down (keel or plank keel) and another near the upper limit of the hull (a thickened upper strake in an open boat or securely fastened decking) will have the attributes of this desirable I section.

The Egyptian solution to excessive tensile hogging stress was to pass a cable between the ends and this was tightened to become a hogging truss (Casson, 1971: fig. 18). The hull was thereby caused to sag somewhat thus minimising reversal forces (Coates, 1977: 218) and resisting deformation of the hull (Fig. 3.5). The European solution has generally been to design a resilient structure with positive fastenings, but to have a bailer or good bilge pump as some leakage was inevitable – see Christensen (1968: 138–9) for a thirteenth-century law which defined a ship as seaworthy if she did not need to be bailed more than three times in 24 hours.

The hull structure of a longer boat must incorporate proportionally more longitudinal strength and thus be more resistant to deformation and working at the seams (Coates, 1977: 218). Although longitudinal bending strength increases quite rapidly with increases in length and slenderness, length is an advantage if high speed is an aim. But, as Corlett (1977) has pointed out, increased length is generally expensive and may introduce structural weakness, as for example the mid-keel scarf in the Ferriby boats.

THE BOAT AS A SHELL

An alternative way of examining the stress imposed on a boat in a seaway, and of investigating how this might best be combated, is to consider the boat as an elastic shell (Gordon, 1978: 293), one which has to be kept watertight and which is moving through and interacting with a wave system of infinite variability. Inadequate knowledge of all the factors involved prevents a full solution of the problem using a structural dynamics approach (Rawson and Tupper, 1976: 203) but certain aspects may be considered.

Such a shell when subject to compressive, torsional and racking forces may part or buckle locally or even over the whole length; these effects are best minimised by reinforcing the shell with transverse frames and bulkheads and longitudinal stringers. It is generally desirable for boat shells to have weight efficient structures combining sufficient strength and rigidity with a certain amount of flexibility so that any distortion under stress is minimised and both stresses and deflections are kept within limits to avoid fracture and excessive working of joints. Under stress the ideal structure absorbs energy by progressive deformation (Gordon, 1978: 91–2, 135); this requires a compromise between stiffness and strength on the one hand and resilience on the other. Such a compromise is difficult to achieve; if weight saving and flexibility are taken to extremes, deflections may become excessive, the shell might buckle or failure of a single component may lead to gross overloading of others, followed by total failure. Concentrated loadings are inevitable in a boat, especially one under sail (Ch. 12); to avoid fracture such point loads must be spread over several members and safely diffused into the shell. Thus the use of frames, beams, horizontal (lodging) and vertical (rising) knees, keelsons and similar timbers which link two or more structural members and thus disperse point loads at fastenings and thrusts from masts, over a wider area. In a good design members will recover their original

Figure 3.5 A 1:10 reconstruction model of Ferriby boat 1 with a hogging truss and a top strake of hide. (Photo: NMM Greenwich.)

form after deformation providing they are not individually overloaded. The boatbuilding tradition of Viking Age Scandinavia seems to have had these virtues in its heyday.

An alternative to the resilient flexibility approach is to increase the scantlings (dimensions) of individual timbers to provide a larger bearing area within the joints where friction without slip will carry the high stresses (Gordon, 1977: 223). Due to the areas of wood required this leads to a massive style of boatbuilding which has some advantages where heavy imposed loads do not greatly increase with hull mass, but none if they do.

It might be thought that if a wooden shell has difficulties in combating these stresses how could a hide boat cope: can leather stand up to the continually varying tensile, compressive and torsional forces without splitting? The answer is that much of the strength of the hull is provided by the framework with the leather skin providing a watertight envelope. Woven basketry framework – the basis for many types of hide boat (Ch. 10) – has somewhat flexible joints where energy is absorbed in friction and dissipated in heat and sound, thus giving the structure toughness with resilience (Gordon, 1978, 103). The practical limits of strength of hide boats do, however, restrict their size.

Shearing forces

The varying tensile and compressive stresses acting longitudinally in the shell of a wooden boat necessarily generate others at ± 45°: these are *shearing* stresses which tend to make one part of the shell slide past the next (Gordon, 1978: 245). Reversing shear stresses can seriously weaken fastenings and start the planking in an edge-fastened boat, and can cause the caulking to drop out or be forced out in a non edge-joined structure. When a boat twists the shell and the deck or thwarts shear. As the shear stiffness of the thwarts of an open boat tends to be much less than that of the shell, the fastenings between thwarts and shell tend to be strained due to hull torsion. The torsional stiffness of most open boats depends mainly upon the stiffness of those fastenings in resisting rotation in the horizontal plane. Hence the requirement to have strong positive fastenings (Ch. 8) such as clenched nails, or treenails with head at one end and wedge at the other; and to have joints in the hull which combine strength with resilience, the strength being such that the joint withstands innumerable reversals of loading. A joint will only withstand such reversals if it does not slip under the loading imposed.

FREE SURFACE EFFECTS

Leakage through strained seams increases with draft. As well as reducing the reserve of buoyancy, damaging cargo and generally making life on board unpleasant it can have a more insidious effect on stability. This *free-surface* effect is especially marked in a flat-bottomed boat (e.g. the Brigg 'raft') but can occur wherever a body of water is not firmly restrained in all dimensions. In the worst cases, if bilge water is allowed to accumulate, a boat's transverse stability can be destroyed (Coates, 1981: 263–4, fig. 4.3.2). Unrestrained bilge water will move across the boat as she heels, and the weight/volume of water transferred to the low side opposes the righting moment generated by the increase of buoyancy on that side with a consequent reduction in metacentric height. This effect is greatest at small angles of heel and in this condition there may be a severe reduction of transverse static stability. A similar effect may occur in the longitudinal plane during pitching, where it is greatest at level trim; but this is not so likely to be a problem for high (L/B) ratio boats (the normal shape).

As this loss of transverse stability is proportional to the cube of the free surface breadth (Rawson and Tupper, 1976: 147) the effect may be decreased by fitting longitudinal bulkheads or sills and thus converting one free surface into several smaller ones. Dunnage will also restrict the movement of free water, and Coates (1981: 266) has shown that closely-packed sticks/branches, to the depth of bilge water could reduce by half the stability loss induced by free-surface effects in the Brigg 'raft'. The systems of cleats on the bottom planking of the Ferriby and Brigg boats (Figs 4.12 and 8.7) would act as incipient longitudinal sills, especially in the latter where ten relatively large cleats projecting upwards from each strake form a discontinuous longitudinal sill impeding the thwartships movement of bilge water. The potential dangers due to bilge water emphasise the requirement to keep bilge water to a minimum by bailing or pumping out.

BEACHING

During operations from beach landing places boats may be run aground bow or stern first, or be allowed to take the ground on a falling tide. Both of these evolutions impose unusual localised loading on the structure. A boat left between high and low water marks may be subjected to severe pounding before she finally settles on her bottom; similar stresses may occur before she lifts clear on the flood (rising) tide.

A boat standing on the beach will generally be less well-supported than when afloat in moderately calm water. This is especially true of round-hulled forms which need reinforcing timbers to oppose the deformation of the hull which may occur due to contact with the beach being over a much smaller area than when she is supported by buoyancy forces afloat; this tendency will be increased by asymetric stresses imposed when a boat has to lie with a list to one side, or at an angle on a steep beach. Flat-bottomed boats and round-hulled forms with little dead rise from keel to bilge are better able to take these stresses, but even here the ends may need extra support to prevent hogging.

LAUNCHING

Launching from a beach site or from a slip also imposes unusual longitudinal loadings on a boat during the period she is partly afloat. These loadings are increased the steeper the launch way. Modern slips have slopes of *c.* 1 : 20 (3°) but slips of the Classical Mediterranean have been recorded with slopes as steep as 1 : 4 (15°) (Flemming, 1980: 169). The design of boats long enough to be affected by considerations of longitudinal bending (see, for example, the Greek trireme – Coates and McGrail, 1984) must cater for such extra loadings by increasing the longitudinal strength or by use of carefully positioned cradles on the slip.

APPENDIX. CALCULATIONS FOR THE POOLE LOGBOAT

1. *Hull weight*: Calculated from the volume of timber in the reconstructed boat. The specific density of the wood, oak (*Quercus* sp.) is taken to be 800 kg/m^3.
   Volume = 1.077 m^3
   Weight = 1.077 × 800 = 862 kg
   By comparison with other logboats for which the vertical position of the centre of gravity (VCG) has been calculated, it is estimated that the centre of gravity of this boat is 1/3 depth of hull above lowest point = (0.5)/(3) = 0.17 m. This estimate is confirmed by calculation

Hull elements	*Weight (kg)*	*Lever (m)*	*Moment (kg m)*
Body	752.8	0.16	120.5
Bow	82.4	0.24	19.8
Stern	26.4	0.25	6.6
	861.6		146.9

$$\text{VCG} = \frac{\Sigma \text{ moments}}{\Sigma \text{ weights}} = \frac{146.9}{861.6} \quad \text{VCG} = \underline{0.17 \text{ m}}$$

2. *Parent log*: An addition of 0.10 m is made to the radius of the minimum tapered cylinder, to allow for sapwood and bark.
   (a) *at butt end*
   radius = 0.76 + 0.10 = 0.86 m
   (b) *at 9 m from butt end*
   radius = 0.50 + 0.10 = 0.60 m
   (c) *mean radius*
   $$= \frac{0.86 + 0.60}{2} = \frac{1.46}{2} = \underline{0.73 \text{ m}}$$
   (d) *girth at butt end* = 2 π (0.86) = 5.4 m
   (e) *volume of log* = π (0.73)^2 × 10.01 = 16.76 m^3

3. *Volume of logboat hollow* = (Volume of displacement (sheer awash)) – (Volume of timber)
   From hydrostatic curves. Volume of displacement (sheer awash) = 4.375 m^3
   Volume of hollow = 4.375 – 1.077 = 3.298 m^3

4. *Freeboard* Taken at the point of maximum beam, in this case the stern of the boat, where half-breadth of boat = 0.76 m
   F = freeboard
   T = draft
   Ø = angle of heel when top edge of logboat is just immersed
   $$\text{Tan } Ø = \frac{F}{0.76} \qquad F = 0.76 \text{ Tan } Ø$$
   (a) Top edge immersion angle of heel at *restricted draft* is given by:
   T = 0.3 m F = 0.2 m
   $$\text{Tan } Ø = \frac{0.2}{0.76} = 0.2632$$
   Ø ≃ 14°
   (b) Top edge immersion angle of heel for *standard draft* is given by:
   F = 0.15 m
   $$\text{Tan } Ø = \frac{0.15}{0.76} = 0.1974$$
   Ø ≃ 11°
   (c) *Minimum freeboard* when Ø = 10°
   F_{min} = 0.76 × 0.1763 = 0.13 m

5. *Loads* These calculations are for the worst case when effective C of G of the crew is their actual C of G and not at their feet or seat.
   State A: *Maximum men*
   Use 18 men (2 standing and 16 kneeling) as theoretical maximum. One man weighs 60 kg; C of G 1.17 m (standing), 0.52 m (kneeling).

Deadweight=	18 × 60 kg =	1080 kg
Hull weight=		862 kg
Displacement=		1942 kg

From hydrostatic curves at this displacement
T = 0.295 m F = 0.205 m (which is greater than minimum)
Calculation of transverse metacentric height (datum = lowest point of boat)

Load	*Weight (kg)*	*Lever (m)*	*Moment (kg m)*
2 men standing	120	1.17	140.40
16 men kneeling	960	0.52	499.20
Hull	862	0.17	146.54
	1942		786.14

$$\text{Thus } \overline{KG}_T = \frac{\Sigma \text{ moments}}{\Sigma \text{ weights}} = \frac{786.14}{1942} = 0.405 \text{ m}$$

From Hydrostatic Curves at a draft of 0.295 m
$\overline{KM}_T = 0.79$ m
Metacentric Height = $\overline{GM}_T$
$= \overline{KM}_T - \overline{KG}_T = 0.79 - 0.405 = \underline{0.385 \text{ m}}$

Thus, this boat can carry the theoretical maximum number of men (18) with a positive metacentric height of 0.385 m.

Note: Should this calculation have shown the boat to be unstable (i.e. with a negative metacentric height) with the theoretical maximum number of men, further calculations would have been made to investigate whether a stable condition can be found when carrying the theoretical maximum crew but ballasted to a deeper draft.
Additional calculations would also be done to find the maximum crew which could be carried without ballast, firstly by using the theoretical maximum numbers but with the men who were standing, now sitting at sheer level. If this failed, the calculations would be repeated with only kneeling men in the crew. If this too proved unstable, the number of men kneeling would be reduced until a stable condition was found.

State B: *Restricted draft*
Draft (T) = 0.3 m Displacement (△) = 2000 kg (from hydrostatic curves)
Freeboard (F) = 0.2 m (which is greater than minimum)
The vertical centre of gravity of the equipment or ballast is assumed to be the same as that of the logboat.
Calculation of transverse metacentric height.

	Weight (kg)	*Lever (m)*	*Moment (kg m)*
2 men standing	120	1.17	140.4
16 men kneeling	960	0.52	499.2
Equipment	58	0.17	9.86
Hull	862	0.17	146.54
	2000		796

Thus $\overline{KG}_T$ = 796/2000 = 0.398 m
From hydrostatic curves at a draft of 0.3 m
$\overline{KM}_T = 0.77$ m
Metacentric height = $\overline{GM}_T$
$= \overline{KM}_T - \overline{KG}_T = 0.77 - 0.398 \text{ m} = 0.372 \text{ m}$

Thus, this boat can carry the theoretical maximum number of men and 58 kg of equipment with a positive metacentric height of 0.372 m.
State C: *Standard Freeboard*
F = 0.15 m T = 0.5 − 0.15 = 0.35 m
The crew in this configuration is taken to be 4 men standing, 2 forward and 2 aft, because of the boat's length and consequent difficulty in manoeuvring. From the hydrostatic curves the displacement at a draft of 0.35 m is 2550 kg.

Crew = 4 × 60 =		240
Hull	=	862
		1102 kg

Weight of cargo = 2550 − 1102 = 1448 kg
Assume 80% of the hollow volume is available for cargo
$= 0.8 \times 3.298 \text{ m}^3 = \underline{2.638 \text{ m}^3}$
Calculate that bulk density of cargo which will result in a zero metacentric height ($\overline{GM}_T$) at this draft. This is the critical bulk density. From hydrostatic curves at draft of 0.35 m
$\overline{KM}_T = 0.73$ m
There will be zero $\overline{GM}_T$ when $\overline{KG}_T = \overline{KM}_T$
Calculations of transverse metacentric height

	Weight (kg)	*Lever (m)*	*Moment (kg m)*
4 men standing	240	1.17	280.80
Hull	862	0.17	146.54
Cargo	1448	X	Y
	2550	0.73	Z

Z= 2550 × 0.73 = 1861.5
Y= 1861.5 − (280.8 + 146.54) = 1434.16

$$X = \frac{1434.16}{1448} = \underline{0.99 \text{ m}}$$

Centre of gravity of this critical load is therefore 0.99 m above the datum.
height of load
= (0.99 − 0.07) 2 = 0.92 × 2 = 1.84 m

volume of critical load

$$80\% \text{ of hollow} \times \frac{\text{(height of load)}}{\text{(internal depth of boat)}}$$

$$\text{critical volume} = 2.638 \times \frac{1.84}{0.43} = 11.288 \text{ m}^3$$

$$\text{critical bulk density} = \frac{1448 \times 0.43}{2.638 \times 1.84} = 128.3$$

The standard loads used in this analysis are:
Stone of bulk density 2500 kg/m^3
Grain of bulk density 680 kg/m^3
Turf of bulk density 435 kg/m^3
Thus the critical load for this boat is less dense than turf, and therefore it will be stable carrying any of the three standard types of cargo. Transverse metacentric height may thus be calculated for a load of turf (i.e. the least dense load).
The height of the load of turf is given by:

$$\text{height} = \frac{\text{Weight of turf} \times \text{internal depth}}{\text{Bulk density of turf} \times 80\% \text{ volume of logboat}}$$

$$\text{height} = \frac{1448 \times 0.43}{435 \times 2.638} = 0.54 \text{ m}$$ (i.e. the top of the load is 0.11 m above the sheer line)

$$\text{vertical centre of gravity} = \frac{(0.54)}{(2)} + 0.07 \text{ m}$$

$= 0.27 + 0.07 = 0.34$ m above datum

	Weight (kg)	Lever (m)	Moment (kg m)
4 men standing	240	1.17	280.80
Turf	1448	0.34	492.32
Hull weight	862	0.17	146.54
Displacement	2550		919.66

Thus, $\overline{KG} = 919.66/2550 = 0.36$ m
from hydrostatic curves, at a draft of 0.35 m
$\overline{KM}_T = 0.73$ m

$$\text{Metacentric Height} = \overline{GM}_T = \overline{KM}_T - \overline{KG}_T = 0.73 - 0.36 = 0.37 \text{ m}$$

Thus, this boat can carry 4 men standing and 1448 kg of turf with a positive metacentric height of 0.37 m. Equivalent loads, by weight, of corn or stone could be carried with greater metacentric heights. Should the critical bulk density have been greater than that of turf, calculations would have been done for grain or stone, as appropriate.

State D *Mimimum freeboard*
F = 0.13 m T = 0.5 − 0.13 = 0.37 m
The crew is again taken to be 4 men standing
From the hydrostatic curves the displacement at a draft of 0.37 m is 2825 kg.
Weight of cargo = 2825 − 1102 = 1723 kg
The standard type of cargo (stone, corn or turf) for which this boat will be stable at this draft can be determined by calculating the critical bulk density as for the standard draft calculation (above). An alternative method is to calculate the transverse metacentric height for a load of turf, and if this proves to be unstable, investigate a load of grain and, should it prove necessary, stone. The following calculation is done by this alternative method.
The height of a load of 1723 kg of turf is given by

$$\frac{1723 \times 0.43}{435 \times 2.638} = 0.64 \text{ m}$$ (i.e. the top of the load is 0.21 m above the sheerline)

vertical centre of gravity

$$= \frac{0.64}{2} + 0.07 = 0.39 \text{ m above datum}$$

	Weight (kg)	Lever (m)	Moment (kg m)
4 men standing	240	1.17	280.80
Turf	1723	0.39	671.97
Hull	862	0.17	146.54
	2825		1099.31

Thus, KG = 1099.31/2825 = 0.389 m
From hydrostatic curves, at a draft of 0.37 m,
$\overline{KM}_T = 0.72$ m

$$\text{Metacentric Height} = \overline{GM}_T = \overline{KM}_T - \overline{KG}_T = 0.72 - 0.389 = 0.331 \text{ m}$$

The positive metacentric height means that the boat is stable in this condition. Thus she can carry 4 men standing and 1723 kg of turf with a positive metacentric height of 0.33 m. Equivalent loads by weight, of grain, or stone could be carried with greater metacentric heights.

4 Trees and wood

Wood is especially important in any study of water transport. It is the principal raw material for log rafts, logboats, bark bundle rafts, bark boats and planked boats, and is used for the framework of buoyed rafts, reed boats and hide boats. Only in reed bundle rafts and frameless hide boats is wood not used, but even in these cases lashings may be made from wood products. Macerated wood can be used as a component of a caulking mix; resins and tars from trees may be used as waterproofing; and split shoots and roots or other light timbers, bast (inner bark) or leaves can be made into ropes for sewing threads or for rigging. Wooden treenails and similar objects have been widely used as plank fastenings or to secure fittings to a boat. Indeed there would be a very limited range of water transport if wood in one of its many forms was not available. Its suitability for, and importance to, boatbuilding is underlined by Hourani's statement (1963: 91) that a traditional Arab sewn plank boat with sails can be made from the coconut tree alone, as *coir*, the fibrous husk of the nut, is fashioned into ropes and sails, while the dried nut kernel yields an oil which can be used to waterproof the planking.

The many species of wood differ in attributes such as density, run-of-grain, average length and girth of bole, strength, resistance to splitting and to tools, durability, and shrinkage. Thus certain species are more suitable than others for particular uses. Individual trees within the species also vary depending on conditions of growth, soil, aspect, rainfall, insect attack, windrock etc. Where there is competition as in forest conditions, trees will tend to have tall, straight boles with few low branches; on the other hand, isolated trees will have a shorter, possibly broader, bole, many branches and curved limbs. Wood is also anisotropic; that is such properties as strength and shrinkage vary according to the orientation of the axis of the cells and this has to be allowed for when choosing the way a particular wooden artifact is to be fashioned from bole, branch or root (Fig. 4.1)

USE BY MAN

Man has long known the properties peculiar to many wood species and has been able to recognise variation in these properties in individual trees and chose that which matched the job in hand: see the wealth of evidence to support this statement in the publications of the Somerset Levels Project (Coles, 1976–84). From early days man has modified a tree's growth pattern by coppicing it to produce large quantities of straight poles of a regular size (Heal, 1978) which originate from dormant epicormic buds in the stump of felled trees of certain species, mainly hardwoods; he has also probably known how to increase the

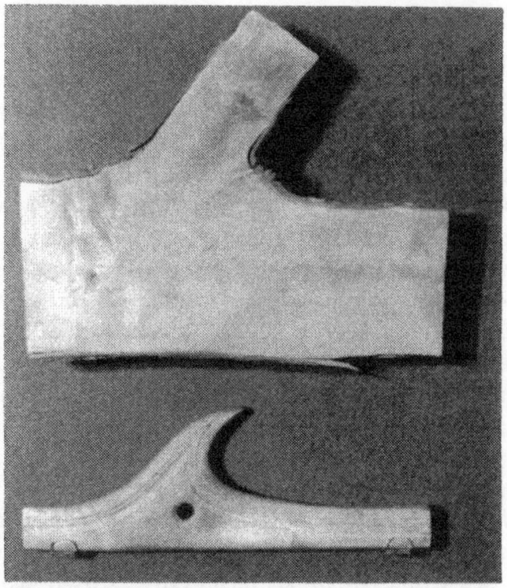

Figure 4.1 This thole for the Greenwich *faering* replica was fashioned, like the original, from a crook to ensure strength. (Photo: NMM Greenwich.)

supply of curved timbers using suitable constraints to branch growth. Different timber species have different specific applications to water transport but the general advantages of its use include:

1. It is generally strong relative to its weight (although weak in tension), free from fatigue and resilient to shock loading.
2. Many species have a density < 1 and therefore have positive buoyancy and, as much of their weight is due to water within the wood even those of density > 1 when growing, will float after drying. Thus teak trees are girdled by cutting through the bark and cambium layer around the base of the bole; continuing transpiration depletes the moisture content so that the density becomes < 1 (Jay, 1961: 6).
3. Generally durable when properly used and ventilated.
4. Water resistant to varying degrees and can be treated with waterproofing mixes to increase this property.
5. Relatively easy to work with a simple tool kit and can be converted, fashioned or bent into a variety of shapes, providing due attention is paid to the nature of the timber (e.g. the run of the grain). Individual elements can be fastened together to produce complex structures.
6. Different parts of a tree can be matched to the various strength requirements of different elements of a boat.

STRUCTURE AND GROWTH

Most trees (perennial plants which can attain a stature of 6 m or more on a single stem) are members of one of three particular *Orders* of plants. Firstly the *monocotyledon* order (bamboos, palms etc.) in which the seedling has one seed leaf. Most of these are inward growers and do not produce the quantities of workable timbers of the other groups, thus they are generally used as a whole bole or split vertically or macerated. In good growing conditions the giant bamboo can be 80 ft (24 m) high with a diameter of 1 ft (0.30 m) (Needham, 1971: 394). Secondly the *dicotyledons* in which the seedling has two seed leaves (Oak, Willow etc.). Commercially these are known as *hardwoods* as most, but not all, have wood which is relatively hard and strong (the lightest and the softest known woods are in fact technically hardwoods). They have broad leaves and generally are deciduous, shedding their leaves in autumn. Hardwoods have a complex structure and a wider range of technical properties when compared with the third group of trees in the *conifareles* order (Cedar, Yew, Pine etc.), which are mainly cone-bearing evergreens with needle-like leaves and are known as *softwoods.*

Hardwoods generally have three types of cell. The *vessels* are vertically-disposed, open-ended cells forming continuous tubes which appear in a cross-section of a stem as small holes or pores; these constitute the 'grain' of the wood. The vessels conduct moisture up and down the tree. *Fibres* are long, narrow and relatively thick-walled cells which constitute the bulk of the wood material and give mechanical support to the tree. The third type, the *paranchyma* are living cells; they are block-shaped and relatively thin-walled and occur both as vertical strands and as groups radiating from the pith (centre of the bole) in medullary rays (Fig. 4.2); their main function is the radial transmission of sap but they are also storage tissue holding nutrient reserves.

Softwoods also have paranchyma cells, but the functions of moisture conduction and mechanical support are undertaken by one type of cell, the *tracheid* which are longer than the hardwood's fibres but similarly vertically disposed. Tracheid cells formed late in the growing season give mechanical support, while those in early wood conduct water from roots to leaves. In addition to paranchyma and tracheids, softwoods, and some hardwoods, have ducts containing resins.

Growth originates in a very thin layer of cells known as the *cambium* which lie under the bark (Fig. 4.2). The bole, branches and roots of a tree increase in size during the growing season (spring and summer in temperate climates) by the repeated division of these cambium cells. The new inner cells become wood elements (*xylem*) which at the beginning of the growing season are large and thin-walled but, in many species, those formed at the end of the season are small, thick-walled and darker; thus growth rings (which are generally annual) may be recognised visually. The xylem cells convey root sap to leaves and shoots. The new cells on the outer side of the cambium develop into new *phloem* or *bast* (also known as inner bark) which carries shoot sap down to the roots. In addition, a thin layer of *bark cambium* outside the bast grows a layer of bark cells. Thus the outer surface of the entire tree, roots, bole and branches grows during this period. As the bole and each branch and root grows in length until maturity, seasonal growth may be visualised as a series of cone-shaped surfaces added to the bole, branch or root (Jane, 1970: fig. 112). At felling date,

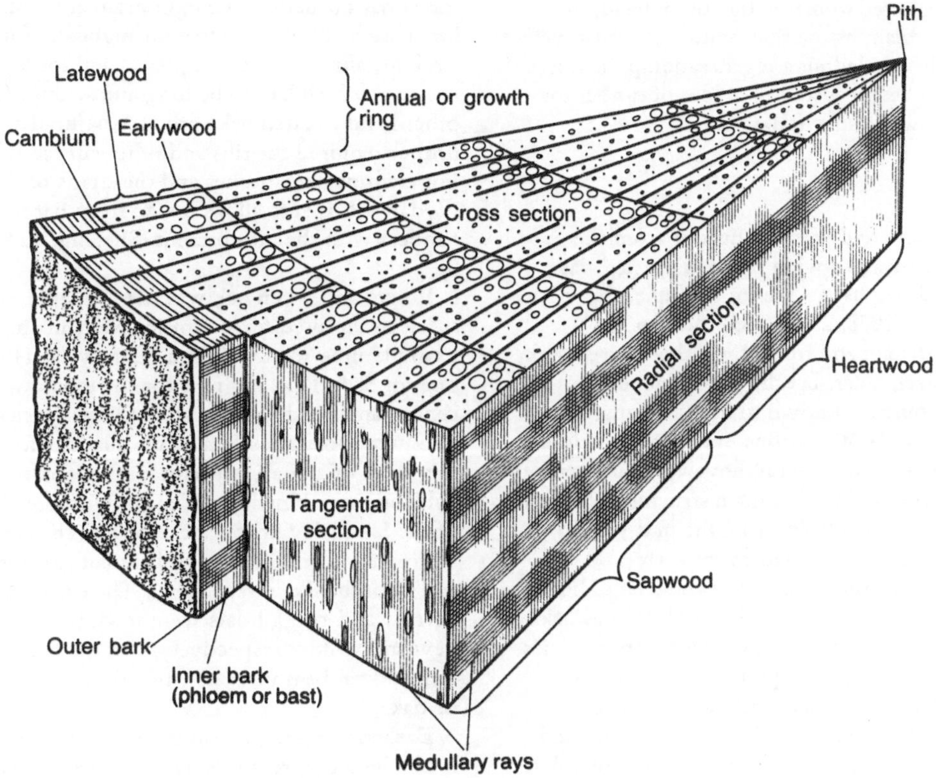

Figure 4.2 Diagram showing the principal structural features of hardwood. After B.J. Rendle: *Growth and Structure of Wood* (BRE Princes Risborough Laboratory: Crown Copyright)

therefore, the outer layer (latest date) of wood will generally be present on all parts of the bole, branches and roots, whereas the early years of growth will only be represented in the lower parts of the bole and branches and in the upper parts of the roots.

Active wood cells in the outer layers of the tree constitute *sapwood*. They function for a limited period and then die and become lignified, a complex chemical process which can include the deposition of tannins in hardwood and resins in softwood, and the vessels may be blocked by tyloses; these cells thus cease being used for conduction and become structural. This inner area of dead cells surrounding the pith or central core of bole, branch or root is known as the *heartwood* which, although susceptible to insect and fungal attack and decay during the tree's life, becomes more durable and generally stronger than sapwood, after felling. Tannins and resins within the heartwood often give it a darker or redder colour which makes it visually distinguishable from sapwood.

One growth ring of sapwood is generally transferred to heartwood each year, thus the number of sapwood rings remains roughly constant at 20 to 30, depending on species. Although the sapwood/heartwood transition generally runs parallel to the bark surface it may cross growth rings (Hughes *et al.*, 1981: 381). For mature northern European oak with more than 120 rings the generally agreed best estimate for the mean number of sapwood rings is 30, with a 95% confidence range of 19 to 50 (Hughes *et al.*, 1981: 389).

SELECTION AND COLLECTION

In the *Heimskringla* the medieval Icelandic author Snorri Sturluson described the building of Olaf Trygvason's ship *Long Serpent* and noted that 'all (timber) that was used was selected very carefully' (Foote and Wilson, 1970: 250). This must have always been the case, other than when very temporary rafts and boats were built. Within the range of timber species regionally available individual trees were selected to match the job in hand; the corollary of this must be that, without a conservative woodmanship including a regeneration policy, a region could be exhausted of grades of timber for which there was a strong demand.

SPECIES

Within northern and western Europe from the Mesolithic onwards a wide range of species was available and was used – see for example, the species listed by Heal (1978: 25) as used by Man in prehistoric Britain and Ireland. Of those listed, oak, ash, elm, hazel, alder, beech, yew, lime, birch, willow and pine are known to have been used in boatbuilding in N.W. Europe in the prehistoric and medieval periods. In general, however, oak seems to have been preferred for the main structural elements whenever it was available, from the mid-second millennium BC Ferriby boats to the early fifteenth-century ship *Grace Dieu* and later. The sample of boat-finds is undoubtedly biased as oak has a greater chance of survival than most other species, nevertheless this apparent preference for oak is confirmed by documentary references; see for example the comments by Caesar (*BG* III.13) and Strabo (*Geog*.4.4.1) on the Veneti's choice of oak for their boats; see also Albion (1952). In assessing which properties of oak (or any other species) made it suitable for building plank boats one must be careful not to assume that prehistoric oaks had exactly the same qualities and grew in the same habitats as today. Certain species may have displayed different characteristics when growing in different climates from that now prevailing, as Rackham has pointed out when discussing the difficulty experienced today in obtaining yew with the right qualities to be used to sew together the oak planking of the Ferriby boats (Wright, 1984). Nevertheless, twentieth-century oak does indeed have properties highly suitable for boatbuilding and it therefore seems probable that similar characteristics were present in prehistoric oaks. As the seventeenth-century diarist John Evelyn wrote, oak is 'tough, bending well, strong and not too heavy, not easily admitting water'. In recent assessments of the relative merits of different timbers oak has been taken as the standard of comparison for the strength properties of many hardwoods, it has 'very good' bending qualities and is in the second of five groups when classified for durability (*Handbook of Hardwoods*, 1969: 166–9). Oak may readily be obtained in elements of different shape and size, depending upon its growth conditions. Thus, tall forest oaks usually of straight grain and without low branches had boles suitable for logboats, for long, straight, almost knot-free planks and for keels and keelsons which had to be long and strong. On the other hand isolated oaks produce naturally curved timber required for ribs and other curved members. For example, the frames and the stems of the *c.*AD 400 Nydam boat 2 are all fashioned from naturally curved timbers with the pith present (Shetelig, 1930: fig. 8, 9).

Where and when oak was not available, other species were used for the major elements in boats, for example lime for the fourth century BC Hjortspring boat of S. Jutland and pine for the medieval boats of central and northern Norway. Other species were also used when they had better qualities for specific members of a boat's structure: ash planking, which evidently combined great length with breadth, in one of the Skuldelev boats; and alder, birch, lime and willow for treenails, ribs and similar fittings (Olsen and Crumlin-Pedersen, 1967). The oak-planked Ferriby and Brigg boats were sewn together with yew and willow respectively, and the latter boat had longitudinal laths of hazel while those of Ferriby were of oak.

Basketry techniques, in which a positive structure is obtained by pre-stressing withies as they are interwoven, are used when building framed hide boats and some other craft. Willow or hazel are suitable for this and the supply of withies can be increased by coppicing.

It is generally considered that the Paper Birch (*Betula papyrifera*) provides the best material for bark boats but many other species have been used where birch was not available, see Table 7.2. The bark of individual birch trees, even those growing in the same vicinity, varies in thickness and quality, and an experienced eye is needed to select those which are suitable for boats. In north-west America in the late nineteenth century, it was considered that birchbark should be over ⅛ in thick, tough, resinous and from a naturally straight bole of sufficient diameter and length to give reasonably large pieces; it was also essential that the bark grain should run horizontally (Adney and Chapelle, 1964: 14–5). The lower part of most trees is unsuitable as the bark there is rough, blemished and thin. From spring to autumn bark can readily be peeled, although late and early in the season the cambium layer comes away with the bark and has to be scraped off before the bark is used. Bark was usually worked soon after being stripped whilst it still had elasticity, otherwise it had to be kept slightly damp.

FELLING TREES

Light timbers, including withies and saplings may be obtained with relative ease, and wind-blown and sea-drifted logs may fortuitously become available, but felling trees, whether by fire or tools, requires more effort and skill (see Adney and Chapelle, 1964: 18 and Johnstone, 1980: 206–7 for methods). Winter has long been felt to be the best time for felling trees and this would have fitted the cyclical nature of medieval and pre-medieval life in northern and western Europe (McGrail, 1974: 39). In addition, deciduous trees can more readily be investigated for soundness and shape of timber and prepared for felling after the leaves have gone, and winter felling of oak means that the acorn crop can be used. Windfalls, branches, saplings, coppiced timber and light timbers generally may be obtained at almost any season although here again an annual cycle of activities would suggest late autumn to early spring.

CONVERSION FOR USE

EFFECTS OF DRYING

Moisture is present in fresh wood in two forms: free water in the cell cavities and absorbed or bound water in the cell walls. Moisture content (m.c.) is generally measured as the weight of water in a specimen expressed as a % of the weight of oven dry wood: thus 100% m.c. means that the wood contains equal amounts by weight of wood material and water. Oak is generally at about 85% m.c. when green, but in several species of tree m.c. can be well over 100%.

After felling a log begins to dry and the reduction in m.c. changes some of the properties of the wood: hardness, stiffness, compressive and bending strength are increased; toughness and resistance to shock are reduced. Thus fresh 'green' oak is considerably easier to work than seasoned oak and is more easily bent, its supple resilience to bending being best at *c.* 25% m.c. (*Handbook of Hardwoods*, 1969: 168 – see also Theophrastus, *Hist. Plant* 5.7.4: 'in shipbuilding, because bending is necessary, wood which is rather green must be used' – quoted by Casson (1971: 205)). On these two counts alone early woodworkers with a simple toolkit preferred to convert and work timber soon after it was felled.

The second principal effect of drying is shrinkage. The wood first loses a large proportion of the free water; only after this is moisture lost from the cell walls. Down to this point, fibre saturation point (f.s.p.), when the cell walls are still saturated, the wood undergoes little if any change in shape or physical appearance. Drying below this point (equivalent to a m.c. of 28% to 30% for most species – *c.* 27% for oak) results in shrinkage. As wood is anisotropic, it shrinks differentially, the relative shrinkages for fresh European oak being, longitudinally 1 : radially *c.* 12 : tangentially *c.* 25. This longitudinal shrinkage in fresh wood is less than 1% and may for most purposes be disregarded; the other measures of shrinkage are significant, however: for example, when fresh oak is dried from the green state to a m.c. of 12%, it shrinks *c.* 7.5% tangentially and *c.* 4% radially (Farmer, 1972: 147). Such differential shrinkage causes stresses in the timber and distortion. In addition, the layers of wood dry at unequal speed, the outer layers drying more quickly than the inner layers can diffuse moisture to the surface. This moisture gradient causes checks and splitting of the wood, especially at the end surfaces.

To avoid this damage, recent practice has been to season timber before use by controlled air or kiln drying until its m.c. is reduced so that it is in equilibrium with its in-use environment; this could be *c.* 20% m.c. at 90% relative humidity (r.h.) or *c.* 12% m.c. at 60% r.h. If the m.c can be reduced carefully and slowly below f.s.p. then splitting and checking is minimised; any that does take place may be removed as the timber is fashioned after drying. Reducing m.c. below 20% has the added advantage that it removes one of the necessary conditions for the growth of rot (McGrail, 1974: 40–1). These conditions are:

(a) wood as food;
(b) more than 20% m.c.;
(c) oxygen;
(d) ambient temperature between *c.* 10° and *c.* 40 °C, the optimum growth range being 17–32 °C.

Thus by reducing m.c. to say 12%, and keeping it there, rot can be avoided.

The dangers of drying – but not of rot – may also be avoided by keeping m.c. *above* f.s.p. By using unseasoned 'green' timber in boatbuilding the m.c. can be maintained at a high level until the boat becomes stabilised with her environment afloat, at 25% m.c. or more (equivalent to a 95% r.h.) which would be sufficiently close to f.s.p. as to minimise splitting, with the timber remaining dimensionally stable. Those parts of a boat out of water may dry somewhat below f.s.p. but where exposed to spray and a salt-laden atmosphere they should accumulate a layer of salt, which, being hygroscopic, should provide a buffer to environmental changes, keeping m.c. relatively high (McGrail, 1974: 39–42).

Fungal spores need still air to settle and thus an open, well-ventilated boat (as were most craft up to the twelfth century AD) would be unlikely to harbour rot; the advantages of using green timber – easier to work and minimum splitting and distortion – thus clearly outweigh the unlikely disadvantage of susceptibility to rot. Archaeological, historical and ethnographic evidence, and recent practice in Shetland, Norway and Britain, strongly suggest that early boats were built of unseasoned timber (McGrail, 1974: 39–41). They were then stabilised at a relatively high m.c. in the boat's natural element, the sea. Careful treatment was necessary when boats were hauled out for the winter: use of an open boathouse near the foreshore and a protective coat of tar (produced by the destructive distillation of timbers of the *Pinacea* family) or resin (from such trees as the spruce (*Picea* sp.) reduce any tendency to dry below f.s.p.

The use of green timber for other maritime structures such as causeway, bridge or waterfront would yield similar benefits. If such timbers, and those from boats, remain wet after deposition and subsequently become waterlogged, they should show no splits or fissures when examined after excavation providing they are kept waterlogged, whereas timber with a similar history but which has been seasoned before use, and not reworked to remove degrade, will display splits and fissures due to drying.

Medieval and post-medieval church and housebuilders who are known to have used green timber (Rackham *et al.*, 1978: 113–14; Hollstein, 1968) must have done so for ease of working rather than to avoid shrinkage problems as, after timbers had been fashioned, they were installed in a relatively dry environment and thus split and distorted once they had dried below f.s.p. – beams of square cross-section may thereby distort to a rhombus shape (Hollstein, 1968; fig. 1b).

CONVERSION OF LOGS

After the crown and the major branches have been removed from a felled tree the first stage of converting the bole for use is to remove bark and sapwood, the latter because it is not as resistant to insect and fungal attack as is the heartwood. However, as discussed in Chapter 8, ancient boatbuilders occasionally left on some sapwood evidently so that they could obtain a greater breadth or a particular shape of worked timber – see the prehistoric Ferriby boats, the Brigg 'raft' and the medieval Graveney boat, and the logboats from Short Ferry (prehistoric) and Llyn Llangorse (medieval).

Whole straight logs of appropriate length and girth are required for keels (oak for strength) and for masts (often pine for resilience). Logboats may also be fashioned from a whole log or from half a log split longitudinally (McGrail, 1978: 122, 310–12). Examples of whole-log logboats are the oak one from Brigg of the ninth century bc and the ash one from Giggleswick Tarn of the fourteenth century ad, while examples of the latter class are the oak logboats from Short Ferry (ninth century bc) and the twelfth/thirteenth-century logboat Warrington 1 (McGrail, 1978).

Planking

Planks have also been fashioned from half logs from early prehistoric times onwards: examples are the oak bottom planking of the Ferriby and Brigg boats, and certain planks (species unspecified) from the causeways (trackways) on the Somerset Levels (Heal, 1978: fig. 15.5). The sequence for conversion (Fig. 4.3) may be summarised as:

(a) Select plane in the log which divides it into two (near) equal halves, and which as far as possible is knotfree.
(b) Remove strips of bark to reveal medullary rays.
(c) Split from alternate sides, using seasoned oak wedges.
(d) Remove remaining bark and outer sapwood by cutting notches at intervals along the log and splitting off wood between them.
(e) Work inner surface of each half log to a (near) plane surface including removal of any rot in the central core (one plank in Ferriby boat 2 was fashioned in a different manner, see Ch. 8). Cut bevels along plank edge (if required) leaving sapwood only where essential to get a particular breadth. Where maximum timber required (as in Brigg 'raft') planks are left with a taper of parent log, otherwise they may be worked to parallel edges along the length.
(f) Hew away the outer heartwood of each half-log to achieve a (near) plane surface (Somerset Levels) or one with upstanding cleat blanks (Ferriby and Brigg).

Figure 4.3 A.B.C. Three early stages in converting a log into a plank for the Brigg 'raft'. (Photos: NMM Greenwich.)

A

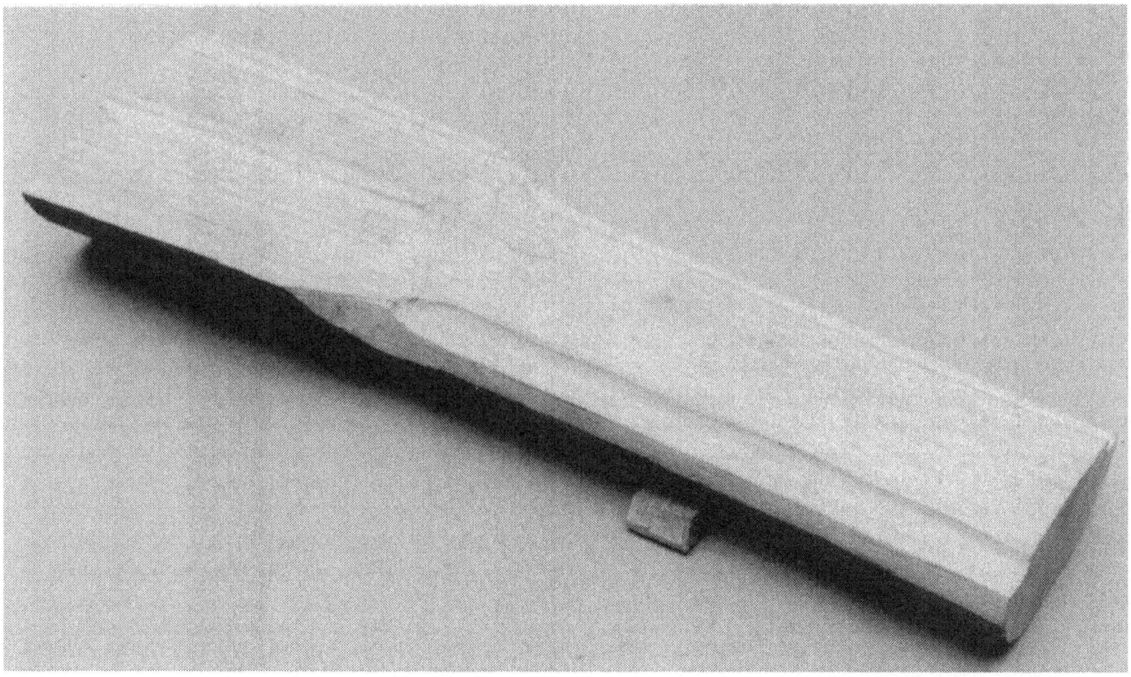

B

C

Figure 4.3 (con't)

(g) Sewing holes bored through plank edges, and transverse holes morticed through cleat blanks (Ferriby and Brigg). Holes for vertical stakes morticed through end of planks (Somerset Levels).
(*References*: McGrail, 1981b: 224–5; Darrah, 1982; McGrail and Denford 1982: 33).

Converting logs in the Ferriby/Brigg manner or as in Heal's fig. 15.5 ensures that maximum breadth of planking is achieved, from near the diameter of the log. Planks in this orientation are more nearly *radial* (growth rings meet the plank faces at (near) right-angles) than *tangential* (rings are (near) parallel to the plank faces) and thus they partake of some of the merits of radial conversion (see below). Such an orientation (Fig. 4.4) is close to the plank orientation which Ewing McGruer, former boatbuilding officer of the Scottish Country Industries Development Trust, identified as the best for boatbuilding. Such a plank has what he called 'neutral grain', that is the growth rings meet the plank faces at an angle between 30° and 60°. McGruer (n.d., 3–4, figs 4c, 5, 7) characterised planks converted from the log in this manner as, 'flexible, stable but not rigid; it deforms readily when required' and 'always tractable. It never splits. It is elastic in the direction of the width, and shrinkage shakes are absent'.

A similar method of conversion was used in medieval times to convert pine logs (*Pinus sylvestris*) for the Skuldelev boats' planking (Olsen and Crumlin-Pedersen, 1967: fig 62, p. 156–7), that of the tenth century Klåstad boat (Christensen and Leiro, 1976) and for the late twelfth century Wood Quay boats. The ash (*Fraxinus* sp) planking of Skuldelev boat 5 and (unusually) one of the oak planks of Skuldelev 3 were similarly fashioned.

The causeway planks from the Somerset Levels illustrated by Heal (1978: figs 15.2, 3, 4) are variants of the conversion process outlined above, but whether or not they could be produced by splitting at an angle to the rays rather than by hewing away surplus wood would depend on the species of timber used. Oak is difficult to split other than radially which it does readily (see below), whereas silver fir (*Abies alba*) splits most easily along the growth rings (Arnold, 1982: 125) as, like most softwoods, the difference between spring and summer wood is well-defined and the long, vertical tracheid cells are

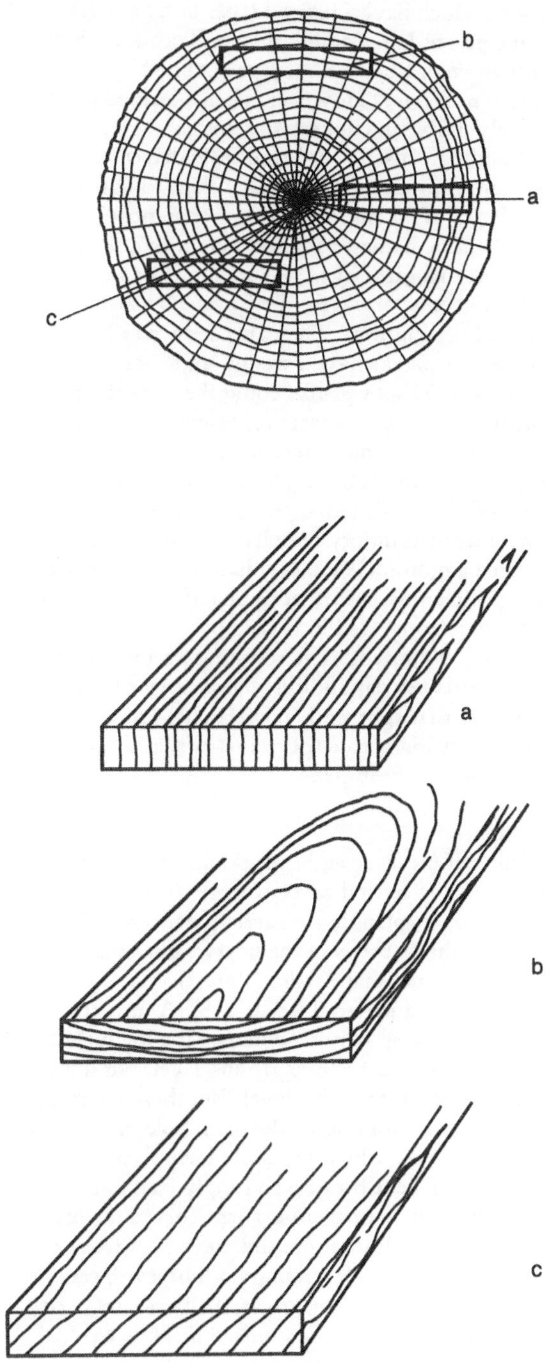

Figure 4.4 Plank orientation in a log. After Ewing McGruer; n.d. (reproduced by permission of the Small Business Division, Scottish Development Agency).
a. radial
b. tangential
c. 'neutral grain'

arranged in regular rows. Ash (*Fraxinus excelsior*) on the other hand, is 70% more resistant to radial splitting than oak, but halving and then quartering is practicable (Arnold, 1982: 122).

Sound and straight-grained green oak logs split well radially; this is because the natural planes of weakness lie along the length of the log, parallel to the medullary rays. Some of the Somerset causeway timbers were produced in this manner (Heal, 1978: fig. 15.1), as was (most of) the oak planking recovered from the five Skuldelev boats, and the oak planking from the Wood Quay boats, the Graveney and Kentmere boats, and all other medieval boat-finds where such aspects have been investigated. There are numerous examples of the use of split logs for non-boat artifacts in Europe from the Neolithic onwards (van de Walle, 1969), and the English royal accounts from Edward III to Henry VIII frequently mention the use of 'clove boards' (Carr Laughton, 1957: 247). This method of conversion results in a stronger oak plank than one fashioned in any other way; one that shrinks less in breadth if allowed to dry below f.s.p.; is less liable to warp, check or split; and is less readily penetrated by fungi (McGrail, 1974: 40, 42–4). In addition, this method minimises the number of planks with strength-reducing knots and it has two distinct advantages when used for clinker boat-building where the lower edge of one plank overlaps the upper edge of the plank below (Fig. 8.5):

1. Dimensional movements due to reaction to changing environmental r.h. are in general proportionally greatest across the thickness of a *radial* plank. This is the plank's smallest dimension and thus the absolute movement is small; it is also a movement in line with the axis of the nail fastenings and is therefore resisted by them, and some of the movement is also absorbed by the caulking in the overlap. Similarly generated movement in a *tangentially* orientated plank is greatest across its breadth, at right angles to the axis of the nails; this is a larger movement in absolute terms and clinker work is not so well able to resist or cushion the effect which results in excessive play in the fastenings.
2. Fastening nails are driven into a radial plank at right angles to the (radial) plane of weakness; whereas nails are driven into a tangential plank parallel to this radial plane and thus have a greater tendency to split the plank.

Saws began to be used for shipbuilding in the medieval period, not later than the fourteenth century

(McGrail, 1981a; 38, 43) and it then became possible and more economic to convert logs in a variety of ways. Methods used in recent times are illustrated in Fig. 4.5. Some of the *plain-sawn* planking, in which the growth rings meet the faces of the plank at an angle of less than 45° (BS 565/1963) are fully tangential planks, while some of the *quarter-sawn* boards (rings meet faces at ⩾ 45°) are true radials. Saws, however, generally part the wood in a straight line regardless of any minor variations in the grain. On the other hand, splitting follows natural lines of cleavage (the grain) without rupturing cells and this results in a stronger board than one in which the grain slopes from one face of a plank to another as can often happen with sawn boards (*Strength of Timber*, 1969: 6).

Splitting oak by cleaving it into a series of diminishing halves produces radial planks of uniform breadth and of triangular or wedge-shaped section (Fig. 4.6) which have a natural taper in length derived from the parent log. Parallel sides and cross sections other than wedge-shaped may be obtained by further hewing if necessary. However, wedge-shaped planks are readily usable in clinkerwork and in housebuilding. Olsen and Crumlin-Pedersen (1968: 158, fig. 62) have estimated that, to obtain radial planks of 30 cm in breadth, a tree must be selected with a diameter at chest height of *c.* 1 m (equivalent to a girth of *c.* 3.14 m at 5 ft or 1.5 m above the ground surface – the forester's usual method of description). They also deduced that, as the Skuldelev planks have a wedge angle of *c.* 11° in cross section, 32 split planks could theoretically be obtained from a log. In practice, however, not all splits will run true and in recent experiments at Moesgård, Denmark Egon Hansen has obtained an average of 20 sound, radial planks per log; boatwrights in antiquity may have done better. In fact Carr Laughton (1957: 247) has noted an entry in the early fifteenth-century accounts of Henry IV that 40 oaks were converted into 1200 boards, and 10 others into 390 boards. The resulting averages of 30 and 39 boards per oak indicate either very competent splitting, or that split timbers were subsequently halved by cutting across the grain, or that these were isolated oaks of great girth.

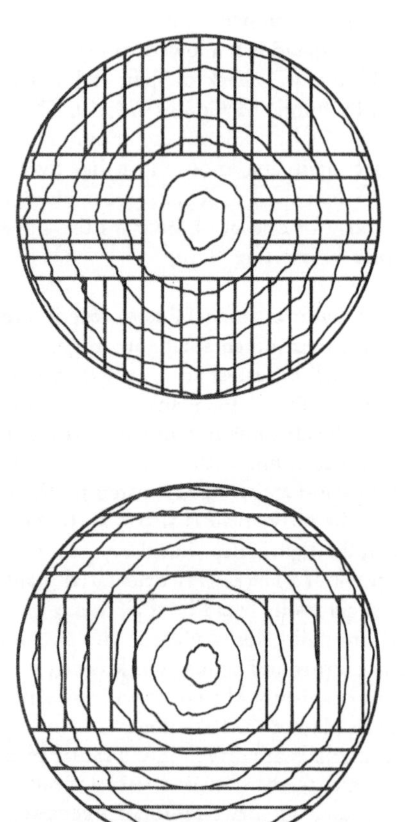

Figure 4.5 Recent methods of converting timber: Lower. Plain sawn conversion produces tangential boards. Upper. Quarter sawn conversion produces boards which are near-radial.

Curved timbers

Long runs of (near) straight planking are needed in quantity for the central parts of larger boats; in small boats, however, planking is rarely parallel-sided (Fig. 4.7) – this is the way the boatwright obtains a three-dimensional shape from (almost) two-dimensional material (see Ch. 8). A turn at one end of the planking may be obtained if a log includes the bole swelling just above ground level, but it is difficult to fell a tree at this level. But there is curved grain wherever a limb joins the main bole, and major limbs themselves, which generally have some curve, may be of sufficient breadth to supply planking. Whether from bole or from branch, curved planks may be split with little more difficulty than from a straight log; boles with *spiral* grain, however, cannot be satisfactorily split – this occasionally occurs in oak, but is frequent in elm. The finished plank is worked so that its curve follows the natural curve of the grain, thus ensuring maximum strength. Ideally there should be no short grain which is a potential weakness, but where the planking is required to have a double curve short grain cannot be avoided if the strake (run of planking from bow to stern) is to consist of only one plank, as in the Gokstad *faering*

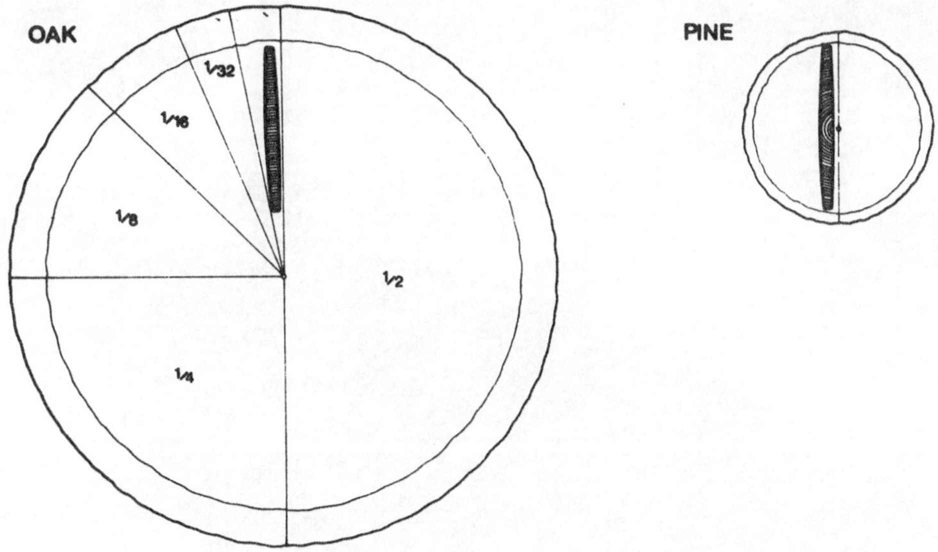

Figure 4.6 Diagram of wood splitting techniques with oak and pine. After Olsen and Crumlin-Pedersen, 1967, fig. 62.

Figure 4.7 The dismantled port side strakes of a small clinker-built boat, showing the two-dimensional curvature required to obtain the three-dimensional shape. (Photo: NMM Greenwich.)

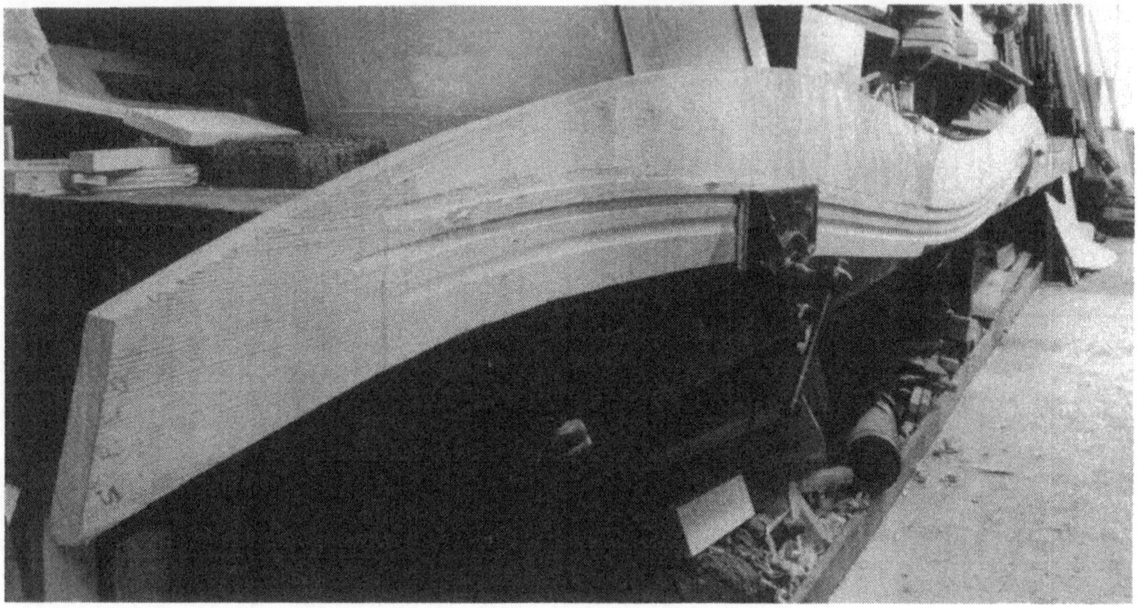

Figure 4.8 A strake in preparation for the Greenwich *faering* replica. (Photo: NMM Greenwich.)

(Fig. 4.8). With two curved planks scarfed together, however, such double curved strakes may be obtained with little, if any short grain.

Curved timbers are also required for transverse members such as ribs (floor timbers), knees and stems. See, for example, Nydam boat 2 (Shetelig, 1930: fig. 6). Curved boles or crooks from the junction of bole and limb (Fig. 4.1) or of two limbs are chosen to match as near as possible the 'design' requirements. An important feature of boatbuilding in the shell tradition (see Ch. 2) is that although the builder may have an ideal shape and size in mind for a particular boat, providing symmetry is maintained, the 'design' is sufficiently flexible to be adjusted within certain limits to match the characteristics (length, curvature etc.) of the timber actually available: thus if the run of grain is not as tightly curved as the 'design' requires, the curve of the stem can be slackened to match that of the timber.

Light timbers and withies

Withies which are to be made into a boat's framework by basketry techniques are today stripped of the bark, dried and stored. Before use they are given an overnight soaking to make them flexible. Whether they were used 'green' in former times is unknown but seems likely. Light timbers which are to be used for structural members which do not require such flexibility may be fashioned from more substantial stems or straight branches: for example the longitudinal laths which covered the seams between adjacent planks of the Brigg 'raft' were split stems or branches of hazel (*Corylus avellana*) with the bark left on and of average diameter 5.5 cm (McGrail, 1981b: 101, 222). The transverse timbers which linked the cleats on the 'raft's' planking had been fashioned from radially split oak, and worked to a rectangular cross section (*c.* 15 cm × 3 cm) in lengths of up to 2.27 cm. Nineteenth-century American Indians converted the two main timber species used for the framing of barkboats according to their specific qualities; thus white cedar (*Thuja occidentalis*) was dried before splitting and then bent and set to shape using hot water; whereas black spruce (*Picea mariana*) was split and bent while green (Adney and Chapelle, 1964: 17).

BARK

North American Indians used bark immediately after it had been stripped or they rolled it up tightly in the direction of growth of the tree and stored it in water to prevent it drying out before use (Adney and Chapelle, 1964: 24). Rolls of birchbark found at the mesolithic site at Star Carr were 1 in to 8 in wide, the largest being only 30 in long (Clark, 1954) thus by recent standards they were inadequate for boatbuilding.

ROOTS AND OTHER MATERIALS USED AS ROPE

Roots have a similar growth to shoot-led stems but it is less seasonal (Edlin, 1970: 4, 5) and the growth rings are less well marked and may be absent or irregular. Roots undergo tension rather than the compression stresses imposed on a bole. Thus roots are less massive, their vessel patterns are different and the fibre cells are often large with thinner walls (Dimbleby, 1978: 102). In good growing conditions roots of even the tallest trees are generally not deeper than *c.* 1 m and thus are relatively accessible. Before use the outer layer is first peeled from the roots and then they are often split. North American Indians used stone knife, clamshell or thumbnail for this, the split being started by biting or by pounding the end with a stone against an anvil (Adney and Chapelle, 1964: 20). The split roots were then kept in water to retain their flexibility; sometimes they were also boiled just before use. Split roots can be used as simple lashings or for continuous sewing without further preparation or they may be laid up and twisted into ropes.

The main materials used for rope in northern and western Europe in recent times have been bast fibres of hemp (*Cannibis sativa*) and of flax (*Linum usitatissimum*), and the leaves of manila (*Musa textilis*) and of sisal (*Agave sisalana*). In prehistoric Britain lime bast, clematis and nettles were twisted into rope (Heal, 1978: 29). The Brigg 'raft' had two split strands of poplar/willow (*Populus/Salix* sp.) for the thread or sewing material used to fasten the planking (McGrail, 1981b: 101); this was badly degraded when excavated but ethnographic analogies and the evidence from the prehistoric withy ropes from Breiddin (Musson p.c.) suggests that the split willow strands had been well soaked and twisted along their length before being laid up as rope. The Ferriby evidence was also difficult to interpret but Wright (1984) has deduced that the sewing thread was single strands of stripped but unsplit yew withies, which had been twisted and cracked to increase flexibility. The rope used to repair a split in the Appleby logboat was two-stranded birch (*Betula* sp.). A rope from the medieval site at Wood Quay, Dublin is also of yew (*Taxus baccata*) – 10 or 15 'yarns' of split or whole withies with bark were bundled into strands and two strands were then laid up, right-handed into a rope of *c.* 10.85 cm (*c.* 4 in) girth. In contrast, a *c.* 5 cm (2 in) girth rope found with the sixteenth/seventeenth-century AD boat from L. Caldecotte, Milton Keynes was made from hemp (*Canabis sativa*) bast and consisted of three right-hand laid strands composed of left-hand laid fibres (J. Squirrell, p.c.)

The properties that all these materials used for rope have in common are:

(a) when laid-up they have good tensile strength, especially when wet;
(b) have very little stretch under load and return to original form;
(c) abrasion resistant;
(d) withstand sunlight and heat. However, if they are left wet they are susceptible to bacterial decay.

DEGENERATION AND DECAY

HEART ROT

A tree may be damaged in several ways during its life: wind, fire, mechanical wear and tear and biological agencies such as rodents, birds, insects, plants and especially fungi and bacteria. Large-sized, and therefore over-mature trees tend to develop a rot known as brittle-heart (Jane, 1970: 228) which generally begins at the butt end extending gradually up the centre of the bole. The precise relationship between extent of rot and age/size of oaks is not known but almost all modern oaks greater than *c.* 2 m in diameter (6.28 m girth) are hollow due to this decay (Mitchell p.c.). Open grown, unpollarded oak can achieve a 6 m girth in something over 200 years (Mitchell p.c.); forest grown oak, on the other hand, grows more slowly in girth but attains a greater height. The 12.2 m bottom planks of the Brigg 'raft' were fashioned from close to the central core of the log (Fig. 4.9) and thus heart rot could not have been extensive in these 170 year-old oaks of *c.* 3 m girth at breast height. The trees used for the Ferriby boats, one of which was at least 260 years old (Hillam, forthcoming) can also have had no substantial heart rot in the sections that have been examined. It is not possible to be so definite about the medieval oaks from which medieval radial planking came, as in their conversion a substantial amount of the central core of the tree is worked away – *c.* 15 cm radius in a 1 m breadth tree (Olsen and Crumlin-Pedersen, 1967: fig. 62) – and this could be sound or rotten.

Heart rot in the lower length of an oak bole would make it unsuitable for hewing into a logboat unless the rotten end were subsequently closed with a fitted transom. Some of the broader British logboats have such transoms, e.g. Brigg whose parent log must have been at least 1.9 m in diameter (5.97 m girth); Poole, 1.72 m; Clifton 1 and 3, both 1.04 m; and

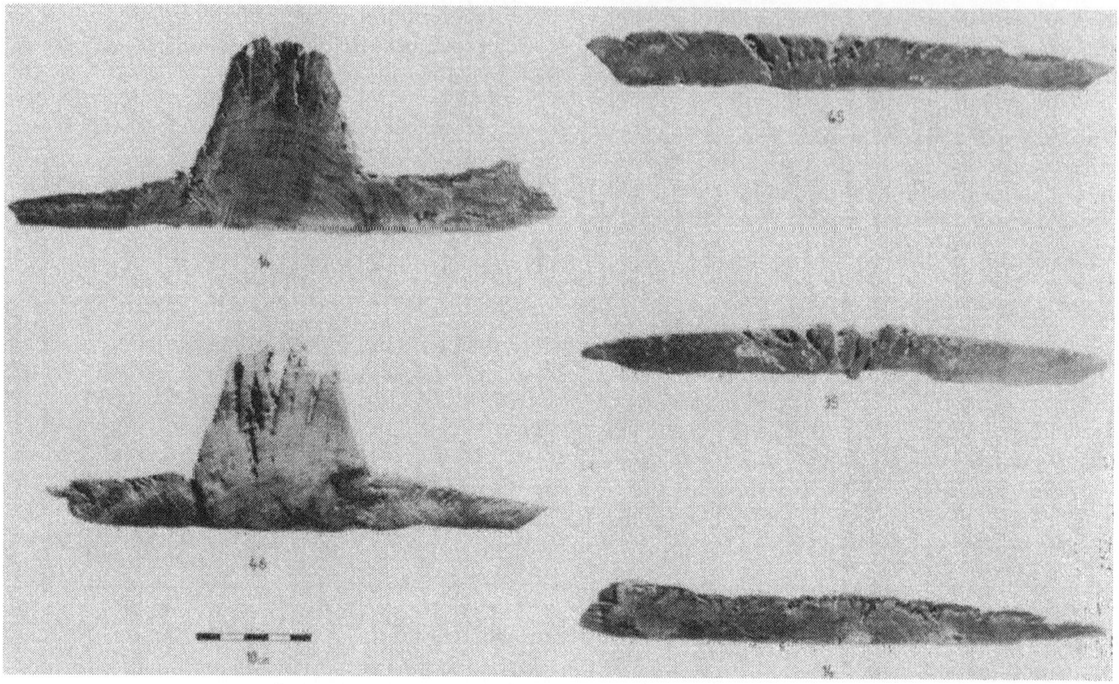

Figure 4.9 Sections through several cleat-units of the Brigg 'raft' showing the ring pattern and the medullary rays. (Photo: NMM Greenwich.)

Preston 1, 0.98 m. The Hasholme logboat (Millett and McGrail, 1985) of *c.* 12 m in length not only has a fitted transom but also a composite bow. Transoms and composite ends may be fitted because of damage sustained during felling or conversion, but it may be that the Hasholme parent log had some heart rot throughout its usable length. On the other hand several other relatively broad boats made from whole logs, notably Hulton Abbey (1.58 m) and Baddiley Mere (1.30 m) had integral ends and thus cannot have had rot in the part of the log used. The parent logs of such logboats as Banks (1.42 m) made from *half* a log split longitudinally may have had a certain amount of heart rot which could have been hewn away during building but leaving sufficient wood for the boat to have two integral ends (McGrail, 1978: 121–2, 310, 317–18).

It seems clear that large, relatively sound oaks could be found in prehistoric Britain, but that some of those of 1 m or more in diameter (3 m in girth) may have had heart rot.

TIMBER DEGRADE

After being felled a tree may be damaged by shrinkage stresses, as discussed above, if dried below f.s.p. During this drying phase the wood is also susceptible to fungi attack as well as weathering and mechanical damage. Sapwood is especially susceptible to fungi, brown rots which attack the cellulose, and white rots which consume both cellulose and lignin.

Timber discarded after use will generally be degraded to humus by bacterial, fungal and insect attack. Timber underwater as in a wreck (or indeed during the useful life of a boat) may also be attacked by marine borers such as the crustacean *Limnoria lignorum*, the gribble, and the mollusc *Teredo navalis*, the shipworm. Fungi which cause the most serious damage to wood generally require oxygen, moisture and an ambient temperature between *c.* 10° and *c.* 40°; thus anaerobic and dry conditions, and high and low temperatures inhibit fungal activity. Ancient wood can therefore survive, albeit in a degraded state, in waterlogged land sites, and on sea, river and lake beds when protected by silts. As this wood becomes waterlogged, the available oxygen is used up by aerobic organisms and further fungal degradation of the cell wall components ceases, except that due to oxygen dissolved in the water. Where low temperatures prevail in waterlogged or underwater sites chemical and bacteriological production of oxygen is reduced to very low rates. Under these conditions some bacterial damage can occur to the wood and sulphate-reducing bacteria may produce hydrogen sulphide (H_2S) which gives waterlogged wood its characteristic smell of rotten eggs. Other bacteria can destroy pit membranes in the wood causing loss of strength and allowing other destructive agents to penetrate the wood material, but, generally, bacterial damage is not so serious as fungal during the deposition stage.

When degraded wood is disturbed from conditions in which it is in (near) equilibrium with its environment (by, for example, excavation) fungal and bacterial action can soon be reactivated. Hence the necessity for extreme care when excavating timber.

The surface of wood that has been decayed by fungi typically shows cuboidal cracking as splits develop across and along the grain. In advanced stages of dry rot decay timber becomes orange-brown in colour and crumbles to dry powder when handled (Jane, 1970: 230–2, fig. 120a). Excavated timber which is not kept waterlogged may be attacked by such rots and the ensuing cuboidal cracking may be interpreted by the inexperienced eye as charcoal produced by the use of fire during the building of a boat; see, for example, Wilson (1851: 32) and Geikie (1879:2).

EXAMINATION OF EXCAVATED WOOD

Much can be learned about the environment, timber management and selection, woodworking technology and boatbuilding by systematic examination of excavated timbers, especially if this can be done in controlled conditions in a post-excavation workshop or laboratory.

ORIENTATION

In addition to identification of timber species, careful examination can reveal facts about the parent tree and the conditions under which it was grown. Logboats and long lengths of planking such as those from the Ferriby boats and the Brigg 'raft' can be especially informative. It is reasonable to assume that in a tapered logboat the broader end is the butt end of the log: except in the case of a most unusual pattern of damage it would be perverse to work the broader end from the narrower (upper) part of a log; the argument may be extended to tapering planks. As growth increments show in a longitudinal section of the log as a series of cones with their apices pointing to the top of the tree (Jane, 1970: 71–2) detailed laboratory examination of cross-sections through a logboat or a long plank could also reveal log orientation. Professor Grimes (1931: 140) pointed out many years ago, however, that branch insertion in an oak tends to be at or above the horizontal: thus if there are several knots in a log or plank at a similar angle to the log's axis the identification of the upper and lower ends can be made with reasonable confidence. Where knot emergence angle is not readily apparent the direction may be obtained by overlaying plans of the outer and inner sides of a plank (McGrail, 1981b: 59).

KNOTS

A knot is that portion of a branch enclosed within the wood as the tree grows. On conversion after felling, a live branch leaves integral knots within the timber; a dead branch leaves a dead knot which may become loose as the wood dries. The normally vertical run of the grain (arrangement of the fibres) is disturbed around a knot thus reducing the strength of the timber, more in tension than in compression or bending. This reduction is proportional to the size of the knot in relation to the size of timber. In general a branch will appear as a strength-reducing knot in many planks of a tangentially converted log, whereas only those one or two radial planks which lie within the angle of the branch will be affected.

In forest conditions the lower, shaded branches of trees die off early and are grown over, leaving a knot only within the first few growth rings. Significant runs of knot-free timber may thus be obtained; the parent log of the Brigg logboat is reputed to have had a clear bole for some 14 m from the butt end; an oak recovered from the peat at Adelaide Bridge in the East Anglian fens and dated to *c.* 2535 bc had a bole of 67 ft (20.42 m) without branches (Willis, 1961: 371; Ashby, 1966: 35). During the post-excavation examination of the Brigg 'raft' few knots were found in bottom planks 1 and 5 (from the same log); in plank 2, although there was a significant number, they were mostly less than 5 cm in diameter and represented lower branches which had died back early as the forest canopy closed and competition increased. Knots of 20 cm and over were, however, found in planks 3 and 4 (from the same log) (McGrail, 1981b: 86, 223).

GROWTH RINGS AND RAYS

The general run of grain and the lie of the growth rings visible in a cross-section of planking can also

Figure 4.10 Hardham 2 logboat in Worthing Museum showing splits in one end radiating from the centre of the log. (Photo: NMM Greenwich.)

show how the parent log was converted (Fig. 4.9). This may also be seen near the ends of logboats. Thus logboat Hardham 2 was made from a whole oak (Fig. 4.10), whereas that from North Stoke was fashioned from one half of a longitudinally split oak (Fig. 4.11). The Brigg 'raft' bottom planking and most of the Ferriby planking was converted from half logs with their underside (outboard) nearer the pith.

The distinctive and characteristic rays can also be seen in the cross-section of an oak plank or logboat. In re-excavated waterlogged wood and in other samples that have been allowed to (partly) dry out, splits radiating from the pith emphasise the alignment of these rays (Fig. 4.10). If such ray patterns are recorded by direct tracing on to drafting film, the partial radii can be projected to give the approximate position of the log's pith at that station in the plank. As ancient boatbuilders were almost invariably seeking to make maximum use of the timber available to them it is reasonable to assume that original edges of the planking or timber are very close to the heartwood/sapwood transition.

Mature oaks today have *c.* 3 in (75 mm) of sapwood and *c.* 1 in (25 mm) of bark (Mitchell, p.c.); a sapwood and bark allowance of 100 mm may thus be added to the distance between pith and heartwood/sapwood transition to give the approximate radius of the parent log of large timbers. If this calculation can be done near both ends of a logboat or plank, an idea of log taper may be obtained. More accurate estimates of parent log size may be made during dendrochronological examination of the ring pattern. The Brigg 'raft' parent logs have been estimated during such examination to have had diameters of 0.64 m to 0.80 m (Hillam, 1981); the corresponding estimates using ray projection were 0.60 to 0.74 m (McGrail, 1981b: Table 2.3.2). It may be more appropriate to use a reduced sapwood and bark allowance for smaller and therefore younger logs and limbs. Thus the parent logs of stem, keel and frames of Nydam boat 2 (Shetelig, 1930: figs 6, 8, 9) which have the pith running through them, would have had diameters of 50 cm, 40 cm and 20 cm respectively, if a 50 mm radius sapwood and bark allowance is used.

The Brigg 'raft' is dated by radiocarbon assay to the mid-first millennium bc, the mean of six dates

Figure 4.11 One end of the North Stoke logboat showing that this boat was fashioned from half a log. (Photo: NMM Greenwich.)

being *c.* 650 bc (McGrail, 1981b), and as no British oak chronology extends back so far Hillam was unable to date the 'raft' dendrochronologically. However, it proved possible to produce a Brigg mean curve covering 160 years and because several samples had *c.* 10 years of sapwood she could estimate that the parent log of plank 4 (which had been fashioned from very close to the pith) had been *c.* 170 years old when felled. Dendro examination also showed that the ring patterns of planks 1 and 5 were so similar that they must be paired halves of the same log; planks 3 and 4 were similarly shown to be halves of another log. This ability to match growth ring patterns from different timbers proved invaluable in the examination of two fragments of timber located some 5–6 m away from, but in line with, the main part of the 'raft' (Fig. 4.12). Dendro-analysis showed that fragment G5 (found in line with plank 5) was from the same tree as planks 1 and 5; and fragment G4 (in line with plank 4) from the same tree as planks 3 and 4. Fragment G4 was too small to compare visually with strake 4, but G5 had a similar cross-section to strake 5. It thus is highly probable that G5 was an integral part of plank 5 when the 'raft' was in use and when deposited (the intervening sections of the plank having been lifted during excavation in 1888). This hypothesis gains additional support from the fact that the girth deduced for fragment G5's parent log, 2.45 m, indicates a steady increase in girth from the 2.01 m estimated for plank 5 at its nearest point to G5. Similar arguments may be advanced to show that plank 4 originally included fragment G4. The minimum length of these planks thus becomes 12.2 m rather than the 6 to 7 m which could only legitimately have been deduced before the dendro examination.

The logs used to build the bottom of the Brigg 'raft' may in summary be described:

> Three standing forest oaks of straight grain were chosen, two of them, trees 2 and 1/5, had few if any low branches, tree 3/4 had some substantial ones. The timber was narrow ringed (0.67 to 2.57 mm) average ring widths being 1.54 mm, tree 1/5; 1.50 mm, tree 2; 1.20 mm, tree 3/4; and probably reflects the crowding experienced from other trees. These three oak boles were more than 12.2 m in length with girths of 2.70 to 3.03 m at breast height (1.5 m); 2.21 to 2.49 m at 5–6 m

Figure 4.12 Vertical view of the Brigg 'raft' during excavation. The main part of the remains consists of five bottom planks and one side strake (top left). The white dots mark some of the holes through which the planking was sewn together. Fragments G4 and G5 are over to the right. (Photo: NMM Greenwich.)

from the butt; and *c.* 2.07 m near their upper ends. It seems likely that all three trees were damaged at their butt ends before conversion, with the damage extending further up the bole of tree 3/4 in the half that became plank 4 (McGrail, 1981b: 84, 224). This description and the details obtained from the parent logs of the Ferriby boats (Hillam, forthcoming) illustrate the value of laboratory examination of excavated timbers even if it involves, as it did for Ferriby and Brigg, sawing samples from ancient timber. The Brigg dendro samples are now being conserved with the remainder of the 'raft' and will be reinstalled in their original position when the remains are reassembled (McGrail, 1981b: fig. 2.4.8).

The Brigg oaks may be compared with other large oaks noted in archaeological publications (Table 4.1).

Slow grown wood has closely-spaced rings (say ≤ 2 mm in width), quick grown has wider ones. In oak very slow growth tends to reduce the wood density compared with more vigorous growth. Narrow and broad rings may be found in the same tree reflecting harder and easier growing conditions. Rings which vary significantly in width on different sides of the bole or branch may be due to:

(a) slope compensation by a tree growing on a hillside.
(b) reaction to lateral crown development caused by prevailing wind: growth rings will be broader on the leeward side.
(c) branches with different widths of rings, above and below; thus in conifers the underneath part of the branch has wider rings whilst the underneath of hardwoods have narrower ones.

Thus measurement of ring widths along only one radius of a log or branch may give misleading information about the general growth rates of the tree. If several radii can be investigated a more reliable overall picture will be obtained. In the case of the Brigg 'raft', measurements were taken along six widely-spaced directions in trees 3/4 and 5/1, and

Table 4.1 Length and girth of large oaks

Oak log	*Length (m)*	*Girth (m)*	*Reference*
R. Went – oak	9.75	4.78	Whitcomb, 1968: 299
Poole logboat	10.01	5.40 near butt 3.77 at 9 m.	McGrail, 1978
Brigg logboat	14.78	5.90 near butt 5.40 at 13.5 m	McGrail, 1978
Hatfield Chase – oak	36.58	11.50 near butt 5.75 at 36 m	de la Prynne 1870; Peck, 1815: 3

three directions in tree 2, and at differing heights in the boles. Thus the conclusion that all three were slow growing is soundly based. Such detailed work has not yet been attempted for logboats, but Ellmers (1973) and Nikkilä (1947: 38) have noted that recent logboat builders orientated asymmetrically grown logs so that the narrower growth rings were uppermost (or sometimes bottom dead centre) so that bias was not built into the boat.

Recognition of the general place of origin of boat's timbers may prove possible where regions have distinctive oak chronologies. The Wood Quay boat timbers were re-used in Dublin waterfront structures dated to the early thirteenth century. Samples from them have been dated by reference to the Dublin chronology which extends from AD 855 to 1306 (Bailey, 1978), resulting in dates between 1089 and 1209. A Bristol master curve for the period 1059 to 1239 has recently been compiled from radially-fashioned boat planking re-used in a waterfront structure (Hillam, p.c.). This curve shows a higher correlation, and individual timbers agree better, with the Dublin curve than with Exeter, Belfast or Ref. 6 (mainly London). As the Bristol curve is entirely from planking from boats which may have been built almost anywhere in the Irish Sea region, it may not be representative of locally grown Bristol timber; and it may be that the Bristol waterfront was built in part from timber grown in the Dublin area. Alternatively, these interim results may be no more than a reflection of similarity in growing conditions for eleventh-thirteenth-century oaks in eastern Ireland and south western Britain. Measurement of further timbers as they become available in S.W. Britain may clarify this.

SHRINKAGE ESTIMATES

If green timber is allowed to dry below f.s.p during building a boat or during use it will shrink differentially; for example, if fresh oak is dried to 12% m.c. it will shrink *c.* 7.5% tangentially and *c.* 4% radially (Farmer, 1972: 147). A circular hole bored radially into a green log has one axis of its cross section in the longitudinal plane of the parent tree and another axis in the tangential plane (Fig. 4.13). Differential shrinkage will change this circular hole into an elliptical one, as its dimensional change in the direction of the longitudinal axis will be negligible, whereas there will be significant shrinkage in the tangential plane. The ratio of elliptical holes' major and minor axes is a measure of the tangential component of this shrinkage. A tangentially bored hole will similarly give a measure of radial shrinkage. Such dimensional changes in fresh wood are reversible and the holes will regain their circularity if m.c. is increased to above f.s.p, providing any splitting experienced has not distorted the holes.

Waterlogged wood, on the other hand, has experienced chemical and biological degradation which has altered its normal structure, thus shrinkage experienced during drying from a waterlogged state will be abnormally high and will be irreversible. Circular holes changed into elliptical holes by differential shrinkage will retain their elliptical cross-section and such holes are frequently found during the post-excavation examination of excavated boat timbers which at some stage have been allowed to dry (McGrail, 1978: fig. 125). The suggestion has been made (McGrail, 1978: 123–4) that, despite the structural differences between fresh and waterlogged wood, the ratio of major and minor axes of such holes will give an estimate of shrinkage experienced between the time the hole was bored and the time it was measured.

The precise effect of differential shrinkage in holes in dried waterlogged wood depends on a number of factors (McGrail 1978: 123–4):

1. original circularity or otherwise
2. orientation of hole
3. local run of grain and nearness or otherwise of knots
4. presence or absence (and if absent the point in the timber's history when it was lost) and the orientation of the fastening associated with the hole
5. timber species (and of treenail if that is the fastening used).

Not only will different boats show different shrinkages depending on their particular history but different parts of the same boat may undergo different rates of shrinkage which may be reflected in the changes imposed on individual holes. The best measure of the overall shrinkage a plank boat or a logboat has experienced is thus considered to be the appropriate *shrinkage index* defined as the mean of the several ratios:

$$\frac{\text{hole dimension in the longitudinal axis of the timbers}}{\text{hole dimension at right angles to longitudinal axis}}$$

for all measurable holes bored into the logboat or the structural elements of a plank boat. Post-shrinkage dimensions are multiplied by the appropriate shrinkage index to obtain pre-shrinkage dimensions. These indices are related to, but not the same as, the

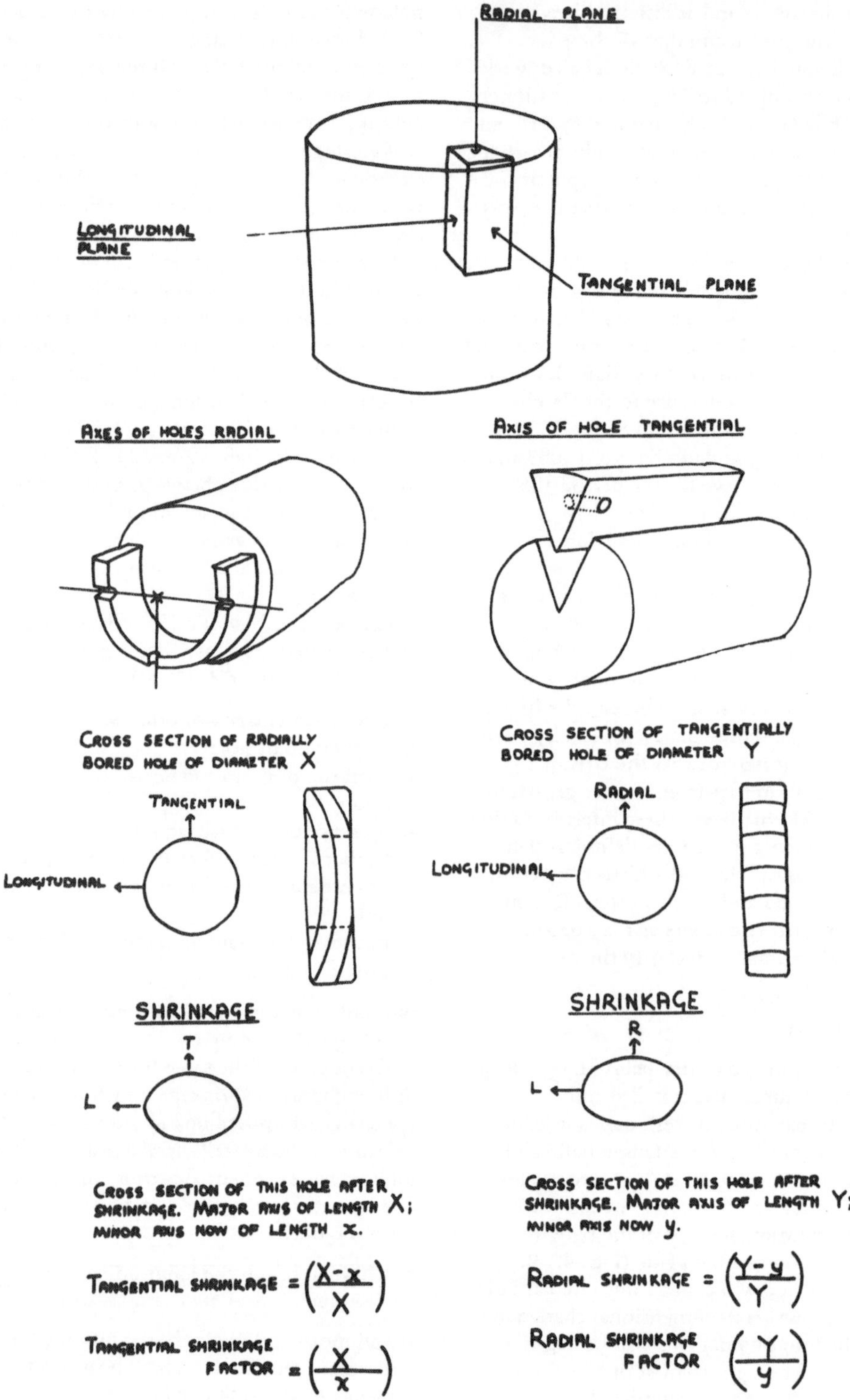

Figure 4.13 Calculation of tangential and radial shrinkage factors. (Diagram: NMM Greenwich.)

measure implied by the normal use of the word *shrinkage*, generally taken to be the difference between post and pre-shrinkage dimensions expressed as a % of the pre-shrinkage dimensions.

When undertaking shrinkage investigations the dimensions of each hole should be measured at right angles to the axis of the hole, and at both faces wherever possible. Any nearby abnormalities in the timber should be noted as should the absence or presence and orientation of any fastening in the hole. Broken holes should not be included. Elliptical holes have been measured in some of the surviving logboats of southern Britain, on the Brigg 'raft' and on some of the Wood Quay timbers, all of which are waterlogged timbers which had been allowed to dry out to a greater or lesser degree. In the Brigg 'raft' 170 holes had a *shrinkage index* (in the plane of the horizontal portions of the planking, twixt radial and tangential) of 1.12, equivalent to a shrinkage of *c.* 11%. Multiplication of the present plank breadths by this index gives pre-shrinkage dimensions which would result in the edges of adjacent planks touching rather than being separated by 2 to 5 cm as they were found to be on re-excavation in 1974 (McGrail, 1981b: 228).

The mean *tangential shrinkage index* for 131 holes in 32 logboats was 1.27 m equivalent to *c.* 21% shrinkage. Individual shrinkage indices were calculated for each logboat, the tangential indices ranging from 1.70 to 1.13. Logboats which do not split asunder and are not distorted by severe rates of shrinkage, shrink in transverse section down the sides and across the bottom, i.e. tangentially. The tangential shrinkage indexes were therefore used to increase post-shrinkage dimensions of breadth and depth to their original values. The radial index was used to obtain pre-shrinkage thickness of sides and bottom (McGrail, 1978: 127–8); where this value was not available for a particular boat 1/3 the value of the tangential index was used. The general effect of applying shrinkage factors to the dimensions of the logboats as they now are in museums is that, whilst the length remains the same, the breadth and depth become *c.* 20% greater and the sides and bottom some 5 to 10% thicker. An excellent demonstration of differential shrinkage may be seen in the Stipelse logboat (Ellmers, 1973: 33). Both logboat and fitted rib have shrunk, mainly tangentially. But the boat's shrinkage has reduced its depth (shrinkage in breadth appears to have been prevented by the rib) whereas the rib's tangential shrinkage has been across its cross-section, and its length (as that of the logboat) has remained the same. Thus a thinner rib protrudes above the boat and the treenail holes in the rib no longer coincide with those in the boat.

The average tangential shrinkage of *c.* 21% is much greater than that of fresh oak (*c.* 7.5%). Individual shrinkages varied from 41% as indicated by measurements of 12 holes in the Smallburgh logboat, to under 12% from 6 holes in a patch on the Oakmere boat. Radial shrinkage ranged from 11.5% to 8.2% whereas for fresh oak this is *c.* 4.5%. Anomolous readings noted range from a tangential ratio of 2.10 from a hole in the North Stoke logboat to one of 0.92 in the Warrington 1. In the latter case the hole had evidently shrunk in the longitudinal axis of the boat.

Support for this method of estimating pre-shrinkage dimensions comes from two directions. The scantlings of those logboats whose dimensions were given in the earliest report on the find were compared with present dimensions. This was an adaption of a method used by Åkerlund (1963: 155); see McGrail (1978: 124) for some practical difficulties. Eighteen logboats gave a mean tangential shrinkage factor of 1.20 (range 1.15 to 1.30), equivalent to the shrinkage of *c.* 16%. The elliptical hole ratios indicate about the same order of shrinkage. Secondly, recent work in the Archaeological Research Centre laboratories at Greenwich has confirmed that the shrinkage factor for naturally dried fresh oak calculated from elliptical holes is a good measure of overall shrinkage; this research is now being extended to waterlogged wood and to investigate the effects of knots and the presence or absence of fastenings (Richard Clarke, p.c.).

5 Log rafts

DISTRIBUTION AND FUNCTION

Log rafts have had widespread use on inland waterways wherever there were trees of buoyant timber species. In addition they have been used for inter-island traffic in, for example, Oceania (Hornell, 1946: 76) and in the coastal waters of S. America (Edwards, 1965: 64–80), Formosa (Taiwan) and India (Hornell, 1946: 61–8, 79–80), and off other countries between latitudes 40 °N and 40 °S where the sea temperatures are suitable for flow-through water transport (Ch. 2). There is no direct evidence for early transocean voyages by rafts, but Heyerdahl (1978: 188–213) has demonstrated that in certain circumstances such conjectural voyages are practicable.

Use as ferries and for cargo and passenger carrying on rivers is known worldwide, as is their use for both river and coastal fishing. The *kola-maram* of Coromandel may spend up to three days at sea and venture out to 25 n.m. from land when in pursuit of flying fish (Hornell, 1946: 62–6). Other catamarans of S. India work at sea in pairs, under sail and towing a trawl (Hornell, 1946: 62–3). An oval tub is fitted abaft the mast of Formosan rafts for fishing tackle, and in Australia a circular fence of upright poles set into the raft platform serves a similar purpose (Hornell, 1946: 72).

In coastal waters in S. America, small log rafts are used as tenders and lifeboats by the crews of larger planked vessels (Edwards, 1965: 64). An echo of such use may be seen in Chapter 7 of the first/second century AD *Periplus of the Erythrean Sea* (Huntingford, 1980) where there is a reference to the unloading of ships by log rafts at *Aualites*, possibly Zeyla in Somaliland.

VARIETY AND SIZE

Hornell (1946: 72) classified the single log with the user sitting astride as a raft, but elsewhere (1946: 17) he called it a tree trunk float. This latter term seems more accurate as the user is partly immersed in the water and thus the log is a personal aid to flotation (Ch. 2). Single logs modified to increase their breadth and hence stability as used in the R. Gascoyne, Australia (Hornell, 1946: 73) where withies are twined around pegs driven horizontally into both sides of the log, may be considered as intermediate between floats and rafts.

The prototype log raft may be taken to consist of two or more logs fastened together by lashings or by wooden or metal fastenings. To give the raft a measure of directional stability it must be longer than broad, a shape which naturally arises when logs are placed longitudinally side by side. Rafts are poorly documented, but the average (L/B) ratio for rafts given measurement by Hornell (1946), Edwards (1966), Eskeröd (1956) and Worcester (1966) is 3.4; the range is from 1.75 to 5.6 with the majority being between 3.0 and 4.0.

These longitudinal logs must be held together so that they cannot move relative to one another – this is so that they form a rigid structure rather than a lattice-type mechanism (McKee, 1983: 51–3). Although this rigidity could theoretically be obtained with spaced-apart logs using diagonal as well as transverse fastenings, the simplest arrangement, and one that is invariably used, is to position them so that they are contiguous along their length and this arrangement requires only transverse fastenings.

McKee (1983: fig. 29) has pointed out that when transverse members or lashings are used to link the longitudinally-laid logs together, a rigid structure may be achieved in two ways:

1. two or more longitudinally-laid, contiguous logs joined transversely by two or more, spaced-out shorter members, or

2. two or more spaced out, longitudinally-laid logs joined transversely by shorter, contiguous logs.

The vast majority, if not all, the world's log rafts seem to have been built by the first method, with the longitudinal logs contiguous. The transverse, spaced-out fastenings, where they are of timber rather than lashings, are almost invariably on top of, or through the thickness of the longitudinal logs; the only known exceptions are in some Solomon Islands rafts where the transverse fastening timbers are underneath, and in the Lacadives where these timbers are above and below (Hornell, 1946: 74–5, 68).

McKee (1983: fig. 30) has also identified seven curved raft structures but these are theoretical constructs derived in retrospect from his conjecture that certain plank boat types were 'developed' from raft structures, and they do not necessarily correspond with reality. In particular the three structures with contiguous short, transverse timbers are not known in the raft world.

Log rafts may be shaped using the natural taper of the log from the butt end, so that they are broader aft than forward. The practice is widespread (for examples in Chinese bamboo rafts (Doran, 1978: fig. 1; Worcester, 1966: 119)): see also the shape of the so-called 'raft' from Brigg, a prehistoric sewn plank boat (McGrail, 1981b). This elongated trapezium shape enhances the raft's waterline shape and hence speed and manoevrability characteristics. A bluntly pointed forward end has also been obtained by using an odd number of logs of varying lengths, the outer ones being set back from the central one (Fig. 5.1): this is also sometimes done at the after end. Such shaping was a feature of the rafts used in S. American coastal waters from at least the sixteenth century (Edwards, 1965: 62, 64, 71); that this was also a feature of pre-Columbian times is suggested by

Figure 5.1 Sixteenth-century log rafts of Ecuador as drawn by G. Benzoni. After Edwards, 1965: pl. 16b (courtesy The University of California Press).

the shape of the model raft excavated from an early site near Arica, Chile (Johnstone, 1980: fig. 2.2). Ends shaped by variable log lengths are known from elsewhere: Marquesas and Solomon Islands and Mangareva (Hornell, 1946: pl. 9, pl. 7AB, 77) and from Japan (Nishimura, 1925: 24).

A log raft may be given a hollow, boat-shape by careful selection and positioning of logs as seen in the rafts from Lobito Bay, Angola (Fig. 5.2). The shape is not obtained by bending or 'springing' the logs as McKee suggests (1983, fig. 30) but by using the natural grown curves and the taper of the logs and by attaching contiguous ones at an angle to each other rather than in the horizontal plane. Certain seagoing rafts of S.E. India are given a hollowed shape, curved both transversely and longitudinally, by similar methods and with the addition of separate upturning bow pieces (Hornell, 1946: 62–6; pl. 10, 11). Rafts of bamboo, when not flat (e.g. Worcester, 1966: 119) are also usually shaped by selection and attachment methods, but that from Formosa (Taiwan) (Fig. 5.3) has a curved up forward end obtained by bending, after the outer, most siliceous layer has been peeled off the bamboos (Doran, 1978: 18). This raft's curved transverse shape is, however, achieved by selection and attachment methods; this curve is supported internally by slightly-curved, spaced-out bamboos which are lashed to the longitudinal bamboos (Hornell, 1946: 79–80).

Log rafts are not just a 'bunch of old logs'; they are built with the workmanship appropriate to the required function, length of life and environmental operating conditions. They range from simple unprepared saplings or limbs temporarily lashed together (Hornell, 1946: pl. 8a; Brindley, 1931: 10) to rafts such as those of South India which Hornell (1946: 61–8) noted were built to 'standardised local designs' with each element having its definitive shape and position, and which had 'very considerable elegance in their proportions and often show(ed) quite pretty lines', and the substantial and durable Chinese rafts used as river homes (Fig. 5.4) or floating market places (Edwards, 1965: 62).

In size rafts have varied from single-man rafts of *c.* 6 × 3 ft (1.8 × 0.9 m) (Hornell, 1946, pl. 8a) to the great rafts used in the eighteenth century and earlier S. American coastal trade, some of which were *c.* 80 × 25 ft (24 × 7.5 m) (Hornell, 1946: 81).

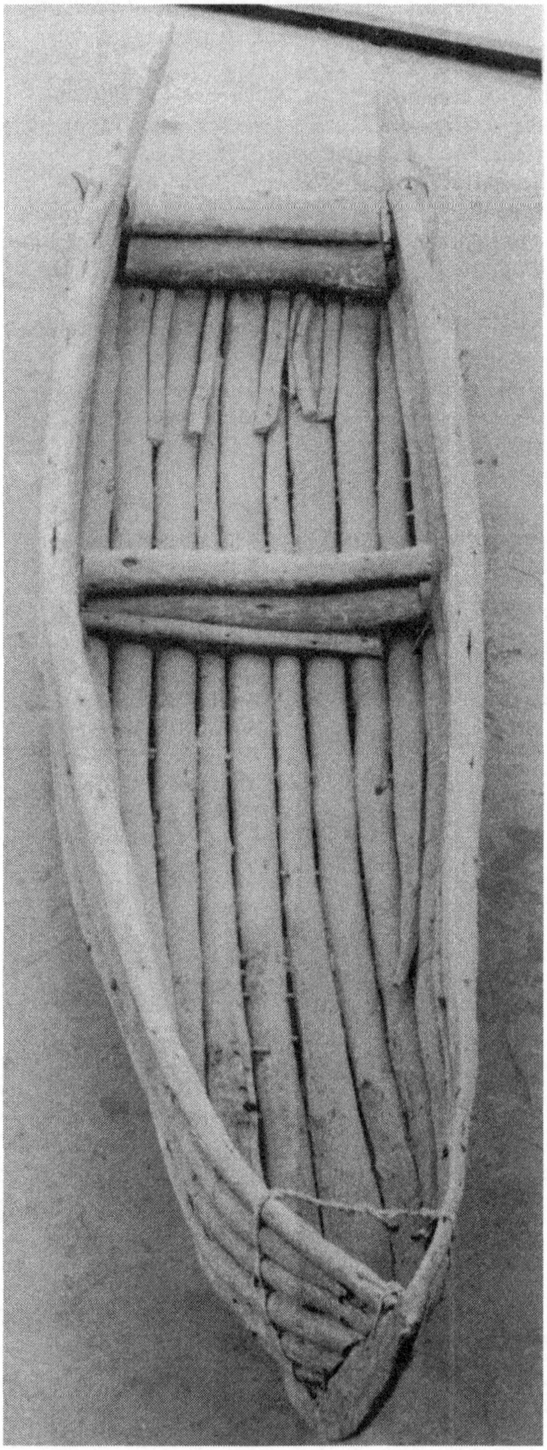

Figure 5.2 A boat-shaped log raft from Lobito Bay, Angola. (Photo: NMM Greenwich.)

CONSTRUCTION

RAW MATERIALS

Rafts have been built from many different species of

Figure 5.3 Model of a log raft from Taiwan (Formosa). (Photo: NMM Greenwich.)

Figure 5.4 A Chinese log raft with river homes (Photo: NMM Greenwich.)

timber, *Ambatch* (*Herminiera elaphroxylon*), Balsa (*Ochroma lagopus*), Bamboo (*Bambusa* spp). Banana, Giant cacti stems, Mangrove saplings (*Rhizophora* spp), Plaintain stems (*Katar bhelā*) (Hornell, 1946: 61–90), to give but a few examples. These species vary in size and in other properties; their one common factor is that they have very low density and correspondingly high buoyancy. Bamboo is very suitable for rafts; the giant bamboo can be 80 ft (25 m) high with a diameter of 1 ft (31 cm) (Needham, 1971: 394) and is of low density. Balsa, from which many S. American, especially Ecuadorian, rafts have been made, generally weighs between 5 and 16 lb/ft^3 (80–250 kg/m^3) at 12% moisture content. This may be compared with oak (*Quercus* sp.) which when green weighs *c* 67 lb/ft^3 (1072 kg/m^3); that is, it has a relative density of *c*. 1 and floats just below the surface of the water; but even when dried to 12% m.*c*. oak weighs 45 lb/ft^3 (720 kg/m^3) thus clearly 3/4 of a log would be submerged when afloat.

Logs or limbs are sometimes used in the round, especially for simple, temporary rafts; where more durable and useful rafts are required the timber is fashioned including the removal of bark and sometimes sapwood. This may result in the use of squared logs as in the Solomon Islands (Hornell, 1946: 74) or even of logs worked to a curved 'boat-shape' as are those of Vizagapatam, India (Hornell, 1946: 67, pl. 12).

BUILDING SEQUENCE

There are three main methods of fastening the longitudinal logs together.

1. *Direct lashing*: The logs are lashed together, often near the ends by, for example, coir rope in India, rattan in New Ireland, 'bark strips' in Australia and New Guinea (Hornell, 1946: 64, 73, 70, 78) and withies in Sweden (Eskeröd, 1956: 60). Details of these lashings are rarely given. However, Hornell (1946) has noted that in the New Ireland rafts the rattan lashing passes through seven series of holes bored transversely through the five logs. He also recorded (1946: 76) that in the New

Hebrides after three or four spaced-out lashings had been bound three or four times round all the bamboo logs, the upper and lower parts of these lashings were bound together by a light line at each interstice.

2. *Transverse timbers on top of the logs or inset*:
   2.1 *lashed* The logs are connected by spaced-out transverse timbers which are lashed to the logs by, for example, vines in S. America (Edwards, 1965: 74) and Fiji (Hornell, 1946: 78). Details of these lashings are seldom given but see Hornell (1946: 61, pl. 8A). In S.E. India grooves are cut in the sides of the logs near the ends of certain types of catamaran, and a transverse, two-horned block of wood is lashed to the logs with coir rope which fits into the grooves (Hornell, 1946: 62).
   2.2 *wooden fastenings* Dampier's surgeon, Lionel Wafer, described seventeenth century Panamanian log rafts in which the longitudinal logs were linked by spaced-out transverse logs fastened by vertically-driven hardwood 'pins of macaw wood' (quoted by Edwards, 1965: 94). Similar fastenings are known from Peru (Edwards, 1965: 95). Other instances quoted by Edwards (1965: 95) are either fastened by horizontal hardwood 'pins' driven through the thickness of the logs (see below) or the descriptions are ambiguous. The raft from Mangskog, Värmland, Sweden published by Eskeröd (1956: 60, fig. 1A) is fastened in this manner at one end. A second Swedish raft from Norberg, Västermanland (Eskeröd, 1956: 60, fig. 1B) has the logs joined by two 'bevelled' timbers set into transverse grooves. A third raft published by Eskeröd (1956: 60, fig. 1C) from Lapland has one end fastened by a thin 'feather' transverse timber set horizontally into the *ends* of the logs.
3. *Transverse timbers through the thickness of the logs*: The method of 'through pinning' or 'skewering' logs by series of hardwood timbers driven horizontally through the thickness of the logs from side to side of the raft, is known from many places: Australia (Hornell, 1946: 72); Oceania (Hornell, 1946: 72, 79, pl. 9; Nishimura, 1925: 88; Nelson, 1961: 166); S. America (Hornell, 1946: 82). The other end of the Mangskog raft described above is fastened by this method. The Lobito Bay boat-shaped raft is fastened together principally by transverse pins driven through the thickness of the logs (Hornell, 1946: 85) but in the raft in the National Maritime Museum's collection (AOA 313) not all such pins go right through all the raft's bottom (or side) logs. In all cases these transverse timbers appear to be held in position by friction generated from an interference fit, as no locking method is apparent.
4. Hornell (1946: 74) states that the logs of the *polat* rafts of the Solomons are 'joined together by three series of hardwood pegs driven tenonwise into their contiguous sides'. He subsequently compares (1946: 75) Polynesian 'pegged-together' rafts with the Moluccan *orembai*, a boat whose planks are fastened by a series of treenails. Hornell (1946: 67) also mentions the Indian rafts from Ganjam and Orissa in which the logs are 'pegged together with wooden pins'.

Mixed fastenings: The Mangskog raft described above has type 2 and type 3 fastenings, and that from Lobito Bay type 1 and 3 (Hornell, 1946: 85). Certain Madras rafts have type 3 fastenings which are supplemented by straw ropes fastened transversely around the projecting ends of the through fastenings (Hornell, 1946: 68). Edwards (1965: 94) has noted that small river rafts in Ecuador have transverse timbers which are not fastened but pressed into the logs by obliquely-driven treenails which bear against them. Lashings to the protruding heads of these treenails reinforce the transverse timber connections.

DURABILITY AND MAINTENANCE

The relatively light timber species chosen for log rafts are not amongst the most durable; bamboo (Hornell, 1946: 69), balsa (Edwards, 1965) and similar timbers are known to become waterlogged relatively quickly (Dick-Reed, 1964: 201, 206). Use of preservatives or other forms of maintenance are only rarely documented, but an almost universal treatment seems to be to haul the raft out of the water when not in use. The smaller ones can be carried; others can readily be dismantled (Hornell, 1946: 67, pl. 12); whilst others can be rolled up the beach with the aid of a sail (Edwards, 1965, pl. 18). Hornell noted that certain Indian log rafts were 'neither painted nor oiled' (1946: 63), but he does not seem to have documented others that were so treated not did he evidently record length of useful life. Doran (1978) has, however, noted that the Formosan (Taiwan) bamboo rafts were formerly soaked in a shark or tung oil, now ashphalt or tar, as a preservative. He estimated that such rafts, if hauled out when not in use should last *c.* 4 years: with constant immersion in teredo-infested water, useful life was reduced to 6–10 months.

Thor Heyerdahl's *Kon Tiki*, built of nine balsa logs remained afloat, safe and seaworthy for 101 days when crossing from Peru to Polynesia (Heyerdahl, 1978: 203). As Heyerdahl pointed out, dry balsa readily absorbs water, whereas *green* balsa fashioned into a raft and launched soon after felling when the moisture content is still high, is very water resistant. He found that, although sea water gradually penetrated the sun-dried outer sections of the logs, the balsa's natural moisture (sap) prevented further absorption. As the outer surface of the logs became soft and spongy the rope lashings were not chaffed as others had feared would happen.

VARIANTS

Four main variants may be identified which are aimed at improving the performance of the basic log raft.

Extension upwards

The aim here is to give the raft more 'freeboard' (structure above the loaded waterline) so that perishable cargo may be kept secure and dry and so that animals and children can be penned in. As with all such extensions, the superstructure obtains buoyancy to counteract the additional weight by forcing the primary logs further into the water; thus draft is increased and the full potential additional freeboard is not achieved. This effect is minimised by using a few light timbers of small scantlings for the superstructure.

The simplest method of extension is to lay a further deck of logs above the main logs as in certain S. American river rafts (Edwards, 1965: 62). At the other end of the spectrum may be cited the seagoing Ecuadorian rafts seen in the seventeenth century by Dampier (quoted in Edwards, 1965: 71–2) where the superstructure rose some 10 ft above the primary raft platform and the weight of it and the cargo resulted in a draft of 2 to 3 ft. The added logs were laid around the ends and sides, not contiguous but spaced apart, leaving a 'hold' amidships into which layers of horizontal poles were inserted to form two 'decks'. The lowest space, being wash-through, held ballast and unperishable cargo; the middle space was for the crew of three or four men, the upper space for perishable stores. Open sided huts for protection from sun and rain are found on rafts from S. America (Edwards, 1965: 62, 74, plate 17), Manila and Fiji (Hornell, 1946: 70, 78, pl. 8B). An analogous effect may be achieved by placing one or two much larger logs under the primary log raft as in New Guinea where they are unattached (Hornell, 1946: 70); here buoyancy is increased without reducing the original freeboard.

Lengthwise extension:

In Australia two shaped rafts are arranged head to tail so that their narrower ends partly overlap. The connection between the two parts of this *kalum* is maintained solely by the weight of upper raft and crewmen (Hornell, 1946: 72; Birdsall, 1977: 139).

Boat-shaped rafts

Boat-shaped rafts remain wash-through (with consequent advantages – see below) but their hollow form gives 'freeboard' and creates a relatively secure space for nets, fish and other equipment and cargo; this shape also improves seaworthiness and gives a better hydrodynamic shape. The principal means of achieving this are:

(a) With light poles pinned or lashed into the required shape as in the Lobito Bay raft (Fig. 5.2).
(b) Logs of larger dimensions are disposed into the shape, and bowpieces and sometimes sternpieces are added, as in the catamarans of S.E. India (Hornell, 1946: 64–7, pl. 10 and 11).
(c) The addition of sewn-on washstrakes as in the craft of Vizagapatam, India (Hornell, 1946: pl. 12). The outer logs are fashioned to the required curved shape, including a rising after end, the washstrake is sewn on and a pegged-on bow piece added. When fitted, the central log has a rising stern and a lashed-on bow piece. The two (or three) logs are then lashed together at the ends.

Logboat rafts

Logboat hulls are rafted downstream for sale in Ecuador (Edwards, 1965: pl. 8A) but the units are only temporarily and relatively loosely linked together – this may be considered a use outside the mainstream of log rafts. The *lakatoi* of the New Guinea inter-island trade had five or more logboats lashed to transverse members which supported a platform of saplings on which huts were erected for people and cargo (Best, 1925: 260–1; Rudolph, 1974b: 199, fig. 147); this is perhaps best thought of as a multi-hull boat, as may the linked logboats used on the R. Dunajec in Poland (see Ch. 6).

Logboats have been combined with logs in an unusual form of water transport in Sweden (Eskeröd,

1956: 65, fig. 84). The logboat has three logs each side and they are linked together by vertically treenailed transverse timbers which pass forward and astern of the boat. This may be thought of as a logboat with several stabilisers (Ch. 6).

PERFORMANCE

The rafts of Torres Is. are known to have carried up to 40 people (Hornell, 1946: 46), Peruvian rafts in the seventeenth century coastal trade could carry 60 or 70 tons of cargo – wine, oil, flour, sugar, cloths, skins and soap, much of it perishable – with a crew of three or four men (Edwards, 1965: 72); at other times horses are known to have been carried (Edwards, 1965: 70).

Speeds are not well-documented although in 1825 it was estimated that Peruvian sailing rafts could achieve 5–7 knots (Edwards, 1965: 77). Rafts of this period are known to have sailed the 600 n.m. from Guyaquil to Lima, often with contrary winds and currents, in *c.* 2 months (Edwards, 1965: 75), giving an average rate of progress of 10 n.m. a day. Hornell (1946:85) stated that the bamboo three-man fishing rafts of Formosa can achieve 5–6 knots when close hauled; however, Doran (1978: 21) in a recent and perhaps more reliable assessment has estimated that such rafts make between 1.72 knots and 3.4 knots depending on the relative wind.

PROPULSION, STEERING AND ANCHORING

Log rafts have been, or are, propelled by all the known methods (Ch. 12): working the tides (Edwards, 1965: 62–3); towed by boat or manpower (Fig. 5.5); by pole (Hornell, 1946: pl. 8B); by paddle (Edwards, 1965: 61–2, plate 16B; Hornell, 1946: 61–2, pl. 8A, 9, 11A); by oar (Eskeröd, 1956; fig. 8; Hornell, 1946: 64–5, 80, 85, 86, pl. 10A, 13A); and by sail (Edwards, 1965; 66–9, 78, pl. 15 to 20; Hornell, 1946: 63–6, 77–8, 81, 84, pl. 11B, 12A, 13A). Uniquely rafts of the Society Islands are said to have been towed by kites (Hornell, 1946: 71–8).

Oars are generally worked standing with the oar pivoted against a thole or wooden crutch lashed onto or driven into a rail or similar timber at the sides raised above the general platform level; sometimes the oar is in a grommet. Square sails (Edwards, 1965: pl. 17), lugsails (Hornell, 1946: pl. 13A) and lateen-type sails (Hornell, 1946, pl. 11B; Edwards, 1965: pl. 15,16) renamed 'crane spritsail' by Doran (1981: 41, fig. 22) have all been used. The seven-log catamaran of Negapatam, S. India has in fact two lateen sails (Hornell, 1946: 66, pl. 11B), the masts being stepped on the leeward outer log. In other cases the single mast is stepped in a block on the central log or between logs (Edwards, 1965: 66–9), or in a transverse timber (Hornell, 1946: fig. 8). Occasionally multi-step fittings have been used as in the Brazilian *jangada* (Hornell, 1946: fig. 8).

As with other forms of water transport, accurate assessments of sailing performance are difficult to make. Peruvian rafts of the early nineteenth century with lugsails are said to have been able to 'beat to windward for many degrees up' and to 'beat to windward like a pilot boat' (quoted by Edwards, 1965: 75–6). Doran (1978) has made a more objective

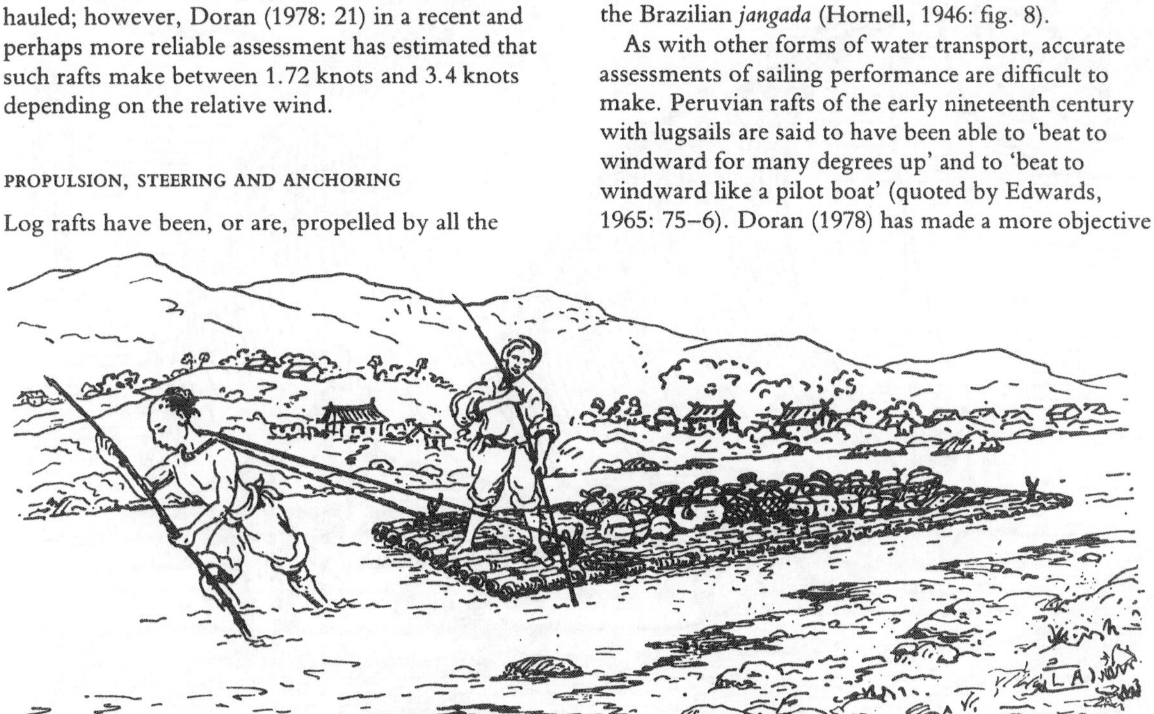

Figure 5.5 Chinese bamboo raft towed from the bed of a river. (Photo: NMM Greenwich.)

assessment of the sailing performance of the Formosan raft with lugsail. He found that they would 'barely sail' when some 3 points off the true wind, made best progress to windward at *c.* 6 points off the wind and reached maximum speed when *c.* 9 points off, i.e. with the true wind just abaft the beam. These results are similar to the limited, reliable information obtained for other types of sailing vessel (Ch. 11).

Sailing performance to windward depends not only on speeds achieved but also on leeway experienced (Ch. 11). Log rafts are especially adaptable to the use of *guares*, adjustable leeboards slipped down between the logs (Figs. 5.3, 5.6) which not only reduce leeway but two or more can be used in combination to obtain sailing balance after a windshift or a sail adjustment or after a course alteration (see Ch. 12). *Guares* are known to have been used off the S. American coast from at least AD 1615 (Edwards, 1965: 69, pl. 16A, 17), and hardwood objects of aerofoil section and with a handgrip at one end which may well be *guares* have been excavated from coastal grave sites in Peru (Heyerdahl, 1978: 102, 107, 205–10), the earliest being dated to *c.* 300 BC. Late eighth century AD sculptures at the Bhuddist temple at Borobodur, Java include one of a log raft with *guares* (Doran, 1971: 129, fig. 7.13). Similar fittings have been used recently in Ecuador, Brazil, Peru (Hornell, 1946: 80, 82, 85, fig. 8; Edwards, 1965: 69, 72–3, pl. 20), Formosa, India (Hornell 1946: 80, 66) and in Vietnam (Edwards, 1965: 98; Doran, 1971: 129, fig. 7.14).

Steering oars, often pivoted on a large block or thole at the stern, are also used in most, if not all, sailing rafts to alter and maintain course; see for example S. America (Hornell, 1946: fig. 8), India (Hornell, 1946: 66), Vietnam (Doran, 1971: fig. 7.14) and China (Worcester, 1966: 119).

Log rafts have used a range of anchoring devices (Ch. 12) – stone (Edwards, 1965: 65, 70), metal (Hornell, 1946: 85) and a mooring pole thrust down between the logs (Edwards, 1965: pl. 13B).

Figure 5.6 Detail of early seventeenth-century drawing by Spilbergen of Paita Harbour, Peru, showing *guares* being used on a sailing lograft. (Photo: NMM, Greenwich.)

ADVANTAGES AND DISADVANTAGES

Log rafts can be built from great logs or from saplings (if used in *bundles* of light poles then analytically they seem best classified as bundle craft – Ch. 9); the only essential characteristic is that the timber species should be of low relative density since their usefulness as water transport is directly related to the buoyancy of each element (Ch. 2).

Rafts represent a lower level of investment than boats, but do not in general have such a long life, nevertheless by correct choice of timber and care in use and when not in use they can last up to four years or more. There is also evidence that they can be kept at sea for periods of up to four months without any serious deterioration.

Log rafts can be operated through surf and off open beaches, even those below cliffs, in conditions which would be very hazardous to boats (Edwards, 1965: 67, 77). They are very seaworthy and their wash-through characteristic means that they do not capsize easily, even in difficult surf conditions (Edwards, 1965: 65; Hornell, 1946: 63). Log rafts can be propelled by any of the means available to boats, and under sail – providing they have an effective means of combating leeway – their windward performance is comparable with that of boats. In general, however, they are large and cumbersome in relation to their payload.

LOG RAFTS IN EARLY N.W. EUROPE

No evidence survives for the use of log rafts in N.W. Europe in the prehistoric period, or if it has survived it has not been recognised. We may, however, tentatively deduce what was technologically possible by considering analogous manufacturing activities for which evidence has survived. The tools and techniques required to build the simplest of log rafts are given in Table 5.1, together with the evidence for their earliest use in non-boat or raft building activities. On this evidence log rafts cannot have been built before the technological state equivalent to that of the Upper Palaeolithic. Whether they were indeed built at a particular time and place would depend upon the local availability of raw materials, whether the idea of applying these tools and techniques to water transport had arisen; and whether the idea of forming multiple floating units by fastening together individual units of buoyant material had been conceived. Before such craft could have been used purposefully methods of steering and propulsion would have had to be devised.

Table 5.1 Earliest technological stages at which the processes required to build the simplest form of log raft could have been accomplished.

Process	*Tools and techniques*	*Earliest evidence for use of analogous techniques*
1. Collect saplings or branches	Windfalls or driftwood	Lower Palaeolithic (Clark, 1977: 26)
2. Collect bindings	Cutting tools	Lower Palaeolithic (Clark, 1977: 25)
3. Bind	Binding and make fast	Upper Palaeolithic (Clark, 1952: 226; Dennell, 1983: table 4)

Notes: 1. The evidence for the earliest use of branches is indirect.
2. Trees of any size were rare in N.W. Europe before the Mesolithic.

Shaping the ends of log rafts by selection of suitably curved timbers or by positioning the shorter ones at the sides (constructive or additive techniques) could have occurred independently of the technological stage. Shaping the ends by reduction (subtractive techniques) was possible from the Mesolithic, although it may have been attempted by controlled fire and scraper in the Upper Palaeolithic (Table 5.2). Shaping using transformation or moulding techniques, as done in recent times in the Formosan raft may have been possible before the Bronze Age, but the earliest known use of these techniques to shape timbers appears to be in the Cheops boat of the mid-third millennium BC (Lipke, 1984). Table 5.2 also suggests that, on present evidence, fastening logs together by means other than lashings (either direct or with transverse timbers) could not have been before the advent of a Neolithic state of technology.

This theoretical study indicates that simple log rafts of light poles could have been used in N.W. Europe from the Upper Palaeolithic onwards; trees which could be converted into large logs were rare in N.W. Europe before the Mesolithic (Dennell, p.c.). It must be assumed that such early rafts were propelled and steered by some form of paddle, or poled in shoal waters, although paddles (not necessarily for use with water transport) do not in fact feature in the archaeological record until the Mesolithic (Clark, 1954). The earliest evidence for rowing in N.W. Europe is not earlier than fifth century BC (Ch. 12) and the earliest evidence for indigenous sail is from the first century BC (McGrail, 1981a: 20–22). It is most unlikely that rafts were used other than on inland waterways (Ch. 2).

Table 5.2 Earliest technological stages for the building of 'improved' log rafts

Process	*Tools and techniques*	*Earliest evidence for use of analogous techniques*
1. Shaping by reduction	By controlled fire and scraper By wedge, hammer and scraper By axe and adze	poss. Middle Palaeolithic (Clark, 1977: 26, 36) Mesolithic (Clark, 1977: 25, 36, 115) Mesolithic (Clark, 1977: 114)
2. Shaping by transformation	Heat and bend Soak and bend	Controlled heat known from Upper Palaeolithic (Clark, 1977: 26) but possibly not applied to shaping timbers until Bronze Age.
3. Fasten by treenails Fasten by tenons Fasten by 'skewers'	Knives and hammers Chisels and gouges Chisels	Neolithic (Clark, 1952: 211–12) Bronze Age (Johnstone, 1976: 51) Neolithic (Clark, 1952: 211)

The earliest documented use of log rafts in N.W. Europe is in Roman times. Caesar (*B.G.*: 6.5) noted that Celtic tribes used them to cross rivers, and there was a specialised group of *ratiarii* in the Romanised world who moved rafts of timber downstream and operated ferries (Johnstone, 1980: 161; Leighton, 1972: 35). Two log rafts dated to the second century AD were found at Strasbourg in 1938 (Ellmers, 1972: 106, figs 83, 84). The larger one consisted of three squared beams 33 cm deep, the central being 12.35 × 0.70 m and the longer, outer ones 13.95 × 0.63 m, giving overall dimensions of 13.95 × 1.96 m. Near the shaped end (bow?) horizontal holes had been bored through the logs, probably to take transverse timbers (type 3) – the precise fastening method is unclear. At the other end there were curved holes near the upper edges where the logs may have been fastened by rope lashings (type 1). The second raft was more fragmentary and consisted of two squared logs each 62 × 33 cm in cross-section and 7.15 m in length. These logs appear to have been fastened (type 2.3) by at least three transverse timbers about 2 m apart set into dovetail-shaped grooves in the upper surface of the logs (Fig. 5.7). The L/B ratios of 7.1 and 5.8 are high and it therefore seems likely that further logs are missing from both these rafts.

Leighton (1972: 134) quotes *Histoire Générale des Techniques* **1**, 445 for a third century AD lograft found at Utrecht, but this may be a confusion with the twelfth-century plank boat Utrecht 1 (Philipson, 1965). Ellmers (1974: 142) has noted that third-century raft shackles, used to force and hold together logs, were discovered in the Roman period fort at Heilbronn-Böckingen.

Ellmers (1972: 106, figs 85–7) has also published three Early Medieval log rafts (Fig. 5.7). One of the fifth/seventh century AD from Vreta Gård, Uppland, Sweden consists of seven logs linked together near the ends by transverse timbers. The other two are Viking period, the larger one from Elisenhof, Schleswig-Holstein consists of four logs, a fifth evidently having been lost; the original overall size would have been *c.* 5 × 1.20 m. These five logs were linked near the ends by transverse timbers through the logs (type 3), which were held in place by vertical pins (keys) through holes in their protruding ends. The smaller raft consists of four logs linked by transverse timbers, fastening method unclear. The L/B ratio for these rafts are 1.7, 4.1 and 1.6 and thus all may be complete.

In one of the very few textual references to the medieval use of log rafts, Bede tells us that the seventh-century monks of Jarrow used to raft timbers for house building down the River Tyne (*Life and Miracles of St. Cuthbert*, Ch. 3). In the late fifteenth century rafts were used on the River Isar at Munich and on the Danube (Ellmers, 1972: 107).

A possible log raft from the R. Ancholme region of S. Humberside was noted by the editor of de la Prynne's diary (1870: 65). He recorded that a 'very primitive "raft" fastened together with wooden pegs' had been found at Greenhoe, Yaddlethorpe near Scunthorpe in the early nineteenth century. Parts of it were said to have been incorporated in nearby farm buildings but these timbers cannot now be traced.

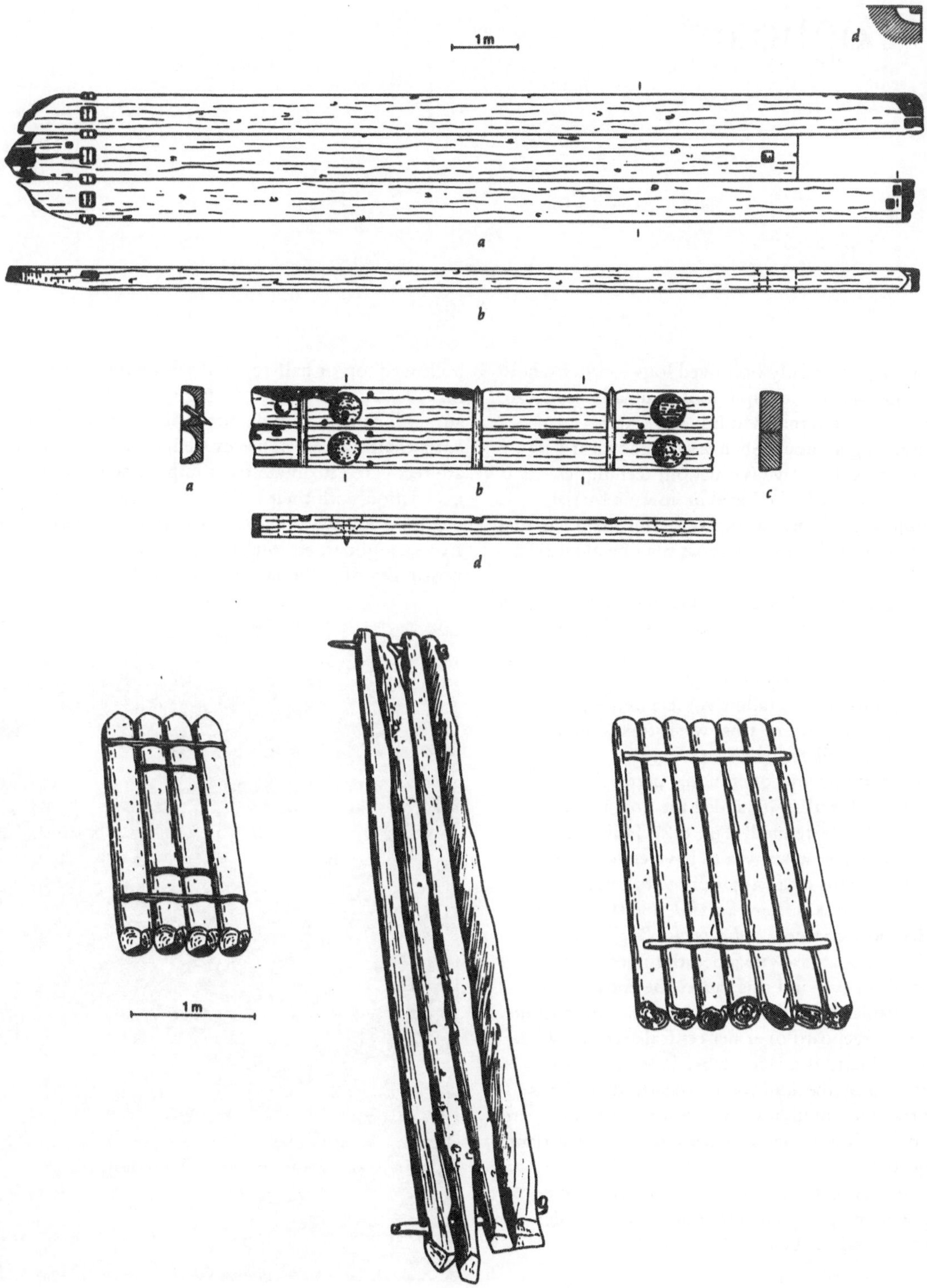

Figure 5.7 Early log rafts. After Ellmers, 1972.
Upper. Remains of two second-century AD rafts from Strasbourg.
Lower. Three Early Medieval rafts from Dalarna, Elisenhof, Schleswig-Holstein and Uppland, Sweden.

6 Logboats

Logboats are essentially hollowed logs – either whole logs or one half of a log split longitudinally. They are usually shaped externally to improve their operational capabilities. Such *basic* logboats are produced by subtractive (reduction) techniques – that is, the raw material is reduced in volume by the hollowing and shaping of the log. Four main variants which elaborate the basic logboat may be identified:

1. To improve stability characteristics (see below) the hollowed log is moulded (transformed) to a new shape, often after heat treatment. Such boats are said to be *expanded* (Fig. 6.1).
2. Using constructive (additive) techniques, extra elements – transoms, ribs, washstrakes, ends, thwarts, stabilisers, etc. – are fastened to the basic logboat to improve operational performance. When washstrakes are added the boat is usually described as 'extended' (Fig. 6.2). However, as logboats are now known to have ends added to them, it seems more appropriate to call logboats with washstrakes, *side-extended*, and those with added bow or stern, *end-extended* (Fig. 6.3).
3. When logboats are expanded the sides are forced apart (Fig. 6.1) and this causes the ends to rise. One result of this deformation is that the boat now has less freeboard over her central section. Where this freeboard is unacceptably low, such boats are extended by the addition of washstrakes. Thus subtractive, moulding and additive techniques are used in this *expanded* and *extended* logboat variant (Fig. 6.4).
4. To further increase stability and capacity two logboats may be joined together side-by-side: these are said to be *paired* (Fig. 6.5).

HOLLOWED LOGS

There are other artifacts which are essentially a hollowed log or half-log and whose remains may be mis-identified as those of a logboat: coffins, irrigation and mill chutes, troughs for cooking, brewing and corn-chaffing, and slipes or sledges. If such artifacts are recovered in an almost complete state or in association with their normal contents or in a working context they may readily be distinguishable from a logboat, although there may remain the possibility of secondary use of a logboat in these

Figure 6.1 The sides of an aspen logboat being forced apart after heat treatment. Finland, 1940. (Photo: courtesy the National Museum of Finland.)

Figure 6.2 Detail of an early nineteenth-century drawing by Paris of Valparaiso harbour, Chile, showing a side-extended logboat. (Photo: NMM, Greenwich.)

non-nautical roles. A complete logboat, especially one of some length, will also probably be recognised for what it is, but there are problems in identifying fragmentary finds. It has been suggested (McGrail,1978a: 19) that in such cases a hollowed-out fragment of wood is probably a logboat if it satisfies two or more of the following conditions:

(a) it is found in or near a (former) watercourse;
(b) it is associated with other nautical artifacts: fishing tackle, anchors, paddles, poles, etc;
(c) at least one end is shaped in one of the well-documented logboat shapes;
(d) it has (vestigial) fittings which are appropriate to logboats, for example, thwarts, tholes, ribs, stabilisers, etc;
(e) it is greater than, say, 3 m in length;
(f) bark and sapwood have been removed.

ARCHAEOLOGICAL DOCUMENTATION

As with other forms of water transport, logboats recovered before *c.* 1960 are generally ill-documented. Some were allowed to dry out, shrink and disintegrate without any recording; while with others which were recorded it is not now clear whether this was done before or after shrinkage which can considerably alter the breadth and depth dimensions of the boat (Ch. 4). It is also unfortunately true that some of the drawings of logboat remains include elements of conjectural reconstruction without this being clearly differentiated.

Dating, with few exceptions, is poor. The special nature of boats and their find-sites means that only rarely can they be dated from their context or by association (McGrail, 1978b: 239). And the study of the form and structure of ancient boats is at too early a stage to attempt precise dating by morphology (indeed, this may never prove possible) or from technological considerations. Thus dating the materials used in the construction of the boat is almost invariably the only reliable method, and this means radiocarbon dating until dendrochronology becomes practicable. Until relatively recently, logboats have generally been presumed to be

Figure 6.3 One of the several parts which make up the bow extension of the Hasholme logboat of *c.* 300 BC. (Photo: NMM Greenwich.)

prehistoric and dates attributed on the flimsiest of stratigraphic or associative evidence. For example, the oft-quoted mesolithic date for the logboat from Perth may be questioned, for Geikie (1879: 1) did not visit the site until a 'number of years' after the boat had been found and by then it had been laid aside on a bank for a long time. The identification of the stratigraphic position on which its date depends was thus not from an *in situ* examination by Geikie but from the memory of the manager of the Friarton brickworks; and in any case, the Tay carse clays in which the boat was said to be lying are not so closely dated as was once thought.

Wilson (1966: 85) was almost a lone voice when he suggested that some of the British logboats could be post-Roman, but his thesis has in fact been confirmed by several recent radiocarbon determinations. Of the 25 dates now available for British logboats, 15 are Romano-British or Medieval, ranging up to *c.* ad 1335 for the Giggleswick Tarn logboat (Q-1245: 615 bp ± 40). Of Continental logboats, 16 are later than 100 bc out of 27 dated.

A rigorous approach to dating has recently emerged in several European countries, and the problems of assessing earlier documentation of logboats and comparing these records with surviving remains to produce a reliable catalogue has been tackled in Britain (McGrail, 1978a; Edwards, 1981), Ireland (MacDowell, 1983) and Denmark (Philipsen, 1983). Similar work has begun in other countries, see, for example, Arnold (1980).

COMPARATIVE EVIDENCE

Logboats, from the simplest to the most elaborate forms, are, or have recently been, in use in many parts of the world, thus there is potentially much information available about logboat building and use. Sadly, many early ethnographers and their predecessors, the travellers and missionaries, appear to have overlooked maritime matters, especially the technological aspects, and detailed accounts which could be of use in interpreting fragmentary and

Figure 6.4 Finnish aspen logboat after expansion and side-extension. (Photo: courtesy the National Museum of Finland.)

fragmented remains did not become available until the mid-twentieth century (McGrail 1978a: 26–7). Nevertheless, by critical study of ethnographic reports and such archaeological evidence as has been rigorously obtained, it has proved possible to compile a general picture, detailed in parts, of how logboats were built and used and thus to interpret many of the features found on excavated remains.

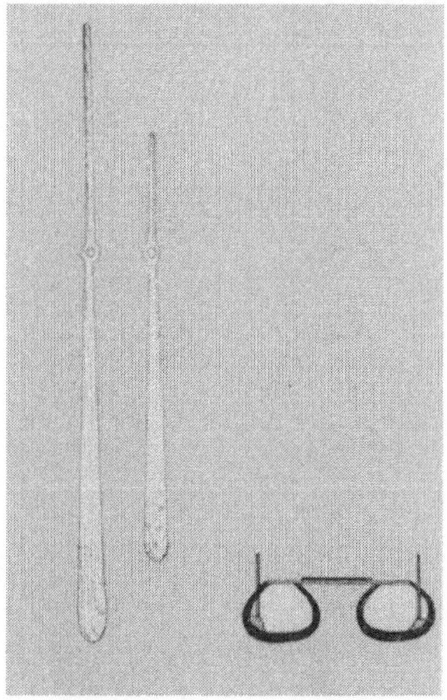

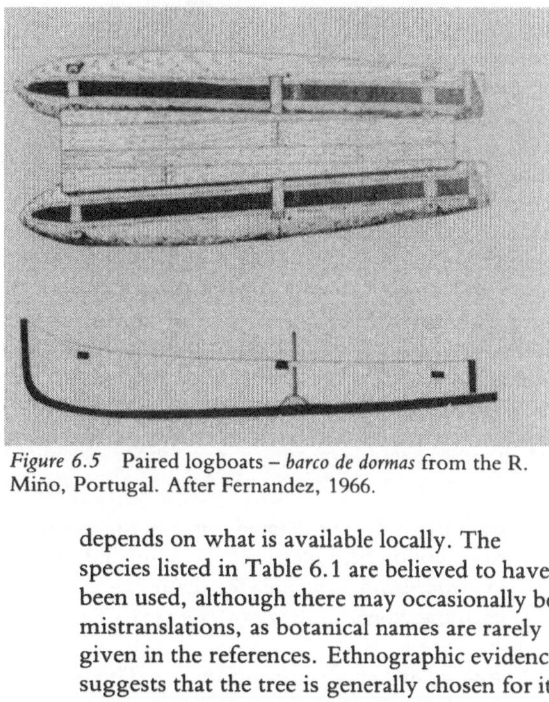

Figure 6.5 Paired logboats – *barco de dormas* from the R. Miño, Portugal. After Fernandez, 1966.

BUILDING A BASIC LOGBOAT

There are several ethnographic descriptions of logboat building from places widely separated in time and space (McGrail, 1978a: 28). Some of these are reports of building expanded or extended logboats, but in the first stages of these activities a basic logboat is made. In addition, Ellmers (1973a: 23–25) has deduced a building sequence from a study of three excavated half-finished German logboats, and from an account of the building of a logboat in 1965 on the Mondsee, Austria.

When these accounts are compared, there emerges a fairly standard procedure with some minor variations. This building sequence can most conveniently be described in stages.

1. Selection of the tree. The timber naturally depends on what is available locally. The species listed in Table 6.1 are believed to have been used, although there may occasionally be mistranslations, as botanical names are rarely given in the references. Ethnographic evidence suggests that the tree is generally chosen for its nearness to water and for its length, girth and straightness of bole, although straightness is not so essential if the log is to form one of a pair. The tree should not be split or have spiral grain which would be difficult to work (Salemke, 1972: 4). A certain amount of decay

Table 6.1 Timber species of ethnographic and archaeological logboats

Species		*Region*	*Excavated (E) or Modern (M)*	*Reference*
Common name	*Botanical name*			
Beech	*Fagus* sp.	Denmark	M	Rasmussen, 1953: 17–18
		Sweden	M	Eskeröd, 1956: 62
		Ireland	16th century AD	Lucas, 1963: 60
Alder	*Alnus* sp.	Denmark	E	Troels-Smith, 1946: 16
Lime	*Tilia* sp.	Denmark	E	Andersen, 1983
Oak	*Quercus* sp.	Sweden	M	Eskeröd, 1956: 62
		Germany	M	Rudolph, 1966: 13
			E	Paret, 1930
		Switzerland	M	Paret, 1930
		France	E	Gomez, 1982
		Italy	E	Castiglioni, 1967: 18
		Britain	E	McGrail, 1978
		Ireland	E	McGrail, 1978
Pine	*Pinus* sp.	Sweden	M	Eskeröd, 1956: 62
		Sweden	E	Crumlin-Pedersen, 1967
		Norway	E	*Radiocarbon*, **17** (1965) 394
		Finland	E	Itkonen, 1941: iv
		Germany	M	Rudolph, 1966: 13
		Germany	E	Paret, 1930
		Switzerland	E	Arnold, 1976
		Austria	M	Paret, 1930
		Netherlands	E	van Zeist, 1957
		Britain	E	McGrail, 1978
Aspen/Poplar	*Populus* sp.	Sweden	M*	Eskeröd, 1956: 62
		Finland	M*	Nikkilä, 1947: 33
		Germany	E	Paret, 1930
		Estonia	M*	Manninen, 1927
Ash	*Fraxinus* sp.	Sweden	M*	Eskeröd, 1956: 62
		Britain	E	McGrail, 1978
Spruce	*Picea* sp.	Sweden	E	Crumlin-Pedersen 1967
		Germany	E	Paret, 1930
		Switzerland	M	Paret, 1930
		Austria	M	Salemke, 1972
Elm	*Ulmus* sp.	Germany	E	Paret, 1930
		France	E	Cordier, 1963: 309
		Britain	E	McGrail, 1978
Chestnut	*Castenea sativa*	Italy	E	Castiglioni, 1967: 18

* Used for expanded logboats

can obviously be tolerated if it is in the section to be hollowed, indeed there is some indirect evidence from Tasmania for the active encouragement of decay (McGrail, 1978a: 30).

2. ***Felling and preparation of the log.*** The felling, which must not be done whilst the tree is still wet from rain according to Salemke (1972: 4) may be by axe (Hornell, 1948: 47) or by controlled fire (Harriott, 1590: 55; Beverly, 1722: 198; Lane-Fox, 1875: 403–4). The tree is then reduced to length; Nikkilä (1947: 34) noted in Satakunta, Finland, that the buttresses (where the bole swells near ground level) were sometimes included in the log. Branches are then cleared by axe or fire, and the bark stripped off (after heating, if necessary, or by soaking). Ellmers (1973a: 4) believes, however, that from the evidence of the three half-finished logboats, some roots and branches are left on at this stage so that the log will not roll during the primary shaping. When roots and branches are not left on, a clean log is held firm on a bed of transverse logs (Eskeröd, 1970: 74–6; Nikkilä,

1947; Salomonsson, 1957; Harriott, 1590: 55). Ellmers (1973a) considers that the log is orientated so that the weather side, which may have narrower growth rings than the lee side, is uppermost (or possibly directly at the bottom) so that no bias is built into the boat. Nikkilä (1947: 38) appears to allude to this practice in Finland. Salomonsson (1957: 293), says that in twentieth-century Poland logs were floated before hollowing, and the axis of the craft determined from the alignment of the floating log.

A recent survey of logboats from southern Britain showed that of 39 boats where the pith of the log can be seen or its position estimated, 18 were made from a whole log (Fig. 4.10) while 21 were from half of a log split longitudinally (Fig. 4.11). It seems likely that when a boat is to be made from half a log, the splitting is done after the branches and any roots have been trimmed.

3. ***Primary shaping.*** There is sometimes a stage of roughly shaping the log to boat-form externally, and possibly internally, on a site near where the tree is felled (Hornell, 1948: 47); it may then be hauled to a waterside site (Longstaff, 1930: 260). However, when a log can be conveniently rolled to a waterside site, this is done before primary shaping (Longstaff, 1930: 260; Eskeröd, 1970: 74–6).

4. ***'Designing' the boat.*** The form of boat is then established either 'by eye' (Longstaff, 1930: 260; Harault, 1970: 65), or by levelling the upper surface of the horizontal log to the height of the future sheerline (Eskeröd, 1970: 74–6; Salemke, 1972: 4; Willis, 1922: 69; Worcester, 1956a: 250; Bidault, 1945: 177; Dick-Reed, 1964: 109). The plan of the boat may then be marked on this surface and transverse sections marked on the ends, using charcoal or red chalk Eskeröd, 1970: 74–6; Salemke, 1972: 4; Hornell, 1948: 47). The log may also be notched to show this outline (Salemke, 1972: 4). Symmetry is normally checked by eye, but may be confirmed by reference to a longitudinally placed line.

5. ***Sequence.*** Ellmers (1973a: 28) states that on the evidence of the three unfinished logboats and from ethnographic parallels, the log was hollowed before the outside was shaped. On the other hand Hornell (1948: 47) and other authors (for example Dick-Reed (1964) and Nikkilä (1947)) state that the exterior is shaped first. Thus, there may well be two distinct approaches to logboat building: external first; and internal first. However, it is probable that with the first method a certain amount of hollowing will be done before the external shape is finalised; and in the latter case, if thickness gauges are used (see below), final hollowing must take place after the bottom has been shaped externally, tó obtain the correct thickness.

   A preliminary shaping, both internally and externally, to reduce the weight of the log, thereby reducing the effort required to move it (either overland to a waterside site, or merely to overturn it to work on the underside), followed by a final shaping externally then internally, would seem to be the general case. This sequence would not be inconsistent with the archaeological evidence of Ellmers's part-finished logboats, nor with the methods used to build logboats in twentieth-century Austria described by Salemke (1972: 4).

6. ***External shaping.*** The external shaping of the boat is done with axes and sometimes by adze. The process may involve rolling the logboat over to work on the underneath surface.

7. ***Thickness gauges.*** After the external bottom is finished it is probable that thickness-gauge holes are bored. Their recent use is well documented from Estonia to Siberia (Manninen, 1927: 12), in Finland (Nikkilä, 1947: 35), Switzerland (Rasmussen, 1953: 34), Pakistan (Greenhill, 1971: 111), and Burma and Siam (Hornell, 1948: 50). Holes are bored into the future bottom of the boat from the outside of the log to the depth of the intended bottom thickness, using a standard gauge to get the precise depth. Manninen (1927: 11) records that in Estonia these holes were bored in close rows right across the bottom from sheer to sheer, but, generally, builders seem to have been satisfied with a single axial row. Normally, these holes are left open during the hollowing phase and are then plugged with a dry treenail flush with both surfaces. However, both Manninen (1927: 11) and Viires (1969: 82) report that in Estonia the holes were sometimes plugged before hollowing with ¾ in to 1 in (19–25 mm) treenails of wood of a different colour from that of the boat. Bidault recorded that in the Amazon basin these holes

were bored to a depth of *c*. 60 mm, gauged by a thread tied around the auger.

Several excavated logboats have had a series of holes which have been interpreted as thickness gauges. Seaby (1973: 1) so interpreted the holes through the bottom of the fragmented find from Bartin's Bay, Lough Neagh, Co. Antrim. Rudolph (1966: 15) states that several finds in the eastern part of lower Germany have a series of plugged holes 10 to 15 mm in diameter along the length of the boat. Beaudouin (1970: 78, 77, fig. 7) has described an excavated logboat (*chaland*) with a series of four holes *c*. 1 in (25 mm) diameter in each of 11 transverse lines spaced at 1½–2 ft (0.46–0.61 m). In this case, and possibly in others, there clearly are difficulties in differentiating between holes for thickness gauges and holes to be used to fasten fitted ribs (see below). An unbroken treenail fitting flush at both surfaces of the bottom is probably a thickness gauge.

Twenty excavated logboats from southern Britain have thickness gauges which are mostly on or near the centre-line of the boat, but Glastonbury 1 has them near the sides (these may alternatively be interpreted as holes for transverse fittings). Fifty-two measurable holes had a pre-shrinkage diameter in the range 12–28 mm. Larger holes which may also have been thickness gauges, have been recorded: Short Ferry with three holes of *c*. 38 mm diameter, Poole with one hole *c*. 93 mm and two of *c*. 51 mm, and Holme Pierepont 1 with a square hole of *c*. 40 mm side.

8. ***The hollowing:*** Excavating the hollow may be done in one of two ways: using tools (often the adze alone) and using tools and fire. The second method is recorded from the Society Islands (Hornell, 1948: 47), Pakistan (Greenhill, 1971: 111), North America (Quinn, 1973: 6; Harriott, 1590) and Denmark (Rasmussen, 1953: 25–26). Salomonsson (1957) found carbonised wood in the hollow of the unfinished logboat he excavated at Tosthult, Sweden (but see Ch. 4). Hollowing using tools alone is recorded in Hawaii (Willis, 1922: 69), Sweden Eskeröd, 1970: 74–6), Finland (Nikkilä, 1947), South America (Lothrop, 1932: 230). North America (Longstaff, 1930), the Guianas (Hurault, 1970: 65), and Austria (Salemke, 1972: 4).

When fire is used it is normally lit on top of the horizontal log and constrained by sprayed water or by wet clay. The charred wood is subsequently adzed or scraped away (Beverly, 1722: 198) to leave a smooth though faceted surface. It is probable that the final shaping of the bottom of the hollow to get the required depth is done without the assistance of fire.

Where tools alone were used, Salemke (1972: 4) recorded that grooves were axed or adzed out across the top of the log and longitudinally along the inside of the future sides of the boat, and billets split off, using beech or maple wedges. Beech wedges were also used to hollow the German and Austrian unfinished logboats (Ellmers, 1973a: 27), the work being finished by adze or possibly chisel. Bidault (1945: 177–8) recorded that the Guianas Indians used a line smeared with charcoal to mark the body of the boat with transverse lines every 0.30 m. A section between two lines was then hollowed down to the thickness-gauge hole; then the next-but-one section was hollowed, followed by the intermediate section: the ends were hollowed last. Eskeröd (1970: 74–6) noted that Swedish logboats were divided into five or six parts before they were hollowed.

Boats that are later to be expanded, are hollowed in a distinctive way (see below).

9. ***Secondary shaping.*** The logboat was then finished externally to the required shape, this could involve rolling over the boat (Salemke, 1972: 4; Willis, 1922: 69). Any 'keel' or 'stems' would be worked in the solid and the sides reduced to the required thickness, sometimes by thickness gauges but other ways were also known: by gently tapping on the boat and judging the sound (Eskeröd, 1970: 74–5; Nikkilä, 1947: 35); or by a finger touching inner and outer sides (Longstaff, 1930: 260; Nikkilä, 1947: 35).

At this stage the boat could be completed as a basic logboat; or it could be extended by the addition of washstrakes, or expanded by heat treatment.

10. ***Intermediate stage.*** Several reports state that the boat was not finished in one season but stored, often under water, until at least the following year. Salemke (1972: 4) stated that the boat, with thicker-than-required bottom and sides, was towed on a sled to the water

near the owner's house and sunk between four posts by loading it with stones in *c.* 2 m of water, where it remained for 15–60 years. Salemke believed that this was to draw decaying substances from the wood. Similarly, logboats of Hasselø, Denmark, were sunk unfinished, and held under water by stones for two years (Rasmussen, 1953: 25), which was claimed to make the logboat timber soft and workable. In the Society Islands, Hornell (1948: 47) reported that after the first hollowing, the logboat was filled with water or submerged for some days to season and absorb water.

Logs destined to be made into expanded logboats in Satakunta, Finland were felled in autumn and seasoned until spring, when the bark was removed. Building did not begin until the summer, and if by then the log had dried too much, it was soaked in water (Nikkilä, 1947: 34).

The storage under water of partly-worked oak split from a recently-felled tree and, therefore, probably still 'green', (that is with a high moisture content) has been recorded from several North European countries from Viking times onwards (McGrail, 1974: 39–40). The reason here was probably to keep the timber fresh and moist above the fibre saturation point, so that it would not shrink (Ch. 4), and so that it could be worked relatively easily in future plank boat-building; it is also theoretically possible that some of the sap would be leached out, thus making the timber potentially more stable in use. There would also be a degree of protection against decay. It seems likely that unfinished logboats were stored under water for similar reasons: when time and timber were available, logboats could be partly built, then stored under water until they were required.

11. ***Final stage.*** If the boat had been stored under water, it was subsequently brought ashore on rollers (Salemke, 1972: 4; Salomonsson, 1957; Itkomen, 1941: iv), placed on its side, covered and allowed to dry. One man could then work a medium size boat to its final shape. It is possible that only at this stage was the bottom hollowed down to the point where the thickness-gauge holes were apparent. Any fittings such as beam-ties and separate transoms, would be fashioned and fitted at this stage.

There is ethnological evidence that logboats were not left in the raw state but were given protective treatment as plank boats are today. The outside could be charred by slight burning (Eskeröd, 1970: 74–5; Longstaff, 1930: 261; Lane-Fox, 1975: 404; Lewis, 1932: 146), the heavy charring scraped away, and the boat polished by rubbing with oil. This charring is said variously to make the timber more readily withstand rot, to ensure that splits would not develop from the sun so easily, and to make the timber insect-resistant. The external finish was beeswax in Ecuador (Lothrop, 1932: 230); tar in Hasselø (Rasmussen, 1953: 20) and in Finland (Nikkilä, 1947: 37); and oil in N.W. America. Knapp (1924: 106) inferred that the Tasmanian logboats were rubbed down with sandstone, and Hornell (1948: 47) stated that hulls were scraped and smoothed by rubbing with coral stone, whilst Lane-Fox (1875: 404) and Longstaff (1930: 261) noted that the Nootka of the N.W. American coast painted the inside of their boats some bright colour. Splits in late nineteenth-century Swedish logboats were treated with warm gum and tar (Eskeröd, 1970: 77), and Rogers (1965: 455) has reported that Ontario Indians coated their logboats with deer, bear and racoon fat to prevent splits.

The proportion of the whole parent log worked away was found to vary between 94.4% and 80.2% in the logboats of southern Britain, while that for half logs varied from 88.6% to 74.6% (or 94.3% to 87.3% in relation to the whole log). This data may be compared with Roop's estimate that, in Amazonia, 85% of the log was worked away (1942: 33), and Beaudouin's estimate of 83.3 to 85.9% for the River Adour region of France (1970: 133). These are large percentages of a log, and it might be assumed that a method of shaping and hollowing would have been evolved by which large fragments suitable for other woodworking skills would be produced, but there is no direct evidence for this.

There is no indication that the choice of either a whole log or half a log was related to the length of log available, nor that this choice is correlated with the date of the boat, or with deduced function. There is also insufficient evidence to suggest that halves of the same log may have been used when building paired boats, although this seems a superficially attractive theory. Bearing in mind the small sample size in the survey of southern Britain, there seems to be a tendency for boats built from half logs to have a semi-rounded transverse section (13 out of 21); and for whole-log boats to have a rectangular, sub-rectangular or flared section (14 out of 18). The reasons for choosing to make a logboat from a whole log or from a half log are thus unclear, although it may have been determined by the extent of any heart rot (Ch. 4).

TIME TAKEN

The time taken to build a logboat obviously varied with the size of boat and the number of men available, the type of toolkit, and, possibly, with the species of timber and the difficulties associated with moving the log to water. Paret (1930: 112) reported that 12 men took 2 days to build a logboat on the Mondsee, whereas Salemke's Mondsee logboats of pine (*Pinus* sp) and 10–12 m in length reached the underwater storage stage in 2–3 days, after 6–12 men had worked on them; the boat could subsequently be finished by one man in 3 days (1972: 4). A roughly finished 15–20 ft cedar logboat on Vancouver Island, British Columbia, could be build by one man, with an assistant for felling and splitting the tree, in about 3 weeks (Longstaff, 1930: 261). Boczar (1966) noted that one man would take 2 days to hollow out a poplar (*Populus* sp) logboat in the River Dunajec region of Poland. Hurault (1970: 73) recorded the data shown in Table 6.2 for building expanded logboats, 4 m to 8 m long, in the Guianas. A larger logboat, some 40 m length, took 28 working days including 2 days to find a suitable tree, interspersed with 7 days rest. It required *c.* 78 man days, and the work team varied from 2 to 10 at different stages of the project (Hurault, 1970: 73–4).

Table 6.2 Time taken to build a logboat

Task	*Men*	*Days*	*Man-days*
Felling tree, hollowing log on site	1	6	6
Haul to river (over 500 m)	4	½	2
Complete hollowing, prepare for heating	1	2	2
Expansion	2	1	2
Final stages	1	5	5
		14½	17

Source: Hurrault 1970

SPECIALISTS OR TEAMWORK

Several ethnographers have reported that logboat building was a communal effort. Salemke's team included a woodworker who, unlike the others, was paid, but it is not clear from the account whether he was in any sense the master boatbuilder. Hornell (1948: 47) records that in the Society Islands a *tohuna* was in charge of the work, and in Hawaii (Willis, 1922: 69) the *kahuna*, who had an apparent priestly function, also planned and measured the inside of the boat and so, presumably, had specialised knowledge. The evidence is insufficient to be certain that there were always specialists in charge of building basic logboats. There can be little doubt, however, that a man with specialist knowledge was in charge of the expansion of a boat, a critical phase when all previous work can be nullified by a minor mistake. See, for example, Nikkolä (1947: 36), Guppys account of a logboat being expanded in the Amazon basin (1961: 184), and the recent experimental building of expanded logboats in Denmark (Crumlin-Pedersen, 1984c: 102).

TOOLS

There is much ethnographic evidence that stone tools can be used to build elegant and useful boats, albeit from tree species other than oak (Lewis, 1932: 146; Knapp, 1924: 103–4; Longstaff, 1930: 261; Lane-Fox, 1875: 403; Beverly, 1722: 198–9; Greenhill, 1976: 27–8).

Boring-tools made from bird bones and set in wooden handles, were used by the American Nootka Indians (Lane-Fox, 1875: 403), whilst on Lake Victoria red hot stilettos were used to bore holes (Worthington, 1933: 159). As an alternative to stone tools, shells have been used as axes, adzes and scrapers (but once again the timber was not oak) (Hornell, 1948: 46; Lane-Fox, 1875: 403–4; Lewis, 1932: 146; Harriot, 1590: 55). Ellmers (1973a: 27) has shown that wedges of beech (*Fagus* sp.) were used in antiquity, and it seems likely that such simple aids as levers, Spanish windlass, slip-knot, half-hitch and toggle were available from an early date.

Thus, as Clark has said, the tools known to have been available from the Mesolithic onwards are adequate for building logboats (1952: 284). Nevertheless, as there is no direct evidence of oak (*Quercus* sp.) being fashioned into a logboat using stone tools, it is possible that only the more easily-worked, less tool-blunting species were used for boats in the pre-metal age. More light may be shed on this matter when further dates for logboats become available.

REPAIRS

There are two main types of repair recorded by archaeologists and ethnographers: those sealing a split and those where a patch or tingle is fitted over a more serious split or hole. Tar and creepers have been used to seal or bind splits in S. American logboats (Lothrop, 1952: 206; Guppy, 1961: 188). Two of the many logboats excavated from Aamosen by Troels Smith (1946: 19, 20) appeared to have splits sewn together. Paret (1930: 106) recorded that the logboat from Vaaler Moor (now in the Schleswig Holsteinisches Landesmuseum) had three double-dovetail clamps (Fig. 6.6) across a split, held in place by vertical treenails. There is a similar fitting across the broader end of a logboat of unknown provenance (code name O = Raftery 5) in the reserve collection of the National Museum of Ireland. A photograph of the now-lost logboat from Loch Laggan, Invernesshire (Maxwell, 1950: 1: pl. XVI) shows what may have been an attempt to pull together the sides of a longitudinal split by a form of sewing with a batten to hold in caulking.

There are several types of patch or tingle, but it is sometimes not clear from reports whether these are used to cover a hole or to hold a split together. Wooden patches fastened by iron nails are known from Ireland: a logboat excavated from Lough Derravaragh, Westmeath (*Journal of the Royal Society of Antiquaries of Ireland*, 101, (1975): 225); and from Scotland, the boat now in Cambuskenneth Abbey, was reported to have had such a repair when it was found in 1874 (Feachem, 1958–9: 117). Metal patches lined with sacking were used by the Wai Wai to repair holes in their boats (Guppy, 1961: 188).

Of the British material, splits in the prehistoric Appleby logboat were lashed together by 2-stranded birch (*Betula* sp.) held firm by oak (*Quercus* sp.) wedges in holes of *c.* 24 mm diameter and about 80 mm apart. In addition, flat double-dovetail clamps of I form (Fig. 6.7) were inserted across the splits. The logboat from Brigg had wooden patches with a caulking of moss, sewn or lashed through 10 mm

Figure 6.6 Recess for a double-dovetail clamp across a split in the Vaaler Moor logboat. The clamp was held in position by treenails. (Photo: NMM Greenwich.)

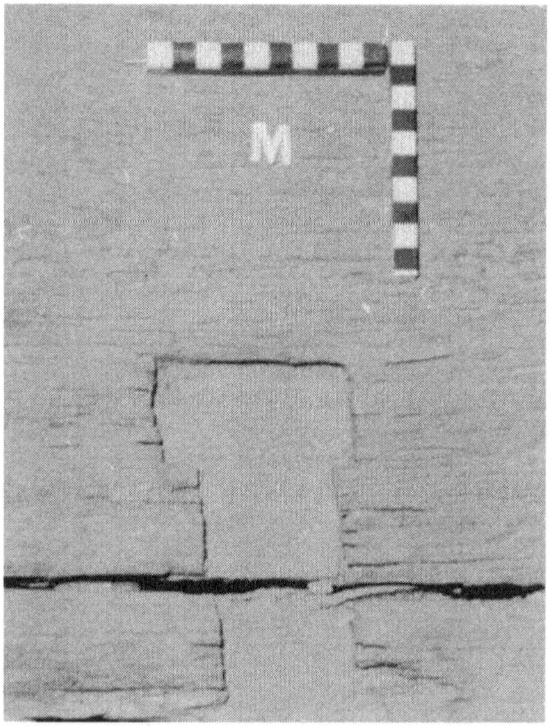

Figure 6.7 Recess cut into the bottom of the Appleby logboat to receive a wooden clamp. (Photo: NMM Greenwich.)

Figure 6.8 The stern of Holme Pierrepont 3 showing an inserted repair block grooved in continuation of the slot for the transom. The dovetail-shaped recess is for a beam-tie. (Photo: Castle Museum, Nottingham.)

diameter holes spaced *c.* 60 mm apart. Brigg's longest patch had three cleats carved in the solid, by which it was wedged on the inside of the split. Lead patches with iron nails were used to repair Banks, Holme Pierrepont 1 and Kentmere 1. A knot-hole *c* 140 mm in diameter in one side of logboat Holme Pierepont 2, had been repaired by iron-nailing a wooden patch into a rectangular recess around the hole with a nailed ovoid-shaped patch inboard. A small split in the same boat had been caulked with a mixture of mosses and other plants. Holme Pierrepont 3 and the Hasholme logboat both have repair blocks let into the bottom of the boat near the stern, and grooved in continuation of the line of the slot for the fitted transom (Fig. 6.8).

The two British logboats with repairs dated to the Bronze Age, Appleby and Brigg, use sewing or lashing techniques; metal is used on the third century bc Holme Pierrepont 1, and on the Kentmere 1 boat of *c.* 1300 AD. The double-dovetail clamps on the Appleby repair (Fig. 6.7) are similar in structure to those used to repair the undated Vaaler Moor logboat (Fig. 6.6) but the latter has vertical treenails to hold the clamps in position. There are, however, differences in the precise shape of the dovetails, for in the Appleby boat they are somewhat rectangular, whereas in the boat from Vaaler Moor they are more truly dove-tailed. It is differences such as these which may identify local boatbuilding techniques, but this possibility cannot be investigated fully until we have more precisely-dated boats.

ALTERATIONS AND ADDITIONS TO THE BASIC LOGBOAT

The principal ways of modifying the basic logboat are aimed at improving the boat's operational performance by (a) improving the boat's transverse stability by increasing effective beam at the waterline; and (b) increasing depth and thereby improving freeboard. These techniques enable the logboat-based craft to escape the limitations imposed by the girth or diameter measurement of the parent log. First to be discussed are three methods of increasing effective beam: by expansion, by pairing, and by stabilisers; then side extension (washstrakes); and finally end extension.

EXPANSION

There are several accounts of the expansion of basic

Figure 6.9 An aspen logboat being heated prior to expansion. (Photo: courtesy the National Museum of Finland.)

logboats and one is well illustrated by Nikkilä (1947). The methods may be summarised:

1. The timber species selected must be one that can be expanded by heat with minimum fear of cracking. A basic logboat is built, but with a complex outline which it is known will give the required shape after expansion (Fig. 6.9). For example, to obtain a level sheerline and a longitudinally-flat bottom after expansion, the sheer of the boat before expansion must be higher amidships than at the ends, and the bottom should be slightly concave. Often, only a long groove, about 0.10 m broad, is cut along the centre line of the top of the log and the inside hollowed out by adze from there. For the logboats of Satakunta, Finland, the walls were reduced to $\frac{3}{4}$ in (19 mm) thickness, and the bottom to $1\frac{1}{4}$ in (32 mm) Nikkilä, 1947: 34–5).
2. Sometimes the boat is soaked in water to soften the fibres (Nikkilä, 1947: 35).

3a. The boat is filled with water, sometimes with oil (Hornell, 1948: 48), and it is heated in one of three ways:
(i) by exposure to the sun on a hot beach, as in S. India (Hornell, 1948: 48).
(ii) heated stones are dropped into the water, as in Burma, Siam, and British Columbia (Hornell, 1948: 50; Bidault, 1945: 175; Longstaff, 1930: 260).
(iii) seal oil in the beech (*Fagus* sp.) logboats of Tierra del Feugo is fired (Bidault, 1945: 176).

3b. In the Guianas and in Brazil the boat is inverted over a carefully controlled, slow-burning fire (Hurault, 1970: 65; Guppy, 1961: 184).

3c. In Finland, the ends of aspen (*Populus tremula*) boats are bound and the inside coated with tar. The boat is then heated to one side of a slow-burning fire for approximately one hour. Whilst it is carefully tarred inside and damped outside the hot side of the boat is stretched by inserting transverse poles of increasing length (Fig. 6.1). The boat is then turned and the process repeated to expand the other side (Itkonen, 1941: viii; Nikkilä, 1947: 35–6).

3d. With some water inside a well-sodden boat in Burma (Hornell, 1948: 50–51) and in Estonia (Manninen, 1927: 8, 9), the boat is heated either directly over a single fire, or between two parallel fires (Lane-Fox, 1875: 405).

3e Aspen (*Populus* sp) boats in Estonia were reported by Manninen (1927: 6) to be made sufficiently flexible for expansion by tarring them internally and externally and leaving them on the rafters in the living room of a house for several days. Nikkilä (1947: 36) was told by a Satakunta, Finland boatbuilder that aspen (*Populus* sp) logs could be made sufficiently flexible for expansion by leaving a hollowed log in the warm sea, held down by stones.

4. When the logboat is sufficiently pliant its sides are slowly forced apart by a variety of means, generally by temporary cross-bars of increasing length, and wedges; but giant 'clothes-pegs' are used in the Amazon basin (Guppy, 1961: 184; Bidault, 1945: 178–9).
5. When the right shape has been obtained, or possibly when there is slightly more expansion than finally intended, and while the boat is cooling, thwarts/beams are inserted to replace the temporary cross-bars, and sometimes knees and ribs are fitted. In Finland, stones are placed on poles laid across the bottom of the boat (Fig. 6.10) to widen the bottom, and sometimes to increase the sheer at the ends (Itkonen, 1941: viii; Nikkilä, 1947: 34–6).

Effects of expansion

By expansion a much beamier boat can be obtained: Bidault (1945: 175) quoted an expansion from 0.9 m to 1.5 m beam; and Hurault (1970, figs 24–40) has recorded changes in beam from 0.31 m to 0.67 m (Fig. 6.11). Increased beam at the waterline leads to improved transverse stability (Ch. 3). At the same time, expansion reduces the freeboard amidships (this may subsequently be regained by the addition of a washstrake) and, generally, the ends of the boat will rise to give it some sheer.

Where the fact is noted in ethnographic accounts, expanded logboats are said to be of rounded transverse section. *A priori* it might be considered that logboats of rectangular section could not be successfully expanded, the operation being so critical even for a boat retaining the rounded section of a log. For example, Hurault (1970: 65) records that one in every four or five logboats built by the Guyanans burst during expansion and became unusable. Similarly, continuous transverse ridges or bulkheads left in the solid of a basic logboat would appear to rule out the possibility of expansion. Manninen (1927: 6), however, records that small cleats can safely be left in the solid of Estonian aspen (*Populus* sp.) logboats which are expanded.

When sufficient remains survive, and providing it has not reverted to its pre-expansion shape during deposition, an excavated, expanded logboat will probably be easily recognised by its form. If the boat has reverted somewhat or has been allowed to

Figure 6.10 An aspen logboat after expansion with temporary crossbeams fitted. (Photo: courtesy the National Museum of Finland.)

become disarticulated after excavation, the fitted ribs and other internal supports will no longer fit, and this fact may support an identification of expansion. However, this effect may be difficult to differentiate from that due to the different shrinkages of the fitted rib and the boat when a find has been allowed to dry. The length of the ribs will scarcely alter (see Ch. 4), whereas the logboat will shrink in breadth and in depth by an appreciable amount, resulting in the ribs no longer fitting the boat.

For fragmented remains the following points should be considered when attempting to recognise expansion:

(a) expanded logboats generally have thinner walls than basic logboats of comparable form and size.

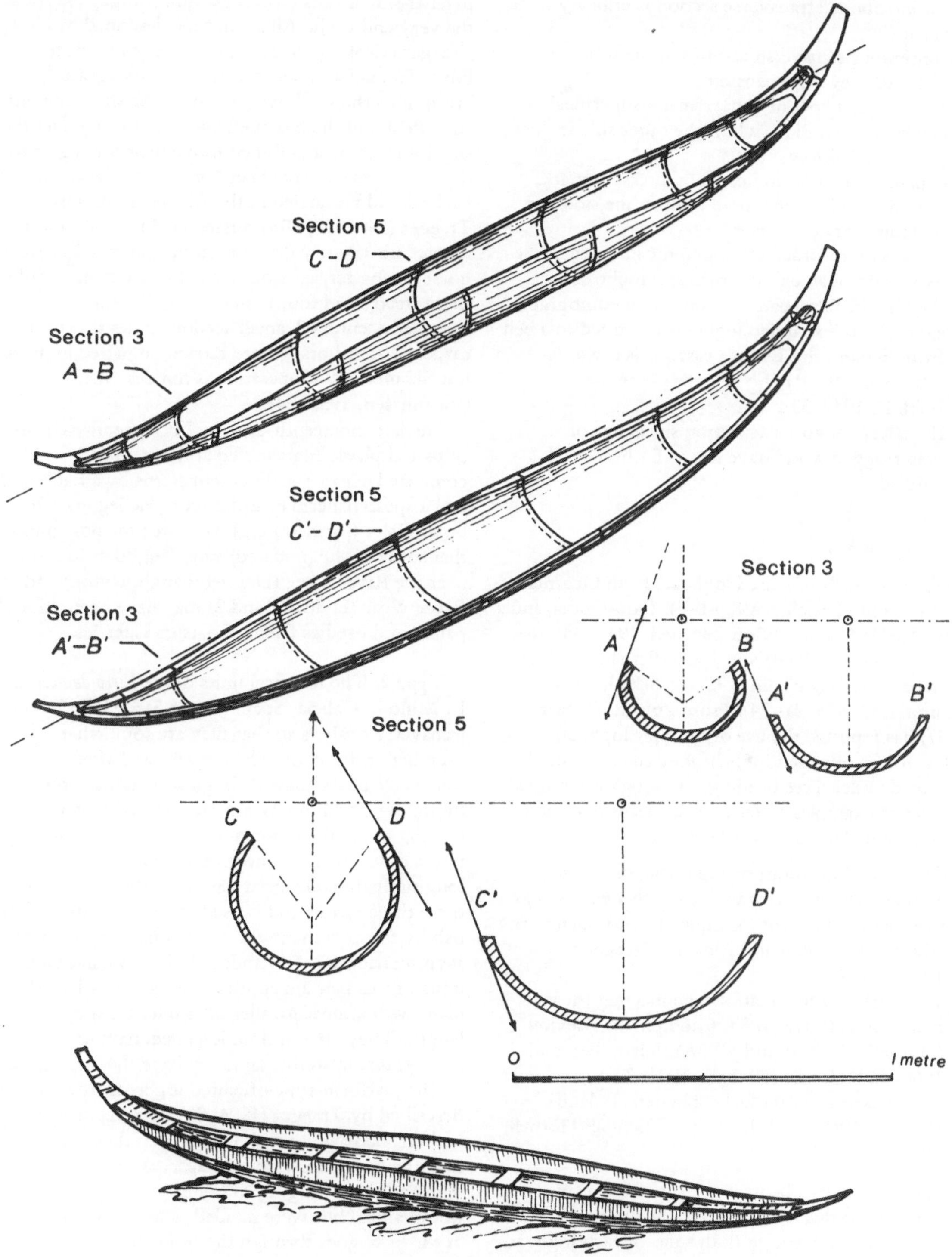

Figure 6.11 A logboat from Guiana showing the effects of expansion. After Hurault, 1970.

(b) a non-rounded transverse section is unlikely to be expanded.
(c) continuous transverse features in the solid are likely to prevent expansion.
(d) an expanded boat may have many superficial cracks on the inside, whilst the outer side is close (but see (h) below).
(e) expansion may be indicated by pressure marks caused by the devices used to open the sides, or by temporary cross-bars.
(f) there is no recorded ethnographic instance of the expansion of an oak (*Quercus* sp.) logboat.
(g) the known European distribution of ethnographic examples of expanded logboats is limited to a belt from Finland and Estonia eastwards towards Siberia and the Pacific Eskeröd, 1956: 68; Nikkilä, 1947: 33).
(h) the effects of post-excavation shrinkage of waterlogged wood have not so far been well studied.

PAIRED LOGBOATS

There are records of paired logboats from Oceania (Haddon and Hornell, 1938: 41–5), Indo-China, India (Edye, 1834: 6) and Ceylon (Hornell, 1943: 48), to Albania (Traeger, 1904) and Spain (Fernández, 1966); and from Russia (Zelenin, 1927) to Scandinavia (Rasmussen, 1953: 21–24). From Poland, Boczar (1927) has reported the use of multiple logboats.

The forms of paired logboat may conveniently be described under three headings, although the limited number of examples so far traced, means that these groups cannot be considered definitive.

1. Those used for transporting cattle, or other animals standing athwartships with a pair of legs in either logboat, for example Hasselø, Denmark (Rasmussen, 1953) and Albania (Traeger, 1904: 27–9).
2. Those used with a platform connecting the two boats: India (Edye, 1834–6 and pl. III), Ceylon (Hornell, 1943: 48 and pl. 5A), Spain (Fernández, 1966) and Albania (Traeger, 1904: 27–9).
3. Those used as individual containers: Poland (Boczar, 1966), and the *donga* of Bengal (Hornell, 1946: 190, 191, fig. 28).

Type 1. The two boats have to be kept a fixed distance apart; this is achieved by two transverse beams which go through both sides of each logboat, one at either end, with a longitudinal beam or plank holding them at a set distance and also providing longitudinal rigidity. The boats are generally parallel-sided, thus ensuring a uniform gap except at the very ends. The Albanian type described by Traeger (1904: 27–9) has some sheer towards the bow. To load animals, the paired logboats are brought to the shallows parallel to the shoreline but still afloat and the beasts encouraged to step over the side into one boat and then move their forelegs into the other. Five or six small horses with packs on their backs could be carried in the Albanian boats used by Traeger (1904: 28). Rasmussen (1953: 23–4) quotes a nineteenth-century description of four to six cattle or horses being ferried from Hasselø with men to hold them steady and four men rowing. An early twentieth-century Danish account states that two carts and four horses were carried on paired logboats, but the method of operation is not described (Rasmussen, 1953: 24).

The late nineteenth-century N. Tyne *trow* consisted of paired plank boats joined at the stem and connected across the divergent sterns by a flat board, and a spear-fisherman stood with one leg in each boat. Osler (forthcoming) has noted the possibility that some of the (undated) single logboats found in or near the River Tyne (Derwenthaugh, Ryton) and the River Wear (Hylton 1 and 2) may have been similarly paired and used as fishing boats or as ferries.

Type 2: The identical units of the *barco de dormas* of R. Miño in Galicia, Spain (Fig. 6.5) are held by three transverse timbers so that they are somewhat closer together at the stern. The forward and after transverse beams pass through rectangular holes in the sides of both boats and are fastened by a wooden locking pin at the outer sides; the central beam slots into a dovetail at the sheer. A platform of longitudinal planks between the boats is fastened to these three beams and is used for equipment and for fish by the crewman who sits in one boat and uses asymmetric oars (Ferńandez, 1966). The individual units of this pair are approximately 'boat-shaped' in plan, with almost parallel sides over most of the length. They are not double-ended, having fixed-transom sterns, and they have sheer at the bow.

The platform-type of paired logboats from Albania described by Traeger (1904: 30) from reports by Liebert, had two units joined so that they touched near the bow and were held apart at the stern. From Traeger's (1904: fig. 8) drawing, the two units were not identical but were handed. The transverse timber at the bow goes through the solid ends of both logboats, the method of fastening not being described. At the stern the boats are linked by a beam passing through horizontal holes in projections from the transom stern of each boat. A platform of transverse planks is fastened to the inner edges of the boats from about midships to the stern, leaving the hollow of each boat usable. However, Nopcsa (1925: 143–4, fig. 105) has recorded paired logboats from the same region with the platform extended across both boats.

Type 3: The third type of paired logboat (or more accurately, multiple joined logboats) is used on the River Dunajec, Poland, more as a raft though attempts are made to make the gaps between individual boats watertight (Boczar, 1966: 219). Four or five of these *c.* 5.7 m poplar (*Populus* sp.) logboats are joined together near the bow by lashing a transverse bar, fitted through holes inside each bow, to a transverse pole laid acrosss the top edge of all the boats. At the stern the boats are lashed together through *c.* 12 mm diameter horizontal holes below the top edges in each side of every boat. This is a somewhat flexible method of attachment, hence the units can independently override obstacles in the shallows. Again, this form of linked logboat is not double ended: in plan, the boats taper slightly from the stern towards the bow.

Humbla (1937: 15) drew attention to the problem of interpreting eight pine (*Pinus* sp.) or spruce (*Picea* sp.) logboats found close together at Skatamark in North Sweden in 1932, and Crumlin-Pedersen (1967) has suggested that they may have been used in a similar manner to those from the River Dunajec. These logboats were alike with one square end; at the other end each had a large horizontal hole through a projecting beak. Individual breadths were only 0.27 m to 0.35 m, and all were roughly rectangular section. A 1.8 m long pole *c.* 80 mm diameter, was found through the horizontal hole in one boat (Ullenius, 1932–3).

Characteristics

There are several combinations of features which could be diagnostic of one unit of a pair. The general form of a single boat is no guide however as there is nothing characteristic about the shapes except for a tendency to be parallel-sided. Almost all reported ones have sheer at least at one end and are not double-ended, (separate or as a pair) which suggests little evidence for roll on roll off type of operations. However, the paired logboats from Hasselø are said to have carried carts and horses and, thus, if the carts were not dismantled, there must have been an ability to manoeuvre a wheeled vehicle on board (Fig. 6.12).

The best evidence of pairing in an archaeological context would be the finding of associated fittings. Two very similar logboats found close together would also be good circumstantial evidence, especially if they were handed. Opposing horizontal holes through both sides of the logboat, probably of a size greater than other holes recorded in the boat, would suggest that the find might be one of a pair. The former presence of a longitudinally-laid platform may not be detectable as it may have been fastened only to the transverse timbers, but a transverse-laid platform should leave some evidence on the boat, possibly only on one side (the inner side when paired). Holes through the sides could also indicate washstrakes or stabilisers, but these are invariably on both sides of a single boat. The N.W. European evidence for early paired logboats (mostly indirect) has been considered by McGrail (1978a: 48–51).

EXTERNAL LONGITUDINAL TIMBERS

There is widespread evidence for the use of longitudinal timbers fastened directly, or close, to the sides of logboats. Their function varies depending on their relationship to the waterline. When at the waterline, these fittings are known as fins or stabilisers, and they enhance initial transverse stability by increasing the waterplane area with little change in displacement. Above the waterline they may have a variety of uses: as rubbing strakes to protect the boat from wear and damage; as spray deflectors reducing the amount of water entering the boat; as stabilisers if the boat were to be loaded beyond its normal draft; to increase large angle stability as one stabiliser entered the water when the boat heeled; and to give longitudinal strength to the boat at this level.

Two drawings by Admiral Paris (1843) show the main characteristics of these fittings. Paris's Plate 75 shows a *banka* from Manila (1843: 150) with a bamboo log fastened to each side, extending almost the full length of the boat. The logs are clear of the water, and are lashed close to the sides through opposing horizontal holes near each end, just below the sheerline. Paris's Plate 130 of a boat from Valparaiso, Chile (Fig. 6.2) shows similar fittings, with the logs extending here for only about half the length of the boat (1843: 68).

There are many accounts and drawings of recent boats in Scandinavia, Finland and other Baltic lands with external longitudinal timbers (Manninen, 1927; Humbla, 1937; Itkonen, 1941; Nilsson, 1943: 126; Clark, 1952; Eskeröd, 1956). Such fittings are also known in Poland (Somalarek, 1972) and an unusual type has been reported from L. Orchid, Albania

Figure 6.12 Paired boats from Teufelsmoor, Bremen in the early twentieth century. These are plank boats, but similar use of logboats is known. (Photo: Detlev Ellmers.)

(Tozer, 1869: 196; Traegar, 1904: 32–3; Nopcsa, 1925: 143).

In the survey of the logboats of southern Britain it was found that several logboats had sequences of horizontal holes along their sides: these could be interpreted as fastening points either for washstrakes or for stabilisers. Freeboard/draft/stability calculations (McGrail, 1978) show that, in general, transverse stability is more of a problem for known British logboats than inadequate freeboard and thus it may be argued that stabilisers are more likely to have been fitted than washstrakes, which have the disadvantage that they increase the height of the boats' centre of gravity. Nevertheless there is no strong evidence for external longitudinal timbers except for the fourteenth-century logboats Kentmere 1 and Giggleswick Tarn, and the tenth/eleventh-century logboat from Stanley Ferry (McGrail, 1978a and b; 1981c).

The Stanley Ferry boat had a series of horizontal holes spaced at *c.* 80 cm over the central section and 30 to 50 cm towards the ends, some 7–11 cm down from the top edge. Small half-logs treenailed along the sides could have compensated for the relatively low side height of this boat; even with these stabilisers at the waterline the boat would have had little freeboard, but ethnographic parallels indicate that this is often the case on inland waterways.

Timbers *c.* 2 m long of wedge-shaped cross-section have survived with the Giggleswick Tarn boat (Fig. 6.13), and they are said in early reports to have been treenailed along the sides near the top edge: the precise method of doing this cannot now be traced in the remains. These timbers were probably for longitudinal strength and to deflect spray, for they would only be immersed at very deep draft.

The Kentmere 1 boat had longitudinal timbers of wedge-shaped cross-section and *c.* 13 cm breadth, treenailed to the sides at the waterline through four horizontal holes of diameter *c.* 18 mm, each side. These holes are not opposing pairs; the intervals being 0.73 m, 1.00 m, and 0.79 m on one side; and 0.91 m, 0.825 m, 0.605 m along the other. The fastening through this last hole also passes through the frame at that station. This extended logboat thus had stabilisers which increased the waterline beam and hence improved transverse stability.

In Scotland, the logboat recovered from Loch Dowalton, Wigtonshire had a plank around the edge of the boat, resting on the top edge and projecting a few inches outboard (Stuart, 1864–5, 118 and pl. XI fig. 1). The plank was fastened to the boat by treenails and seems to have been a gunwale capping, a rubbing strake and a spray deflector. The oak (*Quercus* sp.) logboat from Loch Buston, Scotland (Munro, 1882: 208–9) had a 'sort of gunwale' treenailed to each side near the top edge, which extended from 'within a few feet of the stem till it projected a little beyond the stern'. The sketch by Mrs. Anstruther published by Munro (1882: 207, fig. 191) of the boat *in situ* although not very clear in detail, seems to show a fitting similar to that on the Loch Dowalton boat.

SIDE EXTENSION

Logboats may be extended by adding washstrakes to their sides, thereby increasing freeboard (Fig. 6.2). By varying the angle at which the first strakes are fastened to the logboat, it may also be possible to increase the beam of the extended boat, though this

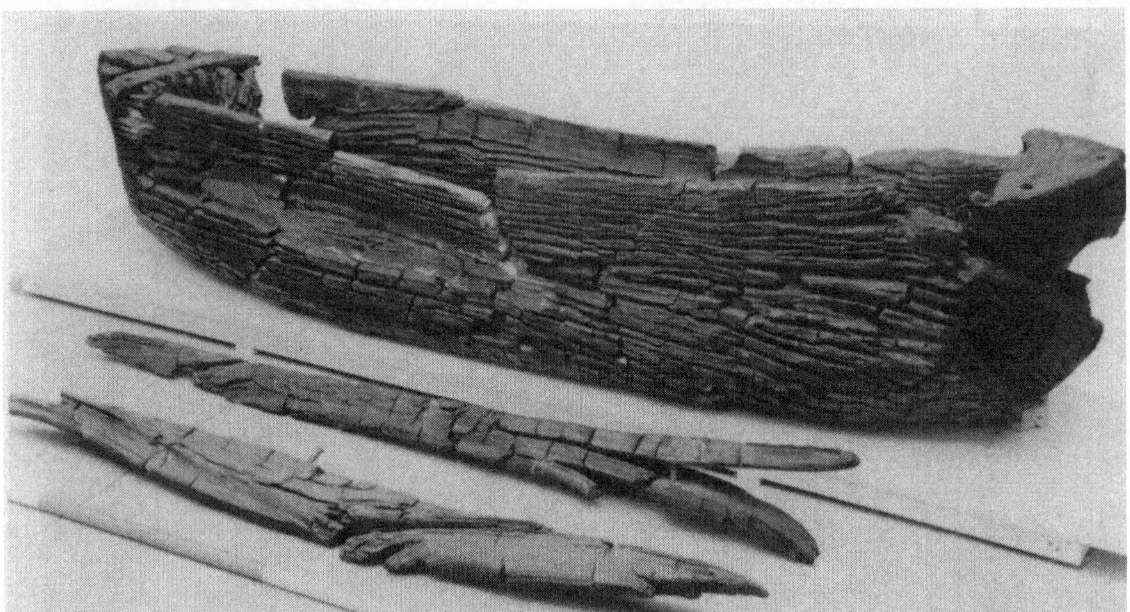

Figure 6.13 The fourteenth-century logboat from Giggleswick Tarn, Yorkshire after treatment at Greenwich. The longitudinal timbers in the foreground are now too distorted to fit the sides of the boat. (Photo: NMM Greenwich.)

will only improve transverse stability if it is effective at the waterline. Both basic and expanded forms of logboat may be extended, and the strakes may be fastened to the logboat in one of the many variant forms of the two main methods: overlapping, as in Pakistan (Greenhill, 1971: 112); or edge to edge as in the Society Islands (Hornell, 1948: 47). The strakes may be fitted before frames are fastened to the boat, for example, as in the Kentmere boat (Wilson, 1966); or after the frames are fitted, as in S. India (Hornell, 1948: 49) and in Finland (Nikkilä, 1947: 37); or they may be fitted when there are no frames – for example the logboats on Lake Atitlan, Guatemala (Lothrop, 1952: 206) or the expanded logboats east of the Urals described by Nikkilä (1947: 42–4).

Closing the ends

The amount of freeboard and beam which may be gained by fitting washstrakes can be seen from Fig. 6.2. These washstrakes are linked at the ends by transoms, the after one being set back from the bow where the logboat has some sheer. Lothrop's drawing of a logboat from Tierra del Fuego (1932: 234) shows that washstrakes here simply made the sheerline more nearly-horizontal amidships between the rising ends, and thus no special fittings is required to close the ends. Washstrakes are similarly used in the Finnish boats from Satakunta described by Nikkila (1947: 37) and Itkonen (1941: 64, fig. 33, viii), to 'fill the gap' caused by the expansion of the basic boat and thus give adequate freeboard over the central section (Fig. 6.4).

Early extended logboats

Zwammerdam boat 3, a 10.5 m oak (*Quercus* sp.) logboat from the Netherlands (Fig. 6.14) had a washstrake of silver spar (*Abies* sp) (de Weerd and Haalebos, 1973: 396). Frames were nailed to the logboat and to the washstrake, but other details of fastenings have not yet been published.

Troels-Smith (1946: 17) recorded a series of horizontal holes on both sides of boat No. 1 excavated from the Verup Complex, Aamosen, Denmark. These holes are reported to be 2 cm diameter, and from Troels-Smith's drawing they appear to be *c.* 20 cm apart, which, from the Greenhill (1971: 112) ratio of 6, would give a watertight fit for a 3 cm thick washstrake.

Figure 6.14 Zwammerdam boat 3 during excavation from a former course of the River Rhine in 1973. (Photo: courtesy the Universiteit van Amsterdam, A. E. van Giffen Instituut voor Prae-en Protohistorie (copyright holders).)

Paret (1930: 106) recorded that a logboat excavated in Gdansk in 1896 had a series of treenail holes for fastening a lapped washstrake. But it is unclear from this report whether the washstrake was excavated or whether its presence was deduced. Ellmers (1973: 56) described a fragment of an extended logboat recovered from the Weser near Rieda in 1953, heavily built, and of rounded transverse section. Here the washstrake was fastened edge-to-edge with the top edge of the logboat, with plaited cow-hair in the seam. A thick batten, almost a plank, was fastened longitudinally over the seam by treenails through the strake and through the boat. Scantlings are not given in the text but from the drawing (1973: 57, fig. 19a) the holes through the logboat appear to be horizontal at about 8 cm down from the top edge and *c.* 5 cm in diameter.

The earliest fully extended logboat from Britain is the *c.* AD 1300 Kentmere 1 boat, which has five clinker strakes each side (Fig. 6.15). Each strake of breadth 8 to 18 cm consists of two planks with *c.* 10 cm overlapping scarf. The garboard strake was fastened to the logboat base by 6 mm diameter, square shank, iron nails clenched over round roves. The radially split planking is now of mean thickness 2.2 cm which would have been *c.* 2.7 cm *pre*-shrinkage. The mean nail spacing is *c.* 20 cm, and thus spacing/plank thickness here is *c.* 7, compared with the ratio of 6 found by Greenhill in Pakistan. There is no 'closing the ends' problem with the Kentmere 1 boat as the strakes merely make the sheerline level between rising ends, the sides of the basic logboat having previously been reduced in height over the central section, perhaps after decay or damage.

Figure 6.15 The thirteenth/fourteenth-century extended logboat from Kentmere, Cumbria. (Photo: David M. Wilson.)

END EXTENSION

Recent examples of extending a logboat lengthwise are known from the Upper Nile (Hornell, 1946: 190–1) where the parts are sewn together, and from Lake Chiem in Bavaria (Paret, 1930: 111) where the two parts are joined by iron clamps.

Pommeroeul boat 2, a first/second century AD logboat excavated from near the River Haine, Belgium has a composite punt-shaped end fastened to the main body of the boat by iron nails (de Boe and Hubert, 1977). The third century AD logboat Zwammerdam 3 also has an extended bow. The *c.* 12 m logboat excavated from Hasholme, N. Humberside, and dated to *c.* 300 BC, has a composite bow. The lower bow timber is fastened to the main hull by two large horizontal treenails; the upper bow is fastened to the lower bow by three vertical treenails, 60mm in diameter (Fig. 6.3).

OTHER FEATURES

RIDGES

Transverse ridges cut in the solid are a feature of many excavated logboats and they are also reported in ethnographic studies. As Paret (1930: 113–4) and Hornell (1970: 187) pointed out, these integral ridges cannot give any significant transverse strength to the boat. Their function is puzzling and this has led Hornell, amongst others, to believe that they were skeumorphs of frames in planked boats (1970: 187). The ridges vary in depth and the deeper ones may be termed dwarf bulkheads. Some of them extend from sheerline to sheerline whilst others are confined to the bottom.

In a discussion of logboat building in the African rain forests Dick-Read (1964: 109) describes how during the hollowing operation by one or two people the boats fill up with rain overnight and have to be baled out before hollowing can recommence. Bulkheads are therefore left at intervals of *c.* 2 ft and thus only a limited amount of baling has to be done between two watertight bulkheads before hollowing can begin. The bulkheads are subsequently removed. Regularly spaced ridges in excavated logboats may be the vestigial remains of such a building process.

However, a curious feature of European logboat ridges noted by Paret (1930: 113–4) and others, is their often irregular spacing. It can be shown, for example, that the ridges in the 8.9 m Federsee logboat 1 (Paret, 1930: 79, fig. 2) are at 11%, 22%, 53%, 83% of the full length of the boat from the bow end: whilst the six ridges in the 15.2 m logboat excavated from the River Lippa at Gartrop-Bühl, Germany, in 1950 (Stampfuss, 1961) were at 7%, 17%, 34%, 52%, 67% and 88%.

In a section of his notable study of the boats of the River Adour, Beaudouin (1970: 76–87) described three excavated logboats each with three unequally spaced ridges. By analogy with documented seventeenth- to nineteenth-century boats, Beaudouin deduced that the two forward ridges were to support thwarts for two oarsmen, and that the third ridge separated cargo and passengers from the helmsman (1970: 86). In the 5.51 m *chaland* in the Musée de la Mer, Biarritz, these ridges are at 36%, 51%, 70% from the bow (1970: 77, fig 7); whilst a *chaland* recorded by the Marquis de Folin in 1892 in his *Souvenirs de Marine conserves* (Beaudouin, 1970: 82, fig. 9) had ridges at 33%, 49%, 70%. Beaudouin's (1970: 86) identification of the function of the various spaces of the *chaland* means that the forward 50% of the boat is for propulsion, the central 25% for cargo and passengers, and the aftermost 25% for command, steering and some propulsion. Beaudouin (1970: 86) emphasised that the high proportion of the boat for propulsion compared with the low proportion for cargo (25%) reflected their relative importance. Boats with different functional requirements could be expected to have a different distribution of space allocation, and if these spaces were marked by ridges, the distribution of these ridges would vary accordingly.

Within this general interpretation of ridges as space dividers, other more specific uses may be considered. A ridge close to one end of a logboat may be used as a foot-timber by a paddler sitting on the flat end. The range for this has been experimentally established as 0.20 to 0.70 m from the edge of the 'seat'. Long sheer-to-sheer ridges can be used to support thwarts as in a logboat from Brazil on display in the New Neptune Hall at the National Maritime Museum: here the thwart is slid into a blind horizontal groove in the ridge. Laver (1926), drawing on his experience of logboats in the Far East, suggested that saplings laid on short ridges would keep cargo clear of the inevitable bilge water, while Clark (1952: 287) argued that these ridges could provide footholds for paddlers, or could be used to give clearance to bottom boards.

Five logboats from southern Britain have integral ridges extending from sheer to sheer, and 23 have them only across the bottom of the boat. Of these latter, eight are in such a position that they could be used as a foot-timber by a paddler sitting on the stern. Details of ridge positions in 12 logboats from southern Britain are given in Table 6.3.

Lethbridge (1951: 230) has argued that ridges were left during the hollowing-out of the Peterborough boat to prevent too large pieces being split off. This suggestion of hollowing by individual sections echoes the building methods described by Biddault (1945: 177–8), Eskeröd (1970: 78) and Dick-Read (1964: 109). The Peterborough ridges were not regularly spaced, and would imply that hollowing sections varied from 10 to 23% of the total length. The ridges across the Clifton boats are also irregularly spaced – those in boat 1 varying from 8 to 23% and in boat 2 from 9 to 15% – and some of the spaces in between the ridges are possibly insufficient for a paddler.

Division into functional spaces appears to be an explanation which could satisfy much of the evidence. The precise function of each space may never be known, although where a ridge can be used as a foot-timber by a man sitting on the end it is probably reasonable to identify the stern sheets as the space for command and steering.

BULKHEADS

Partitions left in the solid which are more substantial than a ridge may be classified as bulkheads. They are known to have been used in the nineteenth and early twentieth centuries in Denmark (Rasmussen,

Table 6.3 Position of logboat ridges expressed as a percentage of overall length

Hardham 1	22 78
Oakmere	32 68
Warrington 2	22 78
Poole	47 77
Baddiley Mere	25 49 71
North Stoke	15 59 90
Brigg	17 26 38
Nocton Delph 1	8 32 48 76
Warboys	31 59 72 81
Peterborough	23 33 52 66 82
Clifton 1	12 22 35 47 58 66 77
Clifton 2	9 19 34 49 63 78 89

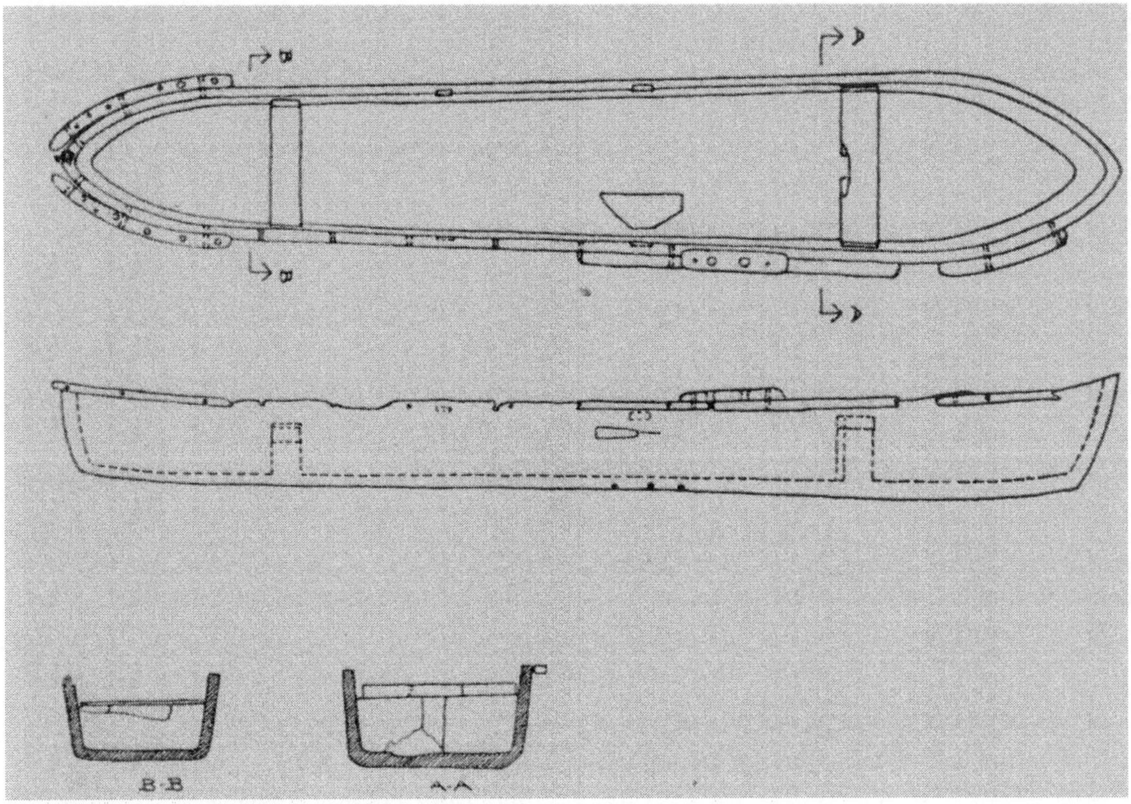

Figure 6.16 A twentieth-century logboat from Hasselø, Denmark. After Rasmussen, 1953.

1953: 18; 26, fig. 8) (see Fig. 6.16). Their task seems to be to divide the boat into functional spaces, and also to provide seats. Rasmussen (1953: 15, fig. 1) figures an 1834 illustration of a man paddling while sitting on a bulkhead, while the Hasselø logboat he describes (1953: 18) had two bulkheads to which thick planks had been nailed to bring them level with the top edge of the boat's sides. The forward thwart had been further modified to take a mast. The two bulkheads divide this boat into small bow and stern sections, and a much larger midship section.

Bulkheads are known in Continental finds from Finland (Itkonen, 1941: 15, fig. 7; 19, fig. 9); Silesia (Paret, 1930: 100, figs 19.5, 19.6; 100, figs 20.4, 20.5, 20.6, 20.7); Germany (Paret, 1930: 104, figs 21.1, 21.3, 21.4, 21.5; Göttlicher, 1971; Ellmers, 1973a: 37, figs 7c, 7d; Timmerman, 1956: 133, fig. 4; 136, figs 11, 12); and Poland (Smolarek, 1972: 18, fig. 1). Of these 20 finds, five have a single bulkhead which divides the craft into two approximately equal parts; one has a single bulkhead dividing the boat in the approximate ratio 1:2; four have two bulkheads forming small compartments at the ends; three have one, forming a small compartment at one end, with a second bulkhead at an intermediate position; four have two bulkheads, dividing the boat into three almost equal spaces; one has two bulkheads, but no surviving ends; and two have three bulkheads, a central one, and two forming small compartments at the ends.

Four boats from southern Britain, Kew, Sewardstone, Waltham Cross and Walthamstow have single bulkheads, and North Stoke may also have had one. The four certain bulkheads are all near or at the midships station, and the one on the Sewardstone boat is broad enough to have been used as a seat.

Some of the considerations discussed about ridges apply to bulkheads, with the obvious difference that the latter are more substantial area dividers, making movement along the boat more difficult. Furthermore, a substantial bulkhead will strengthen the boat. Ellmers (1973a: 36) has suggested that where they are two bulkheads, they separate cargo space amidships from the crew's positions at bow and stern. Alternatively, when fishing is undertaken from a logboat, the crew may have been amidships with tackle and catch at the ends. A compartment formed by a bulkhead, and with holes in its underwater part, may be used as a fish-well (see below).

FITTED FRAMES

Frames are often fitted to modern expanded logboats to help retain their shape – see for example Fig. 6.4. Eskeröd, (1956: 62) and Ellmers (1973a: 31) have pointed out that frames are also fitted in logboats which have not been expanded, and Ellmers interprets this as an unnecessary influence from plank boats. However, there may be other reasons such as:

(a) to support washstrakes.
(b) to support thwarts.
(c) for use as foot timbers.

Fitted frames are generally grown crooks, although Viires (1969: 77–9) has reported that birch (*Betula* sp.) and pine (*Pinus* sp.) roots were used in Estonia. For convenience of analysis, these frames may be divided into four morphological types.

1. ***Single-piece frames with a double curve:*** These are symmetrical and most extend from sheer to sheer. Four ash (*Fraxinus* sp.) frames of this type are fitted in the Kentmere boat, and were fastened to washstrakes (Wilson, 1966). Nine of the original twelve frames fitted to the Vaaler Moor logboat (Akerlund, 1963: 119, fig. 79) survive and are also of this type, as are the frames fitted to the logboat recovered from the Baltic near Gotland and now in the Gdańsk Museum (Crumlin-Pedersen, 1972: 227) (see McGrail, 1978a: fig. 166). Nikkilä (1947: 36–7) noted that young spruce (*Picea* sp) symmetrical crooks were fitted to the logboats of Satakunta, not necessarily at equal intervals, but where they best fitted, subsequently washstrakes were fitted to the protruding frame heads (Fig. 6.4). It may be noteworthy that all these finds are from regions where Early-Medieval plank boats had similar symmetrical single-piece frames.
2. ***Alternating asymmetric frames plus linked side-timbers:*** L-shaped grown timbers with the vertical arms on alternate sides ('long and short work') are fitted to Zwammerdam 3, an extended logboat provisionally dated to the end of the second century AD (de Weerd and Haalebos, 1973: 396). No drawing has been published, but the photograph (1973: 395, fig. 12) appears to show a separate side-timber on alternate sides, forming a composite frame with each L-shaped timber (Fig. 6.14). Washstrakes were fastened to the upper sections of these composite frames.
3. ***Separate half-frames or knees:*** Nilsson (1943: 125, fig. 2) and Eskeröd (1956: 62) figure a logboat from Lane Åsnen, Sweden, with four knees which do not extend to the centre-line or to the top edge of the boat. In a sketch published by Munro (1882: 208, fig. 191), a logboat from Loch Buston, Ayrshire, Scotland, is shown to have fittings of oak (*Quercus* sp) or birch (*Betula* sp) which extend from near the centre-line to the top edge (half-frames) and are treenailed in position.
4. ***Paired alternating asymmetric frames:*** Paired L-shaped grown timbers fastened together by three near-horizontal treenails, were fastened to the logboat recovered from the R. Elbe at Stipelse, Bleckede, Germany (Ellmers, 1973: 31–4) (see McGrail, 1978a: fig. 160). The three fitted frames in a recent, unpublished, oak (*Quercus* sp.) find from Lough Gara, Co. Sligo, Ireland (Wallace, personal communication), are similar to those in the Stipelse boat except that the members of each pair are fastened together by a vertical treenail through a half-joint (see McGrail, 1978a: figs 118 to 120).

Frame spacing
The spacing between frames is only infrequently recorded, but from the limited information available, the frame-spacing generally averages *c.* 0.80 m to 1.20 m; exceptions to this are a Finnish boat recorded by Itkonen (1941: 64, fig. 33) with six frames *c.* 0.50 m apart, and the Pont d'Iena find (Basch, 1972: 16, fig. 8) where the frames appear to be *c.* 1.90 m apart. Frames are not necessarily fitted at regular intervals but where their natural shape best fits the hull form.

Frame fastenings
Two distinct type of fastening are recorded: lashing and nailing. Manninen (1927: 7–8) recorded in Pernau, Estonia, that fitted frames had three grooves around them, and tarred rope was used to lash them to holes in cleats left proud of the bottom and sides of the boat. Projections were also left proud on logboats Manninen (1927: 6) recorded in Wiek, Estonia; these projections had grooves into which the frames were fitted and their heads were then treenailed to the boat, sometimes with additional iron nails. The frames in the Satakunta expanded logboats were iron-nailed to

the boat only near the top edge on either side; they were fitted so that they did not touch the bottom, to allow for shrinkage as the boat dried. The frames fitted to the logboat from Vaaler Moor (Åkerlund, 1963: 119, fig. 79) and to the logboat dredged from the Baltic, now in Gdańsk Museum (Crumlin-Pedersen, 1972: 227) were similarly not fastened to the bottom of the boat but treenailed at their heads.

The L-shaped timbers in Zwammerdam 3 (de Weerd and Haalebos, 1973: 395, fig. 12, 396) were iron-nailed to the boat and to the washstrake, but no details of these fastenings have been published. The paired frames of the Stipelse boat (Ellmers, 1973a: 33–4) are treenailed to the boat bottom at the turn of the bilge and to the side near the top edge, and are iron-nailed through the bevel at the rib head. Thus, the visible signs of missing frames on the bottom of this boat – discounting pressure marks – would be pairs of opposing holes at the turn of the bilge, angled out and down from the inside.

The frame-fastening holes in the Stipelse boat appear to be *c.* 24 mm, while ones in the Vaaler Moor boat are *c.* 30 mm diameter. Other reports do not record sizes but on several drawings the treenails appear to be bigger than those generally used for thickness gauges. We may thus agree with Ellmers (1973a: 31–3) when he interprets the series of three transversely-disposed larger holes across the bottom of the Stipelse boat as the attachment points for a previous set of frames. The frames recovered with the Stipelse boat had additional horizontal treenails in their side-members driven in a fore-and-aft direction (Ellmers, 1973a: 33, fig. 5), and Ellmers has tentatively interpreted these as supports for a thwart.

Transverse rows of holes across a logboat are not necessarily a sign that frames were once fitted there. The *chaland* logboat in the Musée de la Mer, Biarritz, has eleven rows of four holes *c.* 25 mm in diameter (one on either side of the turn of each bilge) with intervals of *c.* 50 cm between the rows (Beaudouin, 1970: 77, fig. 7) and Beaudouin (1970: 76–7) interprets these as holes for thickness gauges rather than for frames. Thickness gauges are generally of smaller diameter than treenails, and they are finished flush with the inner and outer surfaces of the boat's bottom.

Three boats from southern Britain (Banks, Stanley Ferry and Kentmere 1), none with signs of expansion, had fitted frames, but only four degraded frames from Kentmere and a fragment from Stanley Ferry survive. The Kentmere frames are of birch (*Betula* sp), and are fastened to the bottom of the logboat-base by treenails in 2.2 cm diameter vertical blind holes near the centre-line. One frame is further treenailed through the side of the basic logboat and all four are fastened to the fourth strake. The frames are at 20%, 43%, 60% and 79% of the overall length, thus dividing the boat into five roughly-equal spaces, intervals between frames being 1 m, 0.76 m, 0.80 m. These are grown timbers in one piece with a double-curve, the angle at the curve being *c.* 60°. They were used to support the washstrake planking and one of them probably had a thwart associated with it. Three frames (76 mm sided) treenailed to the sides, were found with the Banks logboat, but they no longer survive. Their position has been identified on the surviving remains as at 18%, 50%, and 86% with a spacing of 1.6 m and 1.8 m, and they seem not to have been fastened to the bottom of the boat. Two of these frames may have had thwarts associated with them.

In an early-twentieth century photograph of the tenth-century Stanley Ferry logboat two fitted frames are shown in position near the surviving end. The original frames seem to have been fastened with up to four treenails through the bottom and one or two through each side. The one surviving fragment is part of an oak crook which is now very distorted. As oak crooks with appropriate double curves were probably difficult to find it seems likely that the frames were 'alternating asymmetric frames plus linked side timbers' with the near-vertical arm of the L-shaped, grown timbers on alternative sides. Another possibility that these were closely-paired, alternating asymmetric frames seems less likely. Frame spacings are 32 cm and 45 cm between the first two pairs and then 48 to 50 cm for the remainder. In order to support washstrakes these frames would have to extend above the top edge of the boat and there is no evidence for this in the Stanley Ferry logboat. The least improbable interpretation is that the frames supported thwarts for passengers to sit upon; 50 cm between thwarts is too short for paddlers but adequate for passengers. If this boat had indeed been used to carry passengers, crew with paddles or poles could have been stationed at bow and stern.

TRANSVERSE TIMBERS AND THWARTS

Many logboats, especially unexpanded ones, are used without transverse timbers; for example, the boats of the Zutugil Indians, which are up to 10 m in length and *c.* 1 m to 1.5 m in breadth, have no seats, thwarts or braces (Lothrop, 1952: 206). And Nikkilä (1947: 46) states that whereas expanded boats east of the Urals have transverse timbers, those west of the Urals have only fitted frames. However, some expanded logboats are fitted with transverse timbers during the process of expansion (see Figs 6.1 and 6.10) and sometimes these are left in position permanently, as in the boats of the Kru of Sierra Leone (Hornell, 1929: 234) where bluntly pointed struts are jammed into holes through the sides *c.* ½ in (12 mm) below the top edge, and lashed with split rattan. On the other hand, substantial permanent timbers may replace those used during expansion (Hornell, 1948: 48), and may subsequently be used as thwarts.

Transverse timbers and thwarts have seldom survived on ancient boats but several have fittings which may be interpreted as fastening points for such timbers. Zwammerdam boat 1 (de Weerd and Haalebos, 1973: 391, fig. 7, 392) has rectangular transverse beams at one end, supporting a longitudinally-laid decking to which they were nailed from below. The method of fastening these transverse beams has not yet been published, and it is possible that they are, in fact, cross battens holding the planking together, rather than supporting it.

In several Irish logboats, of which that from Baronscourt, Tyrone (National Museum of Ireland, Reg. No. 1881/536) may be taken as typical, opposing pairs of rectangular projections, with horizontal slots, are left proud of the sides at a height suitable for a thwart. Associated with them are paired blocks left proud of the bottom, and opposing pairs of 'blisters' with vertical holes also left proud near the top edge of the sides; these seem to indicate a thwart, foot-timber (stretcher), thole-pin, combination for rowing.

Åkerlund (1963: 118–120) has argued that similar opposing pairs of blisters with vertical holes, approximately midway between the fitted ribs on the inner edge of the sides of the logboats from Vaaler Moor, are for thwarts treenailed in position, with the ribs serving as foot-timbers for the paddlers. In countering an opposing argument that the blisters were for tholes, Åkerlund assumed that the corresponding thwarts could only be fitted at the top of the ribs. However, it may be possible to balance a thwart around the rib at a position below the sheer, and jam it in position against the boat's side. A thwart at this lower height would not only be better for transverse stability but also permit tholes at sheer level to be used.

A logboat recovered from Loch Kilburnie, Ayrshire, in 1830, had a recess in each side at sheer level approximately one-third of the length from each end, which is interpreted on the published drawing (Mann, 1933: 141, 143) as for a thwart. The boat from Loch Glashan, on the other hand, has a short plank resting on blocks left proud of the sides relatively low down in the hull which would be a more stable position for a crewman.

Five logboats from southern Britain had timbers associated with them of such a size and in such a position that they have been interpreted as thwarts. The timbers reported associated with Banks, Brickfield Farm, Tolcarne, and Warrington 2 have not survived. The rectangular ash (*Fraxinus* sp.) plank found with Kentmere 1 boat has survived, but its method of attachment to the boat is not clear; probably it was associated in some way with one of the frames, although it may have been loose. Two oak (*Quercus* sp.) planks, shaped to fit across each end, were also found with this boat; they fit into notches cut into both sides of the top edge at bow and stern. The thwart reported to be fitted across one end of Warrington 2 was a 2 cm thick plank *c.* 0.18 m in breadth, which evidently fitted into notches in the sheer and was held in place by two vertical treenails of *c.* 2.4 cm diameter.

FITTED TRANSOM ENDS

A fitted transom is a vertical board or boards fitted into a transverse slot near one end of a logboat, usually the stern, and constituting the major part of the closure of that end.

The few ethnographic reports of fitted transoms claim that they are used when bad work or defective wood means that one end has to be replaced (Lothrop, 1932: 233). van Nouhuys (1928: 275) claims to have seen them often in the Eastern Archipelago but never in a new or sound boat. Worcester (1956a), referring to the boats of British Guiana, gives a typical account of the modification of double-ended logboats to take an outboard motor, by sawing them through a few feet from the stern, and nailing a transom stern across. Eskeröd (1956: 64, fig. 7b) has published a drawing of a late nineteenth-century Swedish logboat with stabilisers, which apparently had a transom inserted at the stern.

Fitted transoms have either been found or have been inferred in fifteen logboats from southern Britain. Burpham 1 and Haddenham are reported to have had a slot for a fitted-transom, and the published photograph of Derwenthaugh and the early nineteenth-century woodcut of Horsley Deep 2 show a transverse slot near one end. Appleby has a slot which is now *c.* 4 cm deep and *c.* 12 cm wide; Clifton

Figure 6.17 The sterns of Clifton 1 (left) and 2. The latter has a fragment of transom still in position. (Photo: Castle Museum, Nottingham.)

2 had a slot 4.5 cm wide with moss inside (Fig. 6.17). The slot in the Poole boat is 4.6 cm deep × 4.3 cm width, increasing in width towards the sheer, and it formerly contained hide caulking; the slot in Preston 1 was said to be *c.* 2.5 cm wide; the slot on the Short Ferry boat is now impossible to measure. Transom boards, or fragments of them, have been found with six boats, but that from Peterborough was lost before archaeologists were on site. The Brigg transom was probably of oak (*Quercus* sp.) and was 4 to 5 cm thick. It was made in two sections 0.48 m and 0.28 m high, with a breadth of 1.22 m at the top, and was caulked with moss. The Hasholme boat's transom is an oak board measuring *c.* 1.5 × 1.2 × 0.08 m. A fragment of transom was recorded with the Clifton 2 logboat, in a 6.6 cm wide slot, again caulked with moss (Fig. 6.17). Part of the transom of Holme Pierrepont 3 was also recorded: it is estimated by MacCormick to have been 0.61 m broad × 0.46 m high and 4.5 cm thick, and to have had a thicker board above it with dovetailed ends (Fig. 6.8). The late eighteenth-century drawing of Nocton Delph 1 shows a fitted transom.

This group of fifteen contains some of the broadest logboats, including five certainly and four probably, over 1 m in breadth; these were possibly made from trees of great age which had developed heart rot at the butt end and thus had to have a transom fitted (see Ch. 4). On the other hand, other boats in the group were much less broad, including Burpham 1 with a reported internal breadth of only 0.48 m, and three of 0.76 m external breadth (Clifton 1, Clifton 2 and Peterborough). These may also have had rot, for there is no simple connection between diameter of tree, age of tree, and degree of rot; but it is also possible that their transoms were fitted after damage sustained during use or manufacture. Nine of these fifteen logboats had some form of transverse

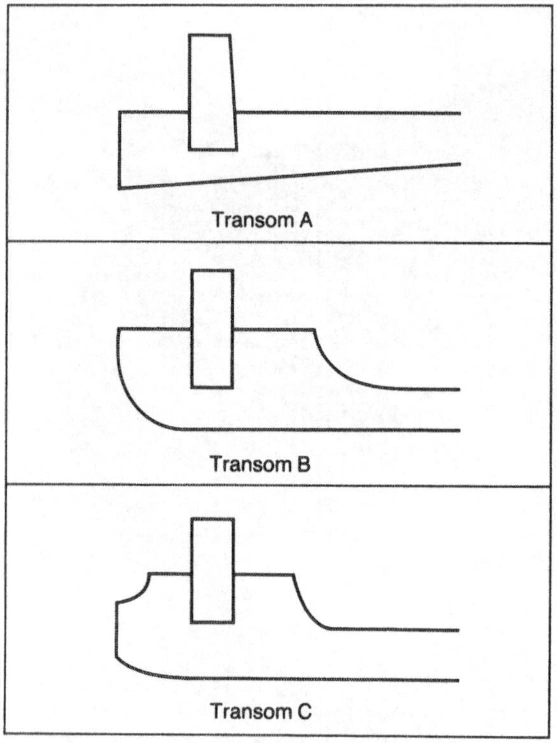

Figure 6.18 Diagram showing three types of fitted transom. (Diagram: NMM Greenwich.)

strengthening fitting at the transom end, either found with the boat or confidently inferred (see below). In the other six boats, insufficient evidence survived to prove or disprove the presence of such a strengthening fitting, but we may postulate that such fittings were probably indispensable.

Three types of fitted-transom stern have been recognised (Fig. 6.18); they differ in the way the boat's bottom is shaped to receive the transom.

Form A includes: Clifton 1, Holme Pierrepont 3, Peterborough; and possibly Horsley Deep 2 and Nocton Delph 1.
Form B includes: Appleby, Clifton 2, Preston 1, Short Ferry; Hasholme; and possibly Derwenthaugh.
Form C includes: Brigg, Poole

There is insufficient evidence to classify Burpham 1 and Haddenham.

These different forms may reflect different degrees of heart rot in the parent log; on the other hand, they may be individual styles of boatbuilding. There is insufficient evidence to resolve this point, at present.

The distribution in southern Britain of logboats with fitted transoms is wide, extending from Dorset and Sussex to North Lancashire and Northumberland; there does, however, appear to be a cluster in the Trent, Humber, Ancholme, Fens regions, where eleven of the fifteen were found.

In her survey of the Scottish material, Edwards (1983) found that twenty-five logboats probably had had fitted transoms, of these nine were from the River Clyde region. Some of the logboats with great breadth (Loch Buston, 1.22 m; Clyde 7, 1.06 m; Clyde 16, 1.50 m; Loch Lotus, 1.52 m; River Tay, 1.30 m) had fitted transoms, but so also did some of the relatively narrow ones (Clyde 6, 12, 26 with breadths of 54, 61, and 60 cm; Loch Castle, 61 cm; two of unknown provenance, No. 86 and No. 87 of 46 and 50 cm). The transom boards only infrequently survived, their former presence being deduced from the presence of a transverse groove near one end. The evidence from elsewhere in Europe is summarised in McGrail (1978: 64–5).

TRANSVERSE FITTINGS AT THE ENDS

There are several cases recorded where transverse strengthening fittings have either been found, or deduced to have been present at the ends of logboats. Possibly, there existed many more than the record indicates, as the upper ends of logboats are only infrequently recovered.

Itkonen (1941: iii) interpreted transverse notches at the ends of two Finnish logboats as seatings for transverse strengthening fittings. Ellmers (1973a: 30) similarly identified the recessed sections with vertical holes at the ends of two logboats in the Hanover Museum (39, fig. 8b; 51, fig. 15a). The shaped horizontal board fastened to the end of Federsee No. 14 (Paret, 1930: 85, 90, fig. 13) would give transverse strength as well as provide a seat. The groove across one end of the Federsee boat 1, and the paired notches in the sheer at the other end may also have been to receive this type of fitting, although no treenail holes are shown in the drawing (1930: 79, fig. 2). The transverse grooves and vertical holes at the ends of the logboat from Kosel, Silesia (Paret, 1930: 100, fig. 19.4), and the paired vertical holes at the surviving end of the boat from Bobernig (1930: 100, fig. 19.3) may have a similar use. The transverse groove with two vertical holes at one end of the logboat from Ockelbo, which Humbla

(1937: 15, fig. 4) argued must be evidence for a paired boat, is more convincingly interpreted as for a transverse tie.

Stuart (1864–5: 118) reported that the logboat from Loch Dowalton had a 'backboard pegged on' above the fitted-transom stern. Hutcheson (1897: 271) calls this a 'thicker top transom'. The drawing in Stuart (1864–5, Pl. xi, fig. i) does not enable us to decide the detailed construction, but it is clear that this fitting gave transverse strength, holding the sides together against the transom.

Lough Eskragh logboat 2 (Collins and Seaby, 1960: 30) had one, possibly two, transverse fittings treenailed across the stern. Somewhat more tenuous evidence on logboat Cahore 1 (National Museum of Ireland Reg. No. 1859: 182) which has a transverse hole 2.5 cm diameter, 3 cm below the top edge, just forward of a groove for a fitted transom, may also indicate a fitting for transverse strength. Iron girth lashings were used to bind both ends of the Hasselø logboat whilst it was still in use (Rasmussen, 1953: 18) and grown crooks were treenailed around the outside of the ends at sheer level (Rasmussen, 1953: 18, 21, fig. 5). A three-arm iron fitting bound the ends of the Mondsee, Austria logboat (Salemke, 1972: 5, 6, figs 3.7, 4).

Of the logboats of southern Britain, eight have direct evidence for strengthening timbers at the ends. One of these Metheringham, is reported to have had timbers fitted in two transverse grooves under the stern. The Short Ferry boat has a transverse slot of dovetail cross-section underneath the stern, which may be inferred to have had a transverse timber (Fig. 6.19); and the inadequately reported Blaenffos boat may have had a similar fitting. Barton and Warrington 2 had natural crooks fastened by vertical treenails around the curve of flat-topped ends. The crook of Warrington 2 was integrated with a thwart; there was probably another crook at the other end of this boat, where three vertical treenail holes survive. The Hasholme logboat has two transverse timbers associated with the transom stern, and fragments of a similar fitting survived with Holme Pierrepont boat 3. This latter timber had dovetailed ends fitting into matching dovetails in the top edge of the boat, aft of the transom and over the open stern (Fig. 6.8). The Giggleswick Tarn boat has transverse fittings at both ends (Fig. 6.13). At the stern a 3 cm thick split-plank, shaped to fit the end, was fastened by three vertical treenails of diameter 2.3 cm. Across the bow end is a double-dovetailed groove 7.7 cm wide by 3.2 cm deep in which a shaped timber was fastened by two vertical, 2.3 cm diameter, treenails. Holes around the

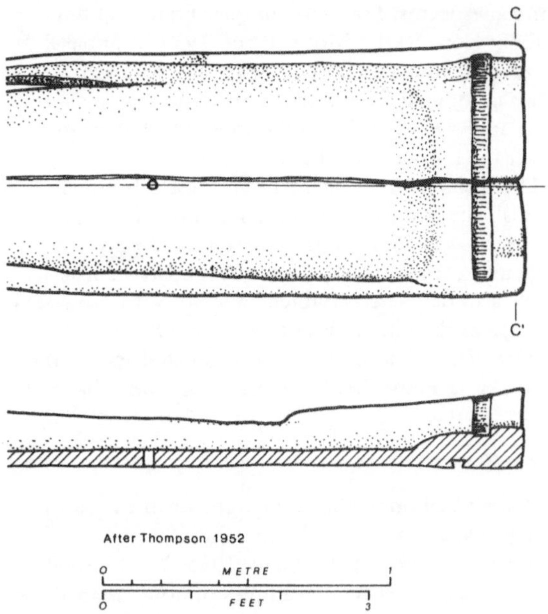

Figure 6.19 The stern of the Short Ferry logboat. (Drawing: NMM Greenwich.)

ends of Shapwick and Warrington 1, 3, 4 logboats suggest that strengthening crooks similar to that fitted to Warrington 2 were fastened there. It is probable that the widened groove at the top of the Poole boat's transom slot held a beam-tie similar to the 'upper transom' of the Holme Pierrepont 3 boat.

SIMULATED KEELS, STEMS AND SIMILAR FITTINGS

A number of modern logboats have 'stems' carved in the solid: for example, the small logboat from Haiti on display in Neptune Hall at the National Maritime Museum; and Hornell (1944: 172) noted that a logboat at Port Amelia, Mozambique, had a simulated stem *c.* 1 in (2.5 cm) wide.

There are a number of other 'false' features in the archaeological record which, like the 'stems' can only be imitations of fittings used in plank boats. The logboat from Lurgan, Addergoole, Co. Galway, now on display in the National Museum, Dublin and probably the largest surviving logboat in these islands, has an 'internal keel' or 'keelson' fashioned in the solid. And some of the logboats recorded by Seaby (personal communication) for the Ulster Museum had the shape of a 'keel' worked externally along the length of the bottom. Giot and Niort (1951–2: 287, fig. 1, 288) recorded a keel outlined along the bottom of a logboat excavated from the R.

Loire at Ancenis, Loire-Atlantique, France, and the logboat from Vaaler Moor (Paret, 1930: 106) has 2 m lengths of simulated keel at the ends. The Loch Glashan logboat from Argyll, Scotland, has a simulated stem worked in the solid which continues under the logboat as a shallow 'keel'.

The boats from Poole and Kentmere 1 both have 'stems', the latter with a horizontal hole through it, at about the waterline. Holme Pierrepont 3 may also have had a 'stem'. It is possible that boats such as Poole also have 'keels' which have not yet been noted because of the difficulties of inspecting their undersurfaces. The Bronze Age logboat-shaped coffin from Loose Howe has both false stem and false keel.

FIGUREHEADS

In British Columbia the large logboats of the Salish Indians have bows carved into a figurehead (Longstaff, 1930: 261; Durham, 1955: 34). Hornell (1946) has given several examples of bows shaped for non-functional reasons. Oculi are frequently associated with such bows, and Hornell has given many examples (1946: 285–289 and 1938).

There are also examples of unusual-shaped ends in the archaeological record. In Scotland there are two logboats which are considered to have had one end carved to represent an animal head. Gillespie (1874: 22) recorded that the logboat recovered from Loch Arthur 1 (Lotus), Kirkcudbright, had a bow 'resembling the outstretched neck and head of an animal' with a 5 in (0.13 m) diameter horizontal hole through the head where an eye might be. This end of the logboat is now on display in the National Museum of Antiquities in Edinburgh and appears to support the interpretation put on it. The second boat, from L. Kinnordy, Angus, also survives, this time in Dundee Museum, but the resemblance here is not so convincing.

A 12 in (0.30 m) wooden model from Roos Carr (Sheppard, 1901, 1902), which is now in the Archaeology Museum, Hull, has been taken to represent a logboat with a crew of warriors, and Piggott (personal communication) has argued for a L.B.A. date, because of the warriors' round shields. One end of this model appears to represent an animal head, with sockets for quartz eyes (Fig. 6.20).

Beaks

Beaks and other protrusions (sometimes with a pierced horizontal hole, which may be for a painter), have also been noted at the ends of excavated logboats. See, for example, Federsee boat No. 2 (Paret, 1930: 83, fig. 6); the boat from Loch Buston (Munro, 1882: 208); the boat from Gartrop-Bühl (Stampfuss, 1961); and a Finnish boat (Itkonen, 1941: 21, fig. 10). Barton, Preston 1, Warrington 5, Irlam, Llyn Llangorse, Oakmere, Warrington 1 and Warrington 11 have, or are reported to have had, beaks. The first three had horizontal holes through the beaks, probably for a painter.

Oculi

Brigg (Fig. 6.21) and Oakmere have projections on both sides of one end, which may be identified as oculi. Two projecting 'bungs' on either side of the narrower end of Holme Pierrepont 3 might be similarly interpreted.

Vertical holes through overhanging ends

A central vertical hole through the forward end of Colombian logboats was used for a painter (Hornell, 1928: 129). While van Nouhuys (1928: 275) recorded that a similarly placed hole, some 2 in (51 mm) in diameter in the Papuan boats of Dutch New Guinea was used as a socket for a mooring pole. He also stated that there was a similar hole in a logboat excavated in 1920 at Terbregge, near Rotterdam, and a thin pole was found lying in the clay immediately below the hole. Roop (1942: 397) has recorded the use of mooring poles in Amazonia, South America. Such holes may also be used as a lead for a fishing line.

Olsen and Sjøberg (1971: 48, fig. 2) have interpreted the rectangular 8 × 10 cm hole through one overhanging end of boat No. 1 from Skäggered, Göteborg, Sweden, as for a painter, but the size and angularity of the hole seem to indicate that a more appropriate use would be for a mooring pole. A similar function may be ascribed to the 10 × 10 cm hole in the logboat from Werie, Gohfeld, Westphalia, (Paret, 1930: 105, 104, fig. 21.8). Zwammerdam boat 3 has a large vertical hole at one end (de Weerd and Haalebos, 1973: 395, fig. 12) and if this is through an overhang it could have had a similar function. There is also said to be a hole through one end of Zwammerdam 1 which the excavators interpreted as a towing point (1973: 392): no drawings have so far been published. Ellmers (1973a, 53–4) has argued that holes through the overhanging ends of two logboats recovered from near Gottingen were attachment points for pairing the boats; use for mooring poles seems to be a more plausible explanation. The near-vertical, 45 mm diameter hole through the bow of Holme Pierrepont 2 may have been used as a socket for a mooring stake, and the boat from

Figure 6.20 A 50 cm long wooden model from Roos Carr, N. Humberside. It is thought to represent a logboat with warrior crew. (Photo: Archaeology Museum, Hull City Museums and Art Galleries.)

Baddiley Mere may have had a similar fitting, though warping makes this difficult to confirm.

FISHING EQUIPMENT

In his description of twentieth-century logboats on the Mondsee, Austria, Salemke (1972: 4, 6, fig. 3) stated that horizontal holes through the sides near the top edge were for fishing nets. His drawing shows two holes to port and one to starboard. Paret (1930: 99, 100, fig. 19.1) suggests a similar use for the irregularly spaced holes through the sides of a logboat recovered from Frauendorf, Silesia. This may also be the function of the puzzling vertical holes in the natural crooks treenailed around the ends of the Hasselø logboat (Rasmussen, 1953: 18, 21, fig. 5).

Fish-wells are compartments between two watertight bulkheads or between a bulkhead and an end, with a series of holes bored below the waterline. They are known to have been incorporated in logboats: in one end of logboats on Lake Aegeri, Switzerland (Paret, 1930: 111); in the bows of boats on the River Havel at Brandenburg (Rudolph, 1966: 14); and in the stern of boats on the Altausseer-see, Austria (Werner, 1973: 44). de Weerd and Haalebos (1973: 392) found 17 holes through one side of Zwammerdam boat No. 1 which they believe may indicate a fish well.

EARLY USE IN N.W. EUROPE

The tools and techniques required to build the simplest form of logboat are given in Table 6.4. The data there suggests that, although suitable tools and techniques may have been available from the Upper,

Figure 6.21 The bow of the Brigg logboat of *c.* 834 bc. (Photo: by courtesy of Lincolnshire County Library.)

Table 6.4 Earliest technological stages at which the processes required to build the simplest form of logboat could have been accomplished.

Process	*Tools and techniques*	*Earliest evidence of use or of comparable techniques*
1. Collect log*	Windfall or driftwood	Upper Palaeolithic or Mesolithic
	Felling by fire	Lower Palaeolithic (Clark, 1977: 26)
	Felling by tools	Mesolithic (Clark, 1977: 114)
2. Shape internally and externally by reduction	Controlled fire and scraper	poss. Middle Palaeolithic (Clark, 1977: 26, 36)
	Wedge, hammer and scraper	Mesolithic (Clark, 1977: 25, 36, 115)
	Axe or adze	Mesolithic (Clark, 1977: 114)

* Trees of a size suitable for logboats were rare in N.W. Europe before *c.*8000 bc (Dennell, p.c.)

Table 6.5 Activities additional to those in Table 6.4 required to build a simple expanded logboat

Process	*Tools and techniques*	*Earliest evidence of use or of comparable techniques*
1. Shaping by transformation	Heat and expand Soak and expand	Controlled heat known from Upper Palaeolithic (Clark, 1977: 26) but possibly not applied to shaping timber until Bronze Age or later.
2. Insert transverse timbers	—	Not known

or even Middle Palaeolithic, there are unlikely to have been suitable logs until the Mesolithic (from say 8000 bc).

On the evidence of radiocarbon assay the Pesse find (van Zeist, 1957) is the oldest in N.W. Europe: Gro-486: 8265 ± 275BP (*c.* 6315 bc). This is uniquely early, but from the mid-fourth millennium onwards, there is a spread of radiocarbon dates for logboats: the earliest from Denmark is Tybrind vig 1 of *c.* 3310 bc ± 95 (K-3557); from France, a logboat from the River Charente of *c.* 2590 bc ± 110 (G1F-5156); and from Britain, a logboat from Locharbriggs, Scotland of *c.* 1804 bc ± 125 (SRR-326).

No dates for an indisputably expanded logboat have been published so far, although shaping by transformation – bending after heating or soaking – may have been technologically possible from the

Table 6.6 Activities additional to those in Table 6.4 required to extend a logboat by fitting washstrakes

Process	*Tools and techniques*	*Earliest evidence of use or of comparable techniques*
1. Fashion plank to required shape	Axe or adze	Neolithic planks known (Coles, 1976: 34, fig. 23) but no evidence of complex shapes.
2. Prepare holes	Chisels and gouges	Bronze Age (Clark, 1952: 213)
3. Fasten to logboat	Sewing or lashing	Upper Palaeolithic (Dennell, 1983: table 4)
	Treenails	Neolithic (Clark, 1952: 211–12)
	Mortice and tenon	Bronze Age (Johnstone, 1976: 51; Clark, 1977: 141)
4. Make watertight	Metal fastenings	Bronze Age?
	Moss	Mesolithic (Clark, 1952: 226)

Bronze Age (Table 6.5). The data in Table 6.6 suggests that extending a logboat by adding washstrakes may have been possible from the Neolithic and several of the early boats do have patterns of holes which may have been for such extension. The woodworking techniques required to extend a logboat lengthwise were certainly available by the time of the Ferriby boats (mid-second millennium), but the Hasholme boat of *c.* 300 BC is probably the earliest example to date of a logboat that is extended lengthwise and also has washstrakes.

7 Bark boats

DISTRIBUTION AND FUNCTION

Table 7.1 shows that there are two bands of latitude which include most of the regions where bark boats are, or are known to have been, used: *c*: 35° to 65°N and *c*. 10° to 55°S. They have also been used just north of the equator in Indonesia and Malaysia (Hornell, 1946: xxiv, xxv), in Guiana and the River Amazon, Brazil (Brindley, 1924: 125–7). Bark rafts are known only from Tasmania: as these are analogous to other types of bundle raft they are discussed in Chapter 9.

Sea voyages by bark boats of advanced design are documented out to *c*. 20 nm from the Australian coast (Edwards, 1972: 9, 10, 35); and the birchbarks of the Newfoundland Beothuk tribe (Fig. 7.1) were used for inter-island voyages of up to 50 nm and possibly from Newfoundland to Labrador (Adney and Chapelle, 1964: 96, 98). The Yahgan and Alacaluf peoples used bark boats in the Chilean archipelago (Fig. 7.2) but these also were complex rather than simple forms of bark boat (Edward, 1965: 21).

Simple bark boats, on the other hand, were widely used on rivers for fishing and hunting, and as ferries: for example in Siberia (Brindley, 1919: 66–7) and in Australia (Edwards, 1972: 7, 41).

Table 7.1 Principal bark boat areas of use

Region	*Approximate latitude*	*Reference*
N. America	35°–65°N	Waugh, 1919; Hornell, 1946: 186; Adney & Chapelle, 1964.
S.E. Siberia	40°–65°N	Brindley, 1919: 66–72, 101–7; Hornell, 1946: 184–6.
Sweden	55°–65°N	Eskeröd, 1956: 70–2; Westerdahl, 1982.
Chile	45°–40°S	Edwards, 1965: 21–5, 90–2; Hornell, 1946: 186.
Australia (N., N.E., S.E.)	10°–40°S	Hornell, 1946: 182–5; Edwards, 1972.
S.E. Africa	10°–20°S	Hornell, 1946: 183; Johnstone, 1980: 20.

VARIETY AND SIZE

The seventeenth–nineteenth century trading bark boats of N. America were probably the most complex of any and could be up to 36 ft (10.97 m) in length (Dunphy, 1979: 80). Bark boats of over 35 ft (10.67 m) have also been noted on the middle reaches of the River Amur, Siberia (Brindley, 1919), but the general run of boats in that region seems to have been *c*. 12–15 ft (3.5–4.5 m) in length (Brindley, 1919: 67). Bark boats of 16–18 ft (4.88–5.49 m) are known from Australia (Edwards, 1972: 42); and 15–25 ft (4.57–7.62 m) ones in S. America (Edwards, 1965: 22–3). The Beothuk seagoing boats were *c*. 20 ft (6 m) long (Adney and Chapelle, 1964: 94).

At the other end of the spectrum are the simple bark boats of Australia, E. Africa and Guyana, which were essentially a single-piece bark trough sometimes with open ends and with little, if any, internal framework (Hornell, 1946: 182–3; Edwards, 1972: 14, 34, 49; Jones, 1976: 239–240). Some of the Siberian boats such as the 9 ft × 3½ ft (2.74 × 1.07 m) bark boats of Kolguev Island were so simple and light that they could easily be carried on a sledge (Brindley, 1919: 104).

The average slenderness ratio (L/B) for twelve types of bark boat for which such data has been recorded is 6, with a range from 5 to 8. One other boat, the large one from the middle Amur (Brindley, 1919: 106) has a ratio of 16.

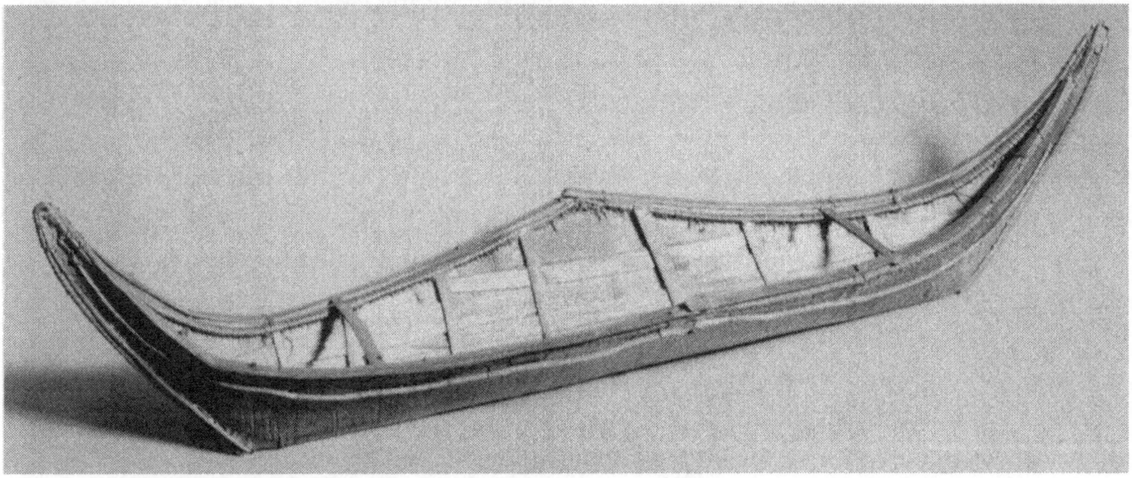

Figure 7.1 An early nineteenth-century model of a Beothuk birchbark canoe. (Photo: NMM Greenwich.)

Figure 7.2 Nineteenth-century drawing of bark boats of Tierra del Fuego. Nishimura, 1931.

CONSTRUCTION

RAW MATERIALS

The most suitable species of tree for boat hulls in N. America was said to be the Paper birch (Waugh, 1919: 23); in Australia it was the River Red Gum (Edwards, 1972: 3). There was a wide range of second-rate species used when the preferred ones were not available: some of these species are listed in Table 7.2. Materials used to fasten the bark by lashing or by sewing are given in Table 7.3; and Table 7.4 lists timber species used for framework.

Table 7.2 Tree species from which bark boats have been made

Region	*Common name*	*Botanical name*	*Reference*
N. America	Paper birch	*Betula papyrifera*	Waugh, 1919: 23
	Button wood	*Platanus occidentalis*	
	Elm	*Ulmus* sp.	Adney & Chapelle, 1964: 14–15
	Cotton-wood	*Populus* sp.	
	Basswood	*Tilia americana*	
	Spruce	*Picea* sp.	
	Hickory	*Carya* sp.	
	Chestnut	*Castenea dentata*	
	Yellow cedar	*Thuja excelsa*	Waugh, 1919: 24
Australia	River red gum	*Eucaliptus camaldulensis*	Edwards, 1972: 31
	Stringy bark	E. *obliqua*	
	Mountain Ash	E. *regnans* or *delagatensis*	
	Blue gum	E. *globulus*	
	White gum	E. *viminalis* or *haemastoma*	
	Mahogany	E. *botryoides*	
S. America	S. Beech	*Nothofagus* sp.	Edwards, 1965: 24
Sweden	Spruce	*Picea* sp.	Clark, 1952: 284

Table 7.3 Materials from which bark boat fastenings have been made

Region	*Common name*	*Botanical name*	*Reference*
N. America	Black spruce roots	*Picea mariana*	Adney & Chapelle, 1964: 15–16.
	White cedar roots	*Thuja occidentalis*	
	E. larch roots	*Larix laricina*	
	Jack pine roots	*Pinus banksiana*	
	White spruce roots	*Picea glauca*	Hansen & Madsen, 1981
S. America	Dried reeds		Edwards, 1965: 21–3.
	Hide		
	Baleen		
Australia	Bark		Edwards, 1972: 9
	Grass		Hornell, 1946: 183
Africa	Bark		Hornell, 1946: 183
Siberia	Hide	*Juniperus* sp.	Brindley, 1919: 67
Sweden	Juniper (roots?)		Clark, 1952: 284

Table 7.4 Materials used in the framework of bark boats

Region	*Common name*	*Botanical name*	*Reference*
N. America	White cedar	*Thuja occidentalis*	Adney & Chapelle, 1964: 17.
	Black spruce	*Picea mariana*	
	Maple	*Acer saccharum*	
	W. larch	*Larix occidentalis*	
	American ash	*Fraxinus americana*	
	Willow	*Salix* sp.	
Australia	Mangrove	*Avicennia officinalis* or *Rhizophora* sp.	Hornell, 1946: 185
Sweden	Hazel	*Corylus* sp.	Hansen & Madsen, 1981

Note: Beam ties were made of bark rope in Australia (Hornell 1946: 183–5; Edwards 1972: 8)

BUILDING SEQUENCE

Selection

A stand of the preferred tree species was searched for those trees which would give a large sheet of bark clear of blemish (Adney and Chapelle, 1964: 14–15, 24; Edwards, 1972: 31). The Paper birch was preferred in N. America because it had a resinous bark which was flexible when green or damp but which did not shrink or stretch unduly. This bark also had a horizontal grain and thus could more readily accept sewn additions, and it did not have a rough surface which in other timbers had to be scraped away to make the bark flexible (*cf.* Edwards, 1972: 7). A bark thickness of between $\frac{1}{8}$ and $\frac{3}{4}$ in (3 to 19 mm) seems generally to have been preferred (Edwards, 1965: 21, 22, 25; Adney and Chapelle, 1964: 14).

Collection

The period late spring/early summer was preferred as the sap was then flowing freely (Adney and Chapelle, 1964: 14, 24; Edwards, 1972: 30) and the bark could more easily be detached; a wet period also facilitated this. Sometimes a whole cylinder of bark was removed from the selected tree by splitting the bark vertically with a stone knife and then levering off with a wooden implement (Adney and Chapelle, 1964: 24); a standing tree was killed by this operation. In Australia, however, half-cylinders seem to have been frequently used, as numerous surviving 'canoe trees' with an elongated barkless scar testify (Edwards, 1972: 32). The rough thick area near the butt end was avoided and the required outline was marked on the bark and cut out with a pointed flint. Ropes were then passed around the tree and bark wedges and sapling levers used to separate bark from tree. The bark segment was then lowered carefully to the ground. Eight to ten men were required for this operation (Edwards, 1972: 30).

Single-bark boats

Simple: The elliptical bark segment used in Australia (Edwards, 1972: 34, 49) or the bark cylinder used in Guyana (Brindley, 1924; Hansen and Madsen, 1981: 8) could be used almost immediately by manipulating the bark to the required shape whilst it was still fresh and pliable (Edwards, 1972: 7). If the tree had had a natural longitudinal curve the bark segment might need no further treatment at the ends; otherwise the ends were blocked with clay or mud tempered with grass (Hornell, 1946: 183). Hornell (1946: 183) states that such boats might have no framework, but the simplest ones described and illustrated by Edwards (1972) have two or three timbers set transversely to stop the bark curling and it seems likely that this was essential. Edwards

(1972: 29) has noted that such a boat could be built in less than half an hour and certainly in less than a day.

Moulded: There are four variant methods:

1. The bark sheet was laid on logs, stones placed inside to keep the bottom flat and the sides supported externally by struts (Edwards, 1972: 28, 33). Small fires were then lit inside causing the sides to curl inwards against transverse timbers which then held the shape; the bark was also kept moist to prevent it splitting under heat (Edwards, 1972: 17). The flexible ends were then bent upwards and pleated, bunched and tied (Edwards, 1972: 9) – see Fig. 7.3; in Rhodesia the ends were skewered (Hornell, 1946: 183). If the bark proved too thick and inflexible at the butt end it was closed with mud and grass. The boat was allowed to set for several days before use.
2. The bark was placed curved side *up* on logs and fires lit underneath. When flexible, the bark was turned the right way up, transverse timbers inserted and the ends forced upwards.
3. The bark was placed curved side *down* on logs and fires lit underneath. When flexible, the bark was turned inside out, transverse timbers or ties inserted, and the ends bunched and tied (Edwards, 1972: 9). This use of bark with its outer side forming the inside of the boat has also been noted in S. America (Edwards, 1965: 23) and in N. America (Adney and Chapelle, 1964: 215; Dunphy, 1979: 78).
4. Edwards (1972: 33) has also noted a simple means of moulding the bark to shape by pressure: a hollow is cut to the required shape in the ground and the bark sheet is placed in it and weighted with mud and clay.

Transverse timbers and ties seem to be the main framework required in such moulded bark boats and there could be up to five different elements in various combinations:

1. Transverse timbers: saplings placed at or just below sheer level (Edwards, 1972: 36).

Figure 7.3 Drawing of an early nineteenth-century Tasmanian lashed bark boat. Lesueur and Petit's Atlas to Peron's *Voyage of Discovery*. (Photo: NMM Greenwich.)

2. Ties: transverse rope ties at or just below sheer level (Edwards, 1972: 6).
3. Ribs: pliant branches curved to shape (Hornell, 1946: 184).
4. Poles: lashed longitudinally at sheer level (Hornell, 1946: 184).
5. Rushwork: (unusual) around the rim (Hornell, 1946: 182–3).

These last two devices stiffened the rim of the bark shell.

Multiple-bark boats

Boats made of more than one sheet of bark are known from Australia (Edwards, 1972: 9–10; Hornell, 1946: 184–5), from S. America (Edwards, 1965: 22–3) and from North America (Adney and Chapelle, 1964). Shapes and sizes of boat can thereby be built that are unattainable with a single sheet. Several sheets are joined together and gores are cut which are subsequently sewn to obtain the required shape, the seams being payed with resin or gum (Adney and Chappelle 1964: 17; Edwards, 1972: 10; Edwards, 1965: 22). S. American boats had either a pithy vine in the seams or straw and reed backed by a longitudinal batten (Edwards, 1965: 21, 25).

The S. American and Australian multiple-bark boats have a framework which is made to fit the bark sheet and hold it to shape; thus the bark shell, which is evidently fashioned by eye or by rule of thumb, determines the shape of the boat. The N. American boats, on the other hand, (at least those that have been documented in the past 100 years or so) have a sequence of construction (Fig. 7.4) which, although primarily shell-first, has aspects of skeleton construction. A framework of two stringers (gunwales) and several transverse timbers (thwarts) is made to the required shape (either of the sheerline plan or the bottom plan of the boat) and used as a former (mould) to shape the bark shell. In the early twentieth century sticks (boat-ells) marked with the main measurements were also used (Adney and Chapelle, 1964: 37); and a shaped building bed was sometimes used to give the required hog or sag to the bottom of the boat (Adney and Chapelle, 1964: 36–57). In more recent times the sheerline (i.e. the upper edge of the bark shell) was determined by the stringer and thwart framework which was forced into the required shape before the bark was attached. By such means a relatively standard shape of boat can be produced and successful 'designs' could readily be transmitted to others.

MAINTENANCE AND DURABILITY

Edwards (1972: 30) gives the maximum useful life of an Australian bark boat as 2 years. However, elsewhere he quotes a mid-nineteenth century description of a simple bark boat which was expected to last months rather than years. The difference may be due to whether or not a boat was given regular protective treatment such as a clay wash or rubbing with grease and ochre (1972: 30), otherwise it would become waterlogged (1972: 79). The seams of the N. American multi-bark boats were payed with spruce gum tempered with animal fat (Adney and Chapelle, 1964: 50). Dunphy (1979: 79) believed that a boat of paper birch bark would last for 10 years if carefully maintained.

VARIANTS

There is some evidence from Sweden and Siberia that boats were built there by fastening bark to a pre-erected framework, unless those who documented this material have misunderstood what they saw. A 1934 find from the River Viskan near Byslätt, Istorp Stoke, Västergotland, and now in the Goteberg Museum, consists of a sheet of spruce bark *c.* 3.5 m × 0.7 m, with fragments of three hazel ribs and juniper lashings, and leather slips inserted between fastenings and bark to prevent chaffing. Holes through the bark have suggested to certain investigators that the bark had been lashed to a framework of ribs and other timbers, rather than that the framework had been inserted into a bark shell (Humbla and von Post, 1937: 11; Clark, 1952: 284; Eskeröd, 1956: 71; Ellmers, 1972: No. 318; Johnstone, 1980: 24; Hansen and Madsen, 1981: 4–5). Pollen analysis suggests a date before AD 1000.

Several reports of late nineteenth/early twentieth century Swedish bark boats also appear to state that the bark was sewn or lashed to a wooden framework: Eskeröd (1956: 7) states that 'bark was sewn to stem, stern and ribs by roots . . . '; Westerdahl (1982) that 'birch bark was sewn to a rib structure and reindeer hair with tar was used for caulking'. When describing the boats of the River Kuntenai region (southern British Columbia, Washington and Idaho), and those of the River Amur region in E. Siberia, Brindley (1919: 106) states that they had a 'wooden frame of "ribs and horizontal slats" on which strips of bark are sewn, the whole structure being caulked'. Brindley heads this section of his paper on Siberian boats 'Canoes of Bark on wooden frame' and thus we must presume that his description of the Amur and Kutenai boats should be taken literally; this suggests a building technique with some analogies with the most widely used method of building hide boats (Ch. 10).

The description of recent Kutenai and Salish bark boat building given by Adney and Chapelle (1964: 168–173) does not confirm Brindley's statement – the method they describe is shell sequence, similar in general to other bark boat traditions in N. America – however they do say that the methods of building the large 'sturgeon-nose' (projecting forefoot) bark boats 'have never been reported' and it may be that Brindley was describing a more northerly type of Kutenai bark boat with affinities to nearby hide boats.

PERFORMANCE

LOADS

Bark boats of S. America and Australia range in size from 2-men boats to ones capable of carrying 10 or 12 (Hornell, 1946: 185; Brindley, 1919: 66; Edwards, 1972: 23). The 'trader' or 'master' bark boats built in N. America after an infusion of European technology could carry up to 20 men (Greenhill, 1976: fig. 80) or 15 men with gear and 4 ton of cargo (Dunphy, 1979: 81).

SPEED

The European 'discoverers' of N. America were impressed by the speed of bark boats (Adney and Chapelle, 1964: 7): a bark boat paddled by 3 men could overtake a 4-oared ship's boat. The bark boat is a lightweight design with a high power/weight ratio and those types with high slenderness coefficients (L/B) and low volumetric coefficients (Ch. 11) can be expected to have high speed capability.

PROPULSION AND STEERING

Bark boats have been poled and paddled (for example see Edwards, 1965: 23; Brindley, 1919: 72; Edwards, 1972: 18, 35) but, as far as is known, not propelled by oars. They are a relatively flimsy construction and generally of narrow beam and thus oar pivots would be difficult, although not impossible, to provide.

Edwards (1972: 35) noted that in Australia single paddlers take station just aft of amidships generally standing, but squatting when hunting or fishing. With a standing paddler such narrow boats can have very little transverse stability (Ch. 3). The transverse timbers are not used as thwarts in the Australian boats – again because of stability requirements – but the kneeling or squatting paddlers sometimes rest their backs against them. Single-bladed paddles are generally used, but double-bladed ones are known (Brindley, 1919: 67; Edwards, 1972: 8). In the larger boats of N. America there is a specialist steersman, who stands and wields a longer paddle; there may also be a bow steersman (Greenhill, 1976: fig. 80). Poles used in Australia often had one forked end which was used to spear fish (Edwards, 1972: 18, 31, 35).

A sealskin sail was used on S. American bark boats but only in light favourable winds; such use of sail was unknown before the late-seventeenth century and may be due to European influence (Edwards, 1965: 23). Adney and Chapelle (1964: 10) also think it unlikely that sail was used on bark boats before Europeans arrived in N. America, after which time they are known to have used blankets, bark, skin and even a bush (Waugh, 1919: 30).

ADVANTAGES AND DISADVANTAGES

Bark boats are built from material which, in certain regions of the world, is readily available and they can be built with a simple tool-kit; there is no need, for example, to be able to fell a tree. Their light weight means not only that they can be carried by their crew, but that they are very buoyant, which in turn means good speed potential. Such a structure has limited strength, however, and can easily be damaged by mishandling or misfortune when under way or when landing, but such damage can be relatively easily repaired by materials ready to hand in 'wilderness travel'.

A bark hull is water resistant to a degree but can become waterlogged after prolonged use (Edwards, 1965: 9). Waterproofing with greasy mixtures delays this and prolongs a boat's useful life.

A range of shapes can be obtained even from a single sheet of bark by moulding after heat treatment. A still wider range of shapes is available when more than one sheet is used, and boats up to 36 ft (10.97 m) in length with a maximum beam of 5 ft (1.52 m) can be built.

The relatively high slenderness ratio of these boats as a general class, except perhaps the river trading boats of eighteenth and nineteenth century N. America, leads to limited transverse stability and the crew need 'a fine sense of balance and agility' in

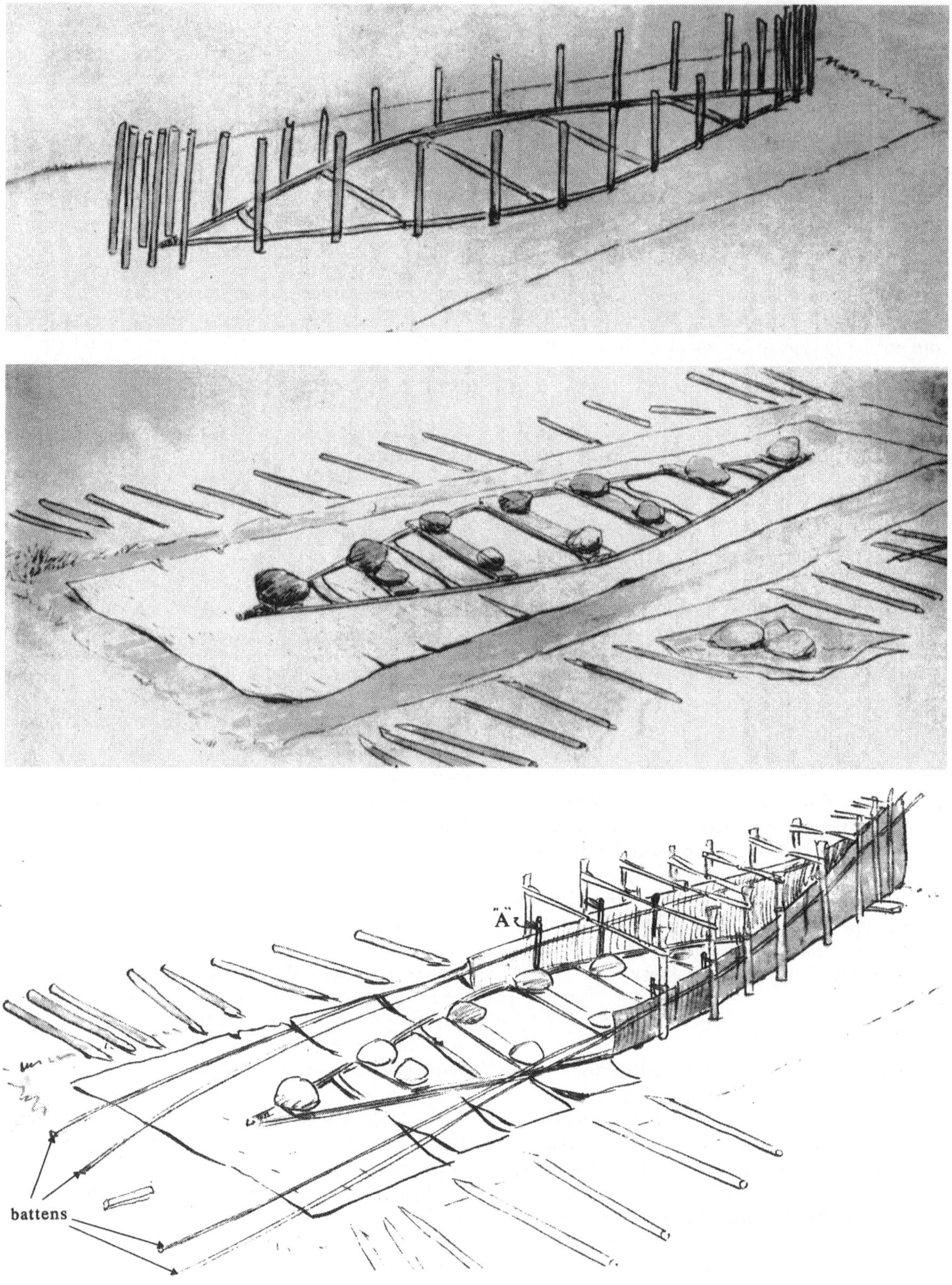

Figure 7.4 Six stages in the building of an American Indian bark boat. After Adney and Chapelle, 1964.

A
A
A
A
B
B

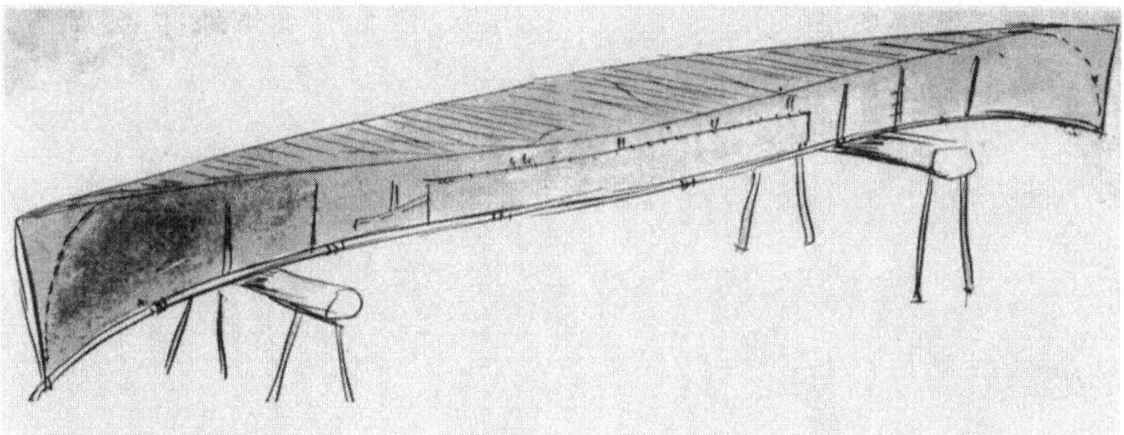

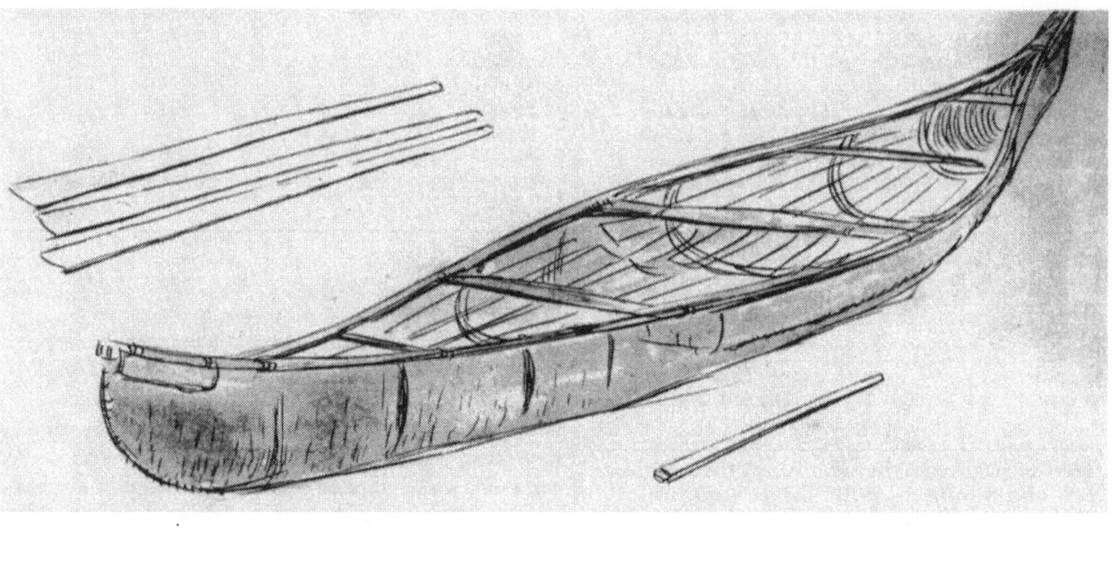

handling their boats (Edwards, 1972: 35); loss of equilibrium could lead to capsize. Although seagoing voyages are known, these have been in the larger, multi-bark boats which also need expert handling in these conditions.

EARLY USE IN N.W. EUROPE

Table 7.5 indicates that the activities required to build a basic bark boat are very simple ones: the substractive (reduction) technique of stripping bark and the additive (construction) technique of stopping the ends of the bark trough. Tools and techniques to undertake comparable tasks were probably available in the N.W. European Upper Palaeolithic stage of technology, but sizeable trees from which a large sheet of bark could be taken may have been rare until *c.* 8000 bc.

Table 7.6 shows that the ability to make a moulded bark boat depended principally upon use of controlled heating which was known in the Upper Palaeolithic but may not have been applied to bending bark until the Neolithic. It may not have been until the Bronze Age, however, that the bark boat of more than one sheet became a possibility (Table 7.7).

Raw material for making simple bark boats existed in N.W. Europe possibly from the late Palaeolithic and certainly by Mesolithic times. The trees available may not have been of the species and size to build first rate bark boats, on the lines of recent N. American ones, but usable bark was there. However, apart from the meagre Swedish evidence (Eskeröd, 1956; Westerdahl, 1982; Anon, 1982) there is no surviving or recent tradition of such use. This may have been because of the early emergence of an expanded and extended logboat (Ch. 6) with the bark boat's advantages and more, as Johnstone (1980: 24)

Table 7.5 Earliest evidence for tools and techniques analogous to those used in building a basic bark boat

Activity	*Tools and techniques*	*Earliest evidence*
1. Strip bark sheet of required shape	Cutting tools Wedges	Tools available from Upper Palaeolithic. Bark use known from Mesolithic but probably derived from Upper Palaeolithic (Clark, 1952: 208)
2. Insert transverse timbers		Not known
3. Stop ends with clay mix		Upper Palaeolithic (Dennell, 1983: Table 4)

Note: Trees of sufficient size to give a large sheet of bark may have been rare in N.W. Europe until after 8000 bc (Dennell p.c.)

Table 7.6 Earliest evidence for tools and techniques analogous to those used in building a moulded single-sheet bark boat

Activity	*Tools and techniques*	*Earliest evidence*
1. Strip bark	As Table 7.5	As in Table 7.5
2. Collect fastenings	—	Upper Palaeolithic (Clark, 1952: 226; Clark, 1977: 101)
3. Shape by transformation	Heat and bend	Controlled heat known from Upper Palaeolithic (Clark, 1977: 26) but possibly not applied to shaping bark until Neolithic (Australia).
4. Fasten ends	Binding	Upper Palaeolithic (Clark, 1952: 226)
	Skewer	Neolithic (Clark, 1952: 211)
	Sewing	E. Bronze Age (Clark, 1952: 209–10)
5. Insert framework	Lashed	Upper Palaeolithic (Clark, 1952: 226)

Table 7.7 Earliest evidence for tools and techniques analogous to those used in building a multi-sheet bark boat

Activity	*Tools and techniques*	*Earliest evidence*
1. Strip bark	As in Table 7.5	As in Table 7.5
2. Collect sewing material		Upper Palaeolithic (Clark, 1977: 101)
3. Fasten sheets of bark	Sewing	E. Bronze Age (Clark, 1952: 208–10)
4. Waterproof seams	—	Mesolithic (Clark, 1952: 226)
5. Insert framework	Lashed	Upper Palaeolithic (Clark, 1952: 226)

has suggested. On the other hand, there are two pointers towards an early bark boat tradition in N.W. Europe. The first is literary: the ninth century Norwegian merchant Ohthere told King Alfred that the Cwenas (a Finnish tribe) carried their small and very light boats overland (Lund, 1984: 21) – these may have been of hide or of bark. The second pointer is the poorly documented Byslätt find (Clark, 1952: 284) which seems to be dated pre-AD 1000 and may be the remains of a bark boat.

8 Wooden plank boats

The aim of this chapter is to bring out some general principles, problems and solutions by posing a series of questions which have to be answered during the building of most, if not all, plank boats, and applying them specifically to those boat-finds from northern and western Europe for which a good record has been compiled. Outline classification schemes are proposed for such timbers as stems, as an aid to the interpretation of individual finds. McKee's schemes (1983) may be more applicable to complete boats and to reconstructions.

HOW TO BUILD TO THE REQUIRED SHAPE?

A boat's performance, measured by such concepts as stability, speed, closeness to wind, leeway, etc., depends on the outcome of several interacting variables, the principal ones being:

hull shape	
form and efficiency of propulsion	discussed in Chapters 11 and 12
match of hull and sail	
capabilities of seaman	

The range of shapes that can be given to a hull depends to a great degree upon the general state of technology and the tools available at a particular time, and on individual inventiveness. But it also depends on the material used; for example planks can only be given so much bend, twist and set; and the angle between adjacent strakes cannot be more than *c.* 20° in both flush-laid (classical) and overlapping clinker (northern) forms of edge-fastened hull, otherwise insufficient wood is left to take fastenings (McKee, 1972: 4).

Which shapes are chosen for a particular boat depends on the qualities it is required to incorporate and the recognition by the builder of how these may be obtained from different shapes. Relationships between shapes and performance are seldom 1 to 1, however, and any boat design is a compromise (Marchaj, 1979: 5). Desirable qualities that might be incorporated into a boat include the following:

Safety attributes:
1. seaworthiness – buoyancy, strength and durability
2. stability

Performance attributes:

3. cargo capacity
4. speed
5. manoeuvrability and controlability
6. closeness to wind and minimum leeway
7. sea-kindliness and dryness

These are not arranged in any order of importance (although without safety there can be no performance) for the required function will determine which attributes should be emphasised in a particular design. Of these seven requirements, shape directly influences 2 to 7; but some of these characteristics are not simultaneously achievable, for example, a design which seeks to maximise speed will thereby impose constraints on cargo capacity, transverse stability and sea-kindliness. The builder has to decide how far reduction in one characteristic is justified by a gain in some other characteristic; the balance chosen depending upon the role the boat is to fulfil. Thus speed and manoeuvrability may be emphasised in a boat for raiding, whilst payload and stability may be the characteristics most desirable in a ferry or fishing boat. Shapes which combine high ratings in the required virtues with acceptable levels in the other characteristics will then be chosen.

BUILDING 'BY EYE'

Boatbuilders in much of northern and western Europe in the recent past have used lines drawings, block models, sets of moulds (patterns) or templates, and sets of key measurements or ratios to give them the shape they require (McKee, 1983; Basch, 1972). It has been claimed that boatbuilding has also been done 'by eye' (see, for example, Christensen, 1972b: 239); the implication here is that the builder approached the building of each new boat armed only with his inherited traditions and acquired skills and that the shape of the boat was developed as the boat grew before his eyes (Christensen, 1972b, 1973). It is true that, except in times of standardised mass production, stock was selected by eye and the supply and character of the timber readily available had to be taken into account in the building of each boat, and undoubtedly this influenced the precise shape and size that could be achieved. Thus the length of the keel would be determined by the length of suitable sound logs available; and floor timbers were not necessarily fastened at right angles to the keel or parallel to one another – they could 'wind' across the hull, their position being determined by the natural curvatures of each log, as in the ninth/tenth century AD Graveney boat (Fenwick, 1978: fig. 1.2.9). The five bottom planks of the Brigg 'raft', even the pairs which came from the same tree, were not worked down in breadth to a standard size determined by the smallest one or to some preconceived plan, but each was fashioned to the maximum breadth of sound wood that could be obtained (McGrail, 1981b: 224–5), and as with many logboats (McGrail, 1978a) the natural taper of the wood was retained.

Thus, each boat had an individuality and was unique, even when the aim was to build a boat similar to another one. Fieldworkers may have been misled by this individual approach (the 'free-arm' method of Ewing McGruer, boatbuilding officer of the Scottish Local Industries Trust in the mid-twentieth century) and by the absence of a pre-erected framework, into thinking that early boatbuilders had no aids to help them obtain the required shape, except the eye. The key to understanding their methods may lie not so much in what they appear to have done, but rather how they taught their apprentices. Harold Kimber, a master boatbuilder from Somerset, whose methods have been documented by McKee (1972) could bore nail holes at precisely 4 in spacing by eye, whereas his apprentice used a ruler until he had gained the experience.

BUILDING AIDS

In those recent cases of 'building by eye' which have been thoroughly documented some form of control based on measurements or 'rules of thumb' has been found to be used to obtain shape. In an unextended logboat and in the very simplest of flat-bottomed boats with rectangular ends (box-shaped), check measures to ensure symmetry may be all that is required, but boats of more curved form (required to improve performance) include parts which are of a relatively complex three dimensional shape (Figs 4.7 and 6.4). It is a skilled task to determine how many and what shape planks are needed, and how each should be fashioned in two planes to form the required three dimensional shapes. Working the changing bevel on a keel to match the varying angle of the garboard (first) strakes also requires great skill.

Rules of thumb

The spatial skills required by the plank boatbuilder and the bark boatbuilder may be compared with those required by a tailor (McKee, 1972). Once acquired and proven they had to be handed on to apprentices as part of their cultural inheritance, as indeed had navigational skills (Ch. 14); such inherited wisdom and experience is encapsulated by preliterate and prenumerate societies in sayings and in rules of thumb. Recent boatbuilders, even in industrialised N.W. Europe, have used similar aids. Thus when determining how to plank the bottom of a rounded bilge boat Harold Kimber used the saying 'quick out to the bilge and slow round it' which determined the relative breadths of the strakes (McKee, 1972: 7); and McKee (1983: 112) doubled the plank thickness to estimate the necessary plank overlap in a clinker-built boat. Certain Welsh coracles have length equivalent to the height of the man (Hornell, 1936: 36) whilst kayaks are approximately three times that (Petersen, 1982: 8).

Families of boats, and boatbuilding traditions which are recognisable principally by shape, may be defined by simple rules such as those for the Somerset turf boats where the ratio length/maximum breadth has a value of 5.5 (McKee, 1983: 107), and thus with a fixed angle and height of sides, the shape of the boat is determined. The shape of Wexford cots is similarly determined: the beam is given by the approximation (L/3 – 2) (Roberts, 1979: 66). The principal dimensions of McGruer's estuary boats were similarly defined in relation to their maximum beam (McKee, 1983: 109). And we find ratios used in the building of seventeenth-century ships when the optimum relationship of length/breadth/depth was said to be 6.5: 2.3: 1 (Salisbury, 1958: 15).

Such traditional rules set the parameters within which individual craftsmen could work and enabled them to translate the idea of a boat-shape into reality. Once a shape had been devised which met certain operational requirements (speed, stability and so on) these rules could be used to build a similar boat even if the original was not to hand to be copied. Alternatively, a variant could be built to meet a different operational specification – 'build me a boat like the one you built for Ola last year, but 2 ft longer and with rather more beam aft' (Christensen, 1972b: 237) – see also McKee (1972: 4) for a discussion of recent methods of modifying an 'ideal' design.

Measuring sticks

The principal characteristics of a boat may be noted in some detail by recording significant breadths and measurements from a base line, for example a rope from stem to stern, at certain stations (e.g. at every (alternate) frame) in the boat, and at specific stages in the construction sequence (e.g. after every (alternate) strake is fastened). The shape of the boat is uniquely defined by such measurements plus the profile of key timbers such as keel and stems. These measurements are, in fact, an analog of a table of offsets which the modern builder uses, and may be recorded in coded form on a boat-ell (Fig. 8.1) as in post-medieval Scandinavia (Christensen, 1972b: 239–40), in twentieth-century Pakistan (Greenhill, 1976: 52), and in the building of bark boats by nineteenth/twentieth century American Indians (Adney and Chapelle, 1964: 37).

Boat levels

A third method is to record for individual strakes the angle of the slope in the transverse plane, which, together with the breadth of the strake (measured between the lines of fastenings in the clinker, overlapping style of building) determine the shape of the hull at a particular station (Christensen, 1972b: 241–7). Such angles may be recorded as marks on a boatbuilder's level (Christensen, 1972b: fig. 2) rather than in degrees (Fig. 8.2). This method is

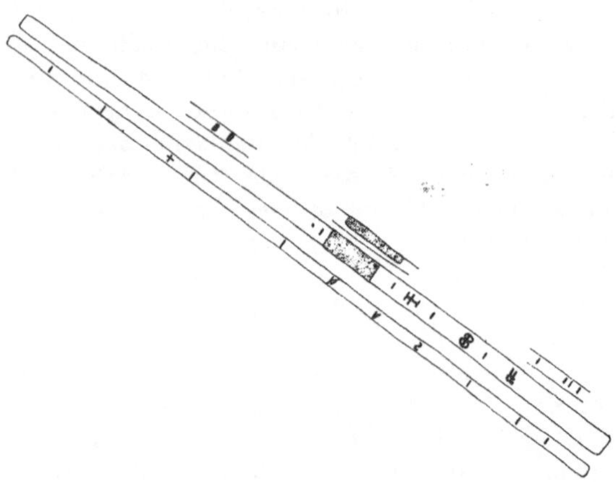

Figure 8.1 A nineteenth-century Norwegian boat-ell. After Christensen, 1972b (copyright Arne Emil Christensen.)

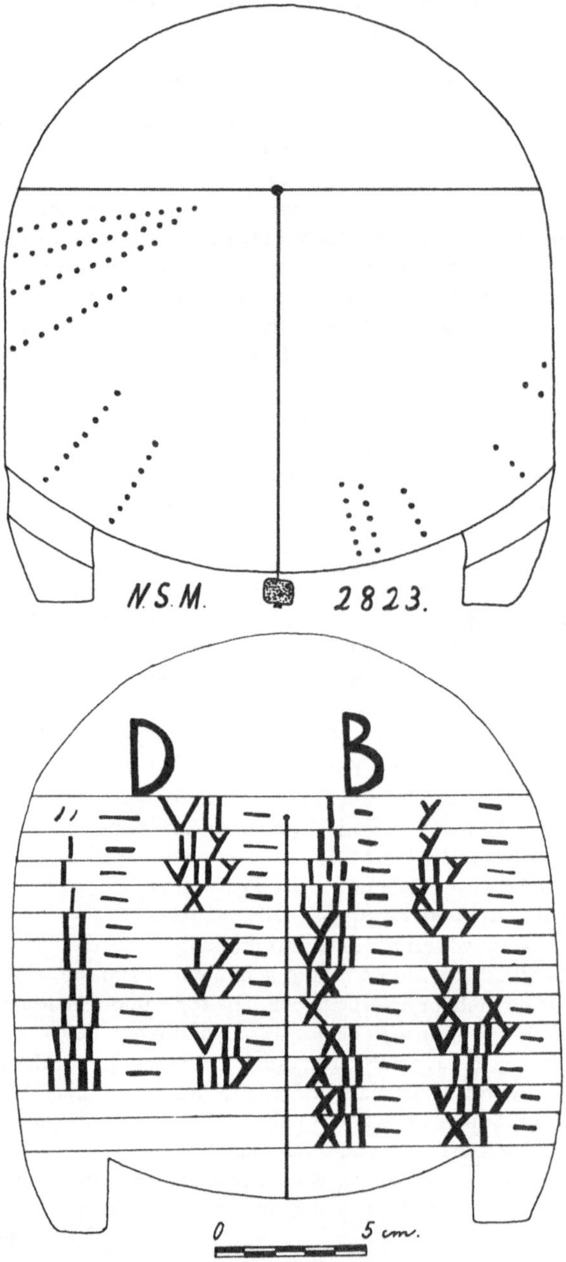

Figure 8.2 A.B. The two sides of a twentieth-century Norwegian boatbuilder's level for a 12-strake boat, with angles and breadths marked in code. After Christensen, 1972b (copyright Arne Emil Christensen).

theoretically applicable not only to N.W. European clinker-built boats but also to those built in the Mediterranean classical tradition, where, although the planks fit flush, edge to edge, the transverse form is also obtained by the angle of bevel applied to the upper edge of each strake, which determines how the next strake will lie (Figs 8.3 and 8.4).

Patterns or moulds

The form of any boat's transverse section at any one station may be recorded by making a pattern: such rib-shaped, wooden templates marked with the run of the strakes are nowadays known as *moulds* (McKee, 1983: 112–4). The outer edge of a mould represents the inner profile of the boat, and a set of these together with some record of keel, stem and sternpost can hold all the information about a boat's shape – once again units of measurement are not essential. Moulds can be permanently in position, athwart the keel, thereby defining the boat's shape throughout the building sequence or they can be used occasionally to check the shape of a boat being built by free-arm methods. When used in twentieth-century shell sequence building they are removed when the hull is finished and replaced by timbers such as frames and floors which give transverse support to the shell. There is analogous use of a building frame of top stringers during the shaping of N. American bark boats (Ch. 7).

Such temporary moulds may be full or half transverse sections of the boat, and they may be used at every or several frame stations or just at one midships station. The shape they convey does not have to be followed exactly – for example the boat may be built broader at certain stations, or the moulds may be spaced out at different intervals (McKee, 1972: 4). *Active* moulds, as defined by Basch (1972), on the other hand, determine the precise shape of the boat and remain in the boat after completion, providing transverse strength throughout her life. Such use of moulds is applicable to the *skeleton* sequence of construction, the transverse moulds forming part of that skeleton.

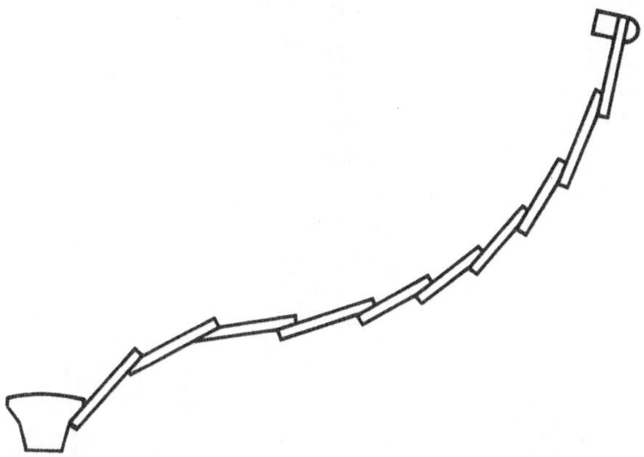

Figure 8.3 Half-section of a clinker-built boat showing how the required section is obtained by varying the angle of bevel along the upper (and sometimes lower) edge of the planking. After McKee, 1983.

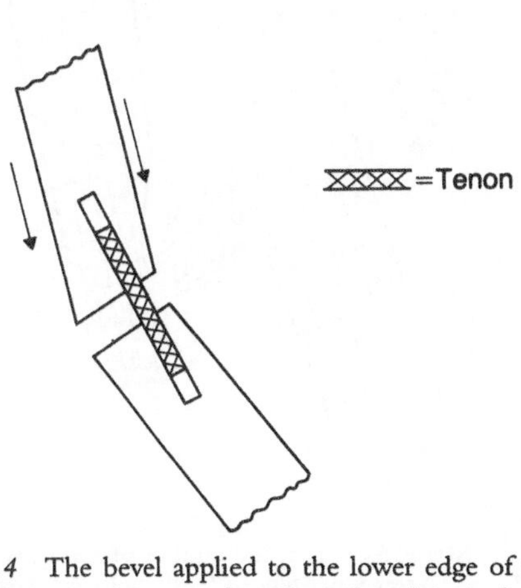

Figure 8.4 The bevel applied to the lower edge of flush-laid planking (in this case, the Classical tradition) also determines the transverse section of the boat.

Worked stems

The methods of determining shape discussed above all require additional information such as details of keel and stems before a complete boat can be specified. Such information can be retained in pattern form or it may be that worked timbers were stored. Partly worked stems found in bogs in Scandinavia (Brøgger and Shetelig, 1971: 41) and in Britain (McPherson, 1877–8) were probably stored under water (which later became bog) to keep the timber moist so that it could be worked relatively easily in future boatbuilding. Landings for the ends of six pairs of strakes had been worked on one of these stems from Eigg and thus the run of planking at that end of the boat would be defined once the stem had been scarfed (joined) to the keel. In this way a key aspect of a boat's design could be recorded for future use.

EARLY METHODS

The foregoing paragraphs highlight the fact that whereas in *skeleton* sequence (frames or moulds first) the form of the boat is necessarily determined by the predetermined shape of the framework of keels, stems and frames, in *shell* sequence (frames inserted after planking has been (partly) completed), the shape may be determined:

(a) purely by eye (if this is indeed possible – see also McKee, 1983: 114); or
(b) by rules of thumb; or
(c) by boat-ell or boat level marked measurements; or
(d) by passive, temporary moulds; or
(e) by a combination of (a) to (d).

(b) to (d): supplemented by stored timbers

The information in (b) to (d) can be stored and re-used by the originator or others.

The construction of active permanent moulds for use in the skeleton building of anything but the smallest boat is much simplified if a model (small-scale or full-size), or drawings or lines plans exist. Little deviation from the planned shape is possible once the moulds have been established, and the planking or hide (for this skeleton sequence predominates in recent hideboat building) must be fashioned to fit the framework. In shell sequence, on the other hand, although a model may exist in the mind's eye, as part of the cultural inheritance, as an innovatory modification, or as an existing boat which is to be copied, the method allows variations to be introduced as the work progresses to match the available timber supply, or to meet a revised requirement or simply because when viewed, the lines are not 'eye-sweet' or fair – in skeleton building the latter requirement is met by fairing the moulds.

Lines plans and other measured drawings do not appear to have been used in northern and western Europe before the seventeenth century (Lyon, 1974). Needham (1971: 409) has pointed out that in AD 1158, Chang Chung-Yan made a small demonstration model before building a ship, which may have been a way of getting the form of the ship from a model, although Needham has also emphasised (1971: 413) that 'the traditional Chinese shipwrights used no templates or blueprints, depending rather upon skill and sureness of eye'. In Europe however, there seems to be no evidence for the use of models as an aid to building before the seventeenth century (Greenhill, 1976: 296).

The use of ratio rules and other rules of thumb may in future be detectable in excavated remains, although the inevitable distortions and incompleteness of this form of evidence makes this difficult to demonstrate unambiguously. The earliest known use of a boatbuilder's level is in the late-seventeenth century (Christensen, 1972: 242), but its use and that of the boat ell and temporary moulds would leave little, if any, evidence on a boat structure. Any of these artifacts *might* have been used as an aid to building in antiquity but only excavation of one from an appropriate context will demonstrate this.

BUILDING IN THE SHELL SEQUENCE

The most positive evidence for shell sequence in

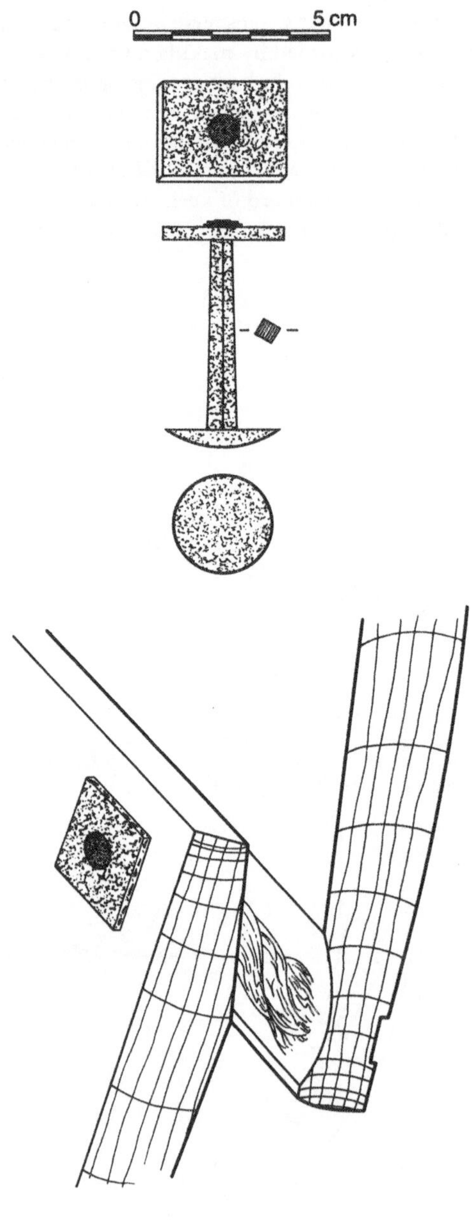

Figure 8.5 Method of fastening clinker planking with clenched nails. After Christensen (copyright Arne Emil Christensen).

clinker work is the existence of clenched strake fastenings (Fig. 8.5) or plank scarf fastenings underneath a frame, as was found in the Viking Age Gokstad 1 (Brøgger and Shetelig, 1971: 78). For flush-laid planking, as in the Classical tradition, where the planks are joined within their thickness (Fig. 8.6) the best proof of shell sequence is the existence under the frames of tenon-securing treenails, driven from the inside or wedged on the inside. However, as Reith (1978: 51) has pointed out, unless such signs are found under every frame it remains a possibility that some frames may have been erected before planking began and thus at least influenced, if not determined, the hull shape. The fact that in the Viking tradition no boats have so far been found with frames fastened to the keel makes this possibility doubtful. Christensen (1973: 139) has made the point that, if frames were previously in position, the edge fastening of clinker or flush-laid planks, which often have to be curved and twisted to get the three-dimensional shape required, would be difficult, if not impossible. This is undoubtedly true if there are many frames but need not overcomplicate building if only one or a few frames were to be prepositioned.

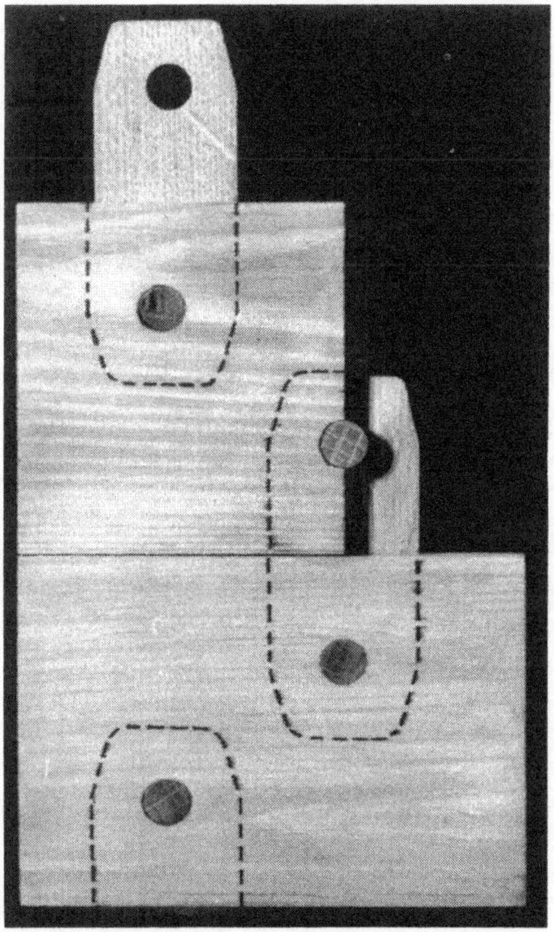

Figure 8.6 Method of fastening flush-laid planking in the Classical tradition. (Photo: NMM Greenwich.)

The case against the use of active frames in the Norse tradition is further strengthened, however, by ethnographic considerations. Scandinavia appears to be one of the few places in N.W. Europe where cultural continuity in boatbuilding can be demonstrated from medieval times until today (*fide* Christensen). Edge-fastened, shell sequence building of at least the lower strakes of a hull is illustrated by Åke Classon Rålaub in a book published in Stockholm in 1695: long-armed clamps and building levels which are also appropriate to this form of free-arm building are also illustrated there. Christensen (1977) has also claimed that in western Norway such continuity of practice may be taken back to the Viking Age on the evidence of boat shape, boatbuilding tools and terminology. As Christensen (1973: 141) has also stated that active moulds were probably not introduced into western Norway for boatbuilding until late in the nineteenth century, and that some boats are still built today without them, their use in Viking times seems unlikely; rather shape was determined and recorded 'by eye', or by rule of thumb, by boat-ell or level, by passive moulds or by some combination of these methods.

BUILDING IN THE SKELETON SEQUENCE

As a shell of planking may be built 'by eye', i.e. from inherited wisdom and individual skill aided by the methods described above, so may a skeleton, framework or mould. Baskets of woven withies are made free-arm, and the framework for hide boats may similarly be built (Herodotus, I, 194; Hornell, 1936, 1937–8). Boats which are members of one specific class of hide boat – Boyne coracle, Donegal paddling curach, etc. – are similar in form but their individual shape and size is determined by their framework which is assembled by eye, possibly aided by rules of thumb and simple ratios. It is theoretically feasible, therefore, to use similar methods to build a framework of keel, stems and ribs (active moulds) for a small plank boat. It is a more open question whether the framework for a *large* boat could be so designed without the aid of drawings or models. An intermediate method may, however, be possible, akin to the seventeenth-century use of one or two moulds or frames (Greenhill, 1976: 65, 72). Longitudinal ribbands or battens were run around the master frames from stem to stern, thereby defining the three-dimensional shape: from this open framework other frames may be derived. Whatever method was used to derive the shapes of individual frames or moulds it seems certain that specialist skills were required; once these moulds were made, skills which were more widely available were then needed to build the hull. On the other hand, special supervisory skills are necessary throughout the shaping of a shell-built hull, although the various aids described above may allow less-skilled boatwrights to produce shapes similar to a proven design.

Earliest use of the skeleton sequence

Hide boats are known to have been built on frames from the fifth century BC (Herodotus, I.194) near the River Euphrates, and probably from the seventh century BC on the evidence of an Assyrian relief from Sennacherib's palace at Nineveh (McGrail, 1981a: pl. 27). Their building and use in northern Europe is described by the fourth century AD Roman author Avienus in his *Ora Maritima* which includes extracts from the *Massaliote Periplus* probably of the sixth century BC (Hawkes, 1977). Rock carvings dated to the Bronze Age in Scandinavia and northern Russia may represent hide boats (McGrail, 1981a: 16), and using these as analogs Ellmers (1984) has attempted to demonstrate the existence of simple, framed hide boats in the late-Glacial period Ahrensburg culture of *c.* ninth millennium BC (Ch. 10), although there are reasons for thinking that the earliest form of hide boat may in fact have been frameless (Ch. 2). The shape of some forms of Iron Age houses was determined by a framework which was subsequently 'clothed' with wattlework (Cunliffe, 1978: 177), as possibly were Neolithic houses (Startin, 1978: 145). Repairs to any form of plank boat could also familiarise boatbuilders with the idea of adding to an existing framework, and logboats may be extended upwards by inserting frames to which planking is subsequently added (Ch. 6), although this technique has not so far been reported on an excavated boat.

Thus the general concept of skeleton or framework building in one form or another was available in many regions of Europe from early times to be applied to plank boats; but when this method of determining shape was first used is not yet clear. In N.W. Europe Romano–Celtic vessels of the early centuries AD were built by frame-based methods (McGrail, 1995). For the Mediterranean, Morrison (1976: 165–6) has pointed out that a form of skeleton building may have been known in the fifth century BC. The seventh-century AD wrecks from Yassi Ada and Pantano Longarini had edge-joined and non edge-joined planking, and shell and skeleton sequences. In all three vessels the skeleton sequence appears to have been used where minimum fashioning of the planking would have been required. Steffy (1982) has deduced that the eleventh century AD vessel from Serçe Liman, which had planking not fastened at the edges, was built on a simple framework of keel, stems and five pairs of frames (active moulds) which defined the shape of the bottom, the bilge and the lower sides over the central 3 m of the 15 m length of the vessel. Strakes 4, 3, 2 and 9 were fastened to this framework and thereby generally defined the shape of the more curved (and therefore more difficult to plank) 6 m lengths at the ends of this vessel. Additional frames were then made to match this shape, and the remaining planking fashioned and fastened to them. If Steffy's deductions are correct this eleventh-century method of boatbuilding was very similar to the seventeenth-century method described above, of building on one or two master frames with ribbands.

Excavated evidence shows, therefore, that a form of skeleton building was used in N.W. Europe during the early centuries AD. Two to three centuries later, Mediterranean shipwrights experimented with skeleton building in those parts of the ship (the near-vertical section of the sides above the waterline, amidships) where only straight runs of planking were required thereby minimising both the problem of determining the shapes of individual planks and the edge-fastening problems. By the seventh century AD, edge fastening in these regions was omitted; and, by the eleventh century, we have a non-edge fastened ship whose shape was derived partly from a pre-erected framework and partly by the run of the planking. The frame-determined regions in these medieval vessels of the Mediterranean were areas of relatively uncomplicated curves and the builders evidently had the ability to translate the shapes required there into a partial framework of timber. The more difficult problem of constructing frames to the shapes of the ends of the vessel, where there were more complex curves, was not attempted directly but by the intermediate step of turning ideas into visible three-dimensional shapes using strakes.

To us it would seem a small step to record in some way the shape of all the frames after they had been installed and thus progress towards *full* skeleton building with the shape of the entire vessel determined by a pre-erected framework. However, *full* skeleton building appears not to have been used in Europe until the 18th century AD (Greenhill, 1995).

MEASUREMENTS

In the methods of recording and reproducing three-dimensional shapes described above, *direct* measurements or comparisons are taken and thus no measurement *scale* in either linear or angular units is required. There is no direct evidence, apart from the stored stems, that these methods were in fact used in medieval or earlier Europe, possibly because surviving evidence has not been identified. Whether or not such methods were used to obtain shape, some form of direct measure or comparison (similar to the use of a boat-ell) would be essential to confirm a boat was being built symmetrically about the fore and aft middle plane. Although boats with intentional asymmetry are known (Chinese crooked stern junk, see Worcester, 1966: 122–6) they are asymmetric over a limited length, usually to counterbalance asymmetric propulsion. Symmetry about the fore and aft middle plane is highly desirable in all forms of boat and indeed rafts (bundle or log) so that there is no biased effect on stability, freeboard, loading, speed and especially direction-holding and course alteration. Checks on symmetry may have suggested methods of recording shape – both can be done without using units of measurement.

Natural units

Natural units of measurement do, however, appear to be deeply ingrained in many cultures – ancient Egyptian, Classical Mediterranean, Anglo-Saxon and recent Indonesian – an obvious relict of this is the survival in several European languages of words denoting units of measurement by *thumbs* or *feet*.

Attempts have been made to evaluate prehistoric units as evidenced in megalithic monuments, in precise modern terms and to demonstrate standardisation over a wide area, but, as Grierson (1972: 5–6) has pointed out, such efforts assume an unwarranted degree of precision. The approximate magnitude of traditional linear units are known, however, and Drillis (1963) has published units which he recorded in Latvia in 1922–24:

1. breadth of thumb (inch, *pirksts*, *le pouce*, *il pollice*).
2. breadth of palm (hand).
3. open index finger to thumb (span).
4. heel to toe (foot, *pedu*, *le pied*, *il piede*).
5. elbow to finger tip (ell, cubit).
6. outstretched fingertip to fingertip (fathom).

Such units are used in Latvia to gauge the lengths of axe handles and other tools, the dimensions of tables and chairs and the rise and tread of stairs. Similar units were used by nineteenth century N. American Indian boatbuilders (Adney and Chapelle, 1964: 36). These are measurements which individuals can take using their own natural units, which are recordable and repeatable by themselves or others, within the standards of accuracy required. Such units were available to the medieval boatbuilder at least in certain parts of northern and western Europe, and theoretically could have been there in prehistoric times; whether they were actually used to record and repeat shapes cannot yet be deduced from surviving evidence.

SHAPING INDIVIDUAL TIMBERS

Timbers may be shaped in three main ways.

1. ***By reduction.*** Comparable with hollowing a log (Ch. 2). The prehistoric Brigg and Ferriby planks are fashioned from half-logs in this way (Fig. 4.3). The Brigg bottom planks have ten cleats fashioned on their upper surface and the plank edges are bevelled underneath. These Brigg planks are near-parallel sided, but the Ferriby planking is more complex; not only have the ends of the boat been given an upward curve by hollowing out the end of the plank-keel but a surviving fragment of side strake has been hewn to give curves in two planes at the turn of the bilge near the end (Fig. 8.7).
   The (near) L-shaped transition strakes between bottom and sides found on many of the Romano-Celtic group of boats (Fig. 8.8) are

Figure 8.7 Ferriby boat 1 on the foreshore of the River Humber during excavation in 1946. (Photo: NMM Greenwich.)

also produced by working away unwanted timber until the required shape is produced. They are sometimes said to have been constructed from a logboat split in two longitudinally to give the L-cross section, but such assertions have not yet been supported by any published drawings of growth ring or ray patterns. *A priori* it is unlikely that a log would be hollowed and then split to produce such strakes; a more economical way would be to fashion such curved strakes from a half-log and there is a tradition of hewing planks in this way extending back to the Iron Age Ljubljana boat (McGrail, 1981b: fig. 4.1.17) and the Bronze Age Ferriby boats (Wright, 1974). The *meginhufr*, the extra thick strake of distinctive cross-section (Fig. 8.9) found near the waterline on many Viking Age boats, must have been fashioned in a similar manner. It may also be thought of as a transition strake where the bottom blends into the sides.

Planking made from timber species which do not readily split radially (Ch. 4) must also generally be produced by hewing away surplus wood. In boats with little deadrise at the midship section (i.e. flattish bottom amidships) the lower strakes twist from near horizontal amidships to near vertical at the stems; where such a twist cannot readily be imposed on a plank or where it is beyond current technical capabilities it may be hewn out of the log rather than achieved by bending. The garboards and the second strakes of the thirteenth-century Sjøvollen boat and the fifteenth-century Folderøy boat were fashioned in this manner (Christensen, 1968a: 137). Twentieth-century boatbuilders in Hardanger Fjord, Norway similarly used axes to shape the garboards of their boats (Wickstead, 1978: 8). It is thus possible that this could be a regional tradition, however medieval finds from this part of Scandinavia are not prolific and in any case this technique has not always been searched for on boat finds, and so may have been overlooked in finds from other places. At least one example has been found in the Mediterranean area, however: the planking of the fourth century BC Kyrenia boat had been similarly hewn to achieve the required curves (Steffy, 1982: 27).

In his detailed recording of the simple boats still to be found in the 1950s and 1960s on the southern shores of the Baltic, Rudolph noted several examples of planked boats with their central bottom member hollowed out from a longitudinal half of a log (1974: figs 2, 3, 4). Stems similarly hollowed out of log sections were also found (figs 4, 5, 6). Generally such hollowing of the inner surface of the stems and the upper surface of the keel is to reduce weight while maintaining external shape and strength, but it may also be done so that the timber is reduced to a thickness through which plank fastening nails can be driven and clenched (see sections below on *keels* and *stems*).

2. ***By deformation or transformation.*** Obtaining a required shape by bending planks may be compared with the *expansion* of a logboat (Ch. 6). There are two principal methods; or a combination may be used:

1. straining to shape

Figure 8.8 Part of the Romano-Celtic boat from Druten, Netherlands with a curved transition strake between flat bottom and sides. (Photo: Rijksdienst voor het Oudheidkundig Bodemonderzoek).

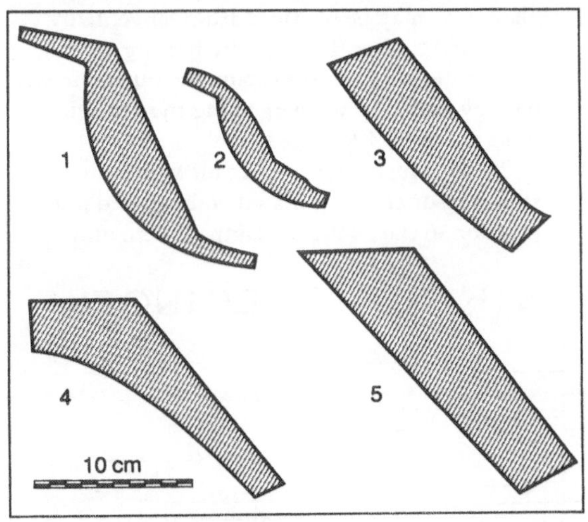

Figure 8.9 Cross-sections of *meginhufr* – distinctive strakes near the waterline – from the Viking Age finds from:
1. Klaståd 2. Åskekar 3. Tune
4. Oseberg 5. Gokstad
After Christensen and Leiro, 1976 (copyright Arne Emil Christensen).

2. steaming or charring
Certain shapes may be obtained by asymmetrically loading a plank so that it is permanently deformed to the required curve, providing the timber species and the plank scantlings will accept this form of stress. Thus the relatively thin and narrow oak strakes frequently found in Viking tradition boats will accept a bend and twist that could not be imposed on the broad thick oak strakes of the Ferriby boats.

Planks may be bent to shape before being offered up to the boat – see Fig. 8.10 for a simple leverage system which permits increasing pressure to be applied to the ends of the planks; or they may be strained to shape whilst they are in position – thus the flat bottom of the Indus punt is forced up at the ends to follow the curve of the bottom edge of the pre-assembled sides, whilst a rockered curvature is applied to the bottom of the River Kabul cargo

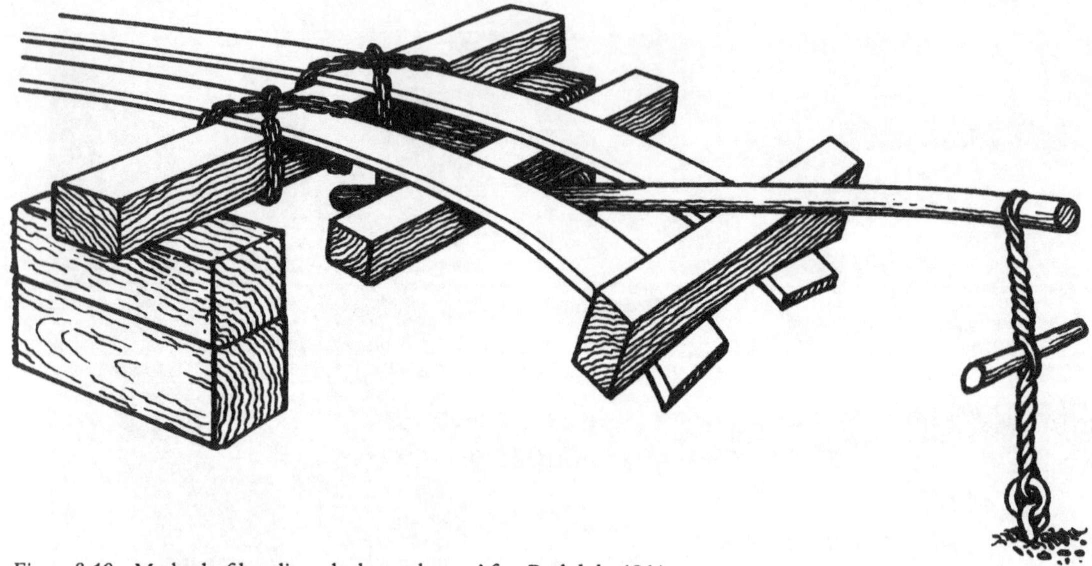

Figure 8.10 Method of bending planks to shape. After Rudolph, 1964.

boats in the Peshawar valley by loading the central sections with stones and forcing up the ends using poles as levers (Greenhill, 1976: 51, 68).

Softwoods used in northern European boatbuilding may similarly be bent to the required shape by struts from roof and floor and by clamps (Christensen, 1973: 142) – see Greenhill (1976: fig. 15) for a similar set-up in Bangladesh in the open air. Where moulds are used they may be used as a fulcrum around which to bend planking.

Fresh green hardwoods bend more readily than when seasoned, and softwoods generally accept bending better than hardwoods. These less docile timbers may have to be heat-treated so that they become malleable. Light (but not heavy) frames and other transverse members may be similarly treated to impose a curvature, or increase a natural curvature, so that they take up the required inside form of the boat. Heat may be applied by heating timbers in a steam box (McGrail, 1974: 8–9) or by charring them on one side while keeping the other side wet: for Britain see Leslie (1892: 47, 52); for Pakistan, Greenhill (1971: 75) and for Japan, Greenhill (1959: 5–6). This charring method was known in the ancient Mediterranean: see V. Flaccus (*Arg.* 1.v. 124, as pointed out by Basch (1973: 332). Curves across the breadth of a plank (*set*) may be obtained from tangentially converted timber (Fig. 4.5) by heating the side nearest the pith whilst keeping the outerside wet thereby inducing warping along the growth rings (McKee, 1983: 53).

Planks which have been made malleable by steaming or charring may then be forced into position in the boat and clamped there until

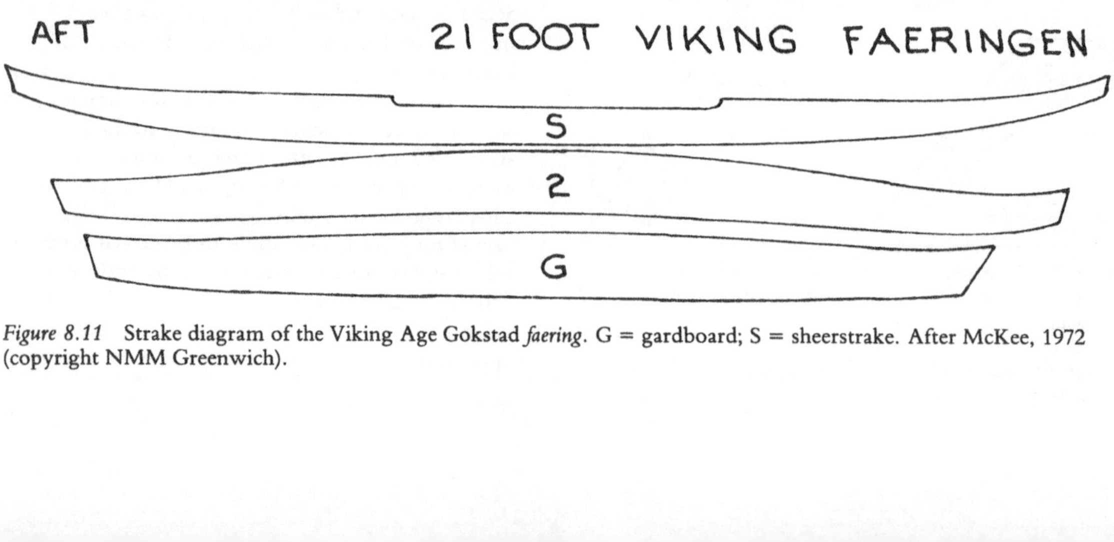

Figure 8.11 Strake diagram of the Viking Age Gokstad *faering*. G = gardboard; S = sheerstrake. After McKee, 1972 (copyright NMM Greenwich).

Figure 8.12 Scarf between keel (to the right) and stem (to left) on the Greenwich replica *faering*. (Photo: NMM Greenwich).

they set on cooling – usually with some slight loss of curvature; thus it may be necessary to apply some force whilst they are being fastened. Where the required shape is especially curved as are the second strakes of the Gokstad *faering* (Fig. 8.11), they may be set to cool against a mould of shaped wedges mounted at intervals on a flat board (McGrail, 1974: 13).

3. ***By addition or construction.*** Shapes may be obtained by adding elements together (Ch. 2) analogous to the extension of a logboat (Ch. 6). Thus keels may be extended to give the necessary curved shape to the forefoot (or toe) of a boat by scarfing them to a curved stem whose run of grain follows the curve required, either direct (Fig. 8.12) or via an intermediate timber. Planks are similarly extended to form strakes and floor timbers to form ribs. These are dealt with in more detail in the appropriate sections below. Here it suffices to draw attention to the simplicity of the scarfs found in the few prehistoric and medieval boat finds that have been recorded in detail; they do not compare at all in complexity either with those generally used in Classical boatbuilding (Fig. 8.13) or with those found in medieval waterfronts, cathedrals and barns (Currie, 1972: fig. 2, No. 2, 4, 8 and fig. 3; Milne, 1982: fig. 2.6 n and o).

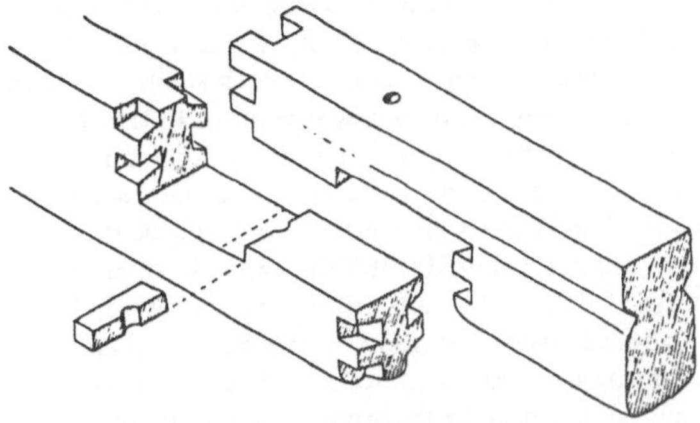

Figure 8.13 Keel scarf from the Classical shipwreck Monaco. After Dubois, 1976.

Shaping strakes

In many designs of boat, strakes are required to be nearer the vertical at their ends than they are over the midships regions of the boat. If a plank of rectangular dimensions is bent into this shape, points along its lower edge will no longer lie in a horizontal plane or in a plane parallel to the run of the keel; a similar effect occurs if a strake has to be bent to match a horizontally curving lower strake or keel. In both cases a gap will appear below the ends of the rectangular plank. A plank that could replace this rectangular plank and also fill the gaps has to have a curved lower edge with the centre higher than the ends (sny). Above the turn of the bilge, however, the ends will need to be higher (hang) (McKee, 1972: 3). Similar considerations apply to the shape of the upper edge of a plank, but the problem is complicated here by the necessity to have sufficient wood at the ends to take fastenings into the stems, and as the planked area is greater around the midships girth than at the ends plank breadths generally taper towards the ends. In addition the ends of the strakes are not necessarily at right angles to the run of the plank but have to match the angle at which they meet the stems. Thus strakes are not rectangular in plan; depending upon how the boat's transverse sectional shape varies along its length so each strake has a corresponding sinuous shape when seen in plan, before bending into position (Fig. 8.11).

Today in Britain boatbuilders find what shape of strake is required by 'spiling' for the lower edge and then estimate the upper edge by adding the varying distances (and overlap in clinker work) between this spiled line and a fair horizontally disposed curve between the required breadths at each stem; the plank is left overlong and subsequently cut to fit the stems. The essence of spiling is that a flexible batten is bent along the length of the boat into the general shape required and offsets are marked on it to give the precise shape required to fill the gap. This marked batten is then used as a template together with plank breadths to mark out the two-dimensional shape on a flat board. Thus a pattern is made in a less rigid material: it would be theoretically possible, for example, for a stretched length of hide to be used to obtain this pattern. (See McKee, 1972: 8, fig. 5 and McGrail, 1974: 12–13 for the spiling method used by Harold Kimber, a master boatbuilder from Somerset).

A method used in Indonesia to spile flush-laid planking has been described by Horridge (1981). The plank to be fitted must first be bent to lie above and in line with the lower strake with a gap between them. A tool (similar to a lap gauge – see McKee, 1972: fig. 5) with a lip and a point is drawn along with its lip tracing out the profile of the lower strake and simultaneously the point scratches a parallel line on the upper plank. This tool is similar to the wooden compasses with one long and one short leg used by Sudanese boatbuilders (Hornell, 1939: 419–20, fig. 1); the short leg is dipped in paint and marks out the required lower edge on the plank to be added.

Greenhill (1976: 52) and Christensen (1972) have described how plank shapes are obtained in Pakistan and Norway 'by eye'. This is essentially a method of trial and error gradually working the lower edge of the strake to match the upper edge of the lower strake and to fit the bevel already cut there. The upper edge of the new strake is then fashioned to a shape determined 'by eye' or by a boat ell measurement of distances from a datum line, or by recorded plank breadths.

Strake diagrams

The set of strake shapes for a boat, clinker-built or flush-laid, produced by whatever means, can fit together in only way one and thus, with patterns of stems and keel, the shapes from one side of a boat (strake diagram Fig. 8.11) define the boat's shape (McKee, 1983: 128–9). In both traditions of building, after the plank has been fastened in position, a variable bevel has to be worked on the upper edge (and sometimes the lower edge – see the section below on *planking*) so that after the next strake has been fashioned to shape and bent into position it will lie snugly all along its length at the required (varying) angle. Greenhill (1976: 80) has pointed out that if the strake diagram of a boat find can be recorded it is not essential, though desirable, to record the bevel angles. As boat finds are rarely complete and are almost always distorted, both sets of data should indeed be recorded, and shape of strake and bevel angle (difficult to measure) may be used to cross check on one another.

Strake diagrams (Fig. 8.11) compiled by tracing planking during the building of a clinker boat or model, or after dismantling a boat find will include the breadths of the overlaps, whereas diagrams taken from flush-laid boats or models or from an assembled clinker boat or boat find or a finished clinker model omit these laps. It would be useful to be able to translate lines plans or construction plans into strake diagrams and *vice versa* but this problem has not yet been solved. Diagrams of strakes may be seen in McKee 1972: fig. 2; McKee, 1983: 128–9, fig. 102 (with some directions on their interpretations); and in

Greenhill, 1976: figs 29 and 31. Lothrop (1932: 246) used a strake diagram 50 years ago to describe the South American *dalca*, a sewn plank boat, whilst Izikowitz (1957: 20) published one for a Solomon Is. *ora*. The late eighteenth-century planking expansion plans produced for some skeleton-built ships of the Royal Navy (Lyon, 1974: 8, fig. 5) are similar.

Early use. Although early boatbuilders in the shell tradition where planking defines the shape, may have been able to break down the mental image of the boat's overall shape into individual strake shapes, no evidence for the use of some form of spiling has survived; nevertheless the possibility of the use of skin patterns to obtain individual shapes should not be overlooked. It seems doubtful that records of strake shapes as in the modern strake diagram were used as an aid to boatbuilding except possibly as a concept behind the 'by eye' method (if this was indeed used). The majority of the information on a strake diagram is more conveniently recorded by boat-ell or boat level and, although these latter records are not so visually obvious, they are readily usable by the boatbuilder.

HOW TO INCORPORATE LONGITUDINAL STRENGTH?

All fore and aft members of a boat contribute to longitudinal strength and, as discussed in Chapter 3 so do the transverse members. In a round-hulled boat the keel is the main centre line strength member but the central bottom plank or strake of a flat-bottomed boat is sometimes of significantly greater scantlings (especially thickness) than the other strakes and is analogous to a keel, see, for example, Ferriby boat 1 (but not the Brigg 'raft'). The strakes at the transition from bottom to side on a flat-bottomed boat, are also strength members, for example the transition strakes or *iles* of some of the Romano-Celtic tradition of boats (see above in section on *shaping timbers*). Stringers, strengthened top strakes and hogging trusses (Ch. 3) also contribute significantly to longitudinal strength (see below in section on *planking*).

Boats without keels, thicker central planks, hogging trusses or other elements giving longitudinal strength are unlikely to have been sea boats, although they may have been used in coastal waters restricted to fair weather conditions.

KEELS

The central longitudinal member or keel, together with stempost and sternpost or transom may be thought of as the backbone of a craft, the lower element of the 'ship girder' (see Ch. 3). The first or garboard strakes are fastened to it or, in the case of the flat-bottomed Romano-Celtic and cog traditions, positioned alongside it. The floor timbers (lowest elements of the frames) are also attached to the keel, directly in the case of the cog, indirectly, for example, in the Norse tradition, thus spreading stresses around the hull. The stresses on the keel depend not only on loading and trim but also on natural variations in buoyancy as the boat passes through different wave and swell formations (see Fig. 3.4). Keels also take the initial impact when a boat takes the ground and thereby protect bottom plank fastenings, particularly important in a sewn boat such as Ferriby. A keel, providing it is of sufficient size and strength to resist the downward loading of mast and sail may (sometimes in conjunction with other timbers such as a keelson) provide a housing (step) for a mast.

Sailing potential

Some historians appear uncertain of the keel/sail relationship and make statements such as: no deep keel, therefore no sail. Masts may in fact be stepped in fittings other than keels, for example in a floor timber as in some of the vessels in the Eastern Baltic variant of the Nordic tradition (Crumlin-Pedersen, 1983). Pole masts may be stepped on an outrigger frame as in the *oru* of Sri Lanka; and bipod masts (pairs of sheers) may be stepped at or near the boat's sides as seen in Egyptian iconography (Casson, 1971: fig. 19), and used in the reed craft of L. Titicaca, S. America.

In N.W. Europe, at least from medieval times, masts in seagoing vessels seem generally to have been stepped in keels or in other fittings mounted on keels. In addition to providing a housing and dispersing sailing stresses around the hull, keels may in conjunction with the shape of the lower hull also reduce leeway (drift downwind). A keel protruding below the boat's bottom acts as an aerofoil in a fluid, thereby producing a sideways force opposing leeway. In general, the more a keel protrudes the greater this reaction. A keel-less boat with mast and sail but without anti-leeway devices such as great deadrise (angle) on the lowest strakes, deep rudders, drop keels, leeboards or *guares* (Ch. 12) will generally be restricted to sailing before the wind (say with the

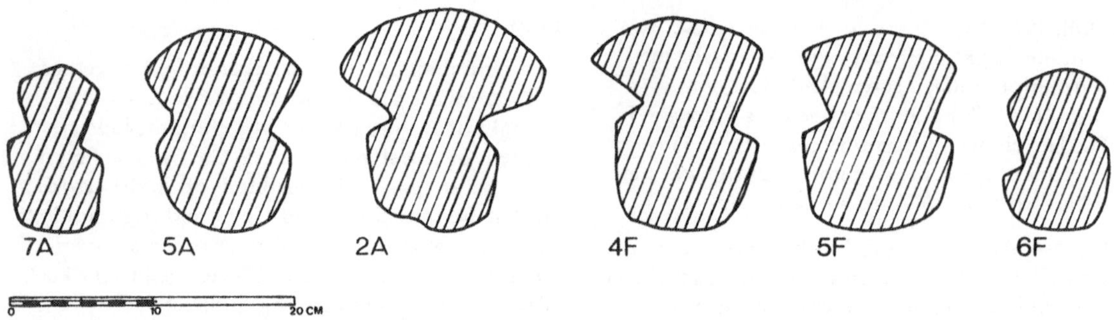

Figure 8.14 Cross-sections of the keel of Skuldelev 1.
2A = Second frame aft of midships
4F = Fourth frame forward of midships
After Olsen and Crumlin-Pedersen, 1967, fig. 18.

wind from a sector 4 points either side of the stern); with other wind directions the leeway would be unacceptably great. In this sense a 'deep keel', protruding below the boat's bottom, improves sailing potential. The actual distance of the keel below the waterline ('deep' = low in the water), except in the sense that it is some measure of the wetted surface of the hull, has no influence on windward performance, but it will determine the minimum depth of water in which the boat can safely be operated.

Keel shapes

In his investigation of recent British boats, McKee (1983: 65, 80) has differentiated between *hogged*, *flanged* and *rabbeted* keels. The latter have longitudinal grooves worked along both sides to take the garboard strakes. A hogged keel has a broader timber (hog) added to the top of the keel to form a T cross-section, while the flanged keel is similar but in one piece. Hogged keels are not known from antiquity, but the other classes are: for example, the Ferriby keel is rabbeted, and the Graveney keel flanged. The keels of Skuldelev boats 1 (Fig. 8.14) 2 and 6 are rabbeted, whereas boats 3 and 5 have flanged keels with a near T-shaped cross section amidships where there is a medium angle of deadrise and through-fastenings of clenched nails are used to fasten the garboards. Towards the ends where there is greater deadrise these two keels become generally triangular or rhomboid and spikes are used as fastenings.

In a flat-bottomed boat such as Ferriby 1 plank attachment requirements do not materially alter keel shape, a suitable landing place for the horizontal outer bottom strakes is all that is required and this is provided by a groove (rabbet) within the thickness of the plank keel. In similar fashion the shape of the Graveney boat's keel changes very little over the central 3 m, with the garboard strakes being fastened under the arms of the T-shaped keel. In other boats, where the deadrise angle of the first strake varies along the length of the keel to match the changing transverse sections of the boat, the cross section of the keel has to vary to match this deadrise (see, for example, the varying cross section of the keel of Skuldelev 3 (Fig. 8.15) (Olsen and Crumlin-Pedersen, 1967: fig. 37).

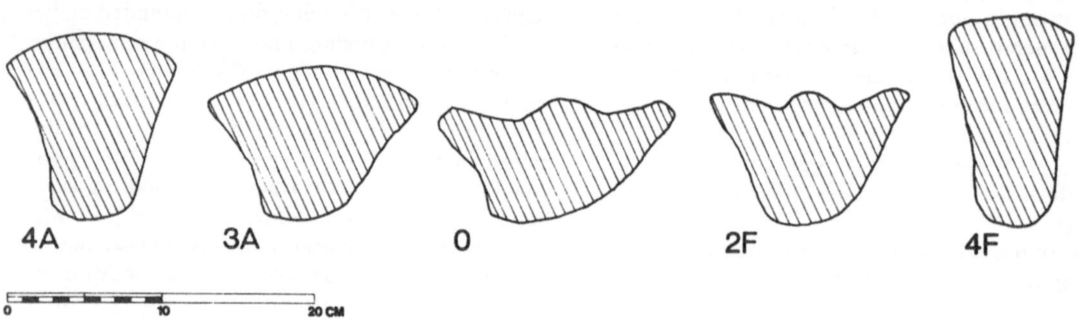

Figure 8.15 Cross-sections of the keel of Skuldelev 3. After Olsen and Crumlin-Pedersen, 1967, fig. 37.

The keels of Gretsedbro, Sutton Hoo 2, Kvalsund 1 and 2 and Oseberg have a T-shaped (non-rabbeted) cross-section amidships and a rabbeted cross-section towards the ends. A T-shaped cross section may in fact be thought of as a rabbet taken to its limit. Such a cross-section, and those of rhomboid shape, give a good seating for the garboards; in addition, the T-shape is a strong yet lightweight section as unwanted wood has been removed. A rabbeted keel, on the other hand, more readily accepts garboards with high deadrise near the ends of a boat.

McKee (1983: 80), following Crumlin-Pedersen (1968: 264), divided keels into *beam* keels which are as deep or deeper than wide, and *plank* keels which are wider than deep. Two points may be made here: firstly as keel cross-sections vary along their length the midship (or as near midships as possible) cross-section should be used when comparing keels. Secondly, this definition seems more applicable to the rabbeted keel and the vertical member of a hogged keel (i.e. not including the horizontal hog) than to flanged keels where the near horizontal arms can give the keel a deceptively low depth/breadth ratio. For example, using the keel midships sections published by Olsen and Crumlin-Pedersen (1967) the ratio of moulding to siding (vertical to horizontal dimension) for Skuldelev 2 is 0.94, 0.87 for boat 3, 0.66 for boat 5 and 0.83 for boat 6. By McKee's definition these should be termed *plank*-keels (m/s<1) but, intuitively they are assessed as (thin) *beam* keels. McGrail and Denford (1982: 50) have proposed that four groups of keels should be recognised with (m/s) ratios of:

(1) < 0.50
(2) 0.50 to 0.70
(3) 0.71 to 1.00; and
(4) > 1.00

The last group corresponds to McKee's *beam* keels; the other groups might be named 3: *thin beam*, 2: *plank*, 1: *thin plank*. Such groupings should be regarded as temporary until sufficient well-recorded keels are available to permit natural grouping to be identified. Keel (m/s) ratios are given in Table 8.1, together with their class by the McKee and McGrail/Denford schemes.

Åkerlund (1963: 98) described the keel of Nydam 1 as an 'almost fully-developed T-shaped keel' although its (m/s) ratio is only 0.33 and thus could be better classified as a thin plank keel of T-shaped cross-section. Gokstad 1 is the most obvious example of a T-shaped keel with a high M/S ratio of 1.84 (beam keel). Skuldelev 3 and 5 and Kvalsund 1 and 2 have more or less T-shaped keels, at least over their central ranges, and (m/s) ratios of 0.75 (thin beam keel) and 0.63 (plank keel).

Keels as an anti-leeway device

Data on cross-sectional shapes of keels are useful in classification studies and in the evaluation of woodworking techniques but are not of immediate use in estimating the effect of a keel on sailing performance, which is related to the shape of the keel protruding below the bottom planking, and the deadrise of the lower strakes. Table 8.1 also lists this data for 25 medieval boats and ships with keels which are well-documented.

There are several factors not considered here which together with the lower underwater shape would determine how suitable a boat was for carrying mast and sail and how good her weatherly performance would be. Nevertheless the angle of the garboards to the horizontal and the relative dimensions of the keel protruding below the bottom do give an indication of the shape of the lowest underwater section near amidships and thus give some measure of anti-leeway characteristics.

Drawing valid conclusions from this small sample is difficult, not least because any apparent correlations and groupings may be due to:

(a) regional/cultural differences
(b) changes over time
(c) differences in the function of the boats and in operational environment
(d) availability of timber.

Nonetheless when the data in Table 8.1 is examined five main groupings may be recognised:

A. Boats from S. Jutland and S.E. Britain of the period AD 400 to 600 have:
   Deadrise: 11° to 24°; (d)/(b): 0.14 to 0.35 — Nydam 1 and 2, Sutton Hoo 2, Gretstedbro.
B. Boats from E and W Norway of the period AD 700–850 have:
   Deadrise: 13° to 40°; (d)/(b): 1.24 to 2.40 — Kvalsund 1 and 2, Klåstad, Oseberg, Gokstad 1.
C. The Graveney boat of *c.* AD 900 has
   Deadrise: 4°
   (d)/(b): 0.14

Table 8.1 Details of selected medieval keels

Boat	*Approx. date*	*Keel cross-section*			*Protruding keel (d/b)*	*Deadrise of garboards*†	*Reference*
		(m/s)	*McKee class*	*Proposed class*			
Nydam 1	400	0.33	plank	thin plank	0.23	22°	Åkerlund, 1963
Nydam 2	400	0.15	plank	thin plank	0.14	11°	Åkerlund, 1963
Sutton Hoo 2	600	0.39	plank	thin plank	0.35	13°	Evans & Bruce-Mitford, 1975: fig. 290
Gretstedbro	600	0.48	plank	thin plank	0.33	24°	Crumlin-Pedersen, 1968b
Kvalsund 1	700	0.75	plank	thin beam	1.80	20°	Shetelig ? Johannessen, 1929
Kvalsund 2	700	0.63	plank	plank	1.25	13°	Shetelig & Johannessen, 1929
Klåstad	800	0.83	plank	thin beam	1.33	40°	Christensen & Leiro, 1976
Oseberg	800	1.00	beam	thin beam	1.27	31°	Brøgger *et al.*, 1917
Gokstad 1	850	1.84	beam	beam	2.40	28°	Nicolaysen, 1882
Graveney	900	0.18	plank	thin plank	0.14	4°	Fenwick, 1978: fig. 10.1.2
‡Skuldelev 1	1000	1.14	beam	beam	0.72	30° (13)	Olsen & Crumlin-Pedersen, 1967: figs 18, 29, 37, 47, 58, 72
2		0.94	beam	thin beam	0.46	57° (*c.* 44)	
3		0.87	beam	thin beam	0.47	47° (35)	
5		0.66	beam	thin beam	0.63	31° (20)	
6		0.83	beam	thin beam	0.74	31° (13)	
				thin beam			
Kyholm	13th century	0.80	beam	thin beam	0.55	55°	Crumlin-Pedersen, 1983a: fig. 65
Kollerup	13th century	0.24	plank	thin plank	0.09	0°	Crumlin-Pedersen, 1979: fig. 2.12
Kolding	13/14th century	0.33	plank	thin plank	0.15	0°	Crumlin-Pedersen, 1979: fig. 2.12
Vejby	1350	0.26	plank	thin plank	0.13	6°	Crumlin-Pedersen, 1979
Q75	1325	0.50	plank	plank	0.36	7°	Reinders, forthcoming
N5	early 14th	0.35	plank	thin plank	0.24	2°	Reinders, forthcoming
NZ.42	14th	0.28	plank	thin plank	0.17	2°	Reinders, forthcoming
NZ.43	14th	0.36	plank	thin plank	0.32	8°	Reinders, forthcoming
M107	late 14th	0.23	plank	thin plank	0.18	13°	Reinders, forthcoming
Bremen Cog	1380	0.32	plank	thin plank	0.19	5°	Ellmers, 1979: fig. 1.6

Note: 1. Data taken from (near) midship station. The keels are often distorted when found: in these cases (e.g. Skuldelev 1 to 6) the figures given are approximate and the mean of port and starboard is given.
† Deadrise is the angle the garboard strake lies to the horizontal.
‡ Olsen and Crumlin-Pedersen (1967: 103, 115, 124, 138, 150) give angles between garboards deduced from the *frames* rather than the keel; their equivalent deadrise angles are given in brackets in column 7.

D. W. Baltic boats of *c.* AD 1000 have:
Deadrise: 30° to 57°
(d)/(b): 0.46 to 0.74 } The Skuldelev boats
The Kyholm boat with 55° deadrise and 0.55 (d)/(b) appears to be a thirteenth-century example of this group.

E. The flat-bottomed (cog-design?) boats and ships from the W. Baltic, N.W. Germany and the Ijssellmeerpolder of the thirteenth and fourteenth centuries have:
Deadrise: 0° to 13°
(d)/(b): 0.09 to 0.36 } Kollerup, Kolding, Vejby, Q. 75, N 5, NZ 42, NZ 43, M 107, Bremen Cog

The pre-twelfth-century boats which are known to have had mast and sail (Skuldelev 1 to 6, Oseberg and Gokstad) have (d)/(b) ratios of ⩾ 0.46 and deadrise angles ⩾ 28°. Klåstad had similar leeway resisting characteristics and on this criterion alone may also have carried mast and sail. Kvalsund 1 and 2 have low deadrise (20° and 13°) but with relatively high (d)/(b) may also have had underwater forms appropriate to sail. Gretstedbro has good deadrise but a very low (d)/(b) and her suitability for sail (solely on the basis of leeway resistance) may be questioned. It has been suggested that Sutton Hoo 2 and the Graveney boat may have carried sail; if so (and the evidence is slight) on the basis of their underwater form as measured by the data in Table 8.1 they were probably restricted to winds from the stern sector.

If we discount the Graveney boat which may, for

other reasons, be not only in a different tradition from her near contemporaries but also have had a significantly different function (McGrail, 1981a: 35), and assume that the fifteen pre-twelfth-century finds are in some sense representative of a relatively homogeneous tradition displaying changes over time we may deduce the following sequence:

Stage A AD 400–600	Sailing (if at all) restricted to conditions with wind from the stern sector, otherwise excessive leeway.
Stage B AD 700–850	Leeway resistance and hence windward performance improved by marked increase in keel ratio and a slight increase in deadrise.
Stage C *c.* AD 1000	Windward performance improved (not necessarily better than stage B but better than A) by combining an increased deadrise angle with a medium keel ratio.

On the other hand Jutland beaches generally have less gradient than those in Norway and beaching may have been more frequent. Thus the combination of medium keel ratio and increased deadrise in the Skuldelev boats (among which are boats of several different functions) may have been a special characteristic of Jutland based boats, and boats in the period 700–850 from this region (of which we have no examples) may have had similar keels and garboards. Equally, Norwegian boats of *c.* AD 1000 may have had the high keel ratios of their forerunners. There are significant gaps in the temporal and regional distribution of finds and the samples of well-recorded finds is too small at present for us to weigh the evidence for and against these two hypotheses. It is of interest to note that within the Skuldelev group of boats, *on this measure of sailing performance alone*, the large warship No. 2 has comparable characteristics to No. 3 the coastal trader, while the smaller warship No. 5 and the fishing boat/ferry No. 6 are not very different from the ocean-going trader No. 1.

Apart from the Kyholm wreck which, although having a cog-like stem, is built in the Norse tradition the post-twelfth-century boats and ships in Table 8.1 are in the cog tradition with only slightly protruding keels and very low deadrise angles. There is direct evidence that most, if not all, these craft had mast and sail; their underwater form as measured by these two criteria was only moderately effective at resisting leeway.

Keel scarfs

Corlett (1977: 229, 230) has emphasised that length is a most expensive dimension in terms of man-hours and materials although, as Wilson (1977: 229) has pointed out, this is only true when the aim is to build a boat longer than the general run of timber available. For this reason and for reasons of strength (a scarfed timber has less than half the strength of an unbroken length – see Olsen and Crumlin-Pedersen 1967: 155) scarfs within the main run of the central longitudinal member were generally avoided in antiquity. The Brigg 'raft' is deduced to have had no bottom planking scarfs in its *c.* 12 m length, but Ferriby boats 1 and 2 did have scarfs near amidships in their plank keels. The two elements in these keels meet in short (F 1, 8 cm; F 2, 14 cm) edge-halved scarfs with square vertical butts; there were no scarf fastenings recorded, only edge fastenings to adjoining strakes.

Medieval keels in the Viking tradition on the other hand are extended at their ends by scarfing them either directly, or indirectly through a curved intermediate timber, to a curving stem so that the natural run of the grain gave an unbroken curve from keel to upper stem. Intermediate timbers (analytically lower stems) were used in the stern of Oseberg (Sjøvold, 1969: 16–17, 20), at both ends of Gokstad (Sjøvold, 1969: 60, 64–5), and at the stern of Skuldelev 2 and at both ends of Skuldelev 5. (Olsen and Crumlin-Pedersen, 1967: 112–13, 134); one of the latter's timbers was 2.7 m in length. The 'stem hooks' of the cog tradition (Ellmers, 1972: 292; Reinders, forthcoming) are functionally the same as the Nordic intermediate timber – they have an angled knee shape rather than a smooth curve.

The scarfs joining Viking Age keels to stems, and to intermediate timbers where fitted, were simple 'vertical' ones, that is they were through-splayed on edge, face-nailed, with the timber faces in the vertical plane (Fig. 8.12). Today woodworkers are recommended to use scarfs with gradients in the range from 1 in 7 to 1 in 12 (14 to 8% slope) – see *Scarf Jointing* publication by TRADA – but medieval keel scarfs seem to have been much steeper, i.e. shorter. The *faering*, Gokstad 3, for example, has keel scarfs with slopes of 1 in 4 (25%), and the average of the keel scarfs found at Wood Quay is *c.* 1 in 3 (36%).

The keel/stem scarfs found on the ninth/tenth-century Graveney boat (Fig. 8.16) and the fourth- to

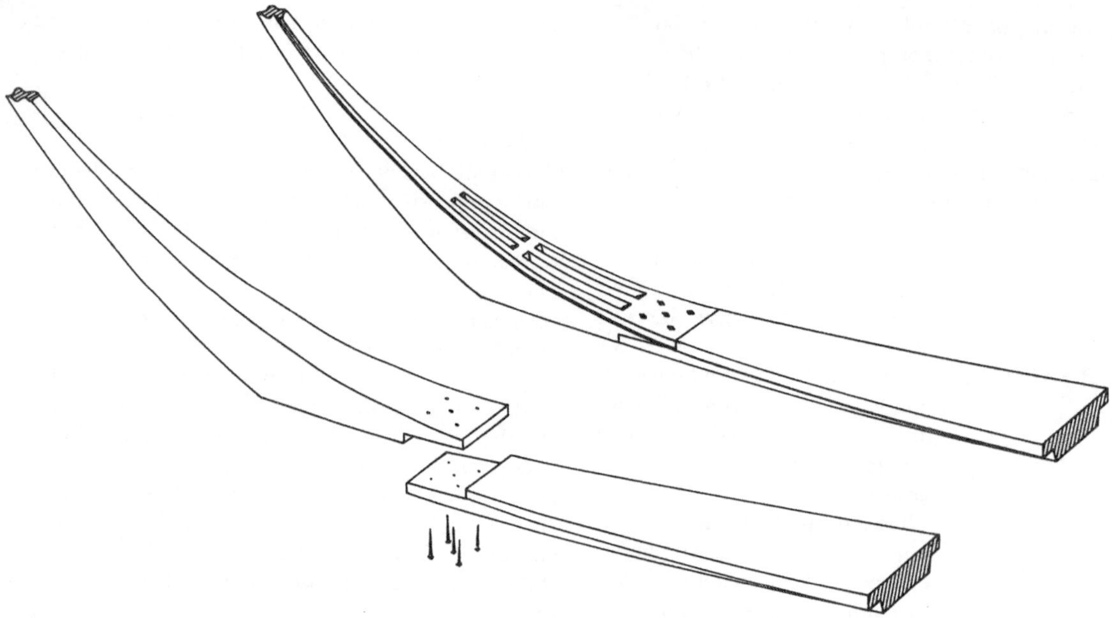

Figure 8.16 Keel/stem scarf of the Graveney boat. (Photo: NMM Greenwich.)

seventh-century finds Nydam 1 and 2, Kvalsund 1 and 2 and Gretstedbro are all stop-splayed on edge, face nailed with the faces in the horizontal plane, and this is also deduced to have been used on the sixth/seventh-century Sutton Hoo 2. The timbers used for the keels of Graveney, Gretstedbro and Nydam 1 and 2 were all broad, shallow planks. Thus the horizontal type of scarf, which gave more room for fastenings in the horizontal than in the vertical face, was more appropriate than a vertical scarf. The keel cross-section ratios for the midships section of the Kvalsund boats are relatively high (0.75 and 0.63) and the ratios at their ends are even higher (1.47 and 1.44), and thus a vertical scarf would seem to be more appropriate. Nevertheless a horizontal scarf was used, possibly because of satisfactory performance on earlier boats.

Table 8.2 seems to indicate that between AD 700 and 800 there was a shift from the horizontal to vertical keel scarfs, necessitated by a change from thin plank-keels and plank-keels to thin beam-keels and beam-keels. With the cog tradition having a plank-keel, the horizontal scarf returned to prominence in the thirteenth and fourteenth centuries (Reinders, forthcoming). From about AD 800 there seems to have been a tendency to use iron nails in scarfs rather than treenails, although both were used in some of the Skuldelev boats, and even as late as the seventeenth century keel/stem scarfs were treenailed in the Llyn Peris boat (Illsley and Roberts, 1980: 345).

Rocker

Keels need not be longitudinally straight but may be curved in the vertical plane, usually with the ends higher than the central section – known as rocker or *spring*. Coates (1984) has suggested a measure of such curvature would be maximum deflection from the horizontal compared with length of keel. Rocker may be achieved by choosing a suitably curved timber, by bending a timber, or by fashioning the curve – in which case there would be some short grain, a source of weakness, at the ends. A keel's rocker may be increased by scarfing on short upward-curving timbers at the ends. Sprung keels have the advantage that directional control is improved and rate of turn increased, and transfer (skid or sideways movement outwards in a turn) and hence radius of turn are decreased. The curvature imposed upon the bottom eases the problem of free surface (Ch. 3) by concentrating bilge water in one place where it can be bailed. Rocker also simplifies beach operations as, in soft mud, the ends can more readily be released by a rising tide, and smaller boats can be pivoted about their lowest point and more easily manhandled. The main disadvantage of rocker is that it complicates the fashioning of the garboard strakes.

Table 8.2 Details of keel scarfs in certain prehistoric and medieval boats

Boat	*Type of keel/stem scarf*	*Gradient*	*Fastenings*	*Reference*
Ferriby 1	Horizontal*	1 in 0.55 (180%)	None (except edge fastenings)	Wright, 1976
Ferriby 2	Horizontal*	1 in 1.36 (73%)	None (except edge fastenings)	Wright, 1976
Nydam 1	Horizontal*	?	?	Åkerlund, 1963
Nydam 2	Horizontal*	1 in 1.58 (63%)	2 vertical treenails	Shetelig and Johannessen, 1929: fig. 6.
Sutton Hoo 2	Horizontal*	1 in 2 (50%)	3 treenails†	Evans and Bruce-Mitford, 1975
Gretstedbro	Horiztonal*	1 in 2.36 (42%)	iron rivets	Crumlin-Pedersen, 1968b
Kvalsund 1	Horizontal*	1 in 2.50 (40%)	?	Shetelig and Johannesson, 1929
Kvalsund 2	Horizontal*	1 in 2.30 (44%)	?	
Oseberg	Vertical	*c.* 1 in 2.00 (50%)	iron rivets?	Brøgger and Shetelig, 1971
Gokstad 1	Vertical	*c.* 1 in 2.00 (50%)	clenched nails and some treenails	Brøgger and Shetelig, 1971: 81
Gokstad 3 (faering)	Vertical	1 in 4 (25%)	clenched nails	Christensen, 1959
Graveney	Horizontal	1 in 3.07 (33%)	5 vertical iron clench nails	Fenwick, 1978: 94, 201, 217
‡Skuldelev 1	Vertical	1 in 2.6 to 1 in 1.6 (39% to 62%)	iron nails	Olsen and Crumlin-Pedersen, 1967
Skuldelev 2	Vertical	—	+ 3 iron nails	Olsen and Crumlin-Pedersen, 1967
Skuldelev 3	Vertical	—	+ 4 iron nails plus 1 treenail	Olsen and Crumlin-Pedersen, 1967
Skuldelev 5	Vertical	1 in 2.8 to 1 in 6 (36% to 17%)	—	Olsen and Crumlin-Pedersen, 1967
Skuldelev 6	Vertical	—		Olsen and Crumlin-Pedersen, 1967
Wood Quay	Vertical	*c.* 1 in 3 (36%)	iron nails	—

* Horizontal scarfs have a horizontal rather than an obliquely-cut face and therefore strictly speaking they do not have a gradient. However, length of scarf compared with thickness of timber which is given in column 3 is a measure of the relative size of the overlap.

† Evans (1975: 398) deduced that Sutton Hoo 2 had three vertical treenails subsequently replaced by clenched iron nails, whereas Fenwick (1978: 221) considers these clenched nails may have been an attempt to bring together the sides of a split – it is difficult to see how these nails could have so acted unless they were driven in at an angle to the vertical i.e. across the split.

‡ Details so far published of the Skuldelev scarfs are general rather than specific, thus only a range of gradients can be given.

Whether an excavated boat had a sprung keel and if so to what degree, may be difficult to recognise due to distortion, damage and settling during deposition or during excavation (Fenwick, 1978: 201, 255). The Graveney's keel rocker was established experimentally on a 1: 10 model; the lowest strakes would not fit the keel and sternpost until the model keel had been sprung so that the ends were *c.* 1 cm higher than the central section (McKee, 1978: 271). It was deduced that, to impart spring to the keel, wedges would have been forced under the ends of the keel as it lay on the stocks. When the required degree of rocker had been obtained the keel would have been fastened to the stocks by treenails driven through three vertical holes in the keel (McKee, 1978: 270–1, 307).

Unlike the Graveney boat, the surviving sidestrake of the flat-bottomed keel-less Brigg 'raft' was insufficiently long to be a control in determining any rocker. The fragmented S.E. ends of bottom planks 4 and 5 were found lying on a gentle slope up to the SE rather than up to the NW as were the main runs of planking (McGrail, 1981b: figs: 1.1.15 and 1.2.1).

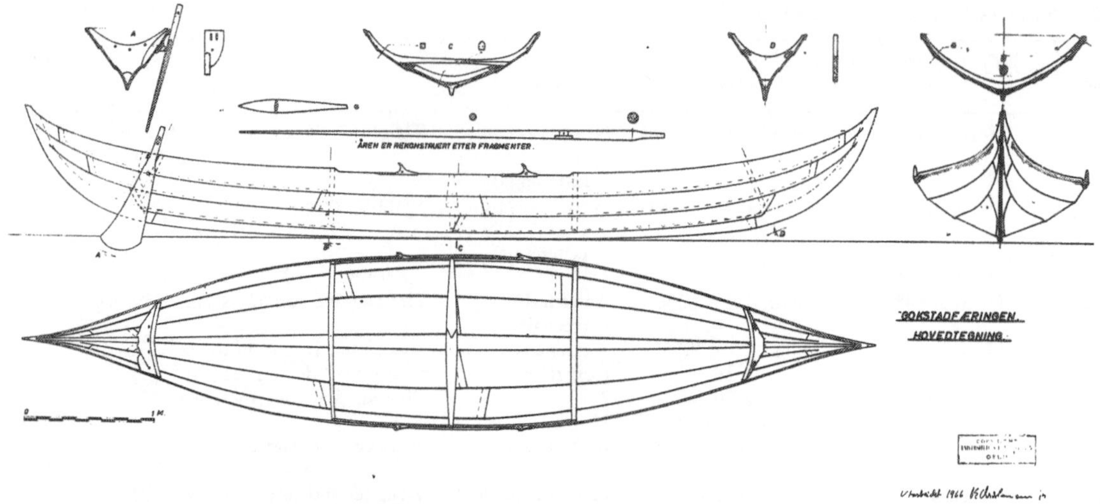

Figure 8.17 Construction plan of the Gokstad *faering*. After Christensen (Courtesy the University Museum of National Antiquities, Oslo, Norway (copyright holders)).

Although some rocker would have been operationally useful to this boat in her deduced role as ferry across the Ancholme arm of the sixth century BC Humber estuary, it was deduced that this change of slope was due to settling on a slightly uneven surface during deposition and thus the minimum hypothesis of a flat bottom was incorporated into the reconstruction drawing (1981: fig. 4.1.23). Although the shape of the fragmentary side strakes might give some control over the bottom curvature of Ferriby boats 1 and 3, the evidence for these plank-keeled boats is difficult to interpret owing to the difficult nature of the work on this inter-tidal site and the fact that Ferriby 1, for which there is the best evidence, could not be recorded in one operation but at irregular intervals using various methods which may not have been successfully harmonised (but see Wright, forthcoming, for a revised interpretation).

The keel-less, flat-bottomed boats of the Romano-Celtic tradition are not all well-documented, but some appear to have rising ends. Other flat-bottomed boats of the medieval period (e.g. Falsterbo) are also under-recorded. However, the one well-recorded find (Fig. 8.20) the twelfth-century Egernsund boat has a sprung bottom, the curvature of which is confirmed by the run of the side strakes (Crumlin-Pedersen, 1977a: 7).

A sprung keel is a feature of the Viking tradition

Figure 8.18 Scenes from the life of St. Nicholas on the late twelfth-century font in Winchester Cathedral. The ship is believed to represent a hulc. (Photo: NMM Greenwich.)

boats (Fig. 8.17). The Oseberg keel has a rocker of 30/1695 = 0.018) the uprising ends running smoothly on into curved intermediate timbers. Medieval cogs had little, if any, rocker if we may judge by the published drawings (Ellmers, 1979: figs 1.6 and 1.8; Crumlin-Pedersen, 1979: fig. 2.13). Iconographic evidence is little help here as not only is there the problem of linking the depicted vessel to a documented typename, but many of them are shown afloat. Earlier representations of flat-bottomed vessels sometimes claimed to be cogs – for example those on the ninth-century Hedeby coins (Ellmers, 1979: fig. 1.7) – have no appreciable rocker.

The shape of the *hulc*, a documented ship type which evidently by the fourteenth century had superseded the cog as the dominant seagoing vessel survives only in representational form which gives the vessel an exceedingly curved hull (Greenhill, 1976: 283–5, fig. 201). It might be thought that such curvature was determined by the circular shape of the town seals and coins on which most of these depictions are found, but the vessels on the rectangular sides of the Winchester (Fig. 8.18) and Zedelgem fonts are also curved and if this may be taken as independent evidence (and not just copied from a widely available model) we may deduce that the hulc had a longitudinally curved bottom which was more sprung than the bottom of the Viking tradition.

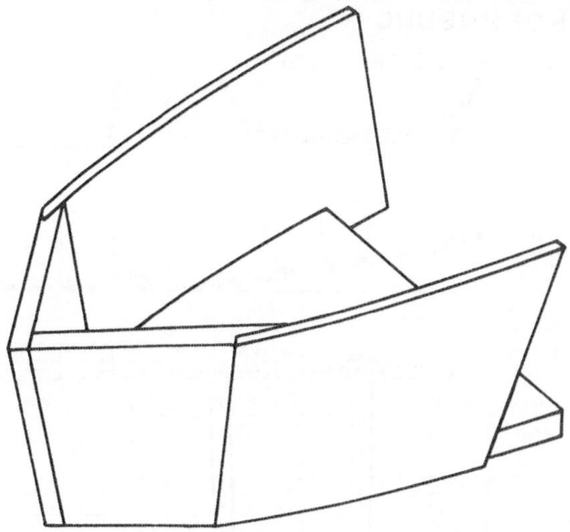

Figure 8.19 The planks of the stemless *Knurrkahn* of Dievenow, S. Baltic meet at the ends. After Rudolph, 1974.

HOW TO CLOSE THE ENDS?

There are four principal ways of building a planked boat so that the ends are securely fastened and so that the whole structure becomes a watertight shell:

1. By curving the side planking of a stemless boat until the sides meet at the ends in a bevel (Fig. 8.19). The planking is fastened along the (near) vertical line of this bevel and the bottom planking also terminates here. The bevel junction may subsequently be supported by a fitted timber. The nature of this bevel means that it is doubtful whether this would be suitable for vessels other than small boats on inland waterways. No ancient boats with this end solution are known.
2. By running the bottom planking up clear of the waterline to end on a horizontal plane with a reinforcing timber, as on the twelfth-century Egernsund find (Fig. 8.20) and in the twentieth-century *Weidling* of the River Aar, Switzerland (McGrail, 1981b: fig. 4.1.7). In McKee's hypothetical reconstruction of the *hulc* (Greenhill, 1976: fig. 39) and in reverse-clinker built boats from the Sylhet area of Bangladesh (Greenhill, 1976: fig. 42) the bottom planking meets a further series of higher horizontal strakes. Analytically the Ferriby boats may be considered as a variant within this classification as the ends of their plank keels are hewn out to give an upturning curve (Fig. 8.21) and terminate above the waterline; the outer bottom planks end where they meet a series of (near) horizontal strakes, but, unlike others in class, this is underwater. Like type 1, boats in this class are stemless.
3. By running bottom and side planking on to a transom, a (near) vertical board, which may be entirely clear of the water as in the bow end of the Norwegian pram (Greenhill, 1976: fig. 38) or partly submerged as in the *galar*, a plank boat of the River Vistula (Fig. 8.22), and in the hypothetical reconstruction of the Brigg 'raft' (McGrail, 1981: fig. 4.1.23) and in several of the Romano-Celtic flat-bottomed boats. The *galar* has one vertical transom and one angled transom forming a ramp; although such a ramp, or even two would have been operationally useful in the Brigg 'raft' the minimum solution of two vertical ends was adopted in the reconstruction.
4. By running bottom and side planking into a post or stem – the several varieties are discussed below.

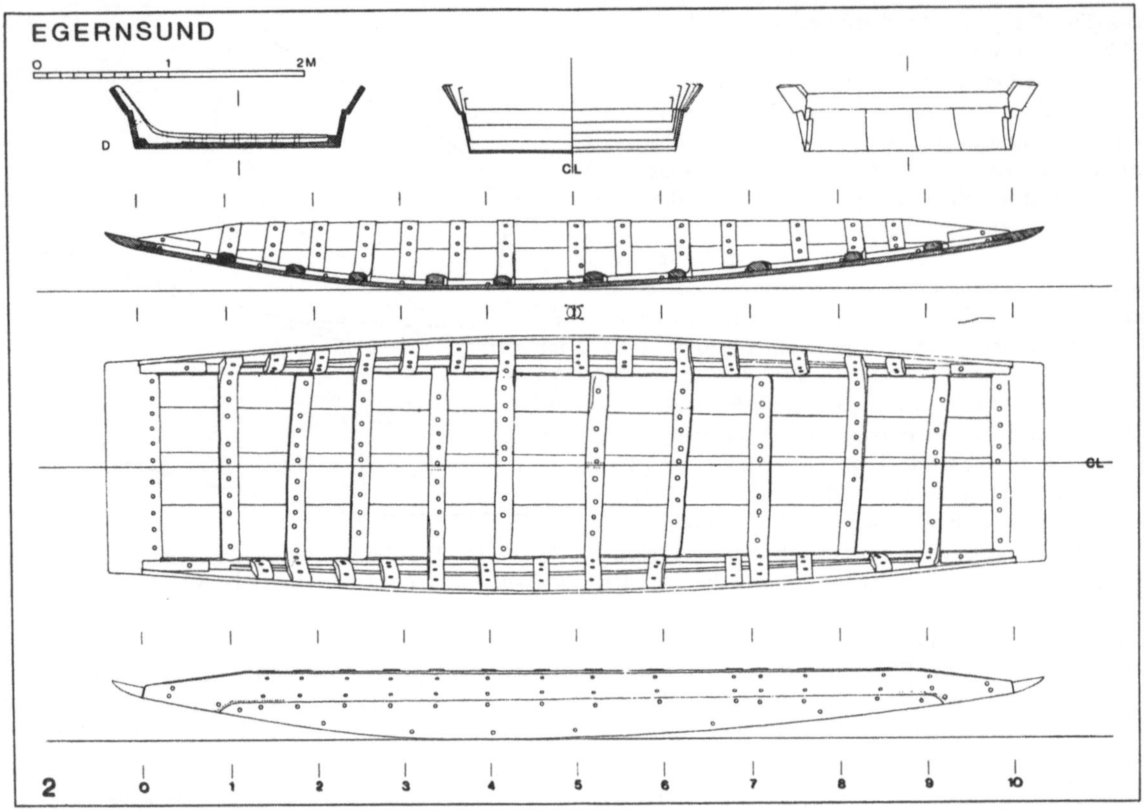

Figure 8.20 Construction plan of the twelfth-century boat from Egernsund, Denmark. After Crumlin-Pedersen, 1977a.

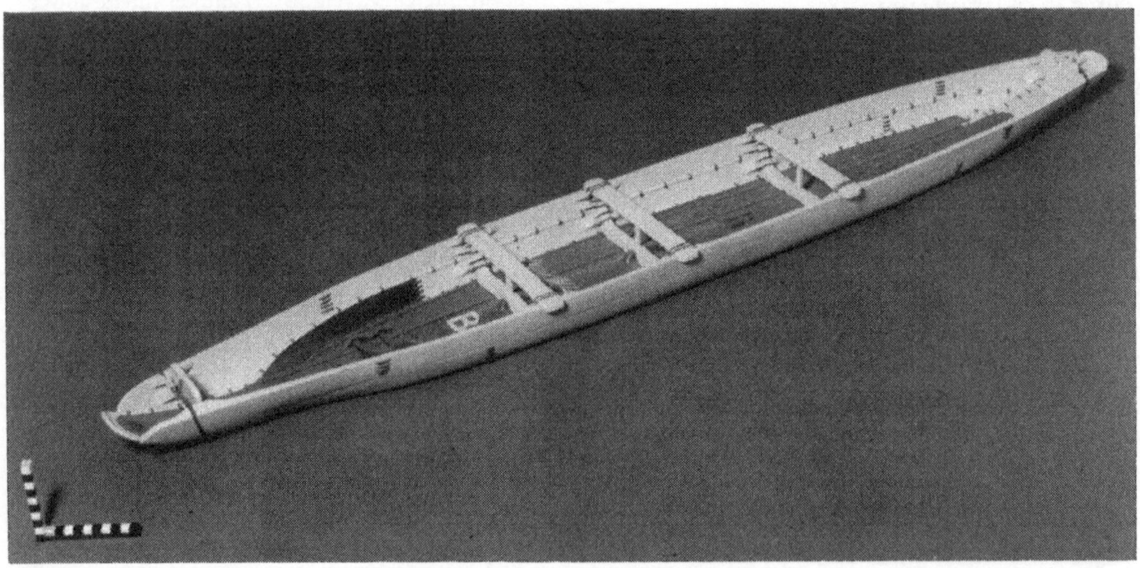

Figure 8.21 A 1:10 reconstruction model of Ferriby boat 1. The black part represents excavated elements; the white is conjectural. (Photo: NMM Greenwich.)

Figure 8.22 A *galar* of Szezvcin, River Vistula, Poland. (Photo: NMM Greenwich.)

Boats may have the same type of end at bow and stern as in the Viking Age tradition of a post at bow (stem post) and stern (stern post). It does not necessarily follow that all such boats will be *double-ended*, i.e. ((almost) symmetrical at the ends, for sharp and blunt ends are compatible with stern posts. Even when the ends are technologically similar they may have different longitudinal profiles, as shown on the seventh-century Anglo Saxon *sceatta* coins (Fenwick, 1978: fig. 8.2b), a seventh-century Merovigian strapend (Joffroy, 1978), twelfth-century Eadwine Psalter (Fenwick, 1978: fig. 8.10b) and on some boats on the Bayeax Tapestry. Dissimilar shaped ends also result when the natural taper of the logs is reflected in the bottom planking as in the Brigg 'raft' (McGrail, 1981b).

On the other hand, technologically different ends are common in boats today – a stem bow and a transom stern readily give the traditional 'boat shape' with a good entry yet broad in the stern where more space is needed for the helmsman and for the base of the sail's sheets. The earliest finds with positive evidence for dissimilar ends in the technological, rather than form, sense are probably the fourteenth-century flat-bottomed boats from Falsterbo, Sweden which have a stem at the bow and a transom stern (Fig. 8.23).

Keel-less boats such as the Brigg 'raft', the Bevaix boat (Arnold, 1975), the Egersund boat (Crumlin-Pedersen, 1977: 7) and the Somerset turf boat (McKee, 1983: fig. 86) and plank-keel boats such as the Ferriby boats, the boats and ships of the cog tradition and many of the boats from the S. Baltic described by Rudolph (1974: types 2.2 and 2.3) may have any type of end, but boats with beam-keels almost invariably have stems, or in recent times a stem terminating in a transom above the waterline (McKee, 1983: 79).

STEM POSTS

Some twentieth-century boats have an 'inner stem' or 'stemson' consisting of a vertical apron timber above a deadwood or knee (McKee, 1983: fig. 37), but early examples have not so far been found unless the cog stem and 'stem hook' are considered analogous, but see above in *Keels* (p. 115). If these are discounted it proves convenient for analysis to classify stems by three groups of attributes:

1. A measure of the timber from which they were formed.
2. Some measure of the shape they gave to the boat:
   2.1 rake (angle to keel)
   2.2 enclosed angle between port and starboard planking at, say, waterline.

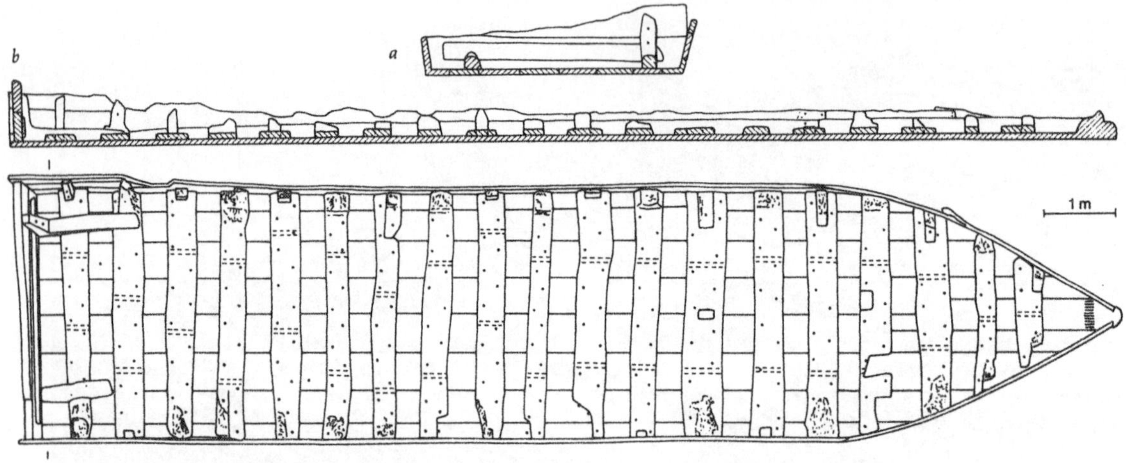

Figure 8.23 One of the fourteenth-century boats from Falsterbo, Sweden. After Ellmers, 1972.

2.3 radius of curvature of underwater profile.
3. Plank fastening methods:
   3.1 rabbeted, bevelled or overlap
   3.2 hollowed or flat inboard surface of stem
   3.3 stepped or continuous on inboard face
   3.4 fastenings, through or blind.

1. Cross-section of the timber
The siding (s = the stem's cross-section dimension in the boat's transverse plane) may be compared with moulding (m = the cross-section dimension in the fore and aft plane) for a representative section, say near the waterline. The medieval stems recorded to date have (m/s) > 1.50 indicating a beam or plank-on-edge stem. Some of the stems described by Rudolph (1974) may have (m/s) values nearer to 1 but dimensions are not available; such stems are sometimes called 'block-stems' but this seems linked to the great absolute values of their cross-section dimensions in relation to the size of boat, rather than to their (m/s) ratio.

2.1 Rake
The shape of the boat at the ends is determined (partly) by the line of fastenings of the strakes to the stem. In most boat finds this is generally parallel to the leading edge of the stem, especially the important underwater section.

Unlike most of the recent boats considered by McKee (1983) ancient stems in the Norse tradition have a variable angle of rake whether measured by the leading edge or by the line of the plank ends (see Fig. 12.34 and the stems illustrated by Evans and Bruce-Mitford, 1975: 411). Rake angle can, however, be given for such straight stems as Graveney, and the Bremen cog. In general a raked stem provides more area of contact between stem and strake than a vertical stem of similar scantlings but this can be a weak, pointed area at an angle to the run of the strake.

2.2 Enclosed angle between planking
In a keel boat this angle will generally decrease lower down the stem. A representative angle may be measured at or near the datum waterline or approximately halfway along the length of the stem. McKee (1983: 79) divided stemmed ends into sharp (enclosed angle < 60°) and bluff (> 120°). The ends associated with medieval stems recorded to date are *sharp*. This is a technological advantage in that planks are more readily fastened to a stem at a sharp end when compared with a bluff end where there would be less wood in contact with the stem, unless the stem was of massive proportions. The modern way to obtain a bluff end is to back a moderate sized stem with apron and deadwood (McKee, 1983: fig. 38) which provide a broader landing for the planking.

2.3 Radius of curvature
Where the leading edge of stems has an almost constant radius of curvature, e.g. Nydam 2 and Skuldelev 3, this may be recorded. Otherwise the radius for the underwater section should be noted. McKee (1983: 78) has postulated a working rule that where this radius is greater than the overall length of the boat the stem may be considered *straight*. If the run of the plank ends is other than generally parallel to the stem's leading edge it should be noted.

The underwater portions of the stems of Oseberg, Gokstad 1 and Nydam 2 are similar. The lower end of Sutton Hoo 2 stem is not unlike these profiles but higher up it takes on a different profile of greater radius which may be due to sagging outwards as the keel scarf opened after (or even during) deposition. Above the waterline Gokstad 1 stem becomes almost vertical whilst Oseberg's changes to a reduced radius and thus bends back on itself. An uprising sheerline at the ends provides a reserve of buoyancy and a degree of spray protection in bow and quartering seas, but a stem which projects as much above the general level of the sheerline as in Oseberg may be for display rather than functional, as the windage would be considerable. The high stem given to Kvalsund 2 by Shetelig and Johannessen (1929) on the other hand may be due to over-enthusiastic reconstruction. The stems of Skuldelev 1 and 3 are also alike underwater, but the former then adopts a reduced radius of curvature and turns inboard beyond the vertical.

3. Plank fastening methods

Methods used to fasten planking to stems may be said to be efficient if they:

(a) provide sufficient plank/stem contact for a firm fastening
(b) do not intrude into the waterflow (McKee, 1983: 64)
(c) protect the end grain of the planking and thus delay onset of rot.
(d) generally a through fastening (e.g. clenched nail) is more secure than a blind one (e.g. spike).

Of the twelve theoretical groups obtainable by combining one of the two states of attributes 3.2 (hollow or flat) with one of the three states of 3.1 (rabbeted, scarfed or overlap and with one of the two states of 3.3 (stepped or continuous), six have known examples (Fig. 8.24).

A. Rabbetted, flat face, continuous Examples are Gokstad 1, Oseberg, the lower part (4 strakes) of the Bremen cog stem, the upper part of the Graveney stem, and Rudolph's group 222 (1974: fig. 7).
The plank ends are protected inside the rabbet, although some planks enter the rabbet over their full thickness while others are half-lapped. As there is no hollow on the inner face the hood ends generally have to be fastened by spikes, but the Oseberg planking is fastened by clenchnails driven from alternate sides through both planks. The Bremen cog planking is further protected by a false stem.

B. Rabbetted, hollowed, continuous Examples are Nydam 2, Gretstedbro, Sutton Hoo 2 and Rudolph's group 221 (1974: figs 4, 5, 6).
The rabbet again protects the end grain, but the hollow allows through fastenings to be used. The

A

Figure 8.24 Six types of stem:
A. Rabbeted, flat face, continuous – *Heuer* of River Oder. (Photo: Wolfgang Rudolph).

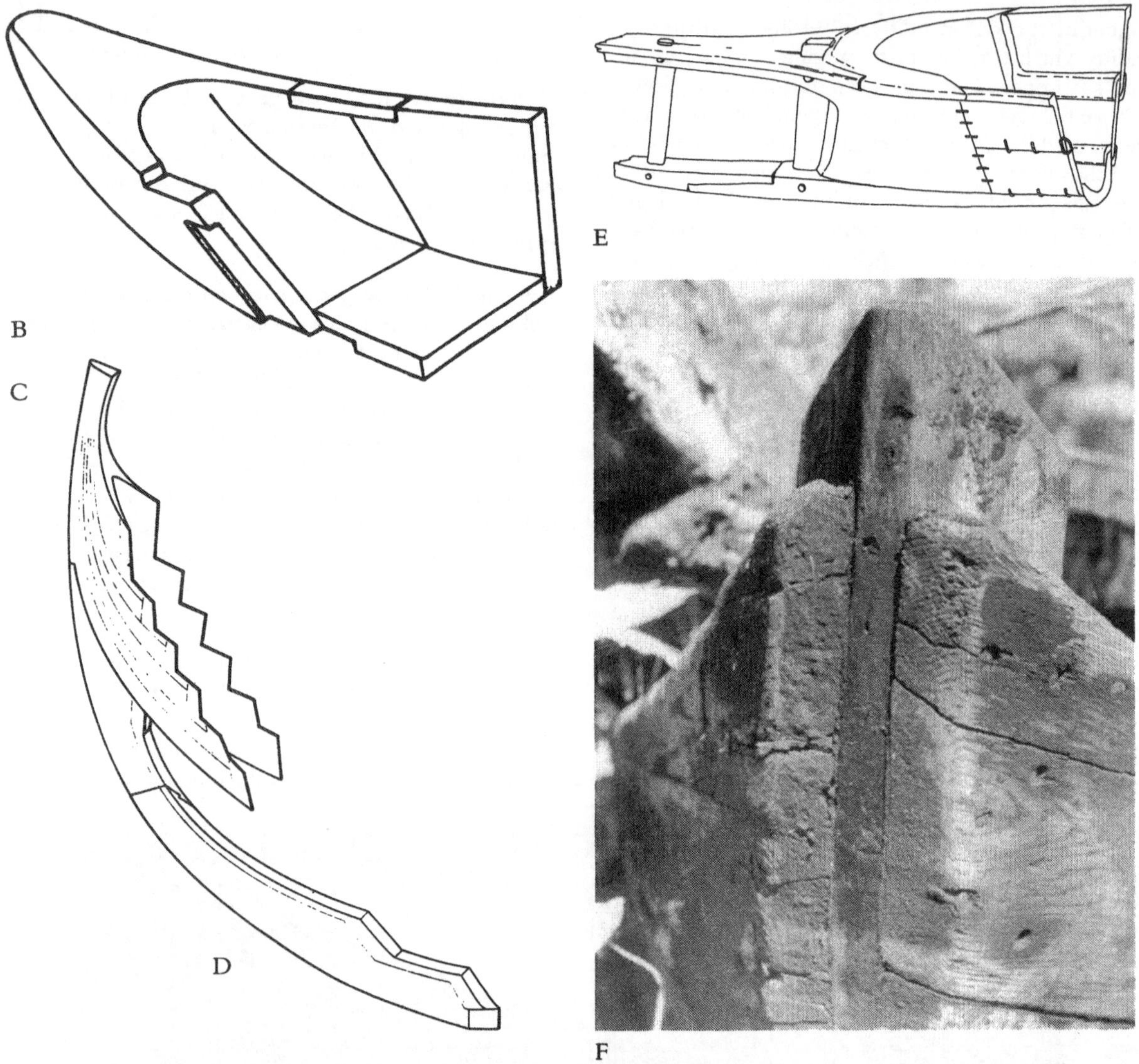

Figure 8.24 (con't)
B. Rabbetted, hollow, continuous – *Kahn* from River Oder. After Rudolph, 1974: fig. 4.
C. Bevelled, hollow, stepped – upper part of Skuldelev 1. After Olsen and Crumlin-Pedersen, 1967: fig. 19.
D. Bevelled, flat face, stepped – lower part of Skuldelev 1. After Olsen and Crumlin-Pedersen, 1967: fig. 19.
E. Bevelled, hollow, continuous – Hjortspring. (Drawing: NMM Greenwich.)
F. Overlap, flat face, continuous – *Waltelle* from Curonian lagoon. (Photo: Wolfgang Rudolph).

lower part of the Graveney stem is a variant of this type, having clenching grooves (Fenwick, 1978: 3.2.2) for the garboard, third and fourth strakes. However, Fenwick (1978: 217) believes that this may be secondary work when these strakes had to be refastened to the stem, perhaps after springing; in this case the whole Graveney stem would originally have been type A. Although Rudolph's group 221 stems are called by him *block stems* as they are hewn and hollowed from logs cut in half, at this level of classification they are analytically the same as those of Nydam 2 and Sutton Hoo 2. Indeed many stems have been fashioned from logs cut in half (e.g. the Graveney stem) but the South Baltic stems are not taken so far from their natural shape.

C. Bevelled, hollowed, stepped Examples are Skuldelev 1 (upper), 2, 3, 5 and probably 6; Gokstad 3 (*faering*); Eigg 1.
The sides of this type of stem are extended inboard

and shaped in steps, the ends of which are bevelled to take the planking. Thus not only is there greater area for fastening each strake than there could be with a continuous type of stem especially a raked one, but through fastenings can be used, and end grain is protected in the bevel.

D. Bevelled, flat face, stepped Examples are lower part of Skuldelev 1; Eigg 2; Danzig-Ohra 1, 2, 3; Sandnes Greenland models; Eltang; Galtabäck; Falsterbo (Olsen and Crumlin-Pedersen, 1967: 165–6). This is similar to type C except that spikes rather than through fastenings have to be used.

E. Bevelled, hollowed, continuous The unusual stem of the Hjortspring boat is of this type with the bevel going forward and inboard. As the breadth of this boat was tied to a hollowed log plank keel and there was no attempt to splay out the strakes, an outward sloping bevel was not required. Sewn-through fastenings were used.

F. Overlap, flat face, continuous The upper part of the Bremen cog stem (8 strakes) and the stems of the Ijssellmeer cogs are of this type. The Bremen cog and some of the Dutch finds additionally have a false stem which not only protects the end grain but nips the plank ends; plank repair and replacement is, however, made more difficult. Rudolph's groups 223 (1974: figs 8 and 9) and 225 (1974: figs 12 to 17) have this type of stem.

PLANK ENDINGS

The bevelled, hollow, stepped stem (type C) tackles three problems: two constructional, the provision of sufficient space on the stem for each strake, and the ability to use through fastenings; and one cosmetic, the apparent continuation of the natural run of the strakes, or, as in the case of Oseberg, the emphasis of the inward curve of the stems above the waterline. Other ways of tackling the first problem are also known:

(a) The 15th and 16th (top) strakes in Gokstad 1 are both scarfed to a single plank about 1.5 m from the stem; the 10th to 13th strakes are similarly scarfed to one broad short plank which becomes their joint hood end. Thus space is only required for 12 strakes on the stem rather than 16. Gokstad 2 (the larger small boat) has an elaborate double stealer for the top three strakes on both sides.

(b) The Oseberg 11th and 12th strakes run out well before the end, and thus only 10 strakes need space on the stem. The 12th and 13th strakes in the Bremen cog similarly run out before the ends.

(c) In the Oseberg and Gokstad 1 finds the lower edges of the hood ends have short reverse curves (Fenwick, 1978: fig. 8.27) thereby increasing the breadth of each plank within the rabbet and allowing more fastenings to be used.

STEM PROJECTIONS

The type E stems of the Hjortspring boat are set back from the ends of the plank keel and each has a horizontal projection turning upwards from its upper surface which parallels the extending projecting plank keel (Fig. 8.24). These two protruding beaks on each end are joined by vertical oak timbers fitting into mortices.

Similar ends are evidently depicted on Scandinavian rock carvings dated to the late bronze age and early iron age (Christensen, 1972: 162). Horizontal projections from the lower part of one end (the bow) are shown on sailing boats depicted on a first-century AD coin (Fig. 12.26) from Canterbury (Muckelroy *et al.*, 1978), and on an unpublished coin from the River Colne at Sheepen, Colchester. Such projections are found today on boats from the Celebes (Hornell, 1946: 210, fig. 39), Lake Victoria (Worthington, 1933), Siberia (Brindley, 1919/20), Oceania and the Indian Ocean (Hale, 1980: 126); and Bonino (1975) has published boat models and engravings from seventh-century BC Italy which also have short projections from the forefoot. A protruding forefoot could be of use when beaching, possibly preventing the bows digging in. In addition, it theoretically increases the boat's speed potential by extending the effective waterline length (Ch. 11). In a short steep sea it is also thought to be effective in keeping up the boat's head (Worthington, 1933: 161).

The Graveney boat's protruding angular forefoot – see Fig. 8.16 and Fenwick (1978: 216–17) – adds to the boat's lateral plane area thus increasing her resistance to leeway.

HOLES

Horizontal holes at or above the assumed waterline have been found through the stems of the Graveney boat (Fenwick, 1978: 213), Kentmere boat 1 (McGrail, 1978: fig. 26), Nydam boat 2 (Fenwick, 1978: fig. 8.7c) and the tenth-century pottery boat models from Lake Paterwolde (Fenwick, 1978: fig. 8.12a). These seem best interpreted as attachment points for securing lines when beached or moored.

The hole well below the waterline on the Graveney stem may be the site of a repair-closing nail as stated by Fenwick (1978: 217). There is also the possibility that such holes were used during the building of the boat.

HOW TO FASHION PLANKING AND MAKE WATERTIGHT?

The planking forms the bottom, bilge and sides of a round-hull boat, and the bottom, transition strakes and sides of a flat-bottomed boat. Certain aspects of planking have been considered above; this section concentrates on other aspects of planking which may be encountered during examination of runs of planking as found, for example, on the Brigg and Graveney sites, or of individual planks, as were many of those from Wood Quay, Dublin and Bristol.

Examination is greatly facilitated by the use of Attribute Lists (McGrail and Denford, 1982) so that nothing significant is overlooked and negative as well as positive evidence is recorded. Seventy-five or more attributes may be listed for planking; when a sufficient number of finds have been recorded and published it should prove possible to search for patterns and identify diagnostic features. As this is not yet possible a few attributes, subjectively selected as important, are discussed below.

DIMENSIONS OF PLANKING

When long plank runs are available and they can be fashioned to the required shape or forced to take up the curves required, then planks running from end to end without scarf joints are preferred, as are found in the bottom of the Brigg 'raft', where there is little if any curvature in the hull. The outer bottom planks of the Ferriby boats were also probably one-piece; the length of side planking is not known here but curvature was achieved by hewing. The Hjortspring boat has side planks of almost constant breadth throughout their length so the boat has little curvature. The Nydam boat is more curvaceous than the boats so far discussed but with the woodworking techniques then available her boatbuilders were able to obtain these shapes using mostly single plank strakes. The top strakes, however, have scarves (Shetelig, 1930: fig. 7), near the 13th and 14th frames from the stern on the port side and the 17th and 18th frames to starboard. The longest plank in these top strakes is *c.* 18 cm. The longest planks in the boat are the fourth strakes which are *c.* 20.30 m (not the 82 ft (25 m) given by Christensen 1972a: 162 or the 23 m given by Fenwick, 1978: 227). The necessity to scarf the top strake may have been due to the lengths of planking of the correct breadth available rather than any difficulty in achieving the shape.

As any curved line consists of an infinite number of infinitely short straight lines, shorter planks can more readily take up the curves found in the more advanced forms of planked boats. For this reason the Sutton Hoo 2 strakes consist of an average of four planks with a range of two to seven (Evans and Bruce-Mitford, 1975: figs 280, 325). These numbers may be slightly inaccurate as the garboards and part of the second strake were too damaged to examine in 1966/7. Gokstad 1 has between five and eight planks per strake, and Oseberg four to nine with the shorter planks in both instances being at the ends. Graveney had at least three planks per strake but it has been deduced that originally there were four or five. Skuldelev 1 generally has four, and Skuldelev 3 and 6 generally three. A significant technological change may be seen between the fourth-century Nydam 2 and sixth-century Sutton Hoo 2 in that the hull of the former boat was built from twelve planks whereas Sutton Hoo's hull had a minimum of 57 planks and probably more than 60, a Nydam/Sutton Hoo ratio of 1:5 when the areas to be planked have a ratio of only 1:3.

We also find that between the Ferriby boats and the Skuldelev boats – a period of some 3000 years – breadths and thicknesses of planking change (Table 8.3). In general terms and based on a very limited sample, from the time of Sutton Hoo onwards planks are thinner and narrower than those used in earlier times. In several finds, however, there are substantial strakes of significantly greater scantlings, at the waterline – see discussion below under *longitudinal strength*.

PLANK SCARFS

There are no scarfs or lengthening joints in the bottom or side planking (in contradistinction to the plank-keel of the Ferriby boats) of the prehistoric finds, and only two in Nydam 2 (Shetelig, 1930: fig. 7). There are unusual tongued-scarfs in the 9th and 10th strakes of Skuldelev 1 (Olsen and Crumlin-Pedersen, 1967: fig. 21). Similar scarfs packed with moss have been found in the planking of thirteenth/fourteenth-century Ijssellmeerpolder wreck 0Z.43 (Reinders, forthcoming) and in the tenth-century Haithabu boat (Crumlin-Pedersen, 1969: 16). Other plank scarfs in these Dutch wrecks

Table 8.3 Breadth and thicknesses of certain prehistoric and medieval boat planking

Boat	Breadth (cm)	Thickness (cm)	Reference
Ferriby	50–54	Bottom 6–10 Sides 6–8	Wright, 1974
Brigg	44–51	Bottom 4 Sides 2–2.5	McGrail, 1981b
Nydam 2*	36–45	2.2–2.5	Shetelig, 1930
Sutton Hoo 2†	25–40 (mean 30)	2.54	See Note†
Gokstad 1	18–22.5	1.8–3.2	Nicoloysen, 1882
Gokstad 3 (faering)	28–38	*c.* 1–1.2	Christensen, 1959
Graveney	≤ 28	2.7–3.3	Fenwick, 1978
Skuldelev 2	24–27	2–3	Olsen and Crumlin–Pedersen, 1967
Skuldelev 3	25–30	2.5–3.5	
Skuldelev 5	25–32	2–3.5	

Notes: 1. Oak planking; breadths taken near midships station where possible.

* Nydam 2 planks were probably originally broader – see Åkerlund, 1963.

† Breadths given by Evans and Bruce-Mitford (1975: 354, table 19) evidently do not allow for plank overlaps which could be up to 2 in (5 cm).

and in the Skuldelev group are through-splayed on edge, faced nailed (vertical). One of the scarfs in Dutch wreck N 5 has a gradient of 1 in 10.6 or 9.4%, otherwise little detailed information has been published about Dutch and Scandinavian plank scarfs, although it seems that such scarfs are generally fastened by two to five clenched nails in addition to those used in fastening the strakes to each other. The Graveney vertical plank scarfs are short ones with steep gradients of 1 in 2.7–3.1) (31–27%) and generally have only one extra nail (McKee, 1978: 97). The twelfth-century planking re-used in the waterfront at Wood Quay, Dublin, on the other hand, has shallower gradients of 1: 6–8 (14.5–12.3%) which, if we allow for the (tangential) shrinkage in thickness which these radial planks experienced while drying out before recording, is close to the gradient of 1 in 7 (14%) recommended today. The planks in the Wood Quay scarfs do not taper to a feather edge as in Skuldelev or Graveney, but have a blunt end some 4–10 mm thick which protrudes slightly from the face of the planking; these are thus stop splayed on edge, face nailed. Similar ones have been noted on Dutch wreck N 5 and they are fastened by four nails (Reinders, forthcoming). The one find in Britain of a boat built in the Classical tradition, the AD third century County Hall boat, had plank scarfs which were stop-splayed on face, edge nailed; that is they were horizontal scarfs and thus different from all other plank scarfs so far reported from boat finds in N.W. Europe.

The outboard edge of plank scarfs generally open aft so that normal water flow does not penetrate the joint. There are exceptions to this however: scarfs in the ends of strakes generally open forward where they meet the forward stem; and planking lengths inserted during repairs are sometimes easier to fit with scarfs facing the 'wrong' way. Some aft-facing scarfs are, however, difficult to explain as repairs: scarfs in two Graveney planks (McKee, 1978: 97), and in the stern area of Oseberg (Sjøvold, 1969: 22) – as Fenwick (1978: 227) has suggested, this may be to do with the sequence or method of planking up, or to eke out damaged timber.

In recent shell boat building it has been the practice to distribute plank scarfs so that they are not close, vertically or laterally: thus a boatwright's rule of thumb might be to have a minimum of two strakes between two scarfs that are within four frame stations of each other. Such rules are difficult to identify, if they exist, in incomplete runs of planking; for example, McKee could not find a rule behind the spacing of the Graveney scarfs, but he did conclude that although they were staggered to avoid too many scarfs near the same station, they were much closer than would be acceptable today (1978: 97). McKee's examination also showed that several of the Graveney plank scarfs had changing angles of bevel on one plank end, possibly indicating that further work had been found to be required when that plank was offered up to the end bevel of a plank already partly in position. The Graveney scarfs, at least those underwater, had a soft caulking of tarred animal hair inserted before they were fastened.

LONGITUDINAL STRENGTH

As stressed in the section on keels, boats require longitudinal strength addditional to that contributed by the keel. Ferriby and Brigg are incomplete and deduction cannot be made about them, but many of the later finds have at least one strake higher up in the hull, which is of greater scantlings than the others or of an unusually stout cross-section. The top strakes of the Hjortspring, Nydam 2, Kvalsund 2 and Sutton Hoo 2 all thicken at their top edge so that their cross section becomes wedge-shaped (Fig. 8.25). Other craft have thick strakes nearer the waterline where they can distribute sailing stresses transmitted to

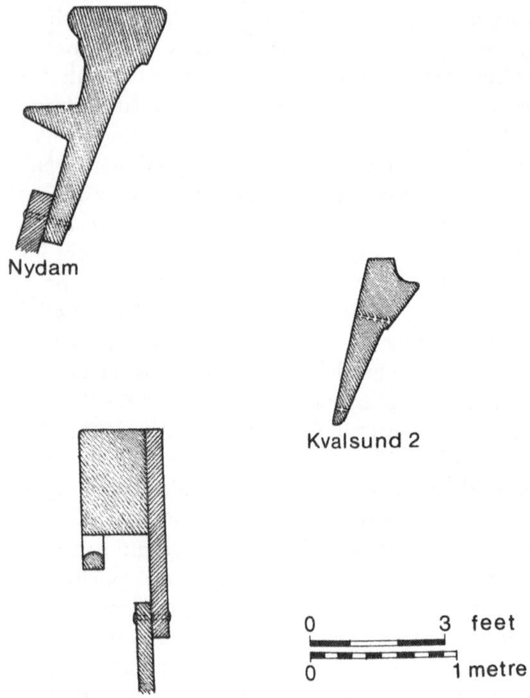

Figure 8.25 Cross-sections of the thickened top strakes of Nydam 2, Kvalsund 2 and Gokstad 1. After Evans and Bruce-Mitford, 1975: fig. 307 (by courtesy of the Trustees of the British Museum).

them through mast partners, mast beams & knees. Such strakes are thought to be the *meginhufr* of medieval Icelandic literature. Gokstad's 10th strake is 4.5 cm thick rather than the general 1.8 to 2 cm, and had a wedge-shaped cross-section, as has the corresponding strake in the Tune find while Oseberg's *meginhufr* is of inverted L-section. The Klåsted and Åskekarr *meginhufr* have S or Z-shaped cross-sections (Christensen and Leiro, 1976) (Fig. 8.9).

Skuldelev 1, 5 and 6 all have thicker strakes near the waterline and the strakes of boat 5 have the inverted L cross-section. The twelfth-century find from Lynaes, Denmark with a total of 19 strakes has a thick and specially shaped 17th strake which would be near the waterline at deep draught. The Graveney boat's 7th, and probably 8th, strakes are also thicker than the others (a mean of 33 rather than 27 cm) making them binding strakes (main distributors of transverse stresses) which over their central sections would be within the range of operational waterlines visualised by Corlett (1978: 303).

In addition to longitudinal strength from keel,

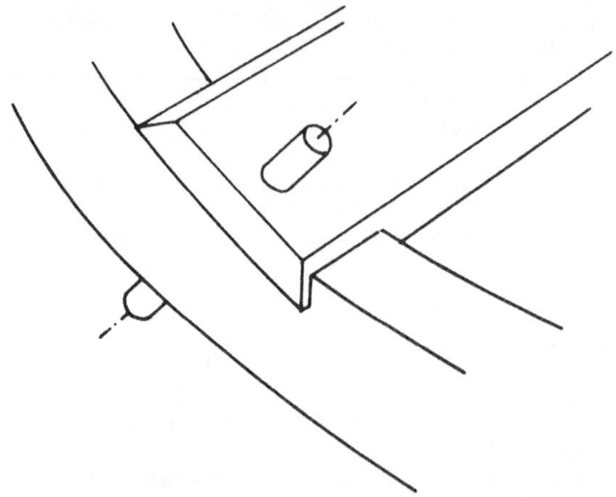

Figure 8.26 Method of fastening stringer to frame in the Graveney boat. (Drawing: NMM Greenwich.)

waterline strakes and topstrakes, some finds have stringers, longitudinal timbers running inside the planking (Olsen and Crumlin-Pedersen, 1967: fig. 15) as on the Skuldelev finds, or let into the frames, as in the Graveney boat (Fig. 8.26).

OVERLAP

Eric McKee (1972: 5) quotes a twentieth-century rule of thumb that overlaps between strakes in the clinker style of boatbuilding should be twice the thickness of the planking. These dimensions can be measured on articulated planking, the thickness being taken near the top edge of the lower plank, below the bevel. On detached planks the overlap can be approximated to the vertical length of the bevel at the top edge; where there is no bevel – as on planking from those parts of a boat where there is no change in the transverse section – it may be possible to see the 'ghost' or impression of the overlap. These data are not available for many finds, but examples are:

	Overlap/thickness
One plank from Blackfriars 3 (15th century)	3.18
Three planks from Wood Quay (12th century)	2.17; 1.35; 1.60
Skuldelev 6 – generally (10th/11th century)	1.07 to 1.60
Nydam 2 – generally (4th/5th century)	2.80 to 3.18.

The few examples calculated to date show little sign of a medieval equivalent of the twentieth-century rule of thumb.

Bevel angle

The bevel along the upper outboard edge of a clinker-laid strake is a measure of the angle between adjacent strakes at any one station (Fig. 8.3). The angle between strakes will vary to match the required transverse section, and thus as the boat has different transverse sections along her length the bevel angle will vary along the length of a strake. In flush-laid building as in the Classical Mediterranean tradition the same principles apply, except that as the planks meet over their full thickness the bevel is worked across the whole thickness of the lower strake (van Doorninck, 1976: fig. 9) – see Fig. 8.4. In the Ferriby style of boatbuilding the bevel angle on the outer edge of the outer bottom planks determines the angle at which the lower edge of the lowest side strake will lie – but as this strake has been given a curved cross-section by hewing, its top edge will lie at a different angle. The joint between the first and second side strakes of Ferriby 1 is an edged half-lap (McGrail, 1982: fig. 3.10). The Brigg bottom planking is butted; there is a bevel between first and second side strakes – an edge splayed lap joint – while the first strake is fastened to the outer bottom plank by a simple overlap joint (McGrail, 1982: figs 3.11, 3.12).

Bevels may also be found on the outboard lower edge of planking as in Graveney (McKee, 1978: 97) where the lower strakes had been rounded where they would otherwise have rubbed when grounded. It is not possible to determine the shape of the inboard top edges of this boat due to damage, but those in the Kentmere 1 boat (Wilson, 1966) had been bevelled, presumably to minimise damage to cargo and to boat. Planking may also be bevelled on its lower inboard edge where the boat has hollow lines, i.e. the transverse section is concave from outboard, and there is a marked angle between adjacent strakes (Fig. 8.3).

CAULKING GROOVES AND CAULKING

Caulking grooves were worked near the lower inboard edges of several Graveney planks – these were within the overlap with the upper strake, along the line of fastenings. Sheep's wool greased with vegetable tar had been laid along this groove before the planks were fastened. Analytically this should be distinguished from caulking forced into seams after fastening to stop a leak, examples of which were also found in the Graveney boat (McKee, 1978: 97). Caulking inserted routinely before fastening is unknown in the twentieth-century clinker boatbuilding using sawn or planed boards, but hewn boards as used in medieval times have an irregular thickness at the edges which would result in slight gaps between strakes; the inserted caulking takes up these irregularities, the tightly-packed greasy fibres preventing water from penetrating the joint by surface tension (Coates, 1977: 223). Caulking grooves have been found in most finds of the Norse tradition, however, in the Wood Quay planking so far examined no grooves have been found, although tarred caulking was undoubtedly used and a pot sherd smeared with vegetable tar suggests that waterproofing or caulking was carried out on site.

The Ferriby boats had caulking within their complex plank edge joints (Fig. 8.27) held in place by a longitudinal lath and the plank fastening stitches. This sequence of caulking, lath, stitch was also found between the edge to edge bottom planking of the Brigg 'raft', and has parallels in the sequences used in later boats, from the Romano-Celtic boats of the late Roman period (Arnold, 1977a) to the late-medieval cogs (Reinders, forthcoming). The Bremen cog and others in that tradition have specialised butterfly-shaped iron clamps (known as *sintel* in Dutch) to fasten lath and caulking to the planking (Ellmers, 1979; Reinders, 1979). In nineteenth-century Switzerland; the *nâcon* of L. Neuchâtel had a moss caulking under a lath fastened by iron clamps (Arnold, 1979b).

A wide range of organic materials has been used in recent times to caulk boat seams, for example, leaves in Oceania, cotton in India, bark in Borneo and moss in E. Siberia (Folkard, 1870: 256, 288, 309, 362). Moss was used in the Ferriby and Brigg boats and in a range of other European boats even as late as the fifteenth-century *Grace Dieu*. Boats of the early Classical tradition are said not to need caulking

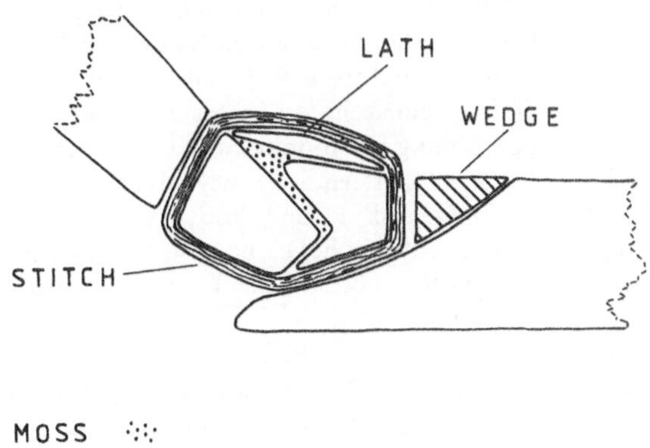

Figure 8.27 Method of fastening lowest side strake to outer bottom plank in the Ferriby boats. After Wright, 1976: fig. 3 (copyright E. V. Wright).

(Casson, 1971: 209; Throckmorton, 1973: 246), the seams being made watertight by good workmanship and the natural swelling of the timber when immersed. None was noticed during the initial examination or during recent work on the County Hall boat (Riley, 1912; Marsden, 1965), which although dated to the late third century AD was built in the earlier style if one may judge by the relatively close spacing of the fastenings (see below). In the later Classical period there was a guild of caulkers in the port of Rome (Casson, 1971: 209) and caulking has been found in a late fourth-century AD wreck from Ause Gerbal, Port Vendres (Liou, 1974b) and the non-edge fastened twelfth-century wreck from Serçe Liman (Steffey, 1982: 20) has a caulking of grass and pitch between keel and garboards. To complicate this simple time progression, however, it seems possible that there was caulking in the third-century BC Marsala (Punic) wreck (Frost, 1973: 42, 44, 49) and in the early second-century BC wreck Jeanne-Garde B from Porquerollo, France (Carrazé, 1977) – but it may have been paying.

The first-century BC Veneti of N.W. France are said to have used seaweed for caulking (Strabo, 4, 195) – E. V. Wright has asked whether this may have been moss. Excavated boats in the Romano-Celtic tradition have a variety of caulking materials – hazel twigs, twisted fibres and moss. Animal hair was generally used to caulk the clinker-built boats of medieval N.W. Europe, but moss has also been widely used. The re-used fragmentary, eighth/ninth-century planking from New Fresh Wharf, London has moss between oak clinker planking fastened by treenails (Marsden, 1979). The closest parallel in time and space for this combination of techniques is the twelfth-century Utrecht boat 1 (Philipsen, 1965) and the possibly eleventh-century Antwerp 7, rather than the eastern Baltic suggested by Fenwick (1978: 224) and Marsden (1979: 86). The eastern Baltic boats with moss caulking and treenailed clinker planking are not well dated but seem to be variants of the western Baltic Viking tradition (Crumlin-Pedersen, 1969: 24–7; 1983). A fifteenth-century wreck (W5) from the Bay of Gdansk, Poland, known as the 'copper wreck' because of her cargo, was hair caulked and clench nail fastened but the plank scarfs had moss inside them (Litwin, 1980).

Three early fourteenth-century finds from the Dutch polders had moss in their clinker seams held in place by laths fastened by iron clamps or treenails (Reinders, 1979) and the Bremen cog of 1380 was similar (Ellmers, 1979, Arnold, 1977a: 294). The clinker-built barge found in a former course of the River Rother beneath Matham Wharf, Kent in 1820 was also caulked with moss, although details were not recorded. Certain features of her structure and some of the objects found with her suggest that she was Dutch, probably of the sixteenth century (Fenwick, 1978: 258–260).

The eastern Baltic and the Rhine estuary thus seem to have been the two main regions where moss was used as caulking in the period AD 800 to say 1600, but it was also used in Britain. Prynne (1973: 231) noted that moss, possibly in caulking grooves, was found during the 1933 investigation of a fifteenth-century wreck in the River Hamble, Hampshire, probably that of the *Grace Dieu*; and the accounts of William Soper, Henry V's Keeper of Ships at Southampton around the time *Grace Dieu* was built, show that moss was indeed being used there for caulking, although oakum (teased-out rope) was also used (Rose, 1982: 73, 112). The use of moss persisted in Britain into the seventeenth century as evidenced by the small clinker built boat Llyn Perris 1 (Illsley and Roberts, 1980); this was excavated from a remote part of N. Wales and may well have been a relict of former use rather than a reflection of continuing mainstream usage.

WATERPROOFING

In addition to caulking seams with organic materials impregnated with vegetable tar, both seams and hull, especially the underwater parts, are often coated with a substance which protects the hull – skin, bark or wood – against biological attack by flora or fauna and also adds to its watertightness. In recent times bitumen has been used for paying skin, reed and plank boats in Arabia and sewn plank boats in S. California. Fish oil was also used for the upper planking and the insides of Arab boats, whilst a mixture of lime, fat and gum was, and is, used on the underwater planking (Hourani, 1963). Lime and oil is used in India (Folkard, 1870: 300) and China; clay, grease and oil for skin floats in S. America; while dung and oil and similar mixes waterproof the basket boats of Indo-China. In Portsmouth Dockyard mixtures of pitch, tar and brimstone or tallow sulphur and rosin were used in the mid-eighteenth century (Upham, 1978: 2). In twentieth-century Norway the mixture is wood tar, raw linseed oil and turpentine. Linseed oil is made from the linseed (flax) plant, whilst pitch, tar, turpentine and rosin (colophony) can all be produced from resinous trees such as the pine. Dry distillation of wood results in a *tar* distillate and a *pitch* residue. If tapped resin is distilled *turpentine* is produced with a *rosin* residue (Hodges, 1964: 165). Moll (1926: 368) has noted that pine roots have a greater tar content (16–20%) than stems (14%). Many of these raw materials were readily available in antiquity and we know that Classical ships were indeed waterproofed/payed with pitch (Pliny *NH* XVI 21, 23). The Nemi barges appear to have been payed with bitumen and the Yassi Ada seventh-century Byzantine wreck with pitch inboard and outboard below the waterline (Casson, 1971: 211). A late-fourth century AD wreck off C. Dramont not only had resin covering her hull, but she was carrying pine resin in Dressel 27 amphorae (Joncheray, 1977). The hull of the early-second century BC Jeanne-Gard wreck B was covered with an unidentified substance (Carrazé, 1977: 303) similar to that found in the third century BC Marsala wreck (Frost, 1973).

The thirteenth-century Norwegian who, in the *Konungs Skuggsja*, advising his son on the essential aspects of merchant shipping, instructed him to tar his ship in autumn (Ross and McLaughlin, 1977: 144–9). A paying of pine tar was found on the planking of the second-century AD New Guy's House boat, and Prynne (1973: 23), has recorded that it was found between the three layers of clinker planking of the fifteenth-century River Hamble wreck.

LEAK STOPPING

There is more excavated evidence for the waterproofing of individual fastenings rather than general waterproofing of the hull. The stitches in both Ferriby and Brigg were probably wedged within their holes although the evidence is not conclusive. The heads of the metal nails fastening the bottom planking to the floor timbers of the flat-bottomed first-century AD, Romano-Celtic boat from Bevaix, Switzerland, were surrounded by moss to pack the pre-bored holes (Arnold, 1975: 123). The projecting heads of the willow (*Alnus* sp) treenails used to fasten the ribs to the planking of the Graveney boat were tarred, and in some cases, possibly during repair, a woollen grommet had been inserted under the head before the nail was driven home (McKee, 1978: 99; Fenwick, 1978: 243). Three of the Graveney plank fastening nails had been 'hardened up' (McKee, 1978: 99) that is loose nails, indicated by a leak, had been driven home and reclenched.

PLANK FASTENINGS

There are three main forms of fastening material – thread, wood and metal. All three have been used on the two basic methods of shell building, edge to edge (Fig. 8.5) and clinker (Fig. 8.6). Hornell (1944: 9) considered that in Madagascar and other remote places sewing had been succeeded by wooden fastenings which in turn had been replaced by metal fastenings; Basch (1973: 336–7) has given further examples of similar sequences. In Europe, both in the Mediterranean (Casson, 1971: 9, 10) and in the north-west, there may have been a similar sequence, but with much overlap. In the Mediterranean there seems to have been an early pre-Classical shift from sewing to wooden plank fastening (Pomey, forthcoming) which persisted beyond the end of the Roman period, although the third-century BC Marsala boat had metal fastenings between garboard and keel (Frost, 1973: 42). Nevertheless, in certain regions, sewn boats continued to be used up to the eleventh century AD (Bonino, forthcoming).

In N.W. Europe there seems to have been a shift from sewing to metal fastenings sometime after the Bronze Age (although this sequence may merely reflect an inadequate and unrepresentative sample of finds) except for the eastern Baltic and the Rhine delta where some wooden fastenings were used in the medieval period. But sewing continued to be used in Scandinavia, Finland and Russia until recent times (Christensen, 1984; Forssell, forthcoming). The tenth/eleventh-century Skuldelev boats 1 and 5 had some treenail plank fastenings amongst mainly metal ones, as did the fifteenth-century 'copper' wreck from the Bay of Gdansk. Such mixed techniques are not unique, for the thirteenth-century Chinese boat from Quanzhou had her layers of planking (in places single, double or treble) fastened by wooden tenons and metal spikes and clamps (Keith and Buys, 1981: 124); and in E. Africa sewing and treenails are used in the same boat (Chittick, 1980).

It is necessary to distinguish between fastenings used to hold caulking in position and those with the principal function of fastening planking. The former may well link two adjacent planks as in the Bremen cog or the nineteenth-century L. Neuchâtel *naçon* (Arnold, 1977a and b) but they are not plank fastenings; indeed analytically the naçon is non-edge fastened, despite this apparent bonding between the planking. On the other hand, all forms of edge fastening, to a greater or lesser degree, help to retain caulking between the planks.

Positive fastenings are ones which are locked at both ends; they will not pull out, but may sheer. Continuous sewing or lashing with wedges may be considered as positive. In wood, the Classical draw tongue joint locked by treenails (Fig. 8.6), and the single treenail with a head at one end and a wedge at the other (Fig. 8.39) or two wedges, are also positive fastenings; as is the metal nail when its pointed end is clenched in one of several ways.

Sewn fastenings

There is worldwide evidence for the early use of sewn fastenings in plank boats. In Europe this technique was used in the pre-Classical Mediterranean (Casson, 1963), pre-Roman Scandinavia, Britain, and Yugoslavia (McGrail, 1981a: 17–19, 21–2). Sewn plank boats are also known to have been used in Arabia and E. Africa from the first century AD, Malaysia third/fourth century, India, Indonesia, Oceania, N. and S. America from the sixteenth century, Burma, Siam, Vietnam from the nineteenth century – in all cases first use could have been considerably earlier. Of the major land masses only in Australia does this technique seem not to have been used.

A wide variety of sewing techniques has been used in early boatbuilding. They vary from simple, single lashings between paired holes (the method used in the Ferriby boats – see Wright, 1976: 16), to sewing with a continuous thread, as for example in the Nin sewn boats (Fig. 8.28), or in the Ljubljana boat (Fig. 8.29) as published by Müllner (1892). The drawing by the latter implies that some form of double stitching was used as the space between every pair of holes is crossed diagonally by stitching. Continuous stitching can be (and has been) used with paired holes or with diagonally disposed holes, but single lashings would seem to be logically restricted to paired holes. Modern sewn boats, for example the East African *Mtepe* (Hornell, 1941), have sewing patterns more complex than that evidently used in the Ljubljana boat, with the thread passing two or more times through every hole.

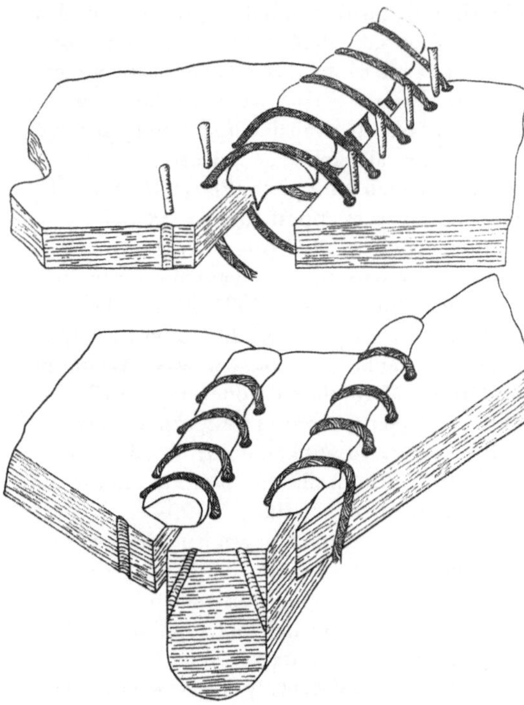

Figure 8.28 Method of fastening the planking of the Nin boat of the first to third centuries AD. After Brusić, 1968: fig. 8.

Ferriby Adjacent planks on the mid-second millennium BC Ferriby boats were interlocked at the edges in a complex joint which can only be described as 'modified tongue and groove' (Fig. 8.27). The planking was fastened together by single strands of yew (*Taxus* sp) stitches, 1 to 1.25 cm in diameter, between paired holes and was not continuous. The holes in boat 1 were generally square 3.2 × 3.2 cm with some long elliptical holes near one end where the lowest side strake meets the plank keel. They had an average spacing of 25–40 cm and were some 3 cm in from the plank edges. It has not yet proved possible to determine from which part of the tree the yew came, but the general internal structure of this wood makes it peculiarly adapted to storing energy (Gordon, 1978: 78). In recent experiments Wright (1984) found difficulty in selecting and preparing yew and threading it to make a tight lashing of $2\frac{1}{2}$ turns, as was found on the boats. Rackham has suggested that some form of forestry cultivation might have encouraged the growth of suitable shoots, but coppicing seems unlikely as, like most softwoods, yew coppices feebly.

The Ferriby boats had a longitudinal lath inboard between the moss caulking and the stitches. Wright (1984) has suggested that the planking was first stitched and wedged, then moss was forced into the seam, and finally the laths, *c.* 1.5 m in length, were forced under the stitches to tighten them. Apart from the doubtful practicality of such a technique and the possibility of disturbing the caulking and overstressing the stitches, there is ample ethnographic evidence that continuous stitching is tightened over lath and caulking as the final act before waterproofing with a tar or similar substance (Hornell, 1941a; Severin, 1982): there is thus a strong possibility that the same sequence was used with the Ferriby non-continuous stitching.

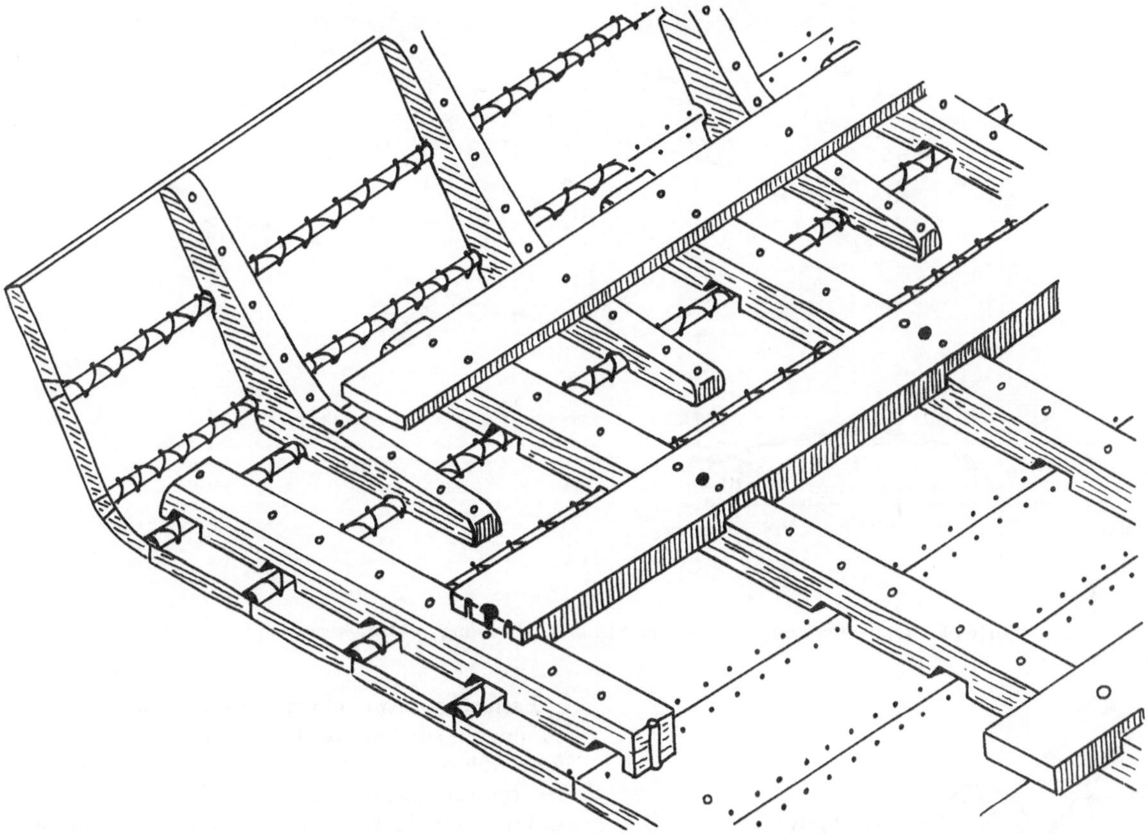

Figure 8.29 Constructional details of the Ljubljana boat of the second half of the first millennium BC. After Müllner, 1892. (Drawing: NMM Greenwich.)

Brigg The evidence from the Brigg 'raft' of the mid-first millennium BC is difficult to interpret as certain details were not recorded when the boat was first exposed in 1888, and much of the plank edges and the stitching were very degraded when re-excavated in 1974. The most likely interpretation is that the planking was sewn with single stitching with a two-stranded thread of split willow threaded 'continuously' through holes which were diagonally disposed to one another. Holes, *c.* 1 cm diameter, were bored almost vertically through the inner edges of planking, where the wood was at least 12 mm thick, some 20 mm from the edge and at an interval of 5.5 cm ± 0.7 (Fig. 8.30). On the outboard sides of the outer bottom planks the sewing holes are bored at an angle so that they emerge through the sides (McGrail, 1981b: fig. 4.1.19). Most holes were parallel sided, but some were hour-glass shaped, having been worked from top and bottom faces. Moss and longitudinal laths were used to make the seams watertight, with the laths inboard.

Hjortspring In this fourth century BC boat from Denmark, the upper strakes overlap the lower strakes by 2 cm in a bevelled lap (McKee No. 21 – see Fig. 8.31). The planks were fastened by lime-bast cord through small holes outside the lap, thus there was only one hole in each plank per stitch. The holes which were at 7 to 8 cm intervals were stopped with resin. The sewing details are mainly known from marks left in this resin and these are as compatible with single lashings as with continuous stitching (Prins, forthcoming).

Halsnøy The overlapping planking of this 2nd/3rd century AD Norwegian boat (Myrhe, 1980) was fastened by stitches within the lap – thus there were two holes *c.* 1 cm apart per stitch in each plank (Fig. 8.31). The holes are now only 2 to 3 mm in diameter and the interval between pairs of holes is *c.* 5 cm. Roots (Shetelig, 1903) or bast (Myrhe, 1980) were used for sewing thread and the caulking was a finely woven, reddish-brown cloth impregnated with tar or pitch (Magnus, 1980).

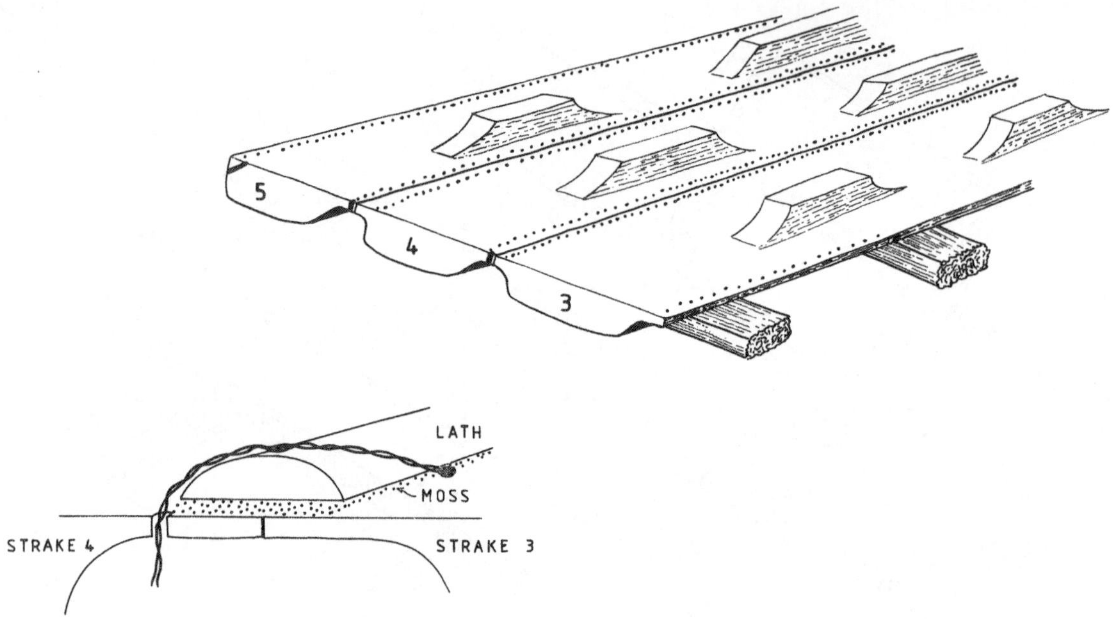

Figure 8.30 Method of fastening the Brigg 'raft' bottom planking. (Drawing: NMM Greenwich.)

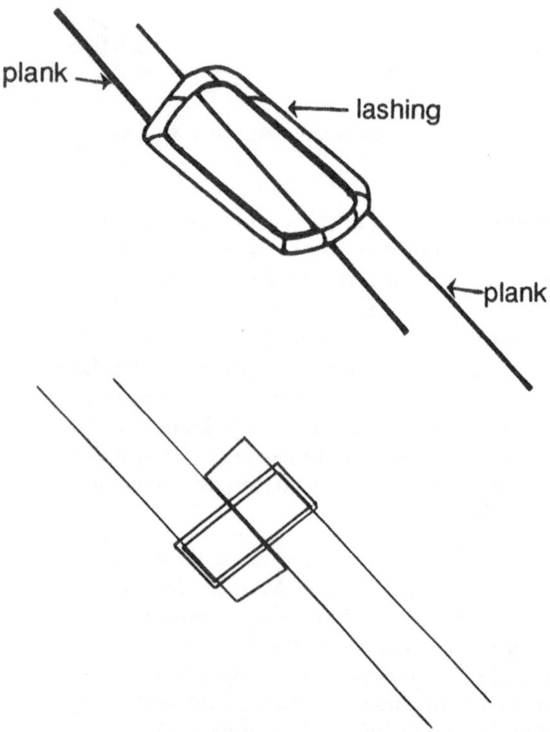

Figure 8.31 Plank fastening methods: upper, Hjortspring of *c.* 350 BC; lower, Halsnøy of *c.* 200 AD. (Drawing: NMM Greenwich.)

Fastening stitches Individual stitches and continuous stitches have to be tightened by some form of lever before they are (temporarily) wedged; see, for example, the Maori *tanekaha* (Best, 1925: 79) and the Cook Is. *keke* (Buck, 1927: fig. 112) and the lever used in the sewing of the E. African *mtepe* (Hornell, 1941: 61). Christensen (1984) has drawn attention to a seventeenth-century drawing of a Lapp boatbuilder, probably with a more flexible material than yew, using his mouth, hands and feet to tighten the stitching. In his experiments Wright (1984) used as a lever a copy of a wooden artifact found near Ferriby boat 2 in 1948.

In recent examples of sewn planking, wedges are sometimes left in position in each stitch hole, otherwise all but the final one of a sequence are removed in succession. Wooden wedges were not found with the stitches during post excavation examination of the Ferriby boats, and Wright found in his experimental work that 'once threaded through and pulled as tight as possible the withy showed very little tendency to loosen and may not have needed wedging': however, unwedged, non-continuous stitching would soon work once the boat was afloat.

Wedges also reduce leakage through the sewing holes and in modern sewn boats wedges or plugs are found in the underwater holes. They were also found in the sewn repair on the prehistoric Appleby logboat (McGrail, 1978a: 148). No wedges or traces of resin were recovered on re-excavation of the Brigg 'raft' but in view of the history of this find their use cannot be entirely ruled out.

Protection of stitches The stitching in a sewn boat is potentially vulnerable and boatbuilders ancient and modern have sought to protect it, especially the outboard portions. This is especially important for boats operating off foreshore or river bank where they regularly take the ground and may also be dragged across the strand. Two well-documented methods of protection are:

1. to countersink the stitches as in Maori boats (Best, 1925: 72); and
2. to wedge the stitches within the sewing holes and cut off the external section so that each stitch becomes in effect a staple or 'dog', as in the Mtepe (Hornell, 1941a: 61).

The Ferriby stitching does not protrude below the bottom plannking except where there are repairs (Wright, 1976, figs 3, 4, 5). The stitches in the Brigg 'raft' are generally 1.5 to 3 cm above the bottom of the planking due to the way the undersides of the planks are bevelled at the edges. The stitching fastening the first side strake to the bottom planking passes through the thickened side of the outer bottom planks (McGrail, 1981b: fig. 4.1.19).

Repairs: There is much ethnographic evidence for the annual or periodic dismantling of sewn boats for stitching to be renewed: Indian Ocean (Hornell, 1941: 62; Hornell, 1946: 236; Hourani, 1963: 94; Folkard, 1870: 309); Oceania (Folkard, 1870: 291; Malinowski, 1978: 141; Best, 1925: 221). Access to the underside of a large boat may be achieved by floating logs transversely underneath the craft and then letting the set-up dry out on an exceptionally high spring tide. Alternatively, the boat may be dismantled in shallow water and manhandled piecemeal. Sewing breaks experienced between major refits would have to be repaired by *ad hoc* means including much use of extra caulking.

Flat wooden tenons

The third-century AD County Hall boat's flush-laid planking was fastened by the insertion of free or loose tenons (called draw-tongues in the original report by Riley, 1912) into opposing mortices cut within the thickness of each plank; these tongues were then locked in position by treenails driven through them from one face of the planking to the other (Fig. 8.6). These mortices were cut at 6 in (15.2 cm) intervals in planks some 3 in (7.6 cm) thick; the tongues were 5 × 2½ × ¼ in (12.7 × 6.5 × 0.6 cm) and the treenails ⅝ in (1.6 cm) diameter. Such positive fastenings of great craftsmanship resulted in a strong waterproof joint, evidently not generally needing caulking.

The earliest form of this method of plank fastening (Fig. 8.32A) is to be found in the Egyption Dahshur boats of *c.* 2000 BC (radiocarbon dates range from BM-22 = 3530 ± 150 BP to UCR-126 = 3750 ± 110 BP). There are, however, no treenails to lock the tenons in position; thus, although the planking is located in the longitudinal direction and constrained from relative movement inboard or outboard, the tenons do not prevent the seams from opening; double-dovetail clamps were therefore used to constrain the planking in this direction (see below). The tenons in the Dahshur boat now in the Field Museum of Natural History, Chicago have not survived, but the mortices now measure *c.* 6 × 2 cm, whereas Bartlett and Hobbs (1897) recorded them as 3 to 4 in × ¾ in (7.6 to 10.2 cm × 1.9 cm) and 3 to 5 in deep (7.6 to 12.7 cm). They are spaced 20 to 25 cm apart in planking which is 8.89 to 12.38 cm thick. The mortices thus occupy 15 to 21% of the thickness of the planking.

Tenons locked by two treenails in each plank were used in some Vietnamese plank boats recorded in this century by Paris (1955) and Pietri (1949) – see also Basch (1972: 31); and boat fragments excavated in 1953 at Jahore Lama, Malaya, and considered to be eighteenth or nineteenth century in date, had similar fastenings with 3 × 1¼ in (7.6 × 3.2 cm) mortices locked by one treenail per plank, in addition to treenail plank fastenings (Sieveking, 1954: 229–230).

The only other evidence for plank fastenings by flat wooden tenon comes from the Mediterranean in the period fifth century BC to the seventh century AD (Bass, 1972; Throckmorton, 1973: 263; van Dornick, 1976: 112–13). Steffey (1984) has summarised the evidence for the fourth-century BC Kyrenia vessel as:

Planking	*Tenons (cm)*	*spacing (cm)*	*treenails (dia.)*
4 cm thick	16 × 4.5 × 0.5	10–12	1 to 0.6 cm
8 cm thick	20 × 8 × 1	10–13	1.5 to 0.8 cm

The mortices thus occupy *c.* 12.5% of plank thickness.

In the fourth century AD Yassi Ada 1 ship the

tenons are less well-fitting and spaced at *c.* 25 cm. By the seventh century AD, Yassi Ada 2 has tenon spacings varying from 30 to 90 cm and there were no locking treenails; more reliance was placed on other fastenings and on the internal framework.

Sleeswyk (1980) has shown from documentary evidence that the Romans also used this type of joint in the construction of the wooden disc in an oil press and that they were known to Cato as *coagmenta punicana* – Phoenician joints. The locked tenon may thus have been transmitted to the Roman world by the Phoenicians, or indirectly via the Etruscans as Basch (1981: 245–8) would prefer.

Treenails

Treenails may be used either across the plank thickness to fasten overlapping strakes or within the thickness of the planking (hidden or obliquely) to fasten flush laid strakes (McKee, 1976, figs 16, 25, 29). Salaman (1967: 50) has described how treenails were fashioned in the eighteenth century from radially split timber and worked to a round cross section with a moot (rounding plane). They were made overlong and cut to size after being driven. Fenwick (1978: 185, fig. 7.9) has drawn attention to the method used by twentieth-century Portuguese builders of the *xavega* who produce similarly faceted treenails with heads using a knife and wooden disc pressed against the chest, and this may have been the medieval method.

Hidden treenails (Fig. 8.32B) are unknown in Europe as plank fastenings but they have been so used in recent times in Indonesia (Folkard, 1870: 256, 265, 273; Horridge, 1981: 1982). Vietnam (Basch, 1973: 337), Madagascar (Hornell, 1944, fig. 6B), Malaysia (Gibson-Hill, 1954: 162, fig. 7), India (Basch, 1972: 31), Somalia (Chittick, 1980: 302). Oblique ones (the ends of which can be seen in the faces of the planking) have recently been used in S. Arabia (Bowen, 1952: 207, fig. 8B), Lamu (Hornell, 1941: fig. 3) and Korea (Basch, 1973: 337). As with flat mortices which are not locked, treenails by themselves will not hold the seams together under sea conditions. Thus their use as plank fastenings is combined with the use of sewing (as in the Somalia *bedgen* and the Lamu *mtepe*); or the treenails are locked by smaller treenails driven through them (Horridge, 1982: 14) as in seventeenth-century Philippines and twentieth-century Madura and Bali; or the planks are compressed and held together against torsional and bending forces either by lashings between transverse fittings locked against lugs on the planking (Horridge, 1982: fig. 2A) or by tensioning a flexible rib downwards to lie across the planking in a series of lugs on each plank (Horridge, 1982: fig. 2b) (Fig. 8.33).

Treenails were used to fasten overlapping strakes in E. Norway until recently (Christensen, 1979: 14) and some clinker boat fragments recovered from the shore at W. Mersea in 1980 and dated to the eighteenth/nineteenth century have the garboards treenailed to the keel, with supplementary spikes, although the remaining clinker strakes were clenchnailed (Dean, forthcoming).

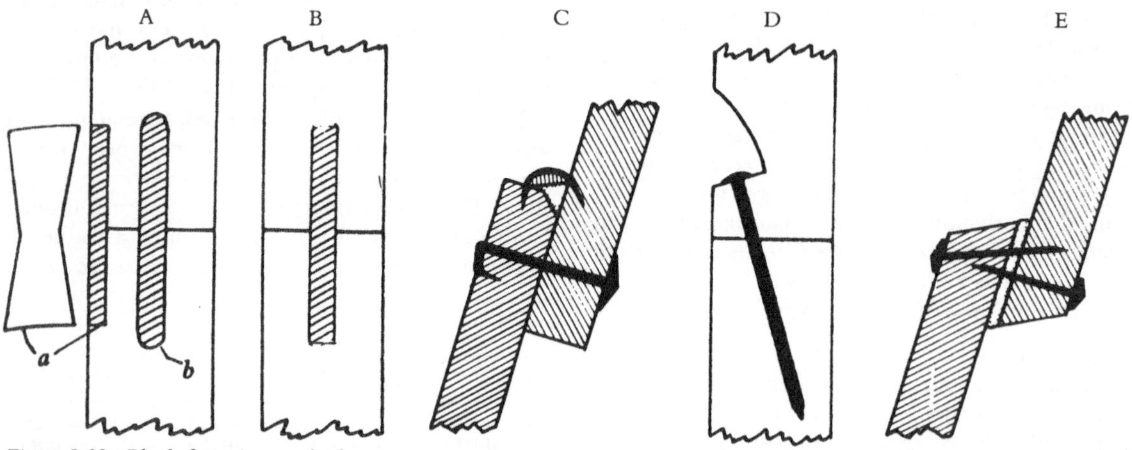

Figure 8.32 Plank fastening methods:
A. Tenons (b) and double-dovetail clamps (a) in the Dahshur boats of *c.* 2000 BC.
B. Treenails in twentieth-century Indian boats.
C. Hooked nails in the Bremen Cog of AD 1380. The metal clamp between the strakes holds in caulking.
D. Obliquely-driven spikes in the Romano-Celtic logboat Pommeroeul 2.
E. Angled spikes on Antwerp boats 1 to 4.

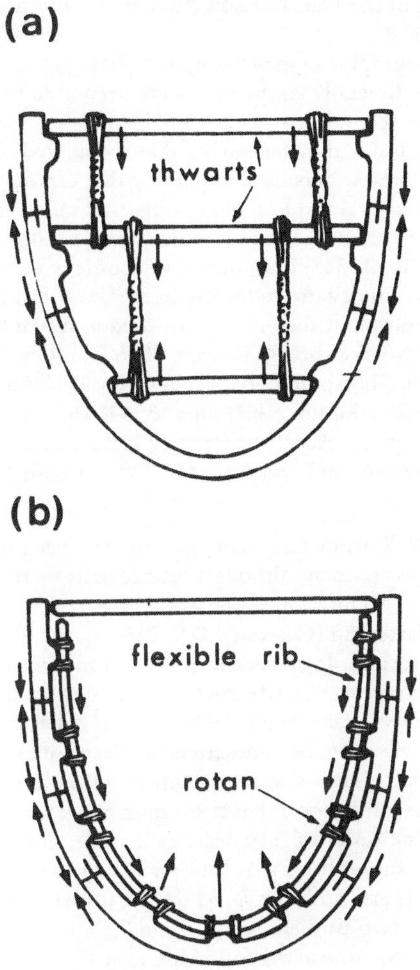

Figure 8.33 Methods of locking treenail-fastened planking:
(a) by lashings between transverse timbers;
(b) by lashing a flexible frame (rib) to the planking. After Horridge, 1982. (Drawing: NMM Greenwich.)

Treenail fastenings have also been used in several medieval wrecks with clinker planking where the principal fastening method was iron nails. Skuldelev 1 has them in part of the 12th strake aft, at an interval of 34–42 cm (more than twice that of the nail fastenings), whilst the after part of Skuldelev 5's top strake is fastened to the 6th strake and to a longitudinal stringer by spikes (see below) and treenails at 45–60 cm pitch. Gdansk Bay wreck W5 (also known as 'copper wreck') has garboards fastened to the keel by 2–2.5 cm diameter treenails, except at the stern (Litwin, 1980: 221). Other Scandinavian finds partly treenail fastened include Hedeby, Schuby, Ralswick (Crumlin-Pedersen,

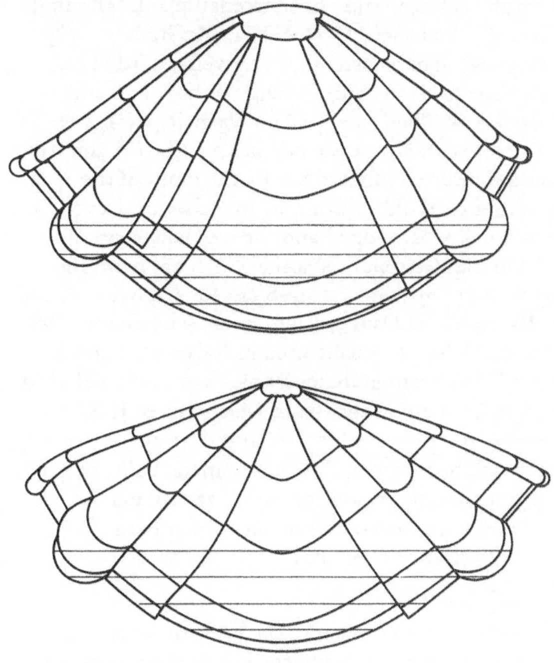

Figure 8.34 The bow (above) and stern (below) of Utrecht boat 1.

1981), Kalmar 1 and Sjøvollen (Christensen, 1968: 140).

Utrecht 1 boat (Philipson, 1965) which is from the twelfth century and Antwerp boat 7, possibly of the eleventh century (Ellmers, 1972) are partly treenailed. Both are extended logboats and Utrecht 1 has three overlapping washstrakes fastened by treenails except at the ends where they are iron nailed to small transoms (Fig. 8.34). The lower washstrakes on Antwerp 7 were similarly fastened. The flat-bottomed Utrecht 2, found in 1974 and provisionally dated to the late-twelfth century has the lowest side strakes treenailed to the transition strakes (*Gemeente Utrecht Archeologische Kronick*, 1974–5: 31–5).

A unique find, so far, in Britain is the eighth/ninth-century clinker oak planking excavated at New Fresh Wharf, London which was fastened by treenails with wedges inboard (Marsden, 1979: fig. 5.2). These New Fresh Wharf treenails are similar in appearance to the *c.* 1 cm diameter treenails used to plug old nail holes in some of the Wood Quay planking before new fastening holes were bored for metal fastenings. Arnold has also reported 1 cm square treenails similarly used in the Bevaix boat (1975: 125–6).

A significant number of Viking Age boatfinds

from the eastern Baltic have treenail plank fastenings (Crumlin-Pedersen, 1969: 1983a; McGrail, 1981a: 34). These fastenings are well recorded on only four finds to date, Danzig-Ohra 1, 2, 3 and Ralswick 4. The Danzig-Ohra treenails are 1.2 cm in diameter with a cone-shaped head outboard and oak wedges inserted inboard across the grain of the planking to avoid splitting it. In Ralswick 4 every 3rd or 4th nail was wedged and the treenails were of 1.2 cm diameter and set some 1–1.5 cm from the plank edge. Spacing was 6–8 cm for Ralswick 4, and 8–10 cm in the Danzig-Ohra boats (Herrmann, 1981; Lienau, 1934). Treenails used in Ralswick 1 and 2 were 1.2–1.4 cm in diameter, 1.2 cm in the Lebafelde find and 1.5 cm in that from Mechlinken. It is noteworthy that plank scarfs in the treenail-fastened Ralswick boats were fastened by metal nails; this may suggest that the reason for using treenail plank fastenings was an economic one rather than technological or structural.

Wooden clamps

Wooden fastenings set into the faces of flush-laid planking were used in ancient Europe but there is less evidence for them than for the other wood fastenings. The few that have been recorded are double-headed, flat, wooden keys, generally of a dovetail shape (Fig. 8.32A), although the earliest one from N.W. Europe, which was used to close a split in the Appleby logboat of the late second millennium BC, is T-shaped (Fig. 6.7). A split in the ninth-century BC Short Ferry logboat has been similarly held together by a single dovetail shaped clamp. Neither of these clamps was locked and so must have relied for their fastening on an interference fit and possibly on differential swelling, that is they were dried before insertion.

Double dovetail-shaped clamps, also not locked, were used to fasten the planking on the 2000 BC Dahshur boats, supplementing the unlocked tenons (see above). They were used, locked by treenails, to close splits on the undated logboats from Vaaler Moor, Schleswig and one now in the National Museum, Dublin (McGrail, 1978a: 37, fig. 162). Details of the Dahshur and Appleby clamps are listed in Table 8.4.

Iconographic evidence suggests that double-dovetail clamps may have been used in the boats of second-century BC India (McGrail, 1981, pl. 33), but it may be sewing that is depicted. Bowen (1956b) and others have suggested that certain post-medieval models, engravings and reliefs may show representations of wooden clamps, although as Basch (1973: 341) has pointed out some may be metal clamps. The evidence for use in medieval Europe is thus slight. But they have been used in recent times on Swiss Lakes boats (Bowen, 1956b) and in several of the flat-bottomed boat types of the R. Adour, France (Beaudouin, 1970: figs 13, 14, 16). The dovetail clamps on the Adour boats were locked by two or four treenails and were spaced *c.* 30 cm apart (Fig. 8.35).

Nails

In N.W. Europe only nails of iron have been found in excavated remains although bronze nails were used in Classical Mediterranean boatbuilding, but not for plank fastening (Casson, 1971: 207).

The terminology currently used in maritime archaeology to describe metal fastenings is not standardised, and objects which could collectively be called nails have been described as *rivets*, *spikes*, *bolts* and *tacks*. Usage is seldom defined but there appears to be an implication that these may be placed in ascending order of size: tack, nail, spike, bolt. There are also suggestions that *bolts* may not be tapered but that nails are, and that *spikes* do not protrude right through two timbers but are 'blind'.

I propose to use the following terms:

Nail. a single arm metal-fastening tapered along its length or pointed.

Clenched nail. The protruding point of a nail is deformed, often over a rove (metal washer) to form what is effectively a second head: this constitutes a positive fastening (Fig. 8.5). Found throughout Nordic boatbuilding, the earliest ones seem to be those associated with third and fourth-century Norwegian boat-houses (Myrhe, 1980).

Table 8.4 Details of prehistoric wooden planking clamps

Boat	*Overall size*	*(L/B)*	*Single or double*	*Locked*	*Splay angle*	*Spacing*	*Clamp thickness/ plank thickness*
Dahshur	19 × 6 cm	3.17	double	no	15°	Pt 75–83 cm St 80–100 cm	0.625
Appleby	18 × 8 cm	2.25	double	no	180°	—	0.317

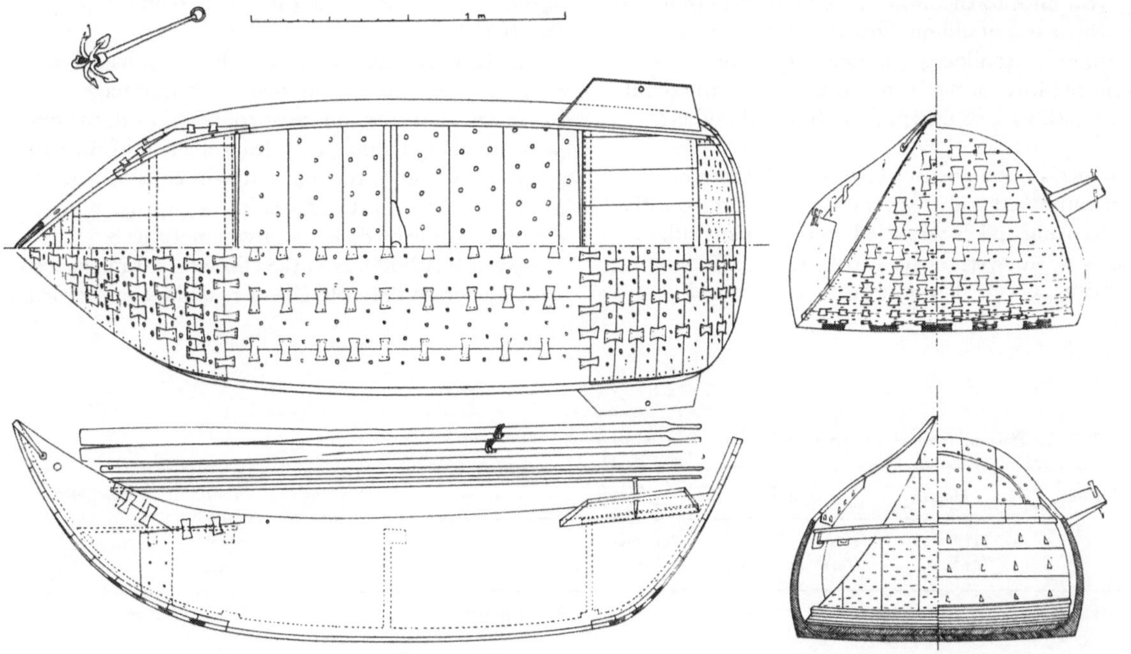

Figure 8.35 Reconstruction drawing of an eighteenth-century *tilhole* of the River Adour. After Beaudouin, 1970 (courtesy Musée de la Batellerie, Conflans–Sainte–Honorine).

Turned nail. The protruding point of a nail is turned through *c* 90° to lie flat along the plank face (McKee, type 20 – see Greenhill, 1976: fig. 26) – a positive fastening. Medieval boat Antwerp 7 had this type of fastening (Ellmers, 1972)

Hooked nail. (Fig. 8.32C). The protruding point of a nail is turned through 180° (sometimes over a rove) to re-enter the planking – a positive fastening (McKee 18 and 19). Found in boats and ships of the cog tradition (Ellmers, 1979; Reinders, forthcoming); hooked nails and caulking clamps of butterfly shape appear to be diagnostic features.

Spike. (or blind nail). A nail which does not penetrate right through the second member to be fastened – not a positive fastening and therefore relies on being held tightly by the wood it displaces, as an interference fit such as a square-section spike in a round hole. Blind nails are generally found in northern and western European boat finds only where a through-fastening (clenched, turned or hooked) cannot be used in the current state of technology: for example, fastening garboards to keel or plank ends to stems where the second element is of such dimensions and cross section that available lengths of nail cannot transpierce. In the Skuldelev finds spikes are also used as plank fastenings where they hold a third timber, a longitudinal stringer.

Bolt. A single arm metal fastening of cylindrical form, without taper, which penetrates through the second timber. The fastenings at the top of the Sutton Hoo 2 ribs (Evans and Bruce-Mitford, 1976: fig. 279) are not 'rib bolts' because they are tapered; by my definitions they are 'nails with circular domed heads of 2.4 cm diameter, round shank of *c*. 0.9 cm diameter and 16.4 cm in length, clenched over a square rove of 3.7 cm side'. Bolts as defined here have not been reported in northern European finds, but in several Mediterranean wrecks (for example Yassi Ada) bolts fastening ribs to planking or keel are reported to have been secured by a metal key passing through a slot near the end of the bolt's shank (Casson, 1971: 207; Liou, 1974a; van Doorninck, 1975: 124; Steffey, 1982: 20–1).

The six types of nail defined above may be used to fasten frames to planking as well as plank fastenings.

Nailed fastenings. Nails of one type or another may be used to fasten both clinker and flush-laid planking. Spikes driven obliquely into a notch in one strake and through into the adjacent strake are used to fasten flush-laid planking in India (Hornell, 1930: fig. 1), Madagascar (Hornell, 1944: fig. 6A), Nubia and Sudan (Hornell, 1946: 193), Chinese-style boatbuilding in Malaya (Gibson-Hill, 1954: fig. 7c), Vietnam, Japan and S. Russia (Basch, 1973), and in

Italy (Bonino, forthcoming). There is one instance of the similar use of obliquely-driven spikes in early European boatbuilding: the elements of the composite bow of the Romano-Celtic Pommeroeul 2 (Fig. 8.32D) are so fastened (de Boe and Hubert, 1977).

As this is not a positive fastening, the method is often combined with a second type. A 'semi-positive' way of using spikes was used on the clinker side strakes of Antwerp boats 1 to 4 (Fig. 8.32E) where alternate nails are driven in from inboard and outboard (Ellmers, 1972: cat. 35a to d and figs 182–3).

The boat nails in northern finds which have been examined are individually made wrought nails hammered hot to a square or round section, the head possibly being formed by hitting the end of the nail against a mould or die. Clench nails constitute the great majority of plank fastenings in northern finds; details of some of them are given in Table 8.5.

Recent boatbuilders in the clinker tradition spaced nails by eye and Harold Kimber, builder of the *faering*

Table 8.5 Details of iron clenched nails from selected medieval boats with clinker planking

Boat	*Holes*		*Shank*		*Heads*		*Roves*		*Spacing (cm)*	*From plank edge (cm)*	*Reference*
	size (cm)	*shape*	*size (cm)*	*shape*	*diameter (cm)*	*shape*	*shape*	*size*			
Nydam 2	—	—	—	—	—	—	square	—	15–16 garboard 16–18 upper	4?	Shetelig, 1930
Sutton Hoo 2	—	—	0.8	round	2.4	round	square	2.4?	15?	—	Evans and Bruce-Mitford 1975
Graveney	0.8	round (t/n)	0.4	rectangular and many faceted	2.5	round	rect.	2.8 × 2.2 × 0.2	15 ± 3		McKee, 1978
Oseberg	0.8	round	—	—	2.5 to 3	round	rect.	2.5 × 3	garboard 19.69 ± 2.68 10th 32.75 ± 5.09	1.5 to 2	Note 1
Gokstad		—	—	—	2.5	round	square	3 × 3	garboard 16.33 ± 2.24 7th 17.36 ± 1.74	1.5 to 2	Note 1
Skuldelev 1		—	—	round	—	—	square	—	16–20	—	Olsen and Crumlin-Pedersen, 1967
2		—	—	—			square	—	12–15	—	
3		—	0.7 to 0.9	round	—	—	rect.	2 × 2.5	12–16	—	
5		—	—	—	—	—	rect.	—	15–20	—	
6		—	0.6 to 0.7	round	—	—	round	—	15–20	—	
Blackfriars 3	1	rect?	0.8 to 1.1	rect.	3.2 to 3.6	round	rect.	4	*c.* 17	—	Note 1
Wood Quay 282	1	rect?	1	round	6	round	square	3.2 × 3.2	*c.* 15	0.7?	Note 1
WQ.TG9–305	1	—	1	square	3.5	round	rect.	3 × 3.5	*c.* 16	2.5 to 3	Note 1
WQ.TG9–306		—	1	square	3–3.5	round	rect.	3 × 3.5	*c.* 17	—	
Eltang		—	—	—	—	—	—	—	12–18	—	Crumlin-Pedersen p.c.
Sjøvollen		—	—	—	—	—	—	—	13–21		Christensen, 1968

Note: Gillian Hutchinson measured representative nail spacings in Oseberg, Gokstad and Blackfriars 3. The Wood Quay spacings were measured by Tony Farrell, Ciaren Campbell and Gillian Hutchinson. I am most grateful to them.

replica (McGrail, 1974) could get a pitch of 3 in in this manner which had a tolerance of ± 0.1 in when subsequently measured. In the original *faering* (Gokstad 3) and in Gokstad 1 a moulding or pattern of lines had been scratched along the upper inboard and lower outboard edges of planking the original purpose of which may have been as a guide for boring nail holes. A nail line was scribed along the lower inner edge of certain Graveney planks (McKee, 1978: 101) in the middle of the lap. On the surviving planks of the Graveney boat nail spacings have a modal value of 16.5 cm, a mean of 15 ± 3 cm, and a range of 2–27 cm (McKee, 1978: fig. 3.2.28). The tail to the left of this distribution indicates that there were many occasions when the builder used a short spacing; the few spacings well above the mean value may be due to ancient error or possibly holes were not found on examination – quite easily done. There is often a closer nail spacing on the lower strakes than on those above the waterline – see data on Nydam 2, Oseberg and Gokstad 1 in Table 8.5. In Nydam 2 (and probably in other boats) the nails were closer at bow and stern (averaging 14–15 cm but as low as 11–12 cm) than elsewhere (15–18 cm) (Shetelig, 1930).

Planking nails driven through pre-inserted treenails have been found in one Wood Quay plank where it is thought to be an individual repair, and in several positions in the Graveney boat where it is considered to have been used generally (Fenwick, 1978: 224). However, the possibility that treenails were inserted during a refastening of a section of the Graveney boat should not be discounted. Jammed fragments of broken stitching in the Brigg 'raft' had to be bored out before resewing and certain holes were enlarged during this boring (McGrail, 1981b: 242). If this happened during a refit of an iron fastened boat such as Graveney the enlarged holes could be plugged with a treenail. On the other hand, McKee (1978: 98–9) recorded that there was no direct sign of such a re-bore.

Whereas the Graveney nails had quasi-rectangular cross-sections, the Sutton Hoo and Skuldelev finds had nail shanks of round cross-section *c.* 0.6–0.8 cm diameter. At Wood Quay both types of cross section have been found, but within specific groups of planking only one type was noted: e.g. TG 7 planks have round nail shanks of *c.* 1 cm diameter, planks in TG 9 have square nails of *c.* 1 cm side. The one plank from Blackfriars 3 that has received detailed examination had rectangular nails sided 0.8–1.1 cm. There is more uniformity in nailheads recorded: they are round and of diameters ranging from *c.* 2.5–3.5 cm. Roves are all (near) quadrilaterals. Graveney's roves were 2 mm thick but this statistic does not seem to have been published for other finds.

Rosenquist (1979) has found that representative clenched nails from Gokstad 1 each weigh 15–20 gm. An outfit of 3500 to 5000 nails gives a weight of 50 to 80 kg for the plank fastenings, which Rosenquist points out corresponds to at least four of the largest 'pigs' of iron known from the Viking Age. A *faering* replica clenched nail weighs *c.* 3 gm. With *c.* 510 nails in the faering the total weight of fastenings is *c.* 1.53 kg. The boat weighs *c.* 89 kg (McKee, 1974: 6, 26) and thus plank fastenings represent *c.* 2% of the total and a significant weight of nails to make for this small, 6.5 m, oared boat.

Metal clamps

Metal clamps or hasps may be used for both overlapping and flush-laid planking. They may be used 'blind' i.e. not penetrating through the second timber (McKee, 27) or they may penetrate and be turned to lie along the planking as in N. China (McKee, 28) or in recent Swedish church boats (Eskeröd, 1973: fig. 99) – these latter clamps are bent from a thin rectangular section bar (11.5 × 0.5 × 0.1 cm), the longer arms measure 5.2 cm and the shorter 3.7 cm, both are pointed, but in opposite directions. Metal clamps have also been used recently in Bangladesh (Greenhill, 1976: 52, figs 14, 16, 17) to fasten rabetted lap (McKee, 5) planking, inset in slots 0.3 cm deep, at an interval of 5 cm.

Basch (1973: 341, fig. 7e) considers that certain medieval iconographic evidence may show metal clamps. However, of the boats he cites as having been fastened by this means Antwerp 1–4 flat-bottomed boats have clamps primarily for holding caulking. Antwerp 7, an extended logboat, also had clamps for caulking but the planking was fastened by turned metal nails and treenails.

HOW TO OBTAIN TRANSVERSE STRENGTH?

The transverse elements in a boat built in the shell sequence serve two main functions:

1. Reinforce the shape-holding qualities of the planking – in certain designs this transverse reinforcement may also play a part in preventing the seams from opening. By thus stiffening the hull transversely and holding the sides in a fixed relationship to one another, the hull's resistance to

longitudinal stresses due to hogging and sagging is increased (Ch. 3).
2. Distribute local loadings and thus disperse stress concentrations.

The main components of transverse strength are *frames* which lie against the inboard faces of the planking and *crossbeams* which run from side to side across the boat. Thus while most frames are generally unobtrusive, crossbeams can seriously interfere with free movement about the boat and can eat into the space available to store cargo. Frames are generally found at fairly regular intervals along the length of the boat, as are the lower beams – those at or below the level of the bilge. In cargo vessels higher beams may only be found at the ends of the boat with one beam for the mast in an otherwise clear central area. In an oared boat, however, upper beams may be used as rowing or paddling thwarts and thus be fitted throughout the length. In certain designs frames and crossbeams are interconnected and supplement the strengthening functions of each other: this is a particularly important feature of the Norse tradition (Fig. 8.36).

Crossbeams would be particularly space-intrusive in a flat-bottomed boat, especially those with low sides. Thus they are not generally found in the Romano-Celtic 'barges' or in twentieth-century inland waterway boats such as the Somerset turf boat (McKee, 1983: figs 84 and 86): these boats do, however, have frames.

Conversely, examples of frameless boats are known, both ancient and modern:

c 2000 BC	Egyptian Dahshur boats (de Morgan, 1895)
1st century AD	craft sailing between India and Ceylon (Strabo, 15.1.15)
20th century AD	the Bangladesh *binekata* (Greenhill, 1971: 89)
	the S. Indian *masula* (Hornell, 1946: 236)
	the R. Nile *nuggar* (Hornell, 1939: 40)
	the Omani *beden seyad* (Bowen, 1956: 286)

All except the craft mentioned by Strabo are known to have crossbeams and it seems highly probable that they also had them. It has sometimes been claimed (e.g. Phillips-Burt, 1979: 45) that satisfactory hulls of some size and seagoing potential could not be made without frames. The *nuggar* and the *binekata* are, and doubtless the Dahshur boats were, rivercraft, but the *beden seyad* and the *masula* are surf boats used in testing conditions. *Masula* sewn boats have frequently attracted favourable comment on their seaworthiness; in the 1904 Somali campaign the Royal Navy used native-manned *masulas* to

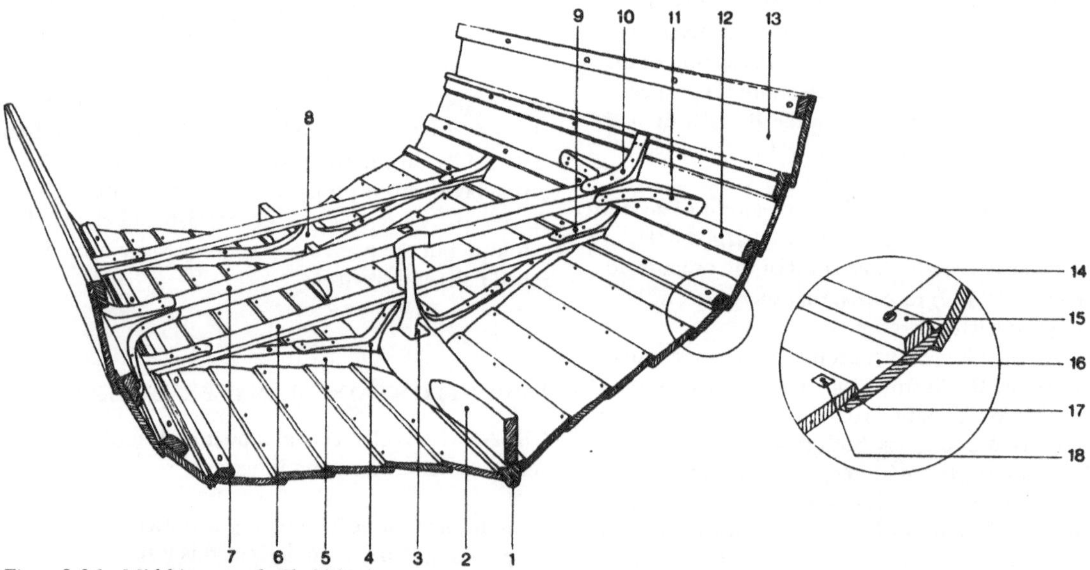

Figure 8.36 Midship part of a Skuldelev boat.
1. Keel. 2. Keelson. 3. Mast step. 4. Knee. 5. Frame. 6. Lower crossbeam. 7. Upper crossbeam. 8. *Snelle*. 9. and 10. Standing knees. 11. Lodging knee. 12. Stringer. 13. Top or sheer strake. 14. Treenail. 15. Stringer. 16. 5th strake 17. Clinker overlap (land). 18. Iron nail clenched over a square rove. After Olsen and Crumlin-Pedersen, 1967, fig. 15 (copyright Ole Crumlin-Pedersen).

Figure 8.37 Part of the remains of Ferriby boat 1 in 1947 showing the transverse timbers passing through cleats in the bottom planking. (Photo: NMM Greenwich.)

disembark men, equipment and even camels from their ships which remained to seaward outside the surf zone – it was found that the sewn plank *masula* 'stood the knocking about better than the man-of-war boats' (Brassey, 1905: 60–1).

The prehistoric British boats from Ferriby and Brigg, built of substantial planks, have transverse timbers linking the bottom planking through integral cleats (Fig. 8.37), but these timbers cannot easily be classified as frames as their primary function is not to reinforce the planking but to:

(a) align the planking during building and during rebuilding after periodic dismantling.
(b) reinforce the plank fastenings against vertical and horizontal movements and thus minimise stitch breakages.
(c) a possible role of holding a dunnage of light timbers or brushwood in the bottom of the boat thereby not only protecting the inboard stitching from accidental damage but also minimising the adverse free surface effects of bilge water (McGrail, 1981b: 236–7, figs 4.1.14 and 4.1.20).

It seems probable that the Brigg 'raft' had no crossbeams, but knees between bottom and sides might have been necessary, although it has been possible to build a 1:10 reconstruction model based on the excavated evidence, without them. An alternative is that movable hurdles were extended across the boat to pen animals, but also giving transverse support to the hull. No crossbeams or similar timbers were found with any of the Ferriby boats, and the indirect evidence is minimal, however they have been incorporated in reconstruction drawings and models (Coates, 1977: figs 16.2, 3, 4) at a level which would allow them to be used by paddlers. The earliest direct evidence in northern and western Europe for both frames and crossbeams is on the *c.* 350 BC Hjortspring boat from Denmark. After this date the surviving evidence suggests that frames and beams were fitted to all round-hull boats.

Although the function and effect of beams and frames cannot be entirely separated and there are timbers intermediate between frame and beam, discussion is clearer if the two elements are treated separately.

FRAMES

A *frame* may be defined as a transverse reinforcement of the bottom and sides of the hull, set against the inboard face of the planking at a particular station. They are generally set vertical to the keel, but near the ends where such an orientation would make them lie obliquely to the run of the planking, they may be inclined so that they lie more nearly across the planking they are intended to support – they are then known as *cant* frames and may be recognised, if separated from the boat, by the asymmetry of the cross section of those parts which cross the centre line. Similarly, the upper parts of frames near the ends of the boat, whether cant or vertical will have an asymmetric cross-section to match the decreasing breadth of the boat.

Examination of the frames may show the number and breadths (without overlap in a clinker boat) of the strakes they span, the angle between the garboard strakes at a particular station (also obtainable from the keel cross sections – see above) and the method and sequence of fastening frames to planking.

Characteristics

From the large number of attributes it is possible to record about a frame the following have been chosen subjectively as a means of describing frames and framing systems analytically and noting significant differences and similarities.

Attribute	*States*
Composition:	single; composite (see below)
Curved form:	bent; grown; sawn/hewn
Span:	topstrake to topstrake; bilge to bilge

The bilge in a round hulled boat is a region where the bottom meets the sides rather than a precise point. It is possible that other states could be added to the span attribute as more evidence becomes available, for example 'waterline to waterline' although this would depend on an agreed definition of 'waterline' (Ch. 11). An attribute measuring asymmetry may also be needed, but at present all frames seem to be symmetrical in that they reach out or up to approximately the same level either side of the keel, although constituent parts of these frames may not individually be symmetrically disposed about the keel.

Composite frames may consist of combinations of the following elements:

Floor timber: spans bottom of boat, i.e. keel and lower strakes. The angle of the arms to the horizontal is a measure of the deadrise of the bottom of the boat. McKee (1983: 79, 81) has proposed the following definitions:

$< 7°$ = flat floor
7°–11° = raised floor
$> 11°$ = Vee floor

Side timber: spans strakes between bilge and top-strake; may be sub-divided into:

(a) *linked*: when they are part of a frame

(b) *interframe*: when they lie between frames.

Futtock: an intermediate timber between floor timber and side timber; not so far identified on pre-thirteenth century finds all of which are relatively small craft.

Asymmetric timber: spans bottom and one side, may be L-shaped, or unequal-arm U or V-shaped.

Half frame: spans one side and half the bottom to the keel.

Knee: spans part bottom and part side in a flat-bottomed boat; may be *linked* or *interframe* (see above).

Frames may be described using these attribute states, for example:

Hjortspring frames are: single; bent; top
Nydam 2 frames are: single; grown; top
Romano-Celtic frames are generally: composite; (L-shaped asymmetric timber with linked side timber); grown; top (Fig. 8.38)
Skuldelev frames are generally: single; grown; bilge; supplemented by interframe side timbers.

Most of the Graveney frames are analytically the same as those of the Romano-Celtic boats, but in his reconstruction drawing, (McKee 1978: fig. 10.1.2) shows two frames at each end which are single; grown; top. Boats such as the Somerset turf boat (McKee, 1983: fig. 84) do not fit so readily into this analytical scheme but her framing system may be described as floor timbers (*rungs*) with interframe knees.

The Hjortspring framing of hazel branches is the only example of frames bent to shape in early N.W. Europe and thus the only ones prestressed to reinforce the plank fastenings by clamping the planks together (Fig. 8.40). Nydam 2 with 19 V- or U-shaped grown frames must have been the most

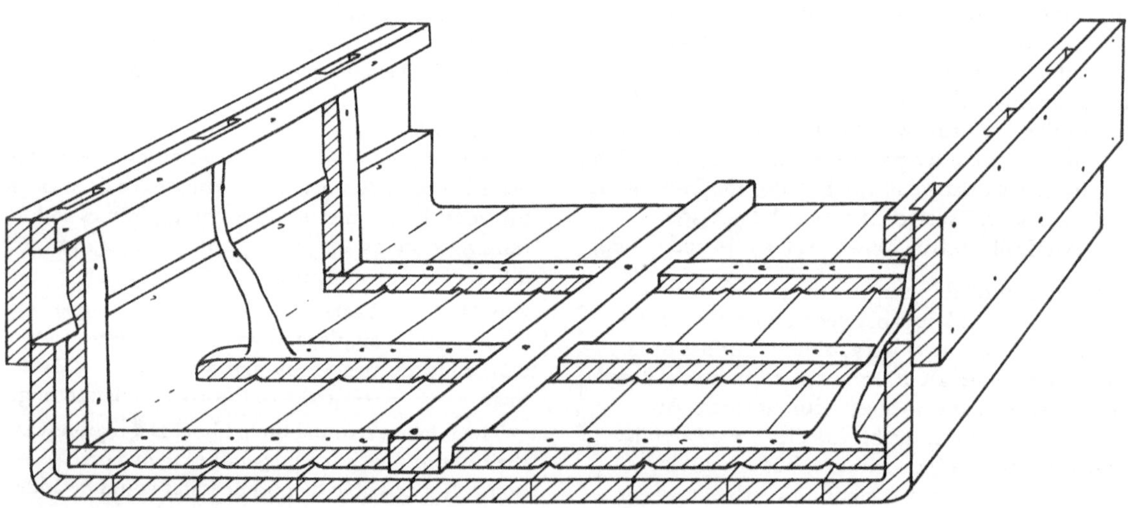

Figure 8.38 Diagram to illustrate the framing of Zwammerdam boat 2. After de Weerd and Haalebos, 1973 (reproduced by courtesy of the Universiteit van Amsterdam, A. E. van Giffen Instituut voor Prae-en Protohistorie (copyright holders)).

demanding in timber selection as each timber had to have a double curve to match the transverse section of the boat from topstrake to topstrake. The evidence at Sutton Hoo does not clarify whether boat 2 had single-piece or composite frames.

There are at least five other aspects of frames or framing systems which should be recorded:

1. ***Joggled or not.*** When joggled the underside of the frame is fashioned in a series of steps so that it fits flush against clinker planking. Frames to fit reverse clinker planking are known, e.g. that from Kastrup, Denmark dated to the thirteenth century (Crumlin-Pedersen, 1981: 46).
2. ***Limber holes.*** Unjoggled frames in a clinker-built boat allow bilge water to move freely fore and aft to a particular low point, usually aft, where it may be bailed or pumped out. Limber holes are cut in joggled frames near their lowest point to permit water flow.
3. ***Fastenings.*** Individual elements of one frame may be fastened to each other or they may just be positioned end to end or overlapping. In some of the Romano-Celtic boats for example, the side timbers are morticed into the L-shaped asymmetric timbers.

Frames are not necessarily fastened to every strake they span, and in the Viking boatbuilding tradition no find has yet been recorded with frames fastened to the keel – a significant point in establishing the building sequence. In Skuldelev boats 1, 2 and 3 the frames are also not fastened to the garboards; the Graveney frames are generally only fastened to alternate strakes (McKee, 1978: 99).

The mid-third millennium BC Egyptian Cheops boats, the oldest plank boat so far found, has 12 frames which are lashed to the planking. Wood and metal fastenings were the norm in Classical boatbuilding (see above), but there is at least one example of frames fastened by sewing: two additional frames were added to the second-century BC wreck Jeanne-Garde B, alongside existing frames fastened by iron nail through treenails; vegetable fibre rope was used to link the new frames to the planking, with the stitching countersunk within the exposed faces of the planking and frame, and the sewing being wedged within the holes with small treenails (Carrazé, 1977: 302).

Ropes were also used in the northern world to lash frames to planking. In the Hjortspring, Nydam, Oseberg and Gokstad 1 vessels (fourth century BC to ninth-century AD) the frames are lashed to cleats upstanding and integral with the planking.

On the other hand the sixth century AD Gretstedbro fragment and the small *faering* which is contemporary with Gokstad 1 had treenail frame/plank fastenings. The frames on Sutton Hoo 2 were similarly fastened but with the topmost fastening being a clenched nail. Graveney frames were fastened by treenails with irregularly-faceted shafts as is common on medieval wrecks, and with round domed heads outboards and oak wedges inboard (Fig. 8.39).

Treenails were principally used in the Skuldelev finds but additional iron nails were used in boats 5 and 6. The Skuldelev treenails were commonly of willow (*Salix* sp) and *c.* 2.5 cm (one inch) in diameter, as were those of Graveney. The treenails used in Skuldelev 2 had conical heads as did many of those found on the Wood Quay timbers, but

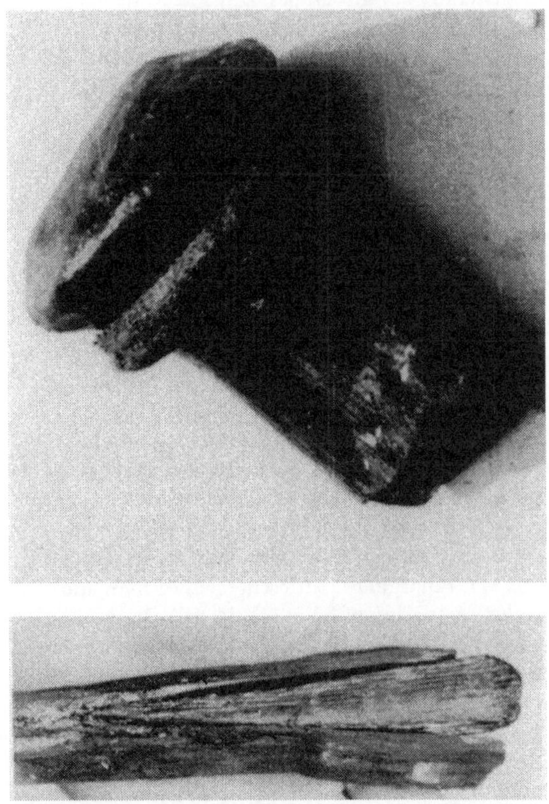

Figure 8.39 Graveney willow treenails: (upper) head with woollen grommet; (lower) shaft with oak wedge. (Photo: NMM Greenwich.)

none were as pronounced as those on the Graveney treenails. Treenail fastenings continued in general use throughout the medieval period. They were used to fasten frames to the 11th century logboat from Stanley Ferry (McGrail, 1981c); and bottom and clinker side planking was fastened to the late fourteenth-century Bremen cog frames by treenail (Ellmers, 1979).

A significant feature of the Romano-Celtic tradition is that the frames are fastened to the planking by iron nails, the emerging points of which are generally hooked, that is turned back into the inboard face of the frame. Bronze spikes, turned nails (Carrazé, 1977: 302; Casson, 1971: 207) or hooked nails were a common way of fastening frames to the planking in the Classical tradition; in many cases these nails are driven through a treenail already fastening frame and plank (Casson, 1971: 207), and this combination is also found on the Romano-Celtic Blackfriars 1 and Bruges finds.

4. ***Frame spacing.*** Although the majority of frames lie generally at right angles to the keel they are often found not to retain this orthogonality along their length but rather wind across the planking according to the natural curve of the timbers (McKee, 1978; Evans and Bruce-Mitford, 1975: 369). Frames are generally fitted at closer intervals towards the ends of a boat, especially at the stern when a quarter rudder is fitted, but over the central part of a boat they are found at near regular spacing; thus there is a range of variation of only 10 to 20 cm in the intervals measured in the Skuldelev finds. The average intervals for boats 1, 3, 5 and 6 range between 91 and 94 cm, whilst Sutton Hoo 2 averages *c.* 91 cm (Evans and Bruce-Mitford, 1975: 369) and the spacing in Nydam 2 is just over 1 m. Other frame spacings are given in Table 11.5.

McKee (1983, fig. 113) estimates that a 30 in (76.2 cm) spacing is the minimum for rowing; this leaves no margin for different sizes of men and styles. A spacing of 36 in (91.44 cm) allows some variation of stroke. From Table 11.5 it can be seen that frame spacings in Nydam 2, Sutton Hoo 2, Oseberg, Gokstad, Tune, Skuldelev 1, 3, 5 and 6 are in McKee's second group. On the other hand in Skuldelev 2 where the average interval was a mere 72 cm only a standard short stroke would have been practicable unless there was only one man on each bench on alternate sides of the boat. As Skuldelev 2 appears to have an underwater shape with good sailing qualities (see above in section on keels), it may be that this method of rowing with only half the complement of oars was sometimes operationally acceptable.

The Graveney boat frame spacing of *c.* 50 cm is altogether too close for pull/sit styles of rowing unless there was only one pair of oarsmen at each end or they were at alternate frames. On the other hand oars or sweeps may have been used at the ends in the push/stand mode (see Ch. 12). This close spacing of such relatively massive timbers strongly suggests that the Graveney boat was designed to take dense cargo in her hold (Corlett, 1978: 303).

An alternative to frame space measurements is useful when assessing boats with irregularly spaced or paired frames. For example, two of the Romano-Celtic boats with single frames have an average of 2.5 timbers per metre of length, whereas two others with paired timbers of similar scantlings have 1.6: this difference may suggest boats designed to carry cargoes of different densities.

5. ***Size and shape.*** The frames in Skuldelev 1, 2 and 6 have a near square cross-section amidships, whereas those in Skuldelev 3 and 5 and the Graveney boat are two to three times as deep (*moulding*) as they are broad (*siding*). Each Skuldelev frame is less deep at the sides of the boat than amidships. The frames in boat 6 also taper in breadth towards their ends, but the frames of boats 1, 2, 3 and 5 have a variable breadth with a broader stretch on either side approximately midway between amidships and their outboard ends.

The Graveney frames taper in depth and in breadth similar to those in Skuldelev 6, but their scantlings are generally much greater than any Skuldelev frames, probably indicating a difference in the bulk density of the cargoes to be carried. In the Skuldelev finds the frames generally become deeper towards the ends of the boat, perhaps influenced by the sharpness of crook required to match tapering transverse sections. The midship frames of the Graveney boat, on the other hand, are relatively deeper than those at the ends, for example frames 1 (near the stern) to 6 (near amidships) have moulded depths of 19, 18, 22, 25, 25, 25 cm; and 7 to 9 measure 25, 22, 20 cm; this pattern may reflect the requirement for extra strength in the midship region of the Graveney boat.

BEAMS

Beams hold the sides of a boat at a fixed distance apart and thereby, in conjunction with the other transverse elements, increase the boat's resistance to hogging and sagging stresses (Ch. 3). Without cross beams or similar continuous transverse members such as fastened deck planking, a boat is an open shell, as are many examples of low-sided, flat-bottomed boats. Beams may duplicate as rowing benches and they may also be used to support a mast. They sometimes rest on, or are notched to take on to, other timbers such as the frameheads or the longitudinal stringers (beam shelves), or they may be lashed to the stringers as has been proposed for the Graveney boat reconstruction (McKee, 1978: 300). They may also be fastened to the planking by knees (grown crooks with arms at near right angles), both horizontal (lodging) and vertical (hanging) knees. These knees bracket the beam and thus resist racking stresses (see Ch. 3) and they disperse local loadings through the hull.

The earliest boat with beams in northern Europe is the Hjortspring boat and she has the combination of beam over frame linked by vertical struts which became commonplace in the Viking Age. She has, in fact, two sets of beams, one at lower strake level and one at upper strake level which doubled as paddling benches. Frames and beams are interlocked at the ends and are also connected to each other by two vertical stanchions (struts). This complex interlocking system (Fig. 8.40) suggests that each frame and its associated lower beam and two vertical timbers must have been assembled together before the frame was lashed to the planking. Nydam 2 has a comparable though less complex arrangement and with only one set of beams which is at top strake level. The method of fastening these beams/thwarts to the hull is not clear (Engelhardt, 1865: 34) but they may have simply been morticed to fit over the frame heads (1865: fig. II.6) and the two vertical timbers.

No evidence survives for the Sutton Hoo 2 beams, and the reconstruction drawing does not show any (1977: fig. 325), however, there is a possibility that at least some of them were removed before the burial chamber was built amidships.

In the Skuldelev finds beams are invariably associated with frames, sometimes at two or three levels. Olsen and Crumlin-Pedersen (1967: 95) call those at the lowest level *bite*, but analytically they are similar to other crossbeams. These lower beams rest on, or in boat 1 are notched to take, the frameheads; in boat 3 they rest on a stringer. The Skuldelev beams are generally fastened in position by knees, and in boat 5 some lower beams have an integral knee, where natural, right-angled crooks have been available. There are vertical stanchions linking frames and lower crossbeams in boats 2, 3 and 5 (Fig. 8.36).

Beams which protrude through the boat's sides are known from Egyptian third millennium BC reliefs (Casson, 1971: figs 14 and 18) and they were used on certain Mediterranean vessels right up to late-medieval times (Nance, 1911: 1913). In Classical ships sometimes all beams protruded and sometimes only three, near amidships and near the ends (Casson, 1971: 210). In northern Europe protruding beams have been found in the *c.* twelfth-century Falsterbo 7, and Kalmar 1 wrecks and in the fifteenth-century wreck W5 from Gdansk Bay, and some of the thirteenth/fourteenth-century Dutch finds have them.

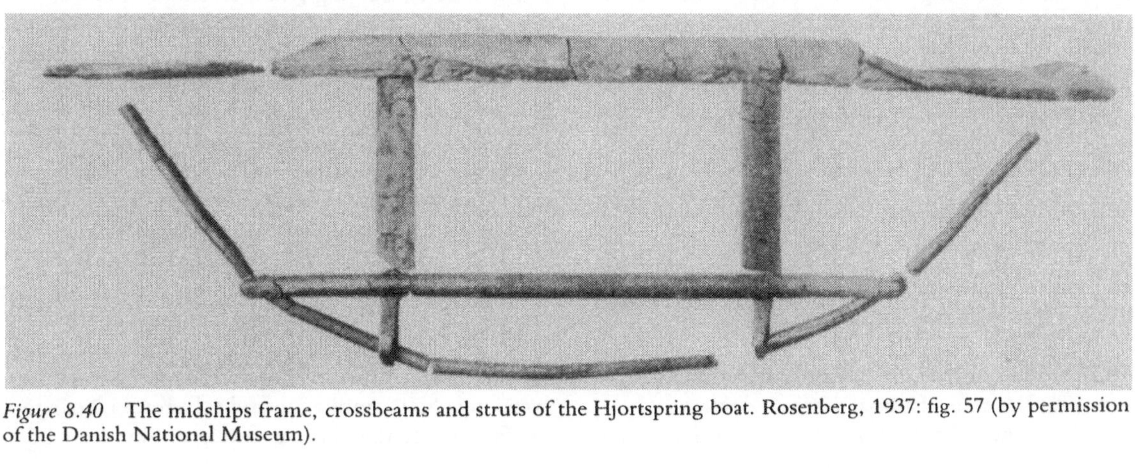

Figure 8.40 The midships frame, crossbeams and struts of the Hjortspring boat. Rosenberg, 1937: fig. 57 (by permission of the Danish National Museum).

Figure 8.41 A bulkhead with knees and a stem from Skuldelev 3. (Photo: Ole Crumlin-Pedersen.)

Figure 8.42 A bulkhead from Wood Quay, Dublin. (Photo: National Museum of Ireland.)

In the late fourteenth-century Bremen Cog four out of five beams protrude through the ship's side (Ellmers, 1979: fig. 1.6). This outwardly distinctive feature seems to be an attempt to obtain a firm seating for the beam ends by notching them over the strakes.

BULKHEADS

A bulkhead of inverted (apex down) triangular shape (Fig. 8.41) was found with Skuldelev 3, and one has been inferred for Skuldelev 2. The boat 3 bulkhead extended from the top strake down to the 6th strake and was nailed through its thickness to strakes 7 and 8; it was also fastened to the boat's sides via the stringers by two horizontal knees each side. A bulkhead from Wood Quay (WQT 351) is of similar shape and dimensions and has a similar hole 8.5 cm in diameter in a central position (Fig. 8.42). This bulkhead is of oak and has been fashioned from timber in close proximity to a fork. Another timber from Wood Quay (WQT 367) has a form intermediate between that of a frame and a bulkhead and has been fashioned from a natural fork. Similar 'dwarf' bulkheads are found at the ends of the Gokstad *faering*. (Fig. 8.43).

Bulkheads combine the functions of frame and beam near the ends of a boat where it could be difficult to find a crook with the necessary acute angle to match the boat's transverse section. They strengthen the planking, hold the sides in a fixed relationship and reinforce the fastenings of the planking ends to the stems. In the Gokstad *faering* the after bulkhead was pierced by holes to which a rope holding the rudder against its boss was fastened; the forward bulkhead had handholds which could be used when manhandling the boat on the foreshore.

THE NORDIC TRADITION OF FRAMING (Fig. 8.36)

The Skuldelev framing arrangements and that of many, if not all boats, in this Nordic tradition may be analytically separated into two parts: a combination of frame, lower crossbeam and knee reinforce the bottom of the boat out to and beyond the bilge; and above this upper crossbeams, knees and stringers strengthen and hold the sides in position. Interframe side timbers further reinforce the upper strakes, but only in the case of Skuldelev 2 is there direct evidence of any overlap in the strakes spanned by the frames and those spanned by the interframe timbers. Thus in the case of boats 1 (Olsen and Crumlin-Pedersen, 1967: 103) and 3 (1967: 125/6) the bottom and the topside are only directly connected by the plank fastenings between the 5th and 6th strakes. Stringers are fastened to the upper planking of Skuldelev boat 1

Figure 8.43 The four-oared boat (*faering*) from Gokstad. (Photo: courtesy the University Museum of National Antiquities, Oslo, Norway (copyright holders).)

(externally), boat 2, boat 3 (four) and boat 5 (two), and are presumed to have been fitted to boat 6; they are interlocked with the frame and beam system, with or without fastenings, and thereby add to transverse as well as longitudinal strength. Wales, internal and external are functionally similar to stringers and are found in other traditions (see above section on longitudinal strength).

This northern European system of frames, beams, knees, stringers and struts, with each individual timber of relatively light scantlings, reinforced 'the planking shell without sacrificing the resilience of the hull in any way' (Olsen and Crumlin-Pedersen, 1967: 124). In general, the upper internal support system is lighter than the lower, thereby keeping the centre of gravity lower.

WHICH TOOLS WERE USED?

Boats have been built in recent times with a comparatively simple tool kit: the Chumash Indians of California built seagoing sewn plank boats with no other tools than flints and whalebone wedges (Durham, 1960: 61; Heizer, 1966); and boats in parts of Oceania were built with stone and bone tools (Greenhill, 1976: 27–8; Johnstone, 1980: 206–9). Tools have been made from materials other than the traditional stone, bronze and iron of European archaeology: the Nootka Indians made boring tools from bird bones (Lane-Fox, 1875: 403); and shells have been used in the axe, adze and scraping roles in Polynesia (Hornell, 1948: 46); Melanesia (Lewis, 1932: 146) and America (Harriot, 1590: 55; Lane-Fox, 1875: 403–4). Furthermore, as Heal (1978) has emphasised, metal and some stone tools require hafts of wood or bone.

In many ethnographic reports of non-metal or stone tools being used for woodworking the timber species involved is not specified; where species are identified they are almost always timbers which are more readily worked than oak. Nevertheless the possibility remains of such use in prehistoric northern and western Europe, especially in working green timber, for even oak is relatively easy to work when freshly felled (Ch. 4).

EXAMINATION OF TIMBERS

Worked timber has survived from the Neolithic Age, and recent experiments have shown that flint and stone tools can be used to fell, convert and fashion oak (Orme and Coles, 1983). From the Bronze Age onwards more woodwork evidence is available, including planked boats, and from all periods and places in northern and western Europe woodworking tools have been excavated. Yet, until relatively recently, few detailed studies had been made of the tools and techniques used in either prehistoric or medieval woodworking. Heal's (1981) evaluation of the Brigg 'raft's' wood technology, Olsen and Crumlin-Pedersen's (1967) preliminary work on the Skuldelev finds, and Christensen's (1982) more generalised paper on the evidence for Viking Age boatbuilding tools and techniques are pioneering studies. Orme and Cole's (1983) publication of their research into wood from the Somerset Levels should lead to useful knowledge of Neolithic, Bronze and Iron Age technology.

As Heal (1981) has pointed out, each operation in the sequence from tree felling to finished artifact obliterates or deforms much of the evidence for the preceding work so that at first sight evidence for only the last, possibly cosmetic, surface treatment appears to be present and even this may be (partly) removed by wear and weathering during use. Nevertheless, much can be learned about tool use by close examination of secondary features such as holes, protrusions, joints, bevels, rabbets, grooves and fastenings for signs which indicate which earlier operations – cutting, splitting, scraping, smoothing, boring, joining, hitting, grasping, marking – had been used on the timber. From this it may be possible to deduce the sort of tool used and possibly the specific tool type.

From the Roman period and later many tools have shapes and constituent parts generally similar to modern ones, and it is not unreasonable to use twentieth-century names for them and to infer that their functions were similar. In earlier periods, however, there is no such straightforward 1:1 correspondence – although earlier archaeologists by giving modern names to classes of tools may have thought so. Some of these names disguise the multi-function of certain tools, for example, some so-called *axes* may also be hafted and used as adzes, and certain socketed tools may be used with a pushing motion (chisel) instead of, or in addition to, a chopping motion (axe or adze) – see McGrail 1977b. Microwear studies as undertaken by Keeley (1980) can show what materials tools have been used on and also the mode of operation. Thus function may in future be attributed to prehistoric tools with more certainty.

Toolmarks were not described in the published account of the Hjortspring boat and as the boat is currently being re-conserved it is not at present possible to say whether any had survived. Numerous toolmarks were visible on the Ferriby boat timbers, but few were measurable or identifiable (Wright, 1976: 21). However, Wright has suggested that a series of 2.5 cm wide square-ended marks on boat 1 may be attributed to known flint tools used as adzes; and a 7.6 cm broad, gently curved toolmark on boat 2 may have been made by a 'flat' or 'flanged' Early Bronze Age tool used either as an axe or adze. No toolmarks on the Brigg 'raft' could be identified after re-excavation in 1974, but by an analysis of the shapes, sizes and angles of the component parts of the structure in conjunction with the tree ring and ray studies, Heal (1981) has been able to identify the woodworking operations and to suggest the sort of tools of stone, bronze or wood which were probably used. For such studies of boatbuilding techniques useful comparative evidence is available from research into the woodworking techniques used in the manufacture of other contemporary wooden artifacts – see McGrail (1982), Coles (1975–84) and Orme and Coles (1983).

ICONOGRAPHIC AND DOCUMENTARY EVIDENCE

The late eleventh-century Bayeux tapestry shows boatbuilders using light felling axes to lop tree branches, T-shaped axes to square-off a balk and to give a final trim to an already fastened strake, and a small hand axe also used to trim planking (Fig. 8.44). Both breast auger and T-handled auger are shown in use to bore holes. Christensen (1982: 329) has suggested that an ill-defined tool in the right hand of one of the Bayeux boatwrights may be a shave (1982: fig. 18.2) used to finish a plank, but the possibility should not be overlooked that this may be an early form of bevel gauge (Christensen, 1972b: figs 8. 11; Salaman, 1967: 27–8, pl. 5) used to check that the correct angle has been worked along the top edge of a plank so that the next strake may be fitted at the required angle (see above, section on *Shape*). Illustrations in manuscripts throughout the medieval period add to our knowledge of tool shape and use –

Figure 8.44 Boatbuilding scene from the Bayeux tapestry (Photo: Photographie Giraudon (courtesy the Musée de l'Evêché, Bayeux, and with the particular permission of the town of Bayeux).)

as for example figured by Olsen and Crumlin-Pedersen (1967: fig. 63) where a hafted wedge is being used to split a log.

Tools are also mentioned in medieval documents but those in inventories or financial accounts add little to our knowledge. There are exceptions, however, Christensen (1982: 329) has drawn attention to a passage in Snorri Sturlason's description of the building of Olaf Trygvasson's late tenth-century *Long Serpent* in which, contrary to modern practice outside W. Norway, an axe is used for the exacting job of finishing the sheerline by trimming the top strake. Recent practice in western Norway is for the axe to be held at the top of the handle close to the blade when undertaking such precise work.

BOATBUILDING TOOLS

Using all forms of evidence at present available an attempt is made below to identify the sort of tools which would have been used in the prehistoric and medieval periods to build a plank boat. In addition rope making tools such as hooks and marline spikes, and the sailmaker's needle and palm would have been needed.

Preparation and felling of the tree

Wind-uprooted trees may become available, possibly at long range when carried by rivers or even across seas as driftwood. Otherwise, trees must first be prepared for felling by lopping off branches using stone or metal blades in the axe mode to cut through the wood fibres. Experimental work, as, for example, reported by Orme and Coles (1983), has shown that stone and bronze tools as well as iron can be used for this sort of work, even on hardwoods such as oak.

Neolithic tools have been used experimentally to fell trees up to 60 cm in diameter (1.88 m girth) and Orme and Coles (1983: 21) have claimed that such tools would be adequate for felling oaks of 1 m in diameter (3.14 m girth). However, there is no direct evidence for this latter statement and Startin (1976: 11) has doubted whether oak greater than 1.88 m in girth could be felled by stone tools.

The 3.14 m girth oak mentioned by Orme and Coles is not exceptional: the logs from which the late Bronze Age Brigg 'raft' was built were *c.* 2.70 m–3.03 m girth at breast height, and the parent logs for the Bronze Age Brigg logboat and the Iron Age Poole logboat had girths of *c.* 5.90 m and 5.40 m (see Table 4.1). It may be that stone tools were indeed used to fell such giants, perhaps using unorthodox techniques. The Maoris felled the *totara* or *kauri*, the toughest of pines, by cutting into the base of the tree with ballista-powered or swing battering ram stone tools (Best, 1925: 50; Johnstone, 1980: 206, fig. 15.3). On the other hand there is much ethnographic evidence for the controlled use of fire (Harriot, 1590: 55; Beverley, 1722: 198; Lane-Fox, 1875: 403–4; Johnstone, 1980: 206). The Somerset Levels trees may have been felled in this manner, as may the trees used to build the Brigg 'raft' (Heal, 1981: 254). Coles (1979: 104) has suggested that large trees may have been felled by first ringing (debarking) them and then felling them after death; this could have made it easier to prepare the tree for felling and reduced the labour involved in removing the crown, but it is difficult to see how this would have made felling easier, especially as living trees are generally less resistant to tools than are dead ones.

In medieval times large heavy axe blades became available for felling and it is likely that felling by burning declined.

After felling, the crown and remaining limbs have to be removed – here again controlled burning would have been possible and there are ethnographic analogies. Manoeuvring logs to the position where they are to be converted, possibly only turning the log into a more accessible orientation, might seem to be difficult, but here again the Maoris evolved effective techniques within their stone age technology (Best, 1925; Johnstone, 1980: fig. 15.4).

Conversion of logs

Oak logs were split by wedge, mallet and lever until pitsaws and sawmills became common in the later middle ages. Wedges of seasoned oak are known from the Neolithic and wooden mallets from the Mesolithic (Heal, 1981: 254). Metal wedges are known from the Anglo-Saxon grave 233 at Sarre, Kent (Wilson, 1968: 150, fig. 2f), and in the Viking Age set of tools from Mästermyr, Gotland (Foote and Wilson, 1970: 180, fig. 19). Hafted wedges are also known from Viking Scandinavia (Crumlin-Pedersen, 1968: 10) and from medieval France (Olsen and Crumlin-Pedersen, 1967: fig. 63). Bronze age tools thought to be flat axes but with butts showing signs of heavy pounding (Heal, 1981: 254) may have been used as hafted wedges or as hammers to drive wedges. Mallets of wood are known from mesolithic times (Heal, 1981: 254); wedges may also have been driven by the butt ends of stone or metal blades – an adze blade from Thetford with a much hammered butt may have been used in this manner rather than in place of a wedge as suggested by Wilson (1968: 143).

Primary shaping

This may have been done, like conversion, on or near the felling site to minimise transport problems, by removing bark and sapwood (and pith, where applicable) and cutting away surplus timber to produce something near the final shape required. Such tasks may be undertaken by stone or metal blades used as axes when a log or part log is to be squared, or used as adzes when a curve or hollow is to be worked. Wedge and mallet may also be used to split off timber between notches cut across the fibres by axes. One of the boatwrights on the Bayeux tapestry is using a T-shaped axe to dress a board – such axes have been excavated from Hurbuck and Crayke (Wilson, 1968: 144, fig. 1) and Milk Street, London (Fig. 8.45). However, during experimental work Crumlin-Pedersen found that replicas of the broad T-shaped adze from the Viking Age Mästermyr find (Arwidsson and Berg, 1983: fig. 12, fig. 26–63) were also excellent for dressing boards after splitting (Christensen, 1982: 334).

Timber-handling equipment

Logs and timber baulks may be moved from forest to building site by relatively simple equipment of wedges, levers, ropes and rollers (see, for example, Best, 1925). On site, additional timber handling equipment such as pinch bars, sheerlegs and tackles have been used recently in cultures with simple technologies; their early use in northern and western Europe have not, however, been documented, although medieval and even earlier use seems likely.

Mauls, hammers and wedges would be required on site to set up the structures of baulks, chocks, and stakes which form the stocks on which the boat is to be built – see for example McKee's (1978: fig. 10.1.4) theoretical design for the Graveney boat or Hasslöf (1972: figs 8, 9) who depicts twentieth-century Scandinavian practice. Sledge hammers were found at Mästermyr (Arwidsson and Berg, 1983: fig 6, 20–69, 21–70, 21–71).

Planks may more easily be bent after they have been made malleable by controlled burning or by steam. Certain species may be bent without that treatment – see the use of rope, wedges and levers in twentieth-century Germany (Fig. 8.10). Planks also have to be clamped together firmly until they are fully fastened: simple V-shaped cramps adjusted and locked by lever (McGrail, 1974: figs 21, 22) or by wedge (Christensen, 1982: fig. 18.4) are in use today, and small V-shaped cramps of horn are known from Viking Age contexts (Simonsen, 1953: fig. 3c) and from medieval Gdansk and Bergen (Christensen, 1982: 334). Other bowsing-in devices are based on the tourniquet or Spanish windlass principle – see for example the method of holding together opposite strakes in the bows of the *faering* replica (McGrail, 1974: fig. 25). Horridge (1982: fig. 7) has recorded the use of a simple device of logs, ropes and wedges to force edge-to-edge planking together in the twentieth-century Island of Bangka, and pointed out how this matches the seventeenth-century description of techniques used in the Philippines.

Figure 8.45 A T-shaped axe from an eleventh-century site in Milk Street, London. (Photo: Museum of London.)

Measuring and marking

It may be – as discussed in the section above on *Shape* – that in certain periods and places in antiquity no tools were used to measure and mark timbers, everything being done by eye, by direct comparison or by use of natural units of length such as thumbs, feet, etc. (Drillis, 1963). On the other hand, tools similar to the post-medieval boat ells, boatbuilding levels and bevel gauges may have been used. Battens to test fair curves and alignments and as straight edges would also have been useful, as would plumb-bobs and set-square or try-squares: these are simple devices to make, but no evidence for such use has been noted until late in the medieval period and clearly their function can be undertaken by the experienced eye.

Evidence for marking off has been found on certain medieval boats: knives or scribing tools were used to incise lines along some of the Graveney planking to mark the inside limit of the overlap and also the nail line (McKee, 1978: 101). Markings have also survived on one frame from the Skuldelev finds where the centre line has been scribed and marks made where limber holes were to be cut and where the frames were to be notched (joggled) to lie against the overlapping strakes; some scarfs were also marked off by knifepoint (Olsen and Crumlin-Pedersen, 1967: 161). Knives are known from tenth to thirteenth-century Dublin (Cat: 162/164/165) and from Mästermyr (Arwidsson and Berg, 1983: figs 26–39, 29–40).

Profile mouldings or decorative lines have been cut along the faces of some medieval planks, near their edges; they are worked, for example, along the upper inboard and lower outboard edges of each strake in the Gokstad *faering* (McGrail, 1974: fig. 26). These are decorative but may also have served to indicate nail lines and overlaps. These marks were cut with a tool known as a moulding or profile shave or scraper which was pulled along the surface of the wood. Examples have been found in Viking Age graves (Christensen, 1982: 330, fig. 18.2c; Arwidsson and Berg, 1983: figs 13, 27–57), and a Roman period find from London (Merrifield, 1983: fig. 32) with a V-shaped projection may also be a moulding iron.

Detailed shaping of timbers

Dressing timbers. The fashioning of planks, transverse timbers, keels, and stems to the final shape required was probably done by stone or metal blades used as axe or adze. In experienced hands such tools can be used with great delicacy to remove mere shavings. Hasslöf (1972: figs 14, 15) has published two contrasting yet complementary photographs: a second-century AD Roman boatwright using an adze to finish the inside of a frame; and a twentieth-century Norwegian using an axe to joggle the outside of a frame.

Olsen and Crumlin-Pedersen (1967: 160) have

identified axe marks on the Skuldelev planking and other timbers, and McKee (1978: 101) axe or chisel marks on Graveney timbers where they had been fitted to one another. Darrah (1982: 222) has described how modern axe marks may be differentiated from those of the adze, and generally Crumlin-Pedersen, McKee and Darrah seem more confident in their ability to make this distinction than Orme and Coles (1983: 32). More experimental work is evidently required here using early tools, and published reports must state tool type, deduced stroke action, timber species and recognition features of tool marks. This sort of information has not always been made available: in the research discussed above, Darrah had used a modern 2 kg axe with a double-bevelled edge, curved in plan; Orme and Coles were seeking to identify tool marks from their, acknowledgely limited, experiments with Bronze Age tools; and Crumlin-Pedersen and McKee identified tool marks by (presumably) comparing them with their knowledge of modern woodworking. Theoretically it would seem possible to use blades as axes to fashion most of the shapes required in boatbuilding including masts and spars, except where very concave hollowed shapes were required as in hollowed garboards, hollowed stems, hollowed upper surfaces of keels and the undersides of some frames, as in the Skuldelev finds (Olsen and Crumlin-Pedersen, 1967: 166), where adze, chisel, or knife would be more appropriate. The experimental building of a replica of Skuldelev 3 now being undertaken by Crumlin-Pedersen at Roskilde may clarify this matter, at least for the Viking tradition of boatbuilding.

Saws. Saws are used today to perform tasks which earlier boatwrights undertook with axe, adze or chisel. Stone tools with serrated edges are known from the eastern Mediterranean and from north-western Europe (Goodman, 1964: 110–1; Jones and Simons, 1960: 12) but these are small and could probably only cut a groove such as would be required to make comb teeth rather than saw right through a timber of any size. Metal saw blades with wider use are known from the Bronze Age and Iron Age (Cunliffe, 1978: fig. 14.4) and saws have been found in Anglo-Saxon (Wilson, 1968: 149) and Viking contexts (Foote and Wilson, 1970: 180, fig. 19; Arwidsson and Berg, 1983, figs 14, 27–42). Two of the saws in the Mästermyr find (Fig. 8.46) have teeth set projecting left and right alternately, thereby clearing a passage through the timber and allowing the saw to be used on push as well as pull strokes. The blades of the two saws are *c.* 45 cm and *c.* 30 cm in length and could have been used for woodworking. Although looked for, saw marks were not found on either the Graveney (McKee, 1978: 101) or the Skuldelev timbers (Olsen and Crumlin-Pedersen, 1976: 161). Evidence for the use of saws has survived on some Mediterranean wrecks: the twelfth-century Serçe Liman boat had sawn hardwood keel and softwood planking; sawmarks were also found on the fourth-century AD Yassi Ada and the fourth-century BC Kyrenia vessels (Steffey, 1982: 20, 23, 27).

The fact that much of the early preparing of timber is obliterated by the final stages means that it cannot be said with certainty that saws were not used at some stage, on some of the timbers of the Graveney and Skuldelev finds; the obvious use of naturally curved timbers makes this unlikely, however. The first direct evidence for the use of sawn boards in N.W. Europe comes from thirteenth- and fourteenth-century cogs (Crumlin-Pedersen, 1983).

Bevels. Bevels and rebates on stems and keels may have been partly worked by axe and then finished by chisel. Wallace (1982a: 269) has found that some grooves were cut in the early thirteenth-century timbers at Wood Quay by first boring holes with a 2 cm diameter spoon auger bit at intervals of *c.* 7 cm and then completing the groove with a chisel. It would be possible to fashion a rebate in this manner, but such work has not been reported.

Plank edges and ends. From at least the time of the Ferriby boats plank edges have had to be worked very carefully in bevels and scarfs to produce good wood-to-wood contact at joins and thus provide the basis for a watertight joint. This fine work could have been done by small axe, adze, chisel or knife – woodworking tools that could have been used with these sort of actions are known from the Neolithic onwards. However, Olsen and Crumlin-Pedersen (1967: 161) found marks of small planes visible on the surfaces within Skuldelev scarfs and similar joints. Planes are unknown from the prehistoric period, but are found in the Roman toolkit. A fifth-century AD plane was found at Vimose, Denmark (Goodman, 1964: 41), a small sixth-century plane with an antler stock (Fig. 8.47) has been found at Sarre, Kent (Wilson, 1968: 149), and a wooden sixth-century plane from the Frisian terps. Planes have not yet been found in Scandinavian Viking context but must have been used by them – not necessarily for boatbuilding – as the word *lokarr* (on loan from Old English) is used in a tenth-century Icelandic poem (Foote and

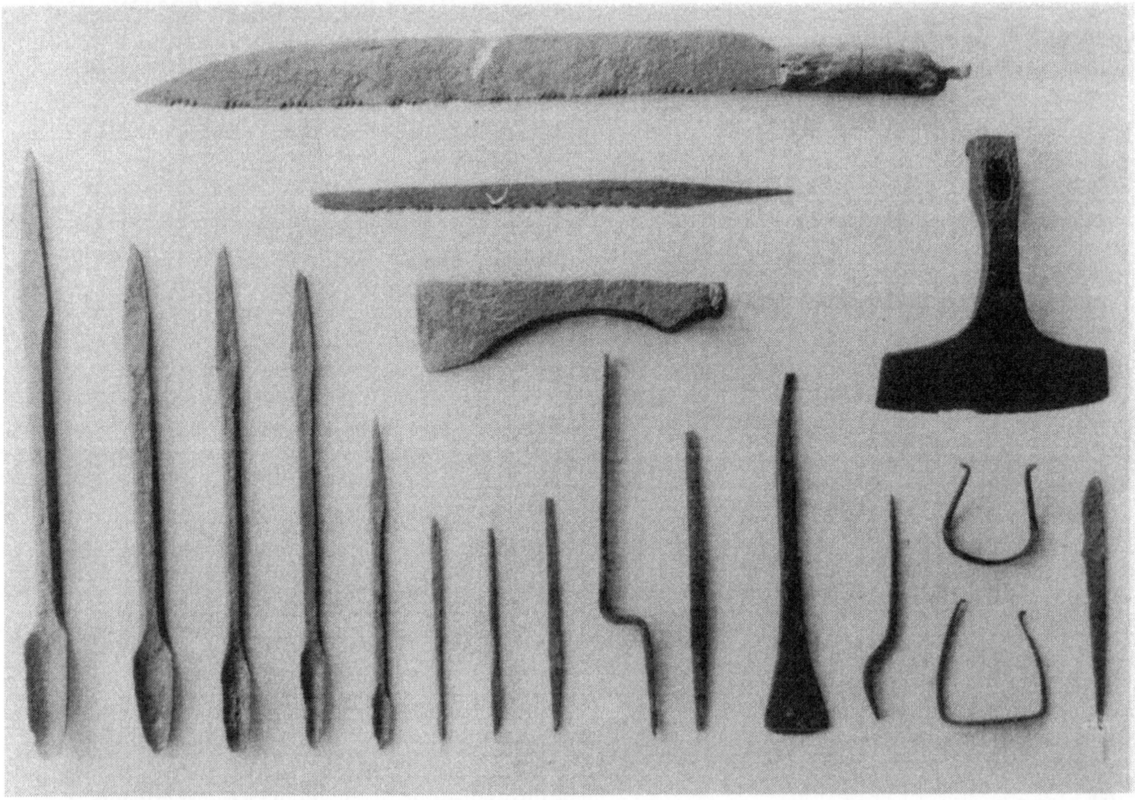

Figure 8.46 Woodworking tools from the Viking Age Mästermyr find. (Photo: Statens Historiska Museum, Stockholm.)

Figure 8.47 A small sixth-century plane from Sarre, Kent. After Wilson, 1968.

Wilson, 1970: 178): the eleventh-century plane from Dublin (Fenwick, 1978: fig. 7.8c) and the one from medieval Bergen (Wallace, 1982: 373) may be grooving or moulding planes.

There is direct evidence for some use of planes in early medieval boatbuilding, but its extensive use cannot be demonstrated and seems unlikely (see next section).

Smoothing. The Brigg and Ferriby oak planks and other timbers seem to have received no other treatment after being worked to shape by blades used as axe or adze, except where timbers were to be joined; the Graveney boat timbers were much the same (McKee, 1978: 101). In the Skuldelev finds the oak planking received no additional dressing after being axed to shape, but the pine planking was smoothed so that axe marks were obliterated (Olsen and Crumlin-Pedersen, 1967: 160–1). Christensen (1982: 331) has noted that several medieval boat-finds have axemarks on the inboard faces of their planking, while the outboard faces have been smoothed. Such smoothing appears to have been achieved by use of the *shave* – see Christensen (1982: fig. 18.2a and b) for examples from Viking Age graves (Fig. 8.48). The shave has handles at right-angles to the breadth of the blade in contrast with the *drawknife*, the handles of which are in line with the blade's breadth.
Unlike the *spokeshave*, the blade of the shave is drawn across the timber with the blade almost at right angles to the surface of the wood, cutting and scraping to achieve a relatively smooth finish but with characteristic marks at intervals of less than 0.5 cm, across the direction of the strokes which are some 2

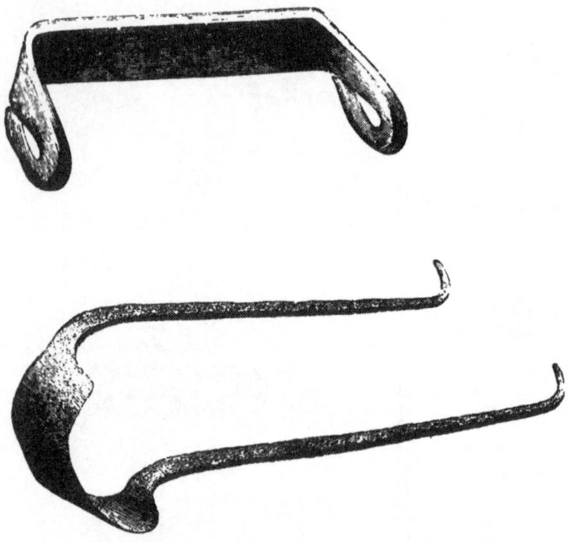

Figure 8.48 Two shaves from Norwegian Viking Age graves. After Christensen, 1982 (courtesy the University Museum of National Antiquities, Oslo, Norway (copyright holders)).

to 6 cm in breadth (Olsen and Crumlin-Pedersen, 1967: 161, fig. 68). Tools which may be shaves include a tenth-century find from Dublin (Cat: 166) and one from Mästermyr (Arwidsson and Berg, figs 13, 27–54).

Files are known from Iron Age Danebury (Cunliffe, 1978: fig. 14.4) and from Mästermyr (Arwidsson and Berg, 1983: figs 7–34, 22–32 and 3, 23–35, 37 and 38).

The boatbuilder's slice is a broad flat chisel used with a pushing action to trim timber, especially where the adze cannot reach (Salaman, 1975: 143). It can also be used to give a smooth finish to timber as Harold Kimber demonstrated when building the Greenwich replica of the Gokstad *faering* (McGrail, 1977: fig. 1) It is possible that certain socketed tools from the early medieval period (for example one from Mästermyr (Arwidsson and Berg, 1983: fig. 12, 26–59)) and originally identified as axes, adzes or hoes, were used in slice fashion to pare off shavings. The slice has the following characteristics:

(a) symmetrical in plan about its longitudinal axis with a slightly curved cutting edge in plan.
(b) longitudinal section is asymmetric about the longitudinal axis, the lower face being flat.
(c) cutting edge is bevelled only on the upper face, a characteristic it shares with the side axe.

Medieval tools which may have been used as slices include ones from:

Thomas Street, London (Manning and Sanders, 1972: 288)
Grave 10, Anglo-Saxon cemetry Sohan (Lethbridge, 1933: 158; Wilson, 1968: 144)
Viking Age tools from Norway usually called 'iron celts' (Petersen, 1951: 159, 517).

Sharpening tools

Whetstones or similar abrasive artifacts would have been required at all stages in boatbuilding to maintain the edge on tools. Such stones have been commonly found in Viking Age context – the Klåstad wreck had a cargo of them (Christensen, 1974), they are known to have been manufactured at Birka (Foote and Wilson, 1970: 210) and one was found at Mästermyr (Arwidsson and Berg, 1983: figs 13, 27–123). In Britain whetstones are known from at least the eighth century BC (Cunliffe, 1978: 32) and an unused ceremonial one was found with the Sutton Hoo burial (Bruce-Mitford, 1975: 190). Sharpening stones have also been excavated from the tenth-century site Winetavern Street and twelfth-century site in Christchurch Place, Dublin (*Viking and Medieval Dublin*, 1973: items 168, 169).

Tools used when fastening timbers

The large holes or mortices cut through the cleats on the Ferriby and Brigg planking are similar to those cut through some of the Somerset Levels planking. Such holes were probably marked out by knife, started by axe – sometimes from both sides – and finished by chisel-shaped tools. The large stitch holes in the Ferriby boats were probably worked by chisel or gouge. The Brigg stitch holes being only *c.* 1 cm in diameter may have been bored by drill – no excavated examples are known from prehistoric northern and western Europe, but there are many ethnographic analogies, operated either by bow (one man) or by leather thong (two men) (Salaman, 1967: 33). Bronze awls, as figured by Heal (1981: fig. 4.2.8) are a possible alternative as may be the type of stone point illustrated by Wright (1976: pl. 2a). Gouges were fairly common in Iron Age Britain (Cunliffe, 1978: 289, fig. 14.4). Some of the small holes in the timbers from the Iron Age settlements at Mere and Glastonbury were evidently burned in by red-hot iron or bronze rod (Orme and Coles, 1983: 40) as were used in L. Victoria in early twentieth century (Worthington, 1933: 159), and Shetelig (1903: 13) considered that the stitch holes in the early medieval Halsnøy boat had been burned. Certain stitch holes had been enlarged on the underside of the Brigg 'raft' and given a truncated cone shape (McGrail, 1981b: fig. 2.1.17): this was

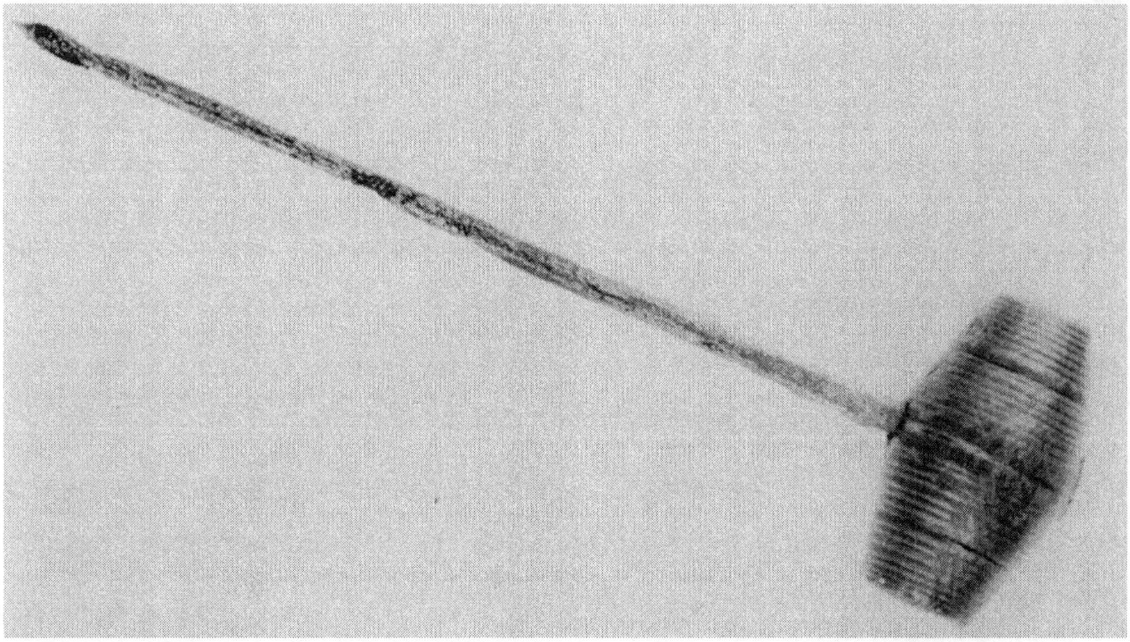

Figure 8.49 Iron gimlet with lathe-turned handle from a thirteenth-century pit in Winetavern Street, Dublin. (Photo: National Museum of Ireland.)

probably done by a boring tool when reaming out a jammed stitch before re-sewing. The other tools required when sewing these early boats would have been a thin blade to split the willow withy used as a thread on Brigg, and a tensioning tool to haul taut the thread before it was wedged in the holes (see above, section on *Plank fastenings*).

Holes for nail fastenings in medieval boats may have been bored by hot metal or by awl, gimlet or auger, either hand-held or breast version – both are seen on the Bayeux tapestry. A 12.6 cm iron awl with wooden handle and 18.7 cm iron gimlet with a wooden barrel-shaped handle have been found in thirteenth-century levels in Dublin (Fig. 8.49) (Cat. Nos. 172, 173) and there was one at Mästermyr (Arwidsson and Berg, 1983: fig. 29–94). Small twist bits are known from Viking Age Scandinavia and Russia (Petersen, 1951: fig. 124; Kolchin, 1953; fig. 87), but the bit most widely found is the spoon shaped bit. Wilson (1968: fig. 2) figures spoon bits from Cambridgeshire, Co. Durham and Norfolk with diameters of *c.* 1 in, 1/2 in and 1/4 in (2.5, 1.3, 0.6 cm). The Mästermyr hoard had six auger spoon bits varying in diameter from 2 to 0.4 cm (Arwidsson and Berg, 1983: figs 28–46 to 51), see Fig. 8.46. A spoon bit of the eleventh/twelfth-century High Street site in Dublin has a diameter of *c.* ½ in (1.2 cm) (Cat. No. 176) (Fig. 8.50) and others from the Wood Quay site have been 1.5 to 3.5 cm (Wallace, 1982a: 269. There are other examples from early medieval Scandinavia (Christensen, 1982: 331) and Hedeby (Wilson, 1968: 147) and from Russia (Kolchin, 1953: fig. 87). A T-shaped handle which would take such bits was excavated from a thirteenth-century context at Mileham, Norfolk (Wilson, 1968: 147). A T-handled auger may be illustrated in the boatbuilding scene on the Bayeux tapestry and there is also another with a swelling or forked crotch at the end of the stock

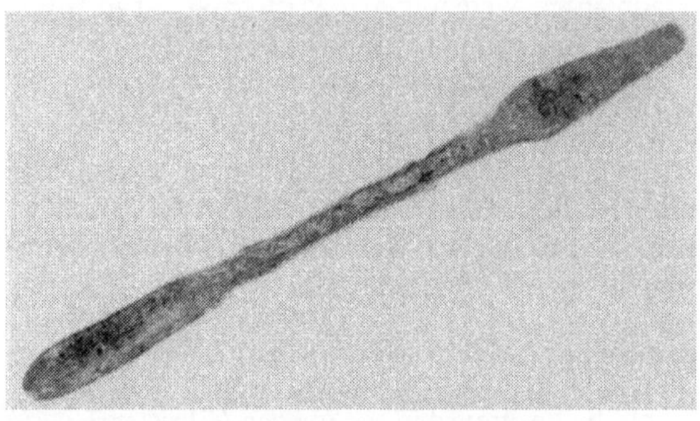

Figure 8.50 A late eleventh/early twelfth-century iron spoon bit from High Street, Dublin. (Photo: National Museum of Ireland.)

similar in form to the breast augers of the later medieval period (Fig. 8.44).

In medieval boatbuilding iron nail holes were of the order of 0.4 to 0.8 cm and treenail holes, *c.* 2.5 cm. The smaller spoon bits from Mästermyr and England, some of the Russian bits (Goodman, 1964: 166–7), the small Scandinavian and Russian twist bits and the known examples of awls or gimlets could have been used for these iron nail holes, while the larger English and Mästermyr and some of the Russian spoon bits could have been used for treenail holes, as Crumlin-Pedersen (1967: 161) deduced from the blind holes with rounded apex found in the Skuldelev timbers.

Hammers to drive nails or to turn and hook them have been found in an Iron Age context at Bulbury, Dorset (Cunliffe, 1978: fig. 14.4). They have also been found in Viking Age graves (Arwidsson and Berg, 1983; Petersen, 1951) but as Wilson (1968: 149) pointed out the butt of an axe blade can also be used. An Anglo-Saxon adze from Thetford (Wilson, 1968: 143, fig.ld) with a much-hammered butt may have been used in this way; use as a wedge as suggested by Wilson seems unlikely owing to its curved longitudinal profile. A roving iron or hollow punch (Salaman, 1967: 46) was found at Mästermyr (Arwidsson and Berg, 1983: fig. 23–52): such tools are used to drive the rove over the point of the nail. Pointed chisels (centre punch), needed to punch a hole through a rove blank, have been found in Viking Age context in Scandinavia (Petersen, 1951) and in Dublin (Cat. No. 171). After the rove has been driven over the point of the nail, the point is turned by hammer, and the excess cut off by hammer and chisel, the stump being deformed by hammer to form a second head on the inside of the boat. During this clenching operation a hand-anvil or dolly, which may be a broken metal tool or the solid end of a roving punch, has to be held against the nail head. Hand anvils were found at Mästermyr (Arwidsson and Berg, 1983: figs 9, 20–73, 20–74, 21–72).

A specialist and previously unexpected tool (the 'nail-hunter') for removing clenched nails during repairs (Fig. 8.51) has been found on the Viking Age site at Pavikan, Gotland (Lundström, 1971/2), and there is a possible one from a Roman London context (Merrifield, 1983: fig. 23).

In many medieval clench-fastened boats longitudinal grooves were cut along the inboard face of planking near the lower edge, so that caulking could be positioned in the overlap (Christensen, 1982: fig. 18, 3b). McKee (1978: 101) deduced that the Graveney caulking grooves had been worked by gouge – gouges are known from Anglo Saxon England (Wilson, 1968: fig. 2h). Christensen (1982: 332) found, on the other hand, that grooves in the Klåstad boat had been worked by shave or scraper. The tar, in which the caulking was usually rolled before being placed in the overlap, may have been in a pot (kettle) and a sherd contaminated with vegetable tar has been excavated from the early thirteenth-century site at Wood Quay, Dublin.

After the change to skeleton building, from say the fourteenth century, and earlier, when leaking seams in shell-built boats had to be repaired (McKee, 1978: 97), caulking had to be driven into V-shaped gaps between the planking after it had been fastened to the frames. In recent times reaming irons were used to remove old caulking, and caulking iron and mallet to drive in fresh caulking (Salaman, 1967: 34–7, 45). McKee (1978: 102) has also discussed the requirement for a scraping tool to rake clean the top edges of clinker planking where cargo remains and other decaying matter may collect.

IRON-WORKING

Smith's tools were required for making nails and anchors and these are well evidenced from the Iron Age (Cunliffe, 1978) and from the medieval period. Hammer, tongs, anvil and bellows were found in a tenth-century AD grave in Bygland, Morgedal, Norway (Foote and Wilson, 1970: 181–2), and hammer and tongs at Mästermyr (Arwidsson and Berg, 1983: figs 7, 22–44). A leverage device or twister and a rod-cutting chisel would additionally be required for anchor chains. The Graveney and the Gokstad roves are of regular thickness and one dimension is fairly constant; this suggests that they were cut from thin metal strips by hammer and chisel or metal shears – such tools were also found in the Bygland smith's grave and at Mästermyr (Arwidsson and Berg, 1983: figs 6, 22–45). A nail-making iron was also found at the latter site (figs 12, 23–86).

MANPOWER AND TIME?

A metal-smith's work can be done at almost any time regardless of season and at any place where there is demand, the main constraints being supply of raw materials and tools and the total demand for his products. Rowlands (1971) concluded from ethnographic evidence that 'full-time smiths are only consistently found when supported by privileged minorities. The ordinary smith rarely seems to be

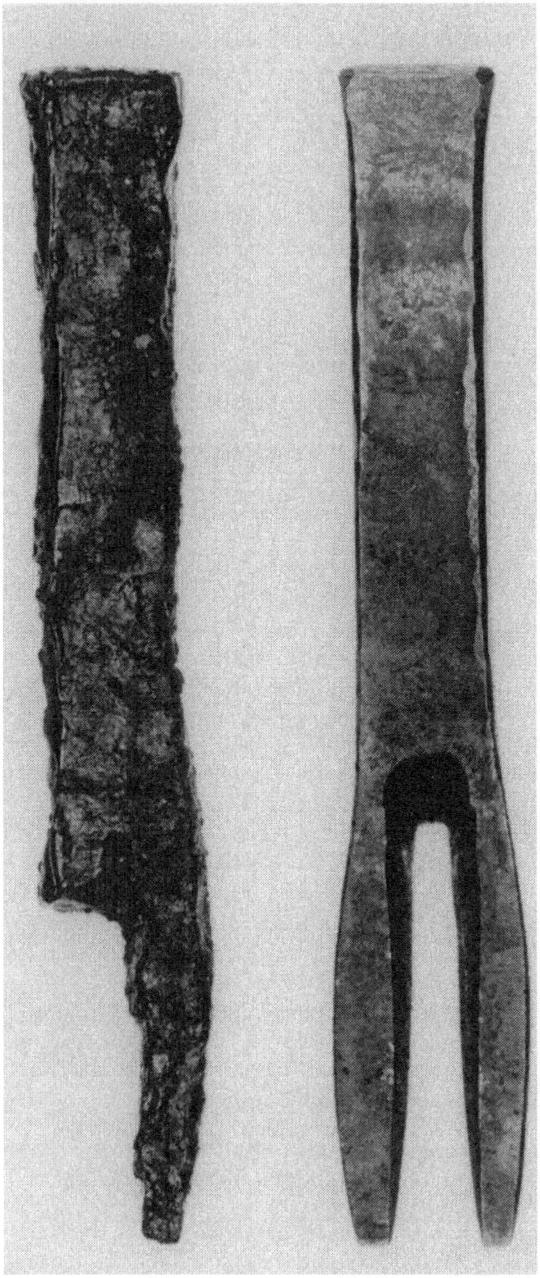

A

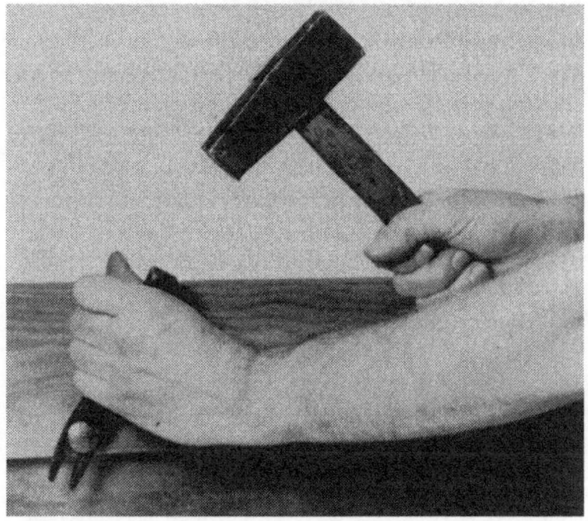

B

Figure 8.51 (A) Nail-removing tool from a Viking Age site at Paviken. Gotland and replica.
(B) Method of use.
(Photo: Per Lundström.)

fully employed because the demand for his product fluctuates'.

SPECIALIST BOATBUILDERS

Boats and boating lend themselves to seasonal building in northern and western Europe where the timing of many activities is determined by the cycle of the seasons. Boatbuilding can readily be undertaken in the winter and early spring when sea-going is not especially attractive and when other activities, such as farming, make few demands. There are also sound reasons for felling trees in winter and using them for boatbuilding (Ch. 4). Unlike a metal-smith's product, therefore, the main demand for boatbuilding skills would be forecastable as a peak in winter. In societies with simple economies and technologies we may expect many people to be capable woodworkers, but the 'design' and fashioning of planks for any boat other than a simple box-like structure and the working of bevels and of effective watertight joints and fastenings calls for specialist ability. Thus *prima facie* we would expect such societies to have specialist but part-time boatbuilders.

Ethnographic evidence on the question of full-time or part-time boatbuilders may be divided into two aspects: that from crofter-type communities where there are individual families exploiting a variety of resources; and groups with a more communal life style as examplified in parts of Oceania and the Indian Ocean.

Eric McKee, in his study (1976) of the small *flatner*

of North Somerset, a boat which is 'not too hard to build', found that one could be built to last 12 years by one man in about 2 weeks. Such a building rate of *c.* 26/year would require a boat population of *c.* 312 (12 × 26) to sustain one full-time boatbuilder. In fact there was insufficient demand for this, and McKee found that in one particular region at any one time there were several men building boats, mostly carpenters, who built anything in wood from 'cradle to coffin', and crofters or fishing farmers. In another part of N. Somerset a more complex *flatner* took longer to build but had roughly the same life span. In this situation a full-time specialist could have been supported, but McKee found that seasonal usage favoured off-season building by user-builders who helped one another when more than one pair of hands was necessary.

The longer the life span and the shorter the building time the larger the population of boats necessary to sustain a full-time boatbuilder. On the other hand there would be a requirement to replace at least one boat each year even in quite small boat populations. For example, with 24 Somerset *flatners* an average of two would require replacement each year. Thus there would be a continuity of techniques and an opportunity to train others and hand on experience during a period of a month or so each year. Maintenance and repair would extend this period; such work is especially important for sewn boats which generally need to be dismantled and re-sewn annually (Hornell, 1941: 62; 1946: 236; Hourani, 1963: 94; Folkard, 1870: 291, 309; Malinowski, 1922: 141; Best, 1925: 221).

In Oceania and the Indian Ocean where more complicated boats were built and where boatbuilding was much more of a communal activity we nevertheless find there was generally one man in charge; for example the *tohuna* in the Society Islands (Hornell, 1948: 47), and the *Kahuna* in Hawaii (Willis, 1922: 69).

The ethnographic evidence suggest that in the Ferriby region in the early second millennium BC there was at least one man who specialised in boatbuilding and repair, who built and maintained boats with the assistance of others during the winter and early spring, and was a member of one of the boat's crew during the operational season.

For the early medieval period there is evidence that on the one hand, there were boatbuilding specialists and, on the other hand, that certain craftsmen worked in several trades. The Mästermyr tool chest and finds from Hordland, Norway and Halleby, Jutland contained both metal- and wood-working tools, while graves at Skovlunde Mark, Denmark and Hurbuck, Co. Durham had woodworking and agricultural tools. Such graves and finds are relatively rich in possessions and suggest the owners had a specialist position and expertise; they may have been makers of tools for several trades or general craftsmen who could make nails, anchors and chains as well as fell trees and build boats; or who could clear ground and prepare the site as well as build a house on it. In support of the case for a somewhat general rather than very specialised capability may be quoted the reference in Aelfric's Colloquy to tenth-century treewrights (*treow wyrhta*) who built boats, houses or other wooden objects (Fenwick, 1978: 188). On the other hand, as Foote and Wilson (1970: 180) have suggested the owners of such tools may have been master craftsmen who supervised the work of men skilled in the several trades needed to build a house or a boat.

Undoubtedly there were builder-users of small boats working alone throughout medieval times, but the building of large boats, and indeed ships, calls for more men, with grades of expertise. There is evidence for this in the later Viking period. Olaf Tryggvason's *Long Serpent* was built at Trondheim during the winter of 998 or 999 by a team of men with four specialisations: tree fellers, labourers, general carpenters or plank-cutters (*filungar*), and specialist stem-smiths (*stafnasmidir*) who were paid twice the wages of plank cutters. One of the stem-smiths, Torberg Skavhogg, was promoted to the position of master boatbuilder (*hofudsmidr*) during the work (Christensenn, 1982: 329; Foote and Wilson, 1970: 250–1). In one Bayeux tapestry scene (Fig. 8.44) we can see one man standing in front of the stem of a partly-built boat, checking the work of men in the boat and confirming that the boat's lines were fair – surely a master boatbuilder.

The specialisation evident when building large vessels in Viking times is reflected in the later medieval period. From the 1422–27 accounts of William Soper, Keeper of the King's Ships at Southampton (Rose, 1982) we can deduce the following table of relative status:

1.	Master carpenter or shipwright	8 pence a day
2a.	*Berders* (carpenters)	6 pence a day
b.	Caulkers	
3a.	Clenchers (carpenters)	5 pence a day
b.	Sawyers (carpenters)	
c.	Fellers (who also split timber)	
d.	Smiths	
4.	Holders (carpenters)	4 pence a day

Davy Owen, the *dyver*, who must also have been a carpenter, was paid 6 shillings and 8 pence as a bonus for repairing *Holigost de la Tour* underwater, when she was lying in the R. Hamble (Rose, 1982: 101). For comparison with shipbuilders' rates of pay, we may note that masters of ships in reserve or under refit were paid 6 pence, and shipkeepers 3 pence, for every day spent onboard.

Some of the shipbuilders or men of equivalent capabilities must also have served at sea to undertake repair and maintenance, as we know from the *Kings Mirror* (Ross and McLaughlin, 1977: 144–9), a document originating in N.W. Norway in *c.* 1250, that nails, axe, gouge and auger and all other tools needed in boatbuilding were to be taken to sea in merchant vessels.

The master boatbuilder's skills were required throughout the shaping of shell-built, edge-joined hulls, although rules of thumb and planking aids might allow him to decentralise some responsibility (see section above on *Shape*). On the other hand, with the skeleton sequence of building large ships, which seems to have become widespread in northern and western Europe during the fourteenth century, specialised skills were required to 'design' the skeleton of keel, stems and frames to which the planking was subsequently fastened (see section above on *Shape*). Once this skeleton was made and erected, less skill was needed to build up the hull. It also became practicable to 'store a design' for future use, perhaps in the form of patterns of the individual parts of the skeleton. Such information could be used many times, independently of the initial skill in 'designing' the shape, and probably speeded the building of subsequent ships of this class.

TIME TAKEN TO BUILD

Man hours or overall time to build are seldom mentioned in ethnographic accounts, and experimental replica building can give only an approximate and excessive guide as there is inevitably a learning process and indeed the work may stop for long periods or even regress whilst thought is given to specific methodological or boatbuilding problems. The Greenwich *faering* replica was begun in June 1972 and was ready for sea by February 1973. In that period of about 8 months one man worked full-time and one part-time (*c.* $\frac{1}{5}$). But, for several reasons, work was not continuous, breaks included, for example, a visit to Scandinavia to study the original *faering* and other medieval boats at first hand. If a second similar boat had been begun immediately afterwards it seems probable that she could have been finished in less than three months. Crumlin-Pedersen's (1968a: 8) estimate of the time required in the Viking period to build one of the *larger* craft is also 3 months.

There can be no valid average time for building a large ship of unspecified size and complexity with an unspecified number of men, unspecified length of daylight, and regardless of whether building was in skeleton or shell sequence. Nevertheless, some idea of the overall time taken to build a medium to large vessel in the later medieval period may be gained from the following data:

1294 Galley built at Southampton in 17 weeks to launch: she then required fitting out for service (Platt, 1973: 61).

1412 Galleys for Henry IV were begun on 20th November and were 'almost complete' by Christmas (Callendar, 1913: 98).

c. 1425 Ballinger *Ane* was built in Southampton between 18 June and 22 October (Rose, 1982: 222).

In addition Carr Laughton (1957: 248) has noted that in 1405 one man could produce 10 to 12 clove boards, 7 to 10 ft in length, in a working day.

Apart from McKee's (1976a) account of the building of a simple *flatner*, which may not have included any estimate for obtaining raw materials, the only detailed ethnographic account known to me, of the time required to build a small boat with a simple tool kit is that published about Guyana expanded logboats by Hurault (1970: 73–4). Boats, 4 to 8 m long, were built in 14½ days elapsed time. For 13 of these days only 1 man was required, 2 on one other day and 4 for ½ day, giving a total of 17 man days (Table 6.2). A larger logboat of some 40 m took 28 days, interspersed with 7 days rest. It required *c.* 78 man days and the team varied from 2 to 10.

An attempt may be made to estimate the man hours required to build a Ferriby boat, the minimum reconstruction variant with two sidestrakes (Fig. 8.21) by summing estimates for individual operations. The estimate below, which includes time to obtain raw materials e.g. fell tree, is based on a method used by John Coates who, from slightly different assumptions, calculated a total of 600 man days.

This is a gross estimate in that no account is taken of whether all operations had to be performed sequentially or whether some could be concurrent. Nevertheless we may make the rough deduction that 2.26 man years of work could be done in 3 months

	man days
7 pairs of planks at 100 man days each	700
3 frames	10
2 transoms	5
3 thwarts and other fittings	12
Fastening boat (including 300 stitches)	74
12 paddles	24
	825 man days

825 man days = *c.* 2¼ man years

elapsed time by 9 men working full time. This assumes that there was a 10th man in charge, that the other 9 were equally skilled and interchangeable, and that there were not many bottlenecks. Coates's estimate requires 6 to 7 men plus boatbuilder.

Hornell (1941) and others have recorded that the sewn *dau la mtepe* of East Africa lasted 3 or 4 years. In N.W. Europe different timber, different operating environments and different usage might suggest a working life of say 6 years for a Ferriby boat. On this assumption each boat would require a maintenance/refit every year for 5 years and be replaced in a 3 or 4 month period of the 6th year. If there were a local boat population of 3 boats a new boat would be required in alternate years.

For comparison we may note that John Coates has estimated that it would take 300 man days to build a hide boat the size of *Brendan*, not counting mast, sails, oars or rudder. From different assumptions, primarily that hide tanning could be done well in advance, and that off-cuts from the hides could be used as lashings, Timothy Severin has estimated 123 days.

9 Bundle rafts and boats

BUNDLE RAFTS

These may be defined as: two or more bound bundles of reed, grass, straw, bark or light poles, lashed together by coiled basketry or similar techniques – bundles are often arranged in a hollowed shape, but rafts with a generally flat upper surface are also known.

DISTRIBUTION

Reed
Bundle rafts of reeds have a wide distribution wherever there are, or were, lakes or swampy flat valley floors, especially near river mouths or confluences; in such places reeds grow abundantly (Hornell, 1946: 39–60; Johnstone, 1980: 9–16). In recent times reed bundle rafts have been used in the following regions:

America (North, Central and South)	Hornell, 1946: 41–6; Edwards, 1965: 1–16, 85–9.
Oceania	Hornell, 1946: 40, 45
Africa	Hornell, 1946: 46–55
S. W. Asia	Hornell, 1946: 56–60; de Graeve, 1981: 89–93
Australia and New Zealand	Hornell, 1946: 40, 60; Birdsell, 1977.
Europe–Ireland	Delaney, 1976; Wakeman, 1872–3
Hungary	Hornell, 1946: 60; Johnstone, 1980: 12
Corfu	Johnstone, 1980: 60
Sardinia	Johnstone, 1980: 12; Hornell, 1946: 60

Earlier use: Pliny (*NH* XIII, XXII, 71–3) described the use of reed rafts with reed sails on the River Nile in Egypt and Strabo (16.1.15) their use in Mesopotamia. Iconographic and textual evidence indicates even earlier use in Egypt (Hornell, 1946: pl. 6) and in Mesopotamia (Hornell, 1946: 56; de Graeve, 1981). Johnstone (1980: 59–60) has also suggested that iconographic evidence and linguistics point towards the use of reed rafts in Greece, Crete and Malta, and Hornell (1946: 55) has argued for their early use in Spain.

Reed rafts were seen in use in S. America when Europeans made first contact in the sixteenth century (Edwards, 1965: 1–16, 85–9) and a small model of a raft excavated from an Atacama burial site in Chile (Johnstone, 1980: 14) suggests that they were in use there at least 2000 years ago.

Poles
Rafts made from bundles of light poles are known from India where sticks of shola (*Aeschynomene aspera* are used (Hornell, 1946: 68); and in Africa (Hornell, 1946: 52–3, pl. 7A) where the rafts are made from bundles of ambatch branches and saplings (*Herminiera elaphroxylan*).

Bark:
Bark bundle rafts are known only from Tasmania (Fig. 9.1) (Roth, 1899: 155; Jones, 1976; McGrail, 1981a: pl. 2) where the 'paper' bark of the teatree shrub (*Melauca* sp) or the 'stringy bark' of the eucalyptus (E. *obliqua*) are used. The former is preferred as it resists waterlogging longer and is lighter – weights of comparable rafts are 350 lb 'paper', 500 lb 'stringy' (159: 227 kg) (Jones, 1976: 245).

BUNDLE BOATS

These are also made by binding and lashing together bundles of reeds but, unlike rafts, they always have a hollow form (elongated or round) which is

Figure 9.1 Early nineteenth-century Tasmanian bark bundle raft. From Lesueur and Petit's *Terre de Diémen*. (Photo: NMM Greenwich.)

strengthened with a light framework; the bundles are then waterproofed with some mastic-like substance. They are known only in Arabia, although they may remain to be recognised in other regions with a readily available supply of resin, tar, creosote, pitch, asphalt or bitumen.

The boat-shaped *Zaima* (Fig. 9.2) and the *jillabie* (Heyerdahl, 1978: 35) are built of reed bundles linked by coiled basketry, with an inserted stiffening of light willows, and coated externally with bitumen. The *quffa* (Fig. 9.3) (McGrail, 1981a: pl. 26; Hornell, 1946: pl. 4B, 17B) is circular but is built and waterproofed in a similar manner. The earliest reference to bundle boats appears to be in an inscription dated *c.* 2300 BC in which Sargon of Akkad claimed to have been placed in the river as a baby in a basket of rushes sealed with bitumen (Anderson, 1978: 49): a similar story is told about Moses (*Exodus*, 2, 3). Round bundle boats similar to the *quffa* (cognate with Akkadian *quppa*) may be depicted on the ninth-century reliefs of Ashur-nasir-pal (Hornell, 1946: 105) which appear to have a smooth skin rather than the patchwork of hides evidently depicted on later reliefs. Babylonian reed boats of the first century AD were described by Strabo (16.1.15); they were built of woven reed and plastered with asphalt.

FUNCTION

Bundle rafts have been used recently in coastal waters off S. America (Edwards, 1965: 1–3; Hornell, 1946: 41–2), off Corfu (Johnstone, 1973a: 6) and off Tasmania (Jones, 1976; Birdsell, 1977). This sea-going use included fishing and acting as tenders to larger boats and ships. Bundle rafts may have been used for coastal voyages from Egypt to Palestine in the eighth century BC, if Isaiah (18, 1–2) has been interpreted correctly. Pliny (*NH* 6.24.82) quoting Eratosthenes clearly states that reed craft were used before the first century AD for coastal voyages from the River Ganges to Ceylon, and Heyerdahl (1978: 28–34) in his experimental reed rafts RA1 and 2 has shown that, in certain circumstances one-way transoceanic voyages may be made in this type of craft.

There is wider evidence for the use of bundle rafts on inland waterways; for example as river ferries and carriers of goods and animals in New Zealand

Figure 9.2 A modern *zaima* reed boat of the southern Iraq marshes. Photo: plate 45 in *Marsh Arabs* by W. Thesiger (reproduced by permission of Curtis Brown Ltd, London).

(Hornell, 1946: 40), S. America (Edwards, 1965: 1–2, 42–3) and Tasmania (Roth, 1899: 155). Reed *boats* have had similar uses in Mesopotamia (Hornell, 1946: 101). Rafts have also been used for fishing and fowling in the reed marshes of Egypt and Arabia from early times and indeed into the present century (Hornell, 1946: 46–8).

VARIETY AND SIZE OF RAFTS

Bundle rafts vary in quality; at one end of the scale are those built from undried reeds as the need occurs (Edwards, 1965: 8–9; Hornell, 1946: 59, 69). Such rafts are generally rectangular in shape with little attempt to shape the ends and they are abandoned or broken up when they have served their immediate purpose. At the other extreme are boat-shaped bundle rafts such as those from Lake Titicaca (Fig. 9.4), Peru (Hornell, 1946: 42–4; Edwards, 1965: 9–11), built with longer term use in mind.

These Lake Titicaca rafts are generally *c.* 17 × 4 × 3 ft (5.2 × 1.2 × 0.9 m). Heyerdahl's *RA2* was bigger at 39 × 15 × 7 ft (12 × 4.5 × 2 m), but, near Basra, he had seen even bigger rafts of *c.* 112 × 16 × 10 ft (34 × 5 × 3 m) made of reeds floated downstream to a paper mill. The largest reed rafts used as water transport (rather than as raw material) are probably those from the Abyssinian lakes which could be up to 35 ft (10.7 m) in length with a 7 ft (2.1 m) beam (Hornell, 1946: 53). The smallest rafts appear to be the one-man 'balsas' from S. America which are *c.* 6 to 8 ft (1.8 to 2.4 m) in length (Edwards, 1965: 2). The mean L/B for 15-bundle rafts, where such dimensions have been recorded, is 3.02 with a range of 1.5 to 5.0. Two others, the Peruvian *cabalitto* described by Hornell (1946: 41) and the grass bundle rafts of the Seri Indians of California (Hornell, 1946: 45–6) are outside this range with L/B of 7.5 and 9. The *cabalitto* is however ridden astride and is therefore strictly not a raft but a float.

VARIETY AND SIZE OF BOATS

Bundle *boats* require a greater investment than is required by the rafts and they are built with a longer useful life in mind, for bitumen preserves as well as waterproofs. The *zaima* which Thesiger (1978: 128) noted on a branch of the R. Euphrates was *c.* 10 × 2.5 ft (3 × 0.8 m) (L/B = 4). The circular *quffa* generally has a diameter of *c.* 13 ft (4 m) and a depth of 7.5 ft (2.3 m), although one of 16 ft 5 in × 10 ft 6 in (5 × 3.2 m) has been noted and small ones of 4.5 × 2.5 ft (1.4 × 0.75 m) are known (Hornell, 1946: 103).

CONSTRUCTION

RAW MATERIALS

Bundle craft have been made from a wide selection of

Figure 9.3 Model of a recent *quffa* from the R. Tigris showing construction of coiled basketry. (Photo: Hornell, 1946, plate XVIIB (copyright Cambridge University Press).)

plants and these are set out in Table 9.1. The common name (reed, bullrush, etc.) is used in some reports and thus it is sometimes not possible to know what botanical species is meant.

In many places the raw material is collected from natural stands of reed, or windfall branches or driftwood; the Tasmanians, for example, collected loose pieces of bark (Jones, 1976: 240). But in Peru (and possibly elsewhere) reed of the right quality is so important to the economy that it is cultivated (Edwards, 1965: 15–6). Cuttings are planted in beds dug down to the water table and the reeds are harvested by cutting six to seven months later. After twelve or so harvests the bed is replanted. Cutting and planting go on almost continuously in adjacent beds.

Bindings for the bundles have been made from a variety of fibrous materials twisted into lengths of rope and some that have been recorded are listed in Table 9.2: the botanical name is seldom given.

BUILDING SEQUENCE

Rafts

A generalised sequence for building a bundle raft may be compiled from the ethnographic accounts;

Figure 9.4 A modern reed bundle raft from L. Titicaca, Peru. (Photo: Exeter Maritime Museum.)

parts of this sequence are illustrated in third millennium BC reliefs in Egyptian tombs (de Graeve, 1981: 91–2; Hornell, 1946: pl. 6B; Phillips-Burt, 1979: fig. 12).

1. Reeds cut and dried for up to three weeks (Edwards, 1965: 10, 15; Johnstone, 1973a: 6).
2. Reeds laid up in bundles and shuffled to get an overlap if a bundle is to be longer than reed length (*c.* 9 ft (2.74 m)); in this way a bundle of up to 18 ft (*c.* 5.5 m) can readily be produced (Edwards, 1965: 5). Much longer ones may evidently be obtained using an adaption of this method (Heyerdahl, 1978: 29–31). For rafts having a tapered shape in plan the reeds are all laid in the same direction; for a double-ended raft the butt ends are placed in the middle to give taper towards both ends. In parts of S. America this laying-up is done between two rows of stakes set some 50 cm apart.
3. These initial bundles are compressed to form a firm heart or core bundle some $\frac{2}{3}$ the original diameter, say 30 cm. This is done by:
   (a) beating with wooden mallets; or
   (b) using leverage devices such as a strap and two levers (Edwards, 1965: 5, 10) or the human foot as seen on Egyptian reliefs (Phillips-Burt, 1979: fig. 12).

   As the bundle is compressed it is tightly bound with lashings which may be as close as 4 in (10 cm). The tighter the bundles the more rigid the resultant raft and the longer she will stay afloat without waterlogging (de Graeve, 1981: 92; Hornell, 1946: 59).
4. More reeds are added to each core bundle to bring the central diameter up to 35–40 cm and this bundle is then bound with a continuous lashing, either spirally or by a series of hitches (Edwards, 1965: 6, 10). Before the ends are bound one is, or both are, given an upward curve. This bending may be assisted by wetting the reeds (Dick-Read, 1964: 206) and the curve may be held firm by a strut underneath until it sets (Hornell, 1946: 47) or by rocks on top (Edwards, 1965: 10). In Sardinia a short, extra bundle is used at the bow end (Hornell, 1946: 60).

Table 9.1 Materials used in bundle rafts and boats

Common name	*Botanical name*	*Region*	*Reference*
Reed	*Phragmites communis*	Morocco S. America	Hornell, 1946: 55 Edwards, 1965: 14
	Scirpus riparius	S. America	Edwards, 1965: 14
Rush	*Typha augustifolia*	S. America	Edwards, 1965: 14
	Typha sp.	New Zealand Tasmania	Hornell, 1946: 40 Jones, 1976: 240
	Typha augustata	Iran, Afghan	Hornell, 1946: 58
	Typha latifolia	India	Hornell, 1946: 59
Sedge	*Cyperus alterifolius*	India	Hornell, 1946: 59
Papyrus	*Cyperus papyrus*	Mesopotamia	de Graeva, 1981: 90
Palm	*Maurita vinifera*	S. America	Edwards, 1965: 13
Fennel	*Ferula communis*	Cyprus	Johnstone, 1973a: 6
Flax	*Phormium tenox*	New Zealand	Hornell, 1946: 40
Straw or grass	?	Mesopotamia	Hornell, 1946: 61
Paper bark	*Melaleuca* sp.	Tasmania	Jones, 1976: 240
Stringy bark	*E. obliqua* *E. regnans*	Tasmania	Birdsell, 1977: 135
Ambatch poles	*Herminiera elaphro-xylon*	Africa	Hornell, 1946: 52
Shola poles	*Aeschynomene aspera*	India	Hornell, 1946: 68

Table 9.2 Materials used as bindings in bundle rafts and boats

Common name	*Region*	*Reference*
Cotton	S. America	Edwards, 1965: 10
Grass	S. America	Edwards, 1965: 10
Vines	S. America	Edwards, 1965: 13
Hide	S. America	Edwards, 1965: 13
Animal hair	S. America	Edwards, 1965: 9, 12
Reed	Iraq	Heyerdahl, 1978: 38
Palm	Iraq	Heyerdahl, 1978: 38; Hornell, 1946: 101
Coir	Iraq	Hornell, 1946: 102
Grass	Morocco	Hornell, 1946: 55
Grass	Tasmania	Roth, 1899: 155
Inner bark	Tasmania	Jones, 1976: 240

5. Bundles are made to different diameters and with different tapers to match the position they will have in the completed raft, and they are arranged relative to each other so that they produce the overall shape required. Examples of such arrangements are:
   (a) Three similar bundles to give a flat-topped raft (Hornell, 1946: 43–46; Edwards, 1965: pl. 3).
   (b) Two similar bundles each with an inner hollowed out portion to give a 'cockpit' near the stern (Edwards, 1965: pl. 2).
   (c) Two similar bundles surmounted by smaller bundles to give a hollowed form, either
       (i) centrally, to give the transverse section tumblehome
       or (ii) outboard, to give a flared transverse section (Hornell, 1946: 43).
   (d) Two similar bundles fastened on the sides of a smaller and shorter bundle to produce a hollowed form (Hornell, 1946: 45).

   More complex arrangements are possible, as for example used by Heyerdahl (1978: fig. 4): two large bundles on either side of a smaller one are themselves surmounted by smaller bundles outboard.
6. These bundles are then joined together using coiled basketry techniques, a light stake may be used to separate the outer reeds to allow the rope to pass through (Hornell, 1946: 43; Edwards, 1965: 4–6, 10).

Early Egyptian and Mesopotamian representations show a truss linking the upturned ends to the floor of the raft (de Graeve, 1981: 93). Heyerdahl (1978: 29) found by experiment that these trusses were to support the raft bottom rather than to support the ends. Such large flexible hulls with weight at the ends are susceptible to hogging stresses (Ch. 3) and this truss opposes these tendencies (Gordon, 1978: 223–5).

Rafts of bark and pole bundles are generally built in the same sequence as those of reed and they are usually tapered at one or both ends and have a hollowed form (Hornell, 1946: pl. 7A, 7 AA). Roth (1899: 155) noted that Tasmanian bark rafts were made of three bundles, two smaller bundles above and outboard of the main central bundle. Jones (1976: 240) noted that one large bundle measured 4.55 × 1 m and the smaller ones, 3.90 × 0.32 m. A nineteenth-century model of a Tasmanian bark bundle raft suggests that each bundle was enclosed in a net of interlaced string and the bundles were joined by a type of coiled basketry. Jones (1976: 242) has estimated that the average time taken to build a bark raft was somewhat less than a day. The *teitei* of Lake Chad consists of two bundles of ambatch poles lashed together endwise so that the thinner ends interlock thus forming a double-dovetail shape (Hornell, 1946: 53), but as the man sits astride with his legs in the water this may more correctly be described as a float rather than a raft.

Boats

Boats are also built of tightly bound reed bundles linked together by coiled basketry techniques. In the building of the *zaima*, Thesiger (1978: 128, pl. 45) noted that the builder first laid a bottom of six small bundles fastened together longitudinally with both ends turned up. A series of small, U-shaped bundles of reeds were then positioned transversely and laced to the bottom bundles to form the sides and ends of the boat. Further reed bundles were fastened longitudinally along the upper ends of the U-shaped bundles to stiffen the 'rim' of the boat, and further hull support was supplied by two or three U-bent willows inserted as ribs. Three transverse wooden members were fastened with bitumen at sheer level. The *zaima* was then coated externally with bitumen.

The *quffa* of Mesopotamia is essentially a huge lidless basket of coiled basketry, made from long cylindrical bundles of grass straw, parcelled and bound to each other by a lacing made from date leaf palmlets. The rim of this circular basket is stiffened by a willow hoop lashed to the top basketry coil, and a framework of ribs (split branches) is inserted and lashed to the sides with coir cord. Two or three transverse members may also be fitted. The outside is then coated with hot bitumen, as is the floor inside (Hornell, 1946: 101–3).

MAINTENANCE AND DURABILITY

Except when building a one-shot, temporary craft, the main aim of reed raft builders is to ensure that the inner core of their bundles is highly compressed; this gives rigidity to the structure and extends the useful life. This core is covered by a shell of reed which decays as it is exposed to sun and water, but which can relatively easily be removed and replaced (Hornell, 1946: 41; Edwards, 1965: 6). While certain types of bundle rafts of S. America last for one month to six weeks, others will last for nine months to one year, and Heyerdahl (1978: 189) noted that a reed raft from L. Titicaca survived for fourteen months on the Pacific coast. Edwards (1965: 4–6) attributes this relatively long useful life to the greater care taken in construction and the higher compression given to individual bundles. It has not been noted, but there may also have been different usage regimes. The generally short life of reed bundle rafts has also been noted in Egypt – about six weeks if left afloat (de Graeve, 1981: 92), and in Iran – up to two months (Hornell, 1946: 59).

After use the smaller rafts are generally hauled out to drain and dry (Edwards, 1965: 2, 5, pl. 4); sometimes they are separated into individual bundles before draining (Edwards, 1965: 2). Thus the onset of decay and waterlogging is delayed. Heyerdahl (1978: 37) believes that reeds cut at the correct season should retain their buoyancy for much longer; in Iraq this is said to be August.

Bundle rafts of bark and poles are also prone to waterlogging and have to be allowed to drain every day (Hornell, 1946: 57). Ambatch rafts of E. Africa are said to have a maximum time afloat of 12 hours (Dick Reed, 1964: 201–6) while Tasmanian bark bundle rafts have to be dried out after 5–6 hours use, paper bark being less quickly waterlogged than stringy bark (Jones, 1976: 246).

Bundle *boats* being coated with bitumen have a longer life. According to Hornell (1946: 103) the bitumen cover of a *quffa* can be renewed and this makes the craft 'nearly as good as new'. Thesiger (1978: 128) on the other hand, was told that the *zaima* lasted for only one year 'as the bitumen could not be renewed'.

VARIANTS

The reed bundle *boat* described above may be thought of as the main variant of the bundle raft. Others may be discussed under four headings:

Superstructure

A superstructure of timber was fastened to the reed rafts recently used on the River Suck in southern Co. Roscommon (Fig. 9.5), principally to provide a rowing thwart and oar pivots (Delaney, 1976). The reed rafts of Lake Tsana have a raised cargo deck of transversely-laid reed bundles (Hornell, 1946: 53).

Strengthening timbers

In some of the New Zealand rafts a longitudinal pole was inserted for strength (Hornell, 1946: 40). Transverse stakes are used for similar reasons in S. America (Edwards, 1965: 12. 13) and in L. Balaton, Hungary (Johnstone, 1980: 12). A sub-variant has been recorded by Edwards (1965: 3, 13) in Brazil where the reed bundles are formed on sticks.

Extension

(a) *Longitudinally*: Two tapered Corfu rafts are sometimes linked by poles stern to stern

Figure 9.5 A reed bundle raft with a light wooden superstructure on the R. Suck, Co. Roscommon in 1962. (Photo: National Museum of Ireland.)

(Johnstone, 1973a: 6). This extension may be what Father Coba noted in western S. America in 1653 but his description (Edwards, 1965: 2) is ambiguous and these rafts may have been paired.

(b) *Paired*: Rafts on Lake Titicaca, Peru may be linked athwartships by poles to carry a greater load (Hornell, 1946: 44).

Built on a frame

Johnstone (1973a: 6; 1980: 60) recorded that the seagoing rafts of Corfu were built on a V-shaped horizontal frame of cypress saplings to which reed bundles were fastened. A similar structure was noted in the mid-nineteenth century on Lough Erne in N.W. Ireland, where 'wreaths of bulrushes (were) tied upon a frame, a raft made of rough branches of trees or saplings' (Wakeman, 1872/3: 75). Such variants are probably best classified as hybrids – reed bundle/log raft (see Ch. 5).

PERFORMANCE

LOADS

Bundle rafts and boats have been used to carry horses and cattle (Edwards, 1965: 2; Hornell, 1946: 101). Macleod noted in Africa that when cattle were transported by bundle raft they were penned in with hurdles (quoted by Hornell, 1946: 54).

Recent *quffas* of Mesopotamia used as lighters are said to have carried 4 to 5 ton, with a freeboard of 6–8 in (15 to 20 cm) (Hornell, 1946: 101). A boat, which may well be reed bundle, is depicted on ninth-century BC Assyrian carvings loaded with a chariot with the wheels and half the shaft outboard (Hornell, 1946: 105). S. American reed bundle rafts were capable of carrying up to 12 men and their baggage or a man and his ox team (Hornell, 1946: 43–4; Edwards, 1965: 2). New Zealand rafts of 22 × 4.5 ft (6.7 × 1.4 m) carried a load of 700 lb (318 kg), while rafts of the Abyssinian lakes of 30 to 35 ft × 7 ft (9 – 11 × 2 m) could carry 7000 lb (3180 kg) of coffee beans (Hornell, 1946: 41, 53). Heyerdahl (1978: 38) claims that the colossal bundle rafts of Lake Chad could carry 40 tons of cattle and other cargo. Many of the Tasmanian rafts were made for one man but others could carry up to 8 people and Jones (1976: 245) has estimated that documented loads of early colonial times were in the region 400 to 1000 lb (182 to 455 kg).

SPEEDS

A New Zealand raft carrying some 700 lb is

estimated to have been paddled at *c* 2 kn. (Hornell, 1946: 41) which seems not unreasonable. Heyerdahl began his 57-day passage under sail across the Atlantic at 4 to 5 kn, but waterlogging and the resulting increased submergence reduced speed to *c.* 2 kn (1978: 37). This may be compared with the 3 kn under sail averaged by the late first-millennium BC bundle rafts which took 20 days on the coastal passage from the River Ganges to Ceylon (Pliny, *NH*, 6.24. 82).

PROPULSION AND STEERING

Bundle craft have been propelled by towing, e.g. in Mesopotamia (de Graeve, 1981: 151) and in Tasmania (Jones, 1976: 244); by paddle, e.g. in Ancient Egypt (Hornell, 1946: 48); by pole, e.g. in India (Hornell, 1946: 59); and by oar, e.g. in Morocco (Hornell, 1946: 55) where a stick was used as the rowing thwart and tholes were inserted in a block of wood lashed to the side bundles. Sail has been widely used: the earliest evidence being from Egypt where a mast of two sheers stepped on the outer bundles was used (Hornell, 1946: 49). Strabo (16.1.19) noted that Egyptian reed craft were propelled by reed sails. A bipod mast was used on some Lake Titicaca reed rafts with a reed matting rectangular sail of high aspect ratio which Hornell (1946: 44, pl. 5A) claims was set *aft* of the mast. This rig could only be used for running downwind or on a very broad reach (Edwards, 1965: 10–11). Johnstone (1973b: 216–7) states that Heyerdahl's *RA* 2 could only manage to sail without leeboards when the wind was within the sector two points (*c.* 22°) either side of dead astern.

ADVANTAGES AND DISADVANTAGES

Bundle rafts have the flow-through advantages of rafts in general – they are almost unsinkable and buoyant, at least until they become waterlogged. But even when waterlogged *RA* 2 floated with the decks awash (Heyerdahl, 1978). Bundle rafts are mostly in use on inland waterways, but some are used in open roadsteads and through 'impassable surf in conditions that would be hazardous to plank boats' (Edwards, 1965: 2, 3, 6), and coastal voyages are well documented.

These craft are made from readily and widely available raw materials and are so light that a mule can carry a man, his bundle raft, fishing net and equipment for a hut (Edwards, 1965: 3). They are relatively much cheaper than comparable plank boats; for example, in 1968 a Lake Titicaca reed raft cost £5 whilst a plank boat cost £30 (C. Sharp, p.c.) Their lashings are, however, vulnerable on beaching, and the reed, bark or poles soon deteriorate so that the longest useful life seems to be about a year, even with conscientious building and maintenance, and many rafts have a much shorter life, perhaps 2–3 months. Operationally it is difficult to make progress against wind and tide.

Bundle *boats* do not have the raft's wash-through advantages, but they have the boat's cargo carrying and freeboard characteristics. Being impregnated with bitumen they are not only waterproofed but have a longer life. In comparison with plank boats, however, they are more easily holed and sunk when deep with cargo on rock-strewn stretches of river (Hornell, 1946: 107).

EARLY USE IN N.W. EUROPE

Table 9.3 sets out the tools and techniques required to build a basic bundle *raft*; the techniques are all construction (additive) ones. The overall shape is obtained by shaping individual bundles and by fastening them together by coiled basketry in a relationship which gives the required shape. Whilst the act of compressing the heart of each bundle is itself a transforming (moulding) technique it is

Table 9.3 Earliest evidence for techniques and tools analogous to those used in building a basic bundle raft

Activity	*Techniques and tools*	*Earliest evidence*
1. Collect reed/grass light poles bark	Cutting tools Stripping tools	Suitable tools were available from Upper Palaeolithic onwards (Clark, 1977: 25) Use of light poles from Lower Palaeolithic (Clark, 1977: 26); reed may have originated in Upper Palaeolithic (Clark, 1952: 226) but first noted in Neolithic (Clark, 1977: 75); bark Mesolithic (Clark, 1952: 208)
2. Make rope. Bind material into bundles and lace bundles together	Coiled basketry	Mesolithic or possibly Upper Palaeolithic origins (Clark, 1952: 226–8) but first noted from Neolithic (Clark, 1977: 75)

Note: Poles and bark may also be windfalls or driftwood.

perhaps best thought of in this context as part of the additive technique or binding bundles.

The insertion of a subsidiary framework (Table 9.4) into a bundle *boat* is additive, although the preparation of the framing may involve subtractive, splitting, techniques. Waterproofing the hull is a transformation of the raw bitumen block by heat into a plastic substance which can be spread as a cover. As the shape of the boat is determined by the bundles (inner elements) and not the waterproofing cover which is added last, bundle boats are built in the skeleton sequence (Ch. 2).

Table 9.3 suggests that, on present evidence, bundle rafts could probably not have been built before the Mesolithic technological stage in N.W. Europe; and from Table 9.4 bundle *boats* could not have been built before the Bronze Age. No bundle *boats* are known, either actual or representational, outside the Mesopotamian region, and, although tree resin or wood tar might have been usable instead of bitumen, it seems unlikely on balance that bundle *boats* were ever used in N.W. Europe.

Table 9.4 Activities additional to those in Table 9.3 required to build a bundle boat

Activity	*Techniques and tools*	*Earliest evidence*
1. Insert framework	Woven basketry	Mesolithic origins (Clark, 1952: 226–31) but first noted in Neolithic (Coles, 1975: 84)
2. Waterproof hull	Transformation of mastic-like substance	Bronze Age (Clarke & Piggott, 1976: 180; Forbes, 1936: 86–7)

Note: An open mesh framework is not so complex as true woven basketry and this may have been used earlier

On the other hand, there is a twentieth-century distribution of bundle rafts in Europe: Ireland, Hungary, Corfu and Sardinia. No archaeological remains have so far been recognised but in view of the almost ephemeral nature of bundle rafts and the fact that they are readily re-used when no longer buoyant (e.g. Edwards, 1965: 4) this may not be significant. There is one scrap of documentary evidence for the early use of reed rafts in N.W. Europe: a tenth-century AD transcription of a Gallo-Roman will has a reference to 'navem liburnam ex scirpo' which Johnstone (1973a: 5) has translated as 'light and speedy reed boat'.

As with lografts, it seems most unlikely that bundle rafts would have been used at sea off N.W. Europe, but their early use on inland waterways should not be excluded from consideration despite the meagre evidence. The raw material was readily and widely available throughout N.W. Europe, and the recent use of simple reed rafts on the River Suck and on Lough Erne is a pointer towards possible similar use in other parts of N.W. Europe from Mesolithic times onwards.

10 Hide boats and float rafts

HIDE BOATS

Some hide boats are skeleton built (C8-RTC; C10-RT) with a waterproofing envelope of hides fastened together and moulded around a pre-erected framework. Similar to them are the nineteenth-century Arabian boats known as *tarada* reported by Layard (1853: 552) which had a woven basketry framework waterproofed by bitumen (C12-T). Other hide boats consist of a watertight shell (a leather bag), which is sometimes reinforced by an inserted framework, and are therefore shell-built (C5-RT; C7-RTC).

DISTRIBUTION AND FUNCTION

Hide boats are known in N. and S. America (Hornell, 1946: 148–54, 163–72; Adney and Chapelle, 1964: 174–211, 219–20), Siberia and Mongolia (Hornell, 1946: 159–63, 172–5; Brindley, 1919; Sauer, 1802; Sinar, 1961: 158–9; Serruys, 1981), Greenland (Petersen, 1986) and the Aleutians (Hornell, 1946: 156–9, 163–5, 169–172), Ethiopia (Hornell, 1946: 21) and India and Tibet (Hornell, 1946: 93–100; Kentley, 1984). In Classical times they are known to have been used in Arabia (Herodotus, 1.194; Strabo, 16.4.19), Italy (Lucan, *Pharsalia* 4.130–8), Spain (Strabo, 3.3.7) and in Britain and Ireland (Caesar, *BG*, 1.54 and others – see Hornell, 1946: 111–48). They are still used in Arabia, Britain and Ireland today.

They have been used on inland waters as ferries and for fishing and the transport of cargo; see for example Hornell's (1938: fig. 4.2) illustration of a cow embarked in a Blasket Is. curach. Off western Ireland they are today used for coastal fishing and inter-island work. In the Arctic waters of N. America, Greenland and Siberia they are used for sea fishing and the hunting of seals and walruses (e.g. Brindley, 1919: 131) as well as for coastal voyages, towing whales and logs and as tenders to ships (Brindley, 1919: 136).

Early seagoing voyages in the Irish Sea and the Channel were described by late-Classical authors (Lucan, *Pharsalia* 4.130–8; Pliny, *NH*, VII: 206; Solinus *Polyhistor* XXIII; Avienus *Ora Maritima*) and in several medieval British, Irish and Anglo-Saxon chronicles (Hornell, 1946: 5–41).

VARIETY AND SIZE

Boats made of a single hide are known (Hornell, 1946: 139, 150; Sinor, 1961: 158) and there are many examples of boats made of several hides sewn together. Some of the single-hide boats are temporary ones, often without any framework. The latter group, and some others with an inserted framework are built in the shell sequence, but most of the recent, better-known hide boats – *coracle, curach, kayak, umiak* – are skeleton built.

There is a wide range of shapes including circular ones (India), elliptical (Arabia, Britain, Ireland), rectangular (Tibet, S. America) and generally 'boat-shaped' ones (Ireland, Arctic). The *kayak* and *baidarka* of the Arctic (Hornell, 1946: 155–175; Brindley, 1919: 130–8) have an elongated, lanceolate shape giving them the speed required in their specialised hunting role. They are also unique in being decked-in with 'tailor-made' openings for one, two or three crew (Sauer, 1802: xv).

The usable portion of an oxhide gives an area of *c.* 4 × 3 ft (1.22 × 0.91 m) (Hornell, 1946: 117; Severin, 1978: 288; Rausing, 1984: 81) and thus single-hide boats are about this size. Table 10.1 gives the known range of dimensions for several types of hide boat.

All hide boats are distinctly lightweight. Typical figures are:

Kayak – 15 to 20 kg (Hornell, 1946: 167–172)

Table 10.1 Dimensions of five types of hide boat

	Smallest (m)	*Largest (m)*	*Group L/B mean (range)*	*Group L/D mean (range)*	*Reference*
Circular	1.52 (dia)	4.27 (dia)	—	3.5 (3.25 to 4)	Hornell, 1946: 96; 1936
Elliptical	1.22 × 0.91	1.83 × 1.37	1.4 (1.1 to 1.7)	—	Hornell, 1936; 1937; 1938
Boat-shaped curach	2.44 × 1.07 × 0.61	7.60 × 1.37 × 0.61*	5.2 (4 to 7.1)	9.2 (6.8 to 12.5)	Hornell, 1936; 1938
Umiak/baidara	6.3 × 1.45 × 0.46	10.67 × 1.37 × 0.61†	5.7 (4.5 to 7.9)	12.1 (10.4 to 13.8)	Adney and Chapelle, 1964: 181–190; Hornell, 1946: 155–163
Kayak/baidarka	4 × 0.50 × 0.24	10 × 0.80 × 0.50	13.1‡ (9.5 to 15)	20 (10 to 40)	Hornell, 1946: 163–75; Adney and Chapelle, 1964: 190–211

* The *Brendan* replica was 36 × 8 × 3 ft (10.98 × 2.44 × 0.91 m) with L/B = 4.5 L/D = 12 (Severin, 1978).
† Others are said to be in the 36–40 ft (11–12 m) range (Adney and Chapelle, 1966: 183) and early Greenland explorers are said to have noted *umiak* of *c.* 60 ft (18 m) in length (Adney and Chapelle, 1966: 175–6).
‡ The Koryac kayak which has a distinctive shape has an (L/B) of only 3.6 (Hornell, 1946: 172). Johnstone (1980: 220) noted a kayak with L/D of only 6.4.

Dee coracle – 16 kg (Hornell, 1946: 127)
Kerry curach – 75 kg (Kentley measurements)

CONSTRUCTION

Raw material The wide range of animal hides used for boat hulls is listed in Table 10.2, as are the few details available about sewing materials. Timber species used for framework are given in Table 10.3 and the materials used to lash the framework together and to lash the hide cover to the frame in Table 10.4

Building sequence Hide boats may be shell built or skeleton built, they may be single hide or multiple

A

Figure 10.1 Five stages in the building of a curach or *navog* in Co. Kerry. (Photos: National Museum of Ireland.)

B

C

D

hide, and they may or may not have a framework. Those with framework may be further classified by whether this is:

(a) Woven basketry i.e. prestressed, interwoven, twilled wooden basketry or interlaced laths which may be:
    (i) closely woven
    or (ii) more open
    Round and elliptical boats are generally in this group today.
(b) Lashed, pegged or nailed framework. These are generally boat-shaped and, except for the twentieth-century curachs of Connemara and Co. Mayo where the longitudinal members touch at the edges (Hornell, 1937: 149; 1938: pl. 1), are open framework. Examples of these various types are given in Table 10.5.

Today, Irish boats have fastened framework and are built upside down, although during the process they are rotated upright or on to their side to gain access to the inside (Fig. 10.1). On the other hand

Table 10.2 Raw materials used for the hulls of hide boats

Material	*Region*	*Reference*
Hull		
Ox	Ireland & Britain	Hornell, 1946: 117ff. Brendan's *Navigatio,* Ch. 4
	India	Hornell, 1946: 94
Horse	Britain	Hornell, 1946: 113, 117
	Mongolia	Sinor, 1961: 158–9
Yak	Tibet	Hornell, 1946: 99
Camel	Mongolia	Sinor, 1961: 158–9
Caribou	Arctic	Hornell, 1946: 166
Reindeer	Arctic	Hornell, 1946: 174
Whale	Arctic	Brindley, 1919: 134
Seal	Arctic	Hornell, 1946: 155; Brindley, 1919: 135
Walrus	Arctic	Brindley, 1919: 131
Sewing hides		
Reindeer sinew	Siberia	Brindley, 1919: 136
Whale sinew	Siberia	Brindley, 1919: 136
Caribou sinew	Arctic	Arima, 1975: 104

Note: Severin used ox hides sewn by greased flax.

E

Figure 10.1 (con't)

Table 10.3 Timber species used in hide boat frameworks

Common name	*Botanical name*	*Region*	*Reference*
Withies	*Salix* sp.	Ireland & Britain	Hornell, 1946: 112–13, 122–4, 133; Fenton, 1972
		Arctic	Adney and Chapelle, 1964: 185
Ash	*Fraxinus* sp.	Wales	Hornell, 1946: 127
Juniper	*Juniperus* sp.	Tibet	Hornell, 1946: 100
Fir	*Abies* sp.	Arctic	Adney and Chapelle, 1964: 185
Spruce	*Picea* sp.	Arctic	(Adney and Chapelle, 1964: 185)
Bamboo	*Bambusa* sp.	India	Hornell, 1946: 94

Notes: 1. Whale bone is used in Siberia (Brindley, 1919: 130–8)
2. Severin used ash with oak gunwales.

Welsh coracles have a woven basketry or interlaced laths (except for a modern Welshpool boat illustrated by Jenkins, 1974: 190) and are built upright (Hornell,

Table 10.4 Raw materials used as lashings in hide boats

Material	*Region*	*Reference*
Lashing frame		
'rope'	Ireland	Hornell, 1937: 149; 1938: 21
nails	Ireland	(Hornell, 1937: 149; 1938: 21)
hide thongs	Arctic	Hornell, 1946: 156–9; Adney and Chapelle, 1964: 178
baleen	Siberia	Brindley, 1919: 131
whale sinews	Arctic	Adney and Chapelle, 1964: 178
whale fins	Siberia	Sauer, 1802 XV
caribou sinews	Arctic	Arima, 1975: 104
Hide to frame		
leather	India	Hornell, 1946: 95–6
	Tibet	Hornell, 1946: 100
withies	Tibet	Hornell, 1946: 100
horsehair	Scotland	Fenton, 1972
balleen	Siberia	Brindley, 1919: 131

Notes: 1. Framework lashings are not essential when the frame is woven basketry.
2. Severin used alum-tawed leather.

1936: 27–8, 264–5). Early description of European hide boats (Caesar, *BC* 1, 54; Lucan 4.130–8; Pliny *NH* VII: 57) seem to suggest that their framework was a woven wickerwork. The seventeenth-century curach drawn by Capt. Thomas Phillips (Fig. 10.2) also has a woven framework as has the nineteenth-century Scottish curach in Elgin (Fenton, 1976). It may therefore be that fastened frameworks are relatively recent in N.W. Europe. Hornell (1938: 35–7) suggested that the Phillips's drawing was inaccurate because it depicts the boat being built upright, contrary to recent Irish practice. It is true that an early seventeenth-century description of curach building seems to imply that the building began with the embryo framework upside down (Hornell, 1938: 29), nevertheless the choice of building upright or upside down may be determined by the particular circumstances rather than by culture or tradition, and size of boat and relative ease of weaving basketry or interlacing laths may influence the decision. The hide boat's framework is light and can be rotated, and indeed is so rotated in all known sequences of building; see for example the Greenland umiak the framework of which is built upright but which is turned upside-down before the hide cover is fastened on (Knuth, 1980; Johnstone, 1980: 34–6).

Frameless boats

In S. America (Hornell, 1941b; 1946: 150–4) Abyssinia (Hornell, 1946: 21) and in Mongolia

Figure 10.2 Captain Phillips's late-seventeenth century drawing of curachs. (Photo: Pepys Library, Magdalene College, Cambridge (by permission of the Masters and Fellows, Magdalene College, Cambridge).)

Table 10.5 Examples of hide boats classified by number of hides and by type of framework

	Type	*Region and boat*		*Reference*
1.1	Single-hide, frameless	Mongolia	*hunt'o*	Sinor, 1961: 158
		S. America	*pelota* (some)	Hornell, 1941b; Hornell, 1946: 150–4
		Abyssinia	*jandi*	Hornell, 1946: 21
1.21	Single-hide, closely woven framework	Scotland	curach	Fenton, 1972
	Single-hide, interlaced framework	Wales	coracle (Figure 10.3)	Hornell, 1936: 5–41
1.22	Single-hide, fastened framework	Ireland	R. Boyne curach*	Hornell, 1937: 154–61
		N. America	bull boat (Figure 10.4)	Hornell, 1946: pl. XXIA
2.11	Multiple hide, closely woven framework	India	*parisil*	Hornell, 1946: 93–7
2.12	Multiple hide, open weave or interleaved lath framework	Wales	coracle	Hornell, 1936: 5–41
2.13	Multiple hide, fastened framework	Ireland	curach (Figure 10.5)	Hornell, 1937: 74–83, 148–175; 1938: 5–39
			umiak (Figure 10.6) baidara	Hornell, 1946: 155–63; Lethbridge, 1937:
			kayak/ baidarka	Adney and Chapelle, 1964: 174–211

Note: 1. The frameless boats are shell-built; some of the framed single-hide boats may also be shell-built, the framing being inserted after the hide cover has been shaped. For example, the boats of the Malecite Indians (Adney and Chapelle, 1964: 219–20) are built by bark boat methods (Ch. 7) and are therefore partly shell-built.

* Discounting the rim reinforcement of closely woven basketry.

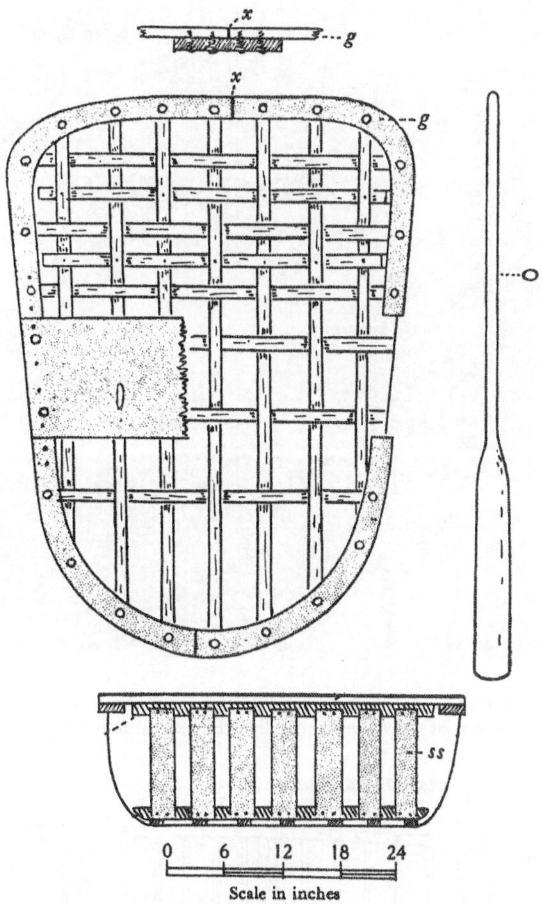

Figure 10.3 Construction drawing of a twentieth century coracle from the River Taf, Wales. After Hornell, 1936: fig. 5.

(Sinor, 1961: 158) certain small boats were built as a watertight shell of hide and used without a framework (Ch. 2). The rim of this 'leather bag' was stiffened either by lashing the four corners or by weaving a thong around the periphery. The hide boats of the Malecite Indians were built in a similar shell sequence and had a light framework inserted (Adney and Chapelle, 1964: 219–20).

Framework

Whatever the shape or method of construction of the framework, the 'rim' of the boat has to be stiffened to give a rigidity to the structure and, for those used under oar, to provide a bed for the oar fulcrum (*c.f.* sheerstrake in a planked boat). This is done by reinforcing the upper basketry as in the Indian *parisii* and Welsh coracle or by having strong gunwale timbers as in the curach/biadara/umiak (Fig. 10.5, 10.6).

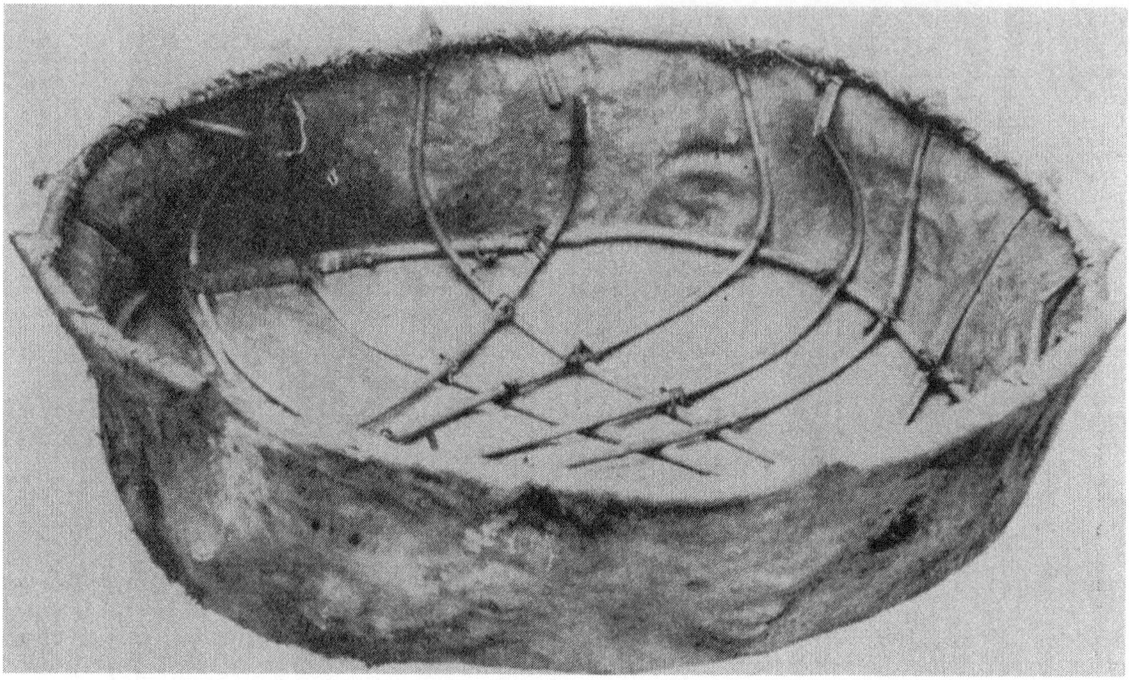

Figure 10.4 A simple form of hide boat (bull boat) used in the twentieth century by Sioux Indians. (Photo: courtesy of the Museum of the American Indian, Heye Foundation.)

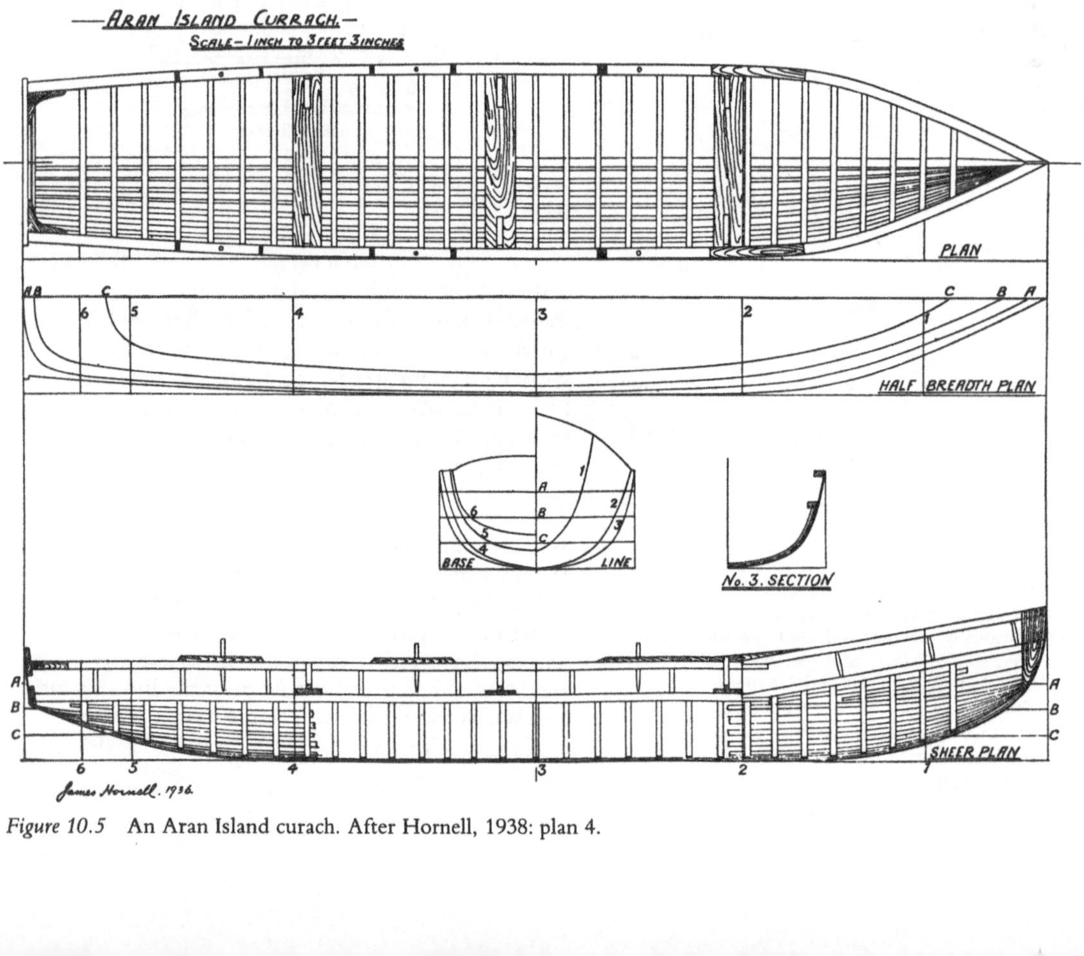

Figure 10.5 An Aran Island curach. After Hornell, 1938: plan 4.

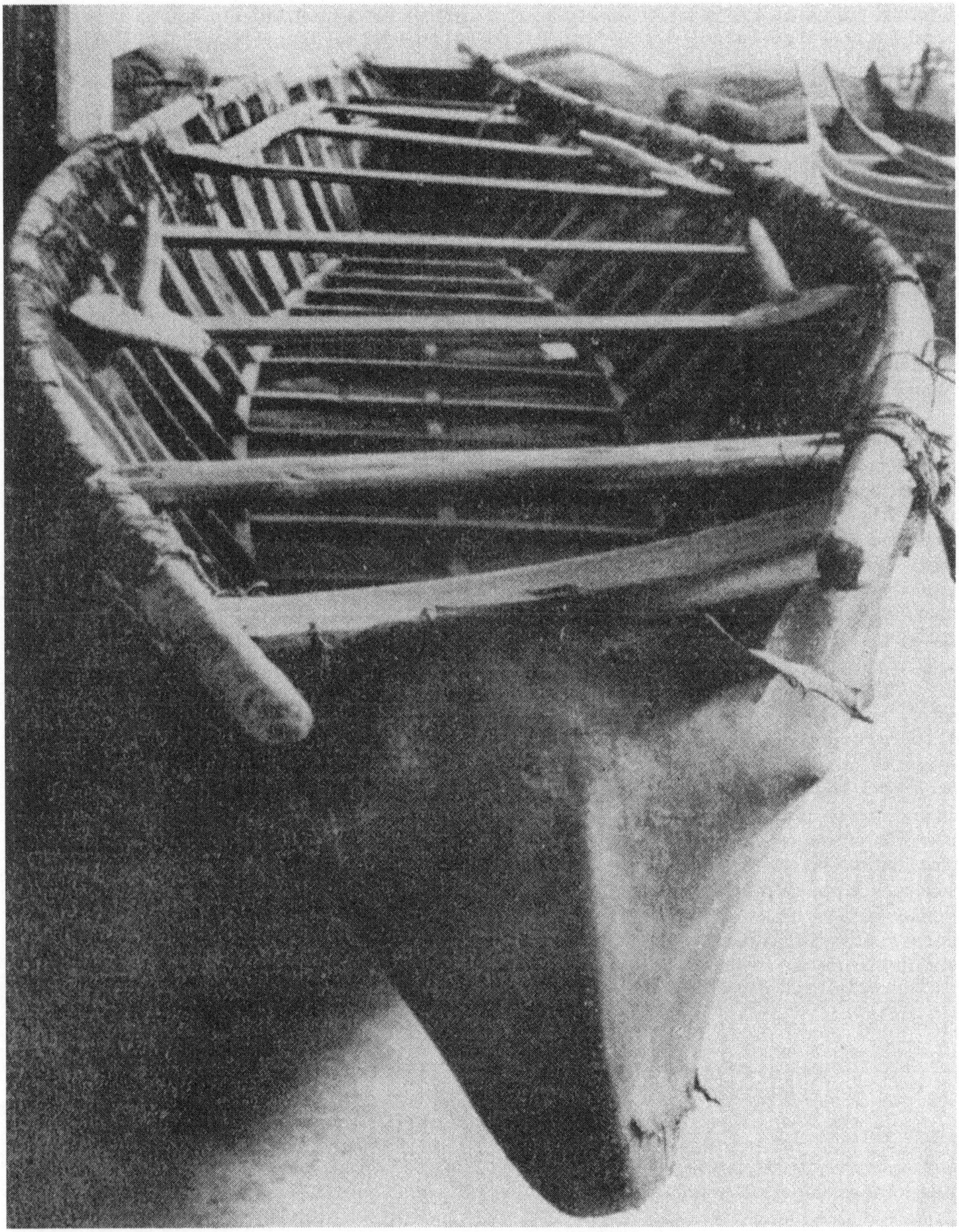

Figure 10.6 A modern Greenland *umiak*. (Photo: Ethnographic Museum, Stockholm.)

The central longitudinal member of the umiak's framework is relatively substantial (Lethbridge, 1937: 12; Johnstone, 1980: fig. 16.3; Knuth, 1980: figs 3 and 10; Adney and Chapelle, 1964: 178), and may be likened to a keel, or perhaps a keelson as it is inside the hide cover. Commodore Billings referred to 'keels' in the baidara he examined during his late-eighteenth century expedition to Siberia (Bauer, 1802: XV). Apart from the Tory Is. curach (Hornell, 1937: 167, pl. IV, fig. 1) twentieth-century curachs generally have no special central longitudinal member, although some, e.g. those from Co. Kerry (Hornell, 1938: pl. VII and VI), have a broader lath. Hornell (1938: 35–7) further criticised Phillips's drawing, which shows a curach with an external keel and stem (Fig. 10.2), claiming that there was no corroborating evidence. In the medieval period, however, Adomonnan in his *Vita St. Columba* and other authors (Marcus, 1953–4: 315) describe curachs with keels, and Caesar (*BC*, 1.54) also referred to British hide boats with keels. It therefore seems probable that pre-nineteenth century British and Irish hideboats had a keel externally like Phillips's boat or internally like the umiak. Severin's (1978: 288) replica had a broad central lath and he fastened an external 'keel-skid' to it through the leather. Some Canadian umiaks have a similar fitting of whalebone (Hale, 1980: 120).

Hide

Animal skins may be treated in several different ways to preserve them and yet make them flexible in use (Hodges, 1964: 148–52). A *rawhide* which has the inner fatty layer scraped off to reduce decay is known as a *pelt*. The hair may subsequently be removed by scraping after the pelt has been sweated by burial or by soaking for a few days. A pelt so treated will, nevertheless, decay in time and, as it dries, it will become stiff. Further treatment to delay, or even prevent, decay turns the pelt into *leather*. There are two main methods:

1. *Dressing* by (a) wood smoke
   (b) oil e.g. tallow
   (c) salt
   (d) alum (tawing)

   Such treatments are reversible to differing degrees and prolonged soaking in water will leach out the preservatives and decay will ensue.
2. *Tanning* with an infusion made from a wide variety of trees and shrubs e.g. oak bark, as used by St Brendan (*Navigatio* Ch. 4). The aim of this treatment is to produce a non-degradable leather which is water resistant, dimensionally stable, and of high tensile strength yet of sufficiently open structure that it will absorb grease (Severin, 1978: 284).

After dressing or tanning the leather is treated with an oil to prevent it drying which would lead to stiffness and cracking. Fat may be used for this (*Navigatio*, Ch. 4 see O'Meara, 1978) or raw wool grease (Severin, 1978: 284–8).

Few ethnographic reports state how hides used in boatbuilding were treated, and those that do appear to present contradictory information. Whereas St. Brendan's boat, certain recent Irish curachs (Hornell, 1946: 135) and other boats had tanned hides, Brindley (1919: 134) states that Siberian *baidarkas* were made of undressed seal skins; and the S. American *pelota* and the Boyne and Donegal curachs are said to have untanned hides (Hornell, 1946: 150; Hornell, 1937: 156, 159). The North American Indians prefer to use green hides rather than cured ones as they can be stretched to shape (Adney and Chapelle, 1964: 186) and fresh hides are used on the Tory Is. curachs so that they shrink to a close fit on the frame as they dry (Hornell, 1937: 164). The resolution of this paradox that certain boatbuilders apparently use untreated and therefore highly degradable hides may lie in the fact that they delay treatment until the boat is finished when it is dressed with oil and/or fats which both preserve and waterproof; see for example the Caribou Eskimos who use a treatment which is said to last for two summers (Arima, 1975: 127–130).

Hides are sometimes used in boatbuilding with their hair still on (Hornell, 1937: 164; 1946: 173; Adney and Chapelle, 1964: 219), but generally it is removed. Hides may then be fitted with their former hairside either inside the boat, e.g. in the N. American bull boat (Hornell, 1946: 149) and the River Spey curach (Fenton, 1972), or outboard as in the S. American *pelota* (Hornell, 1946: 150) and the N. American caribou boats (Arima, 1975: 127). A dry, untreated hide is easier to bend in the natural way (Hornell, 1946: 150) and this may be the reason why certain boats have the former hair-side outboard.

Hide covers are sometimes brought right over the 'rim' of the boat's framework and lashed to an internal member as in the umiak (Adney and Chapelle, 1964: fig. 168), the River Spey curach (Fenton, 1972) and in Tibet (Hornell, 1946: 99); in other boats the hides stop short of the reinforced 'rim': Tibet (Hornell, 1946: 95, pl. XVIA); and India (Hornell, 1946: pl. XVA).

Figure 10.7 Timothy Severin's hide boat replica *Brendan*. (Photo: T. Severin.)

MAINTENANCE AND REPAIR

To preserve the leather and to retain watertightness the hide covers of Indian boats were periodically smeared with a tannin infused from the shrub pobbakka (Hornell, 1946: 96). Tibetan seams were treated with fat as were Eskimo *umiaks* (Hornell, 1946: 99). When a *baidarka*'s skin sagged it was re-oiled (Brindley, 1919: 131). The hides of Severin's replica *Brendan* (Fig. 10.7) were tanned in oak bark, oiled with cod oil and dressed with wool grease (1978: 284–6) and there was no major change in the composition or performance of the leather during the voyage (1978: table 2).

Splits or holes in the hides can be laced together with a leather thong and holes patched, as in the R. Spey curach (Fenton, 1972). A tear in *Brendan*'s hull caused by ice was repaired externally by a sewn patch backed by wool grease and fibre (Severin, 1978: 242). Whenever possible, hide boats not in use are turned upside down to dry out and to preserve their shape.

VARIANTS

Hide substitutes. Waterproof cloth replaced hides in Wales probably in the eighteenth century (Hornell, 1946: 118) and in Ireland by the mid-nineteenth century (Marcus: 1953–4).

Extra freeboard. Adney and Chapelle (1964: 183) noted that certain umiaks had a cloth 'washstrake' of *c.* 20 in (51 cm) breadth, temporarily fastened above the sheerline.

Extra buoyancy. Inflated sealskins are attached amidships at the waterline on *baidara* of Siberia (Brindley, 1919: 136) and Greenland umiaks (Hornell, 1946: 156).

Extra stability. Two Caribou Eskimo Kayaks were sometimes paired (Hornell, 1946: 66).

PERFORMANCE

Loads
The smaller hide boats (Fig. 10.3) are meant for one man, his fishing gear and catch, although one standing passenger can sometimes be carried (Fenton, 1972). Boats of intermediate size such as the 6 × 3 × 5 ft (1.83 × 0.91 × 1.52 m) Tibetan boat could carry five men and their baggage with 1 ft (31 cm) freeboard (Hornell, 1946: 100). Hornell (1938: 28) has noted that a temporary hide boat 26 × 6 × 5 ft (7.9 × 1.83 × 1.52 cm) built to cross the River Shannon in 1602, carried 30 men and towed horses astern. The largest of the Indian round boats carried fifty men or forty bags of grain, and a seventeenth-century report stated that carts were transported across rivers tied by pole and wheels between two such boats (Hornell, 1946: 95). The large *baidara* of Siberia could carry fifteen to twenty men or 3 to 10 tons of cargo, in the latter case they were manned by four paddlers and one steersman (Brindley, 1919: 135–6).

Speed
Kayaks are demonstrably fast boats, speeds of 6 kn under paddle being noted (Ariba, 1975: 138–9). Speeds of umiaks, baidara and curach under oars are not so well documented, but ten oarsmen in *Brendan* could make *c.* 3 kn (Severin, 1978: 289). *Brendan*'s average day's run was 40 nm. (1978: 290). In Force 3 to 4 winds she had a cruising speed of 2 to 3 kn. When attempting to beat to windward, however, the boat made considerable leeway, with no headway when *c.* 5 points off the wind (50–60°), which was the closest she would point.

Propulsion and steering
Hide boats are propelled by pole (Hornell, 1946: 96); by paddle (Hornell, 1946, 96; Fenton, 1972) – double and single paddles are known (Brindley, 1919: 134–5); by oar – pushing against rawhide grommets in Tibet (Hornell, 1946: 100) or pulling in double tholes on a Siberian *baidara* (Brindley, 1919: 135); and by sail, for example the *baidara*. They are also used drifting downstream with a paddler at the bow to keep the boat in the mainstream (Hornell, 1946: 100).

A square, high aspect ratio sail is depicted in use on the three large seventeenth-century curachs drawn by Phillips (Fig. 10.2). Nance (1922) and Hornell (1938: 36–7) have described the setting of the sail abaft the mast of one boat when running, as 'an extraordinary expedient' inferring that the scene must have been carelessly observed. On the other hand Hornell (1946: 44, pl. 5A) has himself drawn attention to the reed rafts of Lake Titicaca, Peru which also set a square sail abaft the mast – possibly as a way of maintaining control in difficult conditions?

Certain *baidara* in the early twentieth century had a centreline rudder (Brindley, 1919) as has the seventeenth-century Irish sailing curach illustrated by Capt. Phillips (Fig. 10.2).

ADVANTAGES AND DISADVANTAGES

The hide boat is in concord with her environment in the circumpolar zone and in such places as the western coast of Ireland where there are few trees, a good supply of hides from land or sea animals, and where life can be sustained by sea fishing and hunting. Such coastlines are often exposed and hide boats such as the curach and the *baidara* are excellent surf boats (Sauer, 1802: XV). These boats can be relatively quickly built; for example, O'Sullivan Beare built a large one in two days (Hornell, 1938: 28–9). They are also relatively cheap – Hornell (1936: 9) noted that in the early twentieth century an Irish hide boat was one-tenth the cost of a comparable plank boat – and are readily repaired.

The frameless, single-hide boat, the essence of simplicity, is readily available for intermittent use on inland waterway crossings and, if dried and dressed, can be re-used many times. Framed boats can be used in more arduous circumstances and if well maintained can also have a long useful life. The lashed framework is resilient yet holds the shape of the boat and gives strength to the structure for least weight. The combination of this framework with a relatively lightweight hide cover is shock resistant, the elasticity of the frame and the ability of the whole skin to move relative to it means that loadings can be absorbed (Adney and Chapelle, 1964: 180). A woven framework is especially able to absorb energy (Ch. 3).

There are further advantages and some disadvantages in such a structure. On the debit side, the hide cover cannot contribute to strength and boats are limited in size (Ch. 3): 40 ft (12.2 m) would seem to be the limit generally reached and the 60 ft (18.3 m) boats reported by early Greenland explorers (Adney and Chapelle, 1964: 175–6) may have been very close to the practical limit. On the debit side too: the skin is relatively easily holed.

Figure 10.8 A curach under oars off the Dingle Peninsula, Co. Kerry. (Photo: NMM Greenwich.)

On the other hand, light weight means that many hide boats are readily portable: a Tibetan 5-man boat, 6 ft in diameter (Hornell, 1946: 100, pl. XVIA) can be carried by one man on his back; a 12-man *baidara* can be carried by four men (Sauer, 1802: XV) and many small coracles weigh only 18–20 lb (8–9 kg) (Hornell, 1946: 116, 127). A Co. Kerry 2- or 3-man curach *c.* 6 m (Fig. 10.8) weighs 75–80 kg whereas the 6.4 m Greenwich replica *faering*, capable of carrying a similar crew, weighs 134 kg (McKee, 1974: 30). A lightweight structure is buoyant with light draft even when loaded (Sauer, 1802: XV) and has good speed potential (*ceteris paribus*). Light draft and good freeboard do, however, mean excessive leeway in a wind, unless means are used to combat this (Ch. 12).

With multi-hide boats a wide variety of shapes can be built to suit different functions. The framework can be shaped to give good sheer at the ends and thus improve seaworthiness (Hornell, 1938: pl. II, 11, plan 2, pl. III, 18, plan 3). Transverse stability can be built in by having a broad boat. The kayak sacrifices this aspect of stability (Ariba, 1975: 139–140) to gain speed, and by having a watertight deck maintains its seaworthiness: techniques – 'the kayak roll' – have been perfected to recover from capsize (Adney and Chapelle, 1964: 223–9).

EARLY USE IN N.W. EUROPE

Table 10.6 indicates that in N.W. Europe the frameless, single-hide boat could have been built by the technology available in the Upper Palaeolithic. Framed single-hide boats with a lashed framework could also have been built in the Upper Palaeolithic, in the Mesolithic if of woven basketry, the Neolithic if fastened with treenails and the Bronze Age if metal fastenings (Table 10.7). The light timbers required in this framework would have been available from the Upper Palaeolithic, or bones could have been used as recently in Siberia (Brindley, 1919: 130–8). Multi-hide boats would have been possible from the Mesolithic (Table 10.8).

Table 10.6 Earliest evidence for tools and techniques analogous to those used in building a frameless single-hide boat

Activity	*Tools and techniques*	*Earliest evidence*
1. Skin animal	Knives and scrapers	Mid-Palaeolithic (Clark, 1977: 25, 26, 36)
2. Collect bindings	Cutting tools	Upper Palaeolithic (Clark, 1952: 226)
3. Fasten (corners of hide)		Upper Palaeolithic (Clark, 1952: 226; Dennell, 1983: table 4).

Ellmers (1984) sought to demonstrate that an antler fragment dredged from the harbour of Husum, Schleswig-Holstein is the remains of a framed hide boat of the Upper Palaeolithic. The evidence seems scarcely to justify this inference, and the fragment is in any case not firmly dated.

Rock carvings in Scandinavia and northern Russia

Table 10.7 Earliest evidence for tools and techniques analogous to those used in building a framed single-hide boat

Activity	*Tools and techniques*	*Earliest evidence*
1. Skin animal	Knives and scrapers	Mid-Palaeolithic (Clark, 1977: 25, 26, 30)
2. Collect bindings and frame material	Cutting tools	Upper Palaeolithic (Clark, 1952: 226)
3. Make framework	Woven basketry	Mesolithic (Clark, 1977: 113)
	Lashed	Upper Palaeolithic (Clark, 1952: 226)
	Pegged	Neolithic (Clark, 1952: 211–12)
	Nailed	Bronze Age (?)
4. Cover frame	Lashing	Upper Palaeolithic (Clark, 1952: 226; Dennell, 1983: table 4)

Table 10.8 Earliest evidence for tools and techniques analogous to those used in building a framed multi-hide boat

Activity	*Tools and techniques*	*Earliest evidence*
1. Skin animal	Knives and scrapers	Mid-Palaeolithic (Clark, 1977: 25, 26, 30)
2. Collect bindings and frame material	Cutting tools	Upper Palaeolithic (Clark, 1952: 226)
3. Make framework	Lashed	Upper Palaeolithic (Clark, 1952: 226)
4. Sew hides	Needles and awls	Upper Palaeolithic (Dennell, 1983: table 4)
5. Waterproof seams	—	Mesolithic (Clark, 1952: 226)
6. Cover frame	Lash	Upper Palaeolithic (Clark, 1952: 226)

include some which probably represent water transport but there has been much discussion about dating and the precise types depicted (Coles and Harding, 1979: 317). In contrast to the skin boat theories of Marstrander (1963: 1976) and Johnstone (1972), Christensen (1972a: 161) believes some of these carvings (dated to the Bronze Age because similar representations appear on early metal swords) represent logboats or planked craft, and similarities with the Iron Age Hjortspring planked boat (Rosenberg, 1937) support his interpretation. On the other hand, some carvings (e.g. the Navestad group illustrated by Marstrander, 1963: pl. 64, figs 15 to 17) have an unusually low length/depth ratio, and thus may represent skin boats rather than planked craft.

The evidence Watkins (1980) adduces to support his hypothesis that grave 2 in the EBA cemetery at Barns Farm, Dalgety, Fife contained a hide boat burial is meagre. The only other possible evidence is the Caergwrle shale boat model from Flint which is generally dated to this period and may represent a hide boat (Denford and Farrell, 1980). The direct evidence for hide boats in Bronze Age N.W. Europe is thus somewhat slight.

A small gold model boat from Broighter, County Derry, is dated by association with other artifacts to the first century BC (Fig. 10.9). The model, now in the National Museum, Dublin, originally had nine rowing thwarts with associated oar grommets through holes in the hull. Mast and yard indicate that the boat could also be sailed, and she appears to have been steered by an oar over the quarter. It is not certain which type of boat is represented, but a hide boat, possibly a curach, seems likely (Farrell and Penny, 1975). The only other reference to excavated material is by Sheppard (1926 who noted a 'coracle-type vessel' containing a skeleton found near the mouth of the R. Ancholme in north Lincolnshire and possibly of Roman date; the remains no longer survive.

Several Roman authors, including Caesar (*B.G.* 1.54) and Pliny (IV: 104, VII: 206) described British boats of wicker framework covered with hides. These are also mentioned in the extracts from the sixth-century *Massaliote Periplus* used by Avienus in his fourth-century AD *Ora Maritima*. Although there are difficulties in identifying the place names it seems probable that this *Periplus* describes hide boat voyages in the seas between Brittany, Britain and Ireland (Hawkes, 1975). References to hide boats on the River Po by Lucan (*Pharsalia* 4: 130–8) and in northern Spain by Strabo (3.3.7) show that their use was not restricted to northern waters.

The continued use of hide boats in parts of Britain is documented by isolated references through the medieval period and later (Hornell, 1946: 9–13, 297–303). Small fishing coracles are still in use in Wales today and the last Scottish curach, thought to be from the nineteenth century, is now in Elgin Museum (Fenton, 1972). Irish hide boats are documented from the sixth century AD onwards, and they continued to be used into this century (Hornell, 1937): 74–83, 148–175; 1938: 5–39; McGrail, 1981a: 26).

Some form of hide boat may have been used on inland waterways in N.W. Europe as ferry and cargo transport and for fishing from the Upper Palaeolithic/Mesolithic period. It is possible to suggest with more confidence that, from at least the later Bronze Age, hide boats similar in form and size to the twentieth-century *baidara* but possibly with some of the structural features of the early twentieth-century curach, may have been in use on the Atlantic seaboard. Such boats would have been capable of coastal and cross-Channel voyages and possibly of extended sea voyages, as for example, to the northern isles. Although Severin's *Brendan* was not an authentic replica of an early seagoing Irish curach, as a large hide boat built generally by pre-industrial age techniques she has provided valuable information about seakeeping qualities. The boat proved to have considerable stability and remained relatively dry in rough Atlantic seas. Despite the continuously varying stresses experienced she was in good shape at the end of the

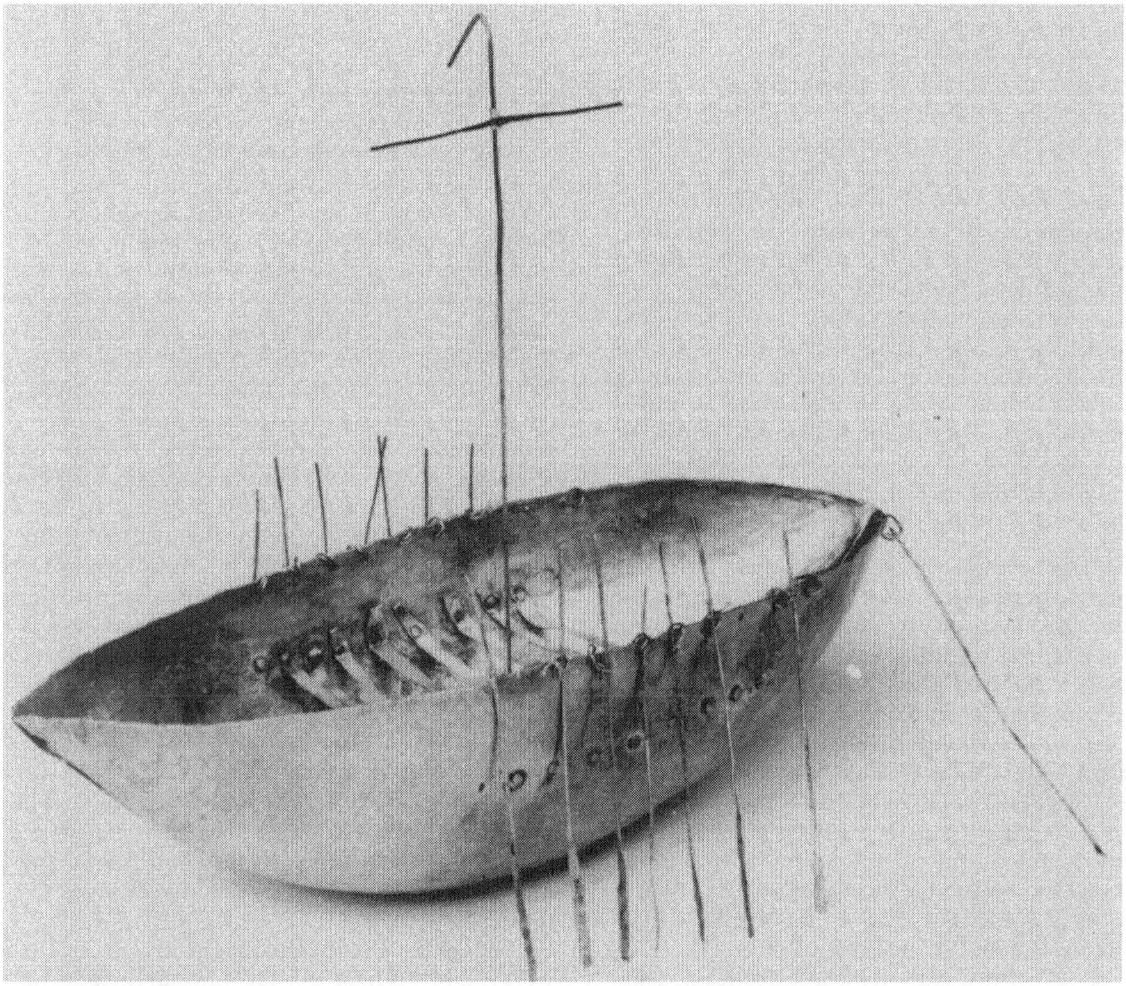

Figure 10.9 Small gold model from Broighter, Co. Derry, Ireland dated to the first century BC. (Photo: National Museum of Ireland.)

voyage, although after 6 months afloat the frame had lost some sheer and the cover of 49 oxhides had developed corrugation.

RAFTS OF HIDE FLOATS AND RELATED TYPES

Hide floats linked together, usually by a framework of light poles, are generally known as *buoyed rafts* (Hornell, 1946: 20) as they are thought of as a subset of the log raft class with extra buoyancy from the floats. However, in most, if not all, cases the log raft element – the framework – is not immersed, and buoyancy is solely derived from the floats. As most known types use inflated bags of hide as floats, these rafts are dealt with here as a subset of hide boats, although this is not entirely satisfactory as related types use gourds, kelp bladders, pots and metal cans (Hornell, 1946: 34–7, 37–9; Edwards, 1965: 93).

DISTRIBUTION AND FUNCTION

Rafts of *hide floats* have been used widely: China, Mongolia and Tibet (Hornell, 1946: 25; Worcester, 1966: 120–2), India and Afghan (Hornell, 1946: 22–4), Arabia, Palestine and Armenia (Hornell, 1946: 20–1, 26–30), Abyssinia and Morocco

(Hornell, 1946: 21, 31–2) and S. America (Hornell, 1946: 32–4; Edwards, 1965: 17–20, 89–90). In Europe they are only known in Albania on the River Drin (Hornell, 1946: 31; Nopcsa, 1925; Traegar, 1904). Classical authors mention their use in Arabia (Pliny, *NH* 6.35; Huntingford, 1980: ch. 27).

Pot float rafts are more restricted in distribution, principally in China, Korea, India and Egypt (Hornell, 1946: 34–7, pl. IVA), although there is some iconographic and documentary evidence for their early use in the Mediterranean region: Etruscan engravings depict such a raft with sail, and they were used to cross the strait of Messina in *c.* 71 BC (Hornell, 1946: 34)

Gourd or calabash rafts have an even more restricted distribution mainly in Egypt, Central and E. Africa (Hornell, 1946: 37–8) and in S. America (Edwards, 1965: 59–60, 92–3), whilst *kelp bladder* rafts are known only from the Chatham Is. (Hornell, 1946: 39).

Use

Float rafts are widely used as ferries and for carrying passengers and goods downstream (Worcester, 1966: 120; Hornell, 1946: 22, 26, 37) and for river fishing (Hornell, 1946: 37). Offshore use is known only in S. America where a specialised form of hide float raft (Fig. 10.10) is used for coastal fishing and as tenders to visiting ships (Hornell, 1946: 33; Edwards, 1965: 18–19). On the Huang Ho river in China rafts carry cargoes of wool, camel hair or tobacco leaf within their hide floats (Hornell, 1946: 25; Worcester, 1966: 120).

In China and Iraq the wooden framework of hide float rafts is sold at the downstream terminal (Worcester, 1966: 120; Hornell, 1946: 30); the floats are dried, deflated and carried back upstream. Thus these rafts are used to deliver light timber; there are analogous uses on the River Nile where the pots of pot-float rafts are sold in Cairo (Hornell, 1946: 36).

VARIETY AND SIZE

Hide float rafts may be divided into two groups:

1. individual floats made from a single hide.
2. individual floats made from several hides – this type is known only in S. America (Hornell, 1946: 32–4; Edwards, 1965: 17–20, 89–90).

The S. American hide float rafts average *c.* 1.25 × 2–3 m: the other types vary in size depending on the job in hand – they can be readily altered in size. A common arrangement in India is to tie something like a wooden bedstead (*charpoy*) across two buffalo hide floats for general ferry duties (Hornell, 1946: 22–3, fig. 3). Even smaller than these is a Mesopotamian raft of only one oxhide float surmounted by four bamboo poles and a light wooden platform some 2.5 × 2.5 ft (76 × 76 cm) (Hornell, 1946: 30). The largest ones in this region are the *Keleks* of Iraq which can be up to 50 × 50 ft (15 × 15 m) with up to 1000 hide floats (Hornell, 1946: pl. 4B). In China hide float rafts vary from very small ones of *c.* 10 × 6 ft (3 × 1.8 m) with 12 to 15 sheephide floats, to freighters of *c.* 75 × 50 ft (23 × 15 m) with 600 to 700 sheep hide floats or 125 to 150 oxen or yak floats, a ratio of *c.* 4.75 sheep to 1 ox or yak. Although several of the rafts described above are rectangular most others for which dimensions are known are square.

Small, temporary *pot float* rafts have been used in China with the pots held together by a framework of four spear shafts (Hornell, 1946: 37). In Korea, by contrast, rafts of up to 80 great pots are known (Hornell, 1946: 36).

In Egypt *gourd* rafts lashed together with canes were *c.* 8 × 4 ft (Hornell, 1946: 37). In Mexico and Peru rafts were generally 1.25 × 1.50 m (Hornell, 1946: 38). A sixteenth-century report described a raft of *c.* 1 × 1 m with 8 calabash (gourds) in a net: an

Figure 10.10 Drawing by Admiral Paris of a nineteenth-century seal hide float raft at Valparaiso, Chile. (Photo: NMM Greenwich.)

Table 10.9 Raw materials used to make and lash hide floats

Material	*Region*	*Reference*
Floats		
Buffalo	India	Hornell, 1946: 22–5
Ox	China	Hornell, 1946: 25
Yak	China	Hornell, 1946: 25
Sheep	China	Hornell, 1946: 25
Sheep	India	Hornell, 1946: 22
Goat	India	Hornell, 1946: 22–5
Goat	Iran	Hornell, 1946: 26
Seal	S. America	Hornell, 1946: 32–3, Edwards, 1965: 17–20
Lashings		
animal sinew	S. America	Hornell, 1946: 32
twisted intestine	S. America	Hornell, 1946: 33; Edwards, 1965: 18
rawhide thongs	China	Worcester, 1966: 121

earlier report mentions a 100 calabash raft, but no size was given (Edwards, 1965: 92).

CONSTRUCTION

Raw materials

1. ***Hide floats.*** Materials used for hide floats and their lashings or thread are given in Table 10.9. The framework often consists of light branches or sticks lashed together (Hornell, 1946: 23, 26) but hurdles or indeed wickerwork beds may be used (Hornell, 1946: 22–3, fig. 3). The Albanian rafts have a carefully made frame of woven osiers (Traegar, 1904: fig. 2) and the two multi-hide floats of the distinctive S. American raft (Fig. 10.10) are linked by a triangular or rectangular framework surmounted by crosspieces covered by skin or reed matting (Edwards, 1965: 18).
   Floats are lashed to the framework by materials such as green withies (Hornell, 1946: 31). In China, and possibly elsewhere the floats are also fastened to one another (Worcester, 1966: 121).
2. ***Pot floats.*** In India nine unglazed pots (*chatty*) lashed to a bamboo platform by withies make a raft some 3 or 4 ft (0.91 – 1.22 m) square (Hornell, 1946: 34). Pot mouths are sometimes left open, closed with a leaf, or stopped with clay (Hornell, 1946: 34–6).
3. ***Gourds.*** Cane platforms link the dried gourds in Egypt (Hornell, 1946: 37). In S. America calabash (*Lagenaria* or *Cucurbita*) are sometimes enclosed in a net which may or may not be given a bamboo framework (Edwards, 1965: 59–60; Hornell, 1946: 38).
4. ***Kelp bladder.*** A number of inflated kelp bladders are fastened to the bottom and sides of a crate-framed raft in the Chatham Islands (Hornell, 1946: 39).

Building sequence

The procedures used in China and in India to make a single-hide airtight bag are very similar (Worcester, 1965: 120–1; Hornell, 1942) and are probably used elsewhere.

The characteristic technique is the skinning of an animal without making a longitudinal slit:

1. The feet and tail, and sometimes the head, are cut off.
2. In India the skin is drawn backwards from the neck; in China the skinner first blows air between skin and flesh through an incision in one leg and pummels the body to separate skin from flesh – the skin is then drawn forward.
3. Hair or wool scraped off, after soaking or burial.
4. Openings are securely fastened on the side which will be inside the float, leaving one hole for inflation.
5. Hides are evidently used untanned (Hornell, 1946: 24) but they are subsequently dressed; pine tar inside, pomegranate infusion outside in India; salt and a vegetable oil in China.
6. The float is inflated, larger floats being inflated from smaller ones (Worcester, 1966: 121).

The male seals used for the S. American multi-hide floats (Fig. 10.10) are skinned in the normal manner and the pelt is then separated into two layers. After softening in water they are cut to size and four or more hides are then joined together to produce one slender tapering float. To join hides, cactus spines or fish bones are passed through the edges of adjacent hides in the manner of toggles, which are then lashed in a criss-cross pattern using soaked seal intestines to produce an airtight seal. When finished, the float is filled with sand or reeds and left to dry in the sun. The float is then emptied and inflated and the seams treated with seal grease and the entire float coated with a mix of clay, grease and oil (Hornell, 1946: 32; Edwards, 1965: 17–20).

MAINTENANCE AND DURABILITY

With good maintenance hide floats can last a long

time, although no precise figures are known. When not in use hides are kept dry and then soaked in water before inflation. After use they are washed and rubbed with a tannin preparation (Hornell, 1946: 29). Floats becoming detached on passage can be retrieved and punctured floats can be replaced and repaired by sewing patches inside and out (Worcester, 1966: 121). Spare floats are carried on the *kelek* of Iraq (Hornell, 1946: 29) and probably elsewhere.

Pot floats require no special preparation, but gourds are dried before use. The Chatham Islands kelp bladders, some 2 ft × 8 in (61 × 20 cm) have their soft internal tissue removed before they are inflated (Hornell, 1946: 39).

VARIANTS

In recent times metal cans have replaced pots in Egypt (Hornell, 1946: 36) and gourds in S. America (Edwards, 1965: 93).

PERFORMANCE

Loads

Herodotus (1.194) states that the fifth century BC hide float rafts of Mesopotamia could carry 5000 talents which Hornell (1946: 29) estimates as 125 to 150 tons, but which has elsewhere been translated as 14 tons. Little other reliable data is available, but Chinese rafts of *c.* 7 × 7 ft (2.13 × 2.13 m) are said to carry six or seven men and ½ ton cargo (Worcester, 166: 122) and an Albanian raft 1.3 × 1.3 m can carry 100 kg (Npocsa, 1925: 141).

Propulsion and steering

When drifting downstream as many of these rafts do, they are steered by bow and stern sweeps (Worcester, 1966: 121; Hornell, 1946: 28). Ferries are often towed or pushed by swimmers (Edwards, 1965: 59; Hornell, 1946: 22–3) and paddled on rivers and in coastal waters (Hornell, 1946: 22, 33) – a double paddle is used in S. America (Edwards, 1965: 17–18).

The *kelek* has one or two large oars each side (Hornell, 1946: 28) and the hide float rafts of Swat, India (Fig. 10.11) has an oar pivoted on a vertical strut.

Small cotton sails were sometimes used on the S. American hide float rafts (Edwards, 1965: 18) and Etruscan engravings depict early pot rafts with sail (Hornell, 1946: 34).

ADVANTAGES AND DISADVANTAGES

Once hide floats, pots or gourds are to hand, rafts are quickly assembled, e.g. a 16 goat-hide float raft of

Figure 10.11 A twentieth-century hide float raft on the River Swat, northern Pakistan. (Photo: Basil Greenhill.)

c. 6 × 6 ft (1.83 × 1.83 m) can be assembled in *c.* 1½ hours (Hornell, 1946: 23). Many of them have a temporary *ad hoc* existence and when in use they can easily be increased or decreased in size. Component parts can readily be stowed away or sold when not needed.

The one-way, downstream rafts of Iraq and China are an economic way of delivering light timber as well as other goods. Float rafts carry a bigger load, when compared with log rafts, yet retain a light draft and they are easier to manoeuvre on river bends (Worcester, 1966: 120).

Hide float rafts are the only form of water transport capable of making a successful passage down the rocky rivers of the Himalayas and in Iraq: they remain afloat and are not delayed even if a few floats are punctured; in this respect they have been compared to a ship with watertight compartments (Hornell, 1946: 29).

The specialised hide float rafts of S. America are seaworthy and can be successfully operated through surf. They have been known to stay at sea for two to three days (Edwards, 1965: 19).

Table 10.10 Earliest evidence for tools and techniques analogous to those used in building a raft of multiple-hide floats

Activity	*Tools and techniques*	*Earliest evidence*
1. Skin animal	Knives and scrapers	Mid. Palaeolithic (Clark 1977: 25, 26 36)
2. Collect bindings and frame material	—	Upper Palaeolithic (Clark 1952: 226)
3. Sew hides	Needles and awls	Upper Palaeolithic (Dennell, 1983: table 4)
4. Waterproof seams	—	Mesolithic (Clark, 1952: 226)
5. Make framework	Lashed	Upper Palaeolithic (Clark, 1952: 226; Dennell, 1983: table 4)

Table 10.11 Earliest evidence for tools and techniques analogous to those used in building a raft of single-hide floats

Activity	*Tools and techniques*	*Earliest evidence*
1. Skin animal (special)	Knives and scrapers	Later than Mid. Palaeolithic
2. Collect bindings and frame material		Upper Palaeolithic (Clark, 1952: 226)
3. Lash openings	—	Upper Palaeolithic (Clark, 1952: 226; Dennell, 1983: table 4)
4. Make framework	Lashed	Upper Paleolithic (Clark, 1952: 226; Dennell, 1983: table 4

EARLY USE IN N.W. EUROPE

Tables 10.10 and 10.11 suggest that multiple hide float rafts would have been technologically possible in N.W. Europe from the Mesolithic. On the other hand single-hide floats which do not have waterproof seams but require a specialised skinning technique may have been possible from the Upper Palaeolithic. Pot float rafts could have been in use from the Neolithic or Mesolithic. Gourds are not indigenous to N.W. Europe, although other plants with similar characteristics may be, as may suitable forms of seaweed.

Rafts of pot floats are known from mid-first millennium BC representations and documentation in the Mediterranean region, but are not otherwise known in Europe. Hide float rafts are known only from twentieth century Albania. The lack of evidence in N.W. Europe may be due to the essentially temporary nature of this form of water transport and the possibility remains of their use in early times on inland waterways.

11 Assessment of performance

Assessing the performance of excavated examples of early boats and ships is difficult but by making certain assumption it is possible to give broad answers to such questions as how fast was this boat? and what was her cargo capacity? Upper and lower bounds may be established, rather than precise answers; for example it may be possible to suggest that a particular boat could carry between 10 and 15 tonnes of cargo depending on the mean density of her cargo and on what her master considered to be a safe freeboard. The value of performance estimates – what they tell us about ancient finds – is directly related to the quality of the data used and the appropriateness of the assessment techniques. These factors include:

(a) Accuracy of excavation documentation and the post-excavation research.
(b) Authenticity of the theoretical reconstruction, the *hypothesis*, or of the model or replica if built.
(c) Validity of values given to such parameters as:
bulk density of 'typical' cargo
safe freeboard
performance of sail material
(d) Relevance of twentieth-century naval architectural methods to the analysis of data from earlier centuries.

Questions which are even more difficult to answer than those about maximum speeds and cargo capacity are those that deal with windward performance and average speeds achieved over long distances; such matters depend on prevailing weather, predominant winds and on the master's seamanship and are thus considered in Chapter 13.

Boat finds are invariably incomplete: the ends, the upper parts and the evidence for the rowing outfit or the sailing rig are rarely excavated. Thus the original form and structure of all boat finds have to be theoretically reconstructed, some more so than others, and the details of the propulsion equipment have to be deduced or 'guesstimated'. Reconstruction drawings, models and replicas of ancient boats abound, but seldom is the reasoning behind them published; the logical link between the excavated remains and the reconstruction or replica is frequently lacking. Informed appraisal may identify anachronistic reconstructions, such as leeboards on a sixth-century BC boat or two masts on a Viking Age craft, but other assumptions made by the reconstructor or replica builder may not be so readily apparent unless the deductive process is published. The smaller proportion of a vessel that survives the more likely it is that there will be more than one reconstruction compatible with the excavated remains. Indeed, single-solution reconstructions seem likely to be rare (McGrail, 1984b: 23–4); yet until recently single solution reconstructions have almost invariably been published.

METHODS OF ASSESSMENT

Once an authentic reconstruction drawing is available the performance of the ancient boat it represents may be assessed in several ways (McGrail, 1985b).

By eye. An experienced seaman, boatbuilder or naval architect may be able to make general assessments of performance, providing he has some familiarity with the particular tradition of boat building.

Using simple coefficients. These are based on the boat's *overall* measurements rather than the underwater form. Thus (L/B) and (B/D) summarise the overall proportion of the boat and, as performance is related to shape, may be used in a general way to give relative assessments of a boat's capabilities.

Using hydrostatic curves. These involve the definition of the waterline(s), assessments of the underwater shape and calculations of displacements and transverse section areas (Ch. 3). Coefficients based upon the underwater geometry of the hull may be calculated and used to give forecasts of performance.

Small-scale models. These may be built for tank, and possibly wind tunnel tests to evaluate performance by empirical methods. This has the additional advantage that, if a constructional model is made as a prelude to a tank model, something may be learned about the structure of the boat and the building techniques used.

Computer work. This can be a cost effective method of investigating certain aspects of performance. It can be especially useful when there are alternative reconstructions to be investigated or the single reconstruction is defined somewhat imprecisely over a range of values. Impractical solutions may be eliminated and promising ones evaluated.

Building a full-scale replica and undertaking trials. Such projects are popular and spectacular but much of the experimental work undertaken to date in Maritime Archaeology has been of slight consequence when measured by the standards of authenticity and cost/benefit. It is most important that, before such work is undertaken, the hypothesis formulation state – the establishment of a theoretical reconstruction – is undertaken rigorously. In many cases hydrostatic curve investigations, small-scale model building and testing, and computer modelling may well be more cost effective than full-scale replica work. Where replica building appears to be justified it is essential that the aim of the project be clearly defined at an early stage, and the complete sequence from obtaining raw materials to method of publication be set out and costed. As such experiments are expensive, and for all practical purposes are unrepeatable, it is especially important that only the most worthwhile are considered and that these should be critically appraised from all angles before being undertaken (McGrail, 1985b). Only in this way can scientifically valuable results be obtained.

GUIDANCE FROM NAVAL ARCHITECTURE

Although replica building has been popular recently, few projects have been published, and the authenticity of most replicas may be questioned. Tank and wind tunnel tests are relatively expensive and most performance estimates to date have been based on simple calculations, including the use of hydrostatic curves (Ch. 3). An understanding of how size and shape of hull determine a boat's cargo capacity and speed should ensure that the assessment of performance by eye, by coefficients, or by simple calculation is undertaken with more confidence. But a boat's performance is not just a function of hull design, it is also determined by the propulsion outfit and by the seaman's ability to use this combination in the prevailing circumstances – these two important factors are considered in Chapters 12 and 13.

The hull shape and dimensions give a boat buoyancy and stability and also influence other characteristics. Which attributes are to be emphasised in any one particular design will depend principally on the role the boat is to fulfil and on her operating environment. Thus, speed and manoeuvrability may be emphasised in a boat to be used for raiding, whereas payload and stability are required in a boat for trading or for fishing. As some characteristics are interlinked and cannot be simultaneously maximised (e.g. speed, capacity, stability) every boat will be a compromise to a degree; a multi-purpose boat more so.

SHAPE OF HULL

The underwater sections of a hull which will minimise power requirements (oar or sail) should have gentle 'eye-sweet' curves offering minimum resistance to passage through the water. The lines should allow the water which has been displaced by the forebody (bow and entry) to move relatively downwards and astern and then to rise diagonally along the buttock (or quarter) lines (the run and stern) back to the surface and to leave the boat with minimum turbulence at the stern.

The general shape of the hull should also conform to the function the boat is to undertake, that is in general to match the buoyancy as given by the various transverse sections along the length of the boat (bow, amidships, stern) with the loadings expected there. The requirements of speed and manoeuvrability should be considered in conjunction with these factors when determining the overall relationship between Length (L) and Breadth (B); capacity and stability considerations should similarly be considered when determining the relationship between B and Depth (D).

Ends

At the bow gentle underwater curves in all three

planes will ease water entry and minimise resistance. Flare above the waterline will improve sea-keeping qualities by helping to keep the boat dry. A gently curving stern has similar advantages. If the stern is roughly the same form as the bow (i.e. the boat is double-ended) the boat may be propelled either way with little loss of efficiency in an oared boat. A narrow stern may not give sufficient buoyancy at this end to match the weight of the steersman, or the varying loadings imposed by a quartering or stern sea. A broader stern gives the steersman more space to manoeuvre and provides a wider base for the sheets to the sail.

Sides

Outward sloping sides (flare) result in a boat being drier in a lop, but they may be more easily damaged than vertical ones. The greatest advantage of flare, however, is that transverse stability and payload increase with increasing draft. In addition, breadth of free-surface bilge water (Chapter 3) is less than that of the flotation plane, and thus transverse stability is not all lost as it can be in a wall-sided boat. The safety of flat-bottomed boats in rough water depends on this effect (Coates, 1984). A measure of assessing relative seaworthiness of this sort of boat is given by:

$$\frac{\text{depth of hull} \times \text{Tan } \emptyset}{\text{breadth of bottom}} \quad \text{(Fig. 11.1) where}$$

$\emptyset$ is the angle of flare measured from the vertical (Coates, 1984).

A boat with a hard bilge (sharp angle between sides and bottom) has more volume available for cargo. However, sharp changes in the transverse section disturb the waterflow and eddies are induced resulting in greater viscous pressure resistance. In contrast, rounded bilges allow the water to flow diagonally astern, resulting in less resistance.

Bottom

A flat bottom gives maximum capacity for minimum depth (and hence draft) of boat. A perfectly flat bottom longitudinally and transversely may be difficult to use from a beach due to mud suction and resistance to being dragged. Some fore and aft rocker and/or transverse camber best suits such boats. In addition rocker and camber help to control the adverse effects of free water by causing it to collect amidships where it is easier to bail out (Coates, 1984).

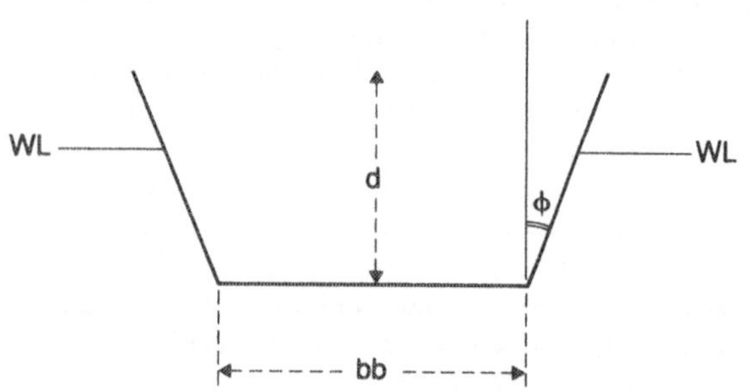

Figure 11.1 Diagram to illustrate calculation of flare index:
WL – waterline d – depth of hull
bb – breadth of bottom Ø – flare

SIZE OF HULL

Length and Breadth

A longer boat (*ceteris paribus*) may achieve a higher speed because the speed at which wave-making resistance becomes significant is increased, and in an oared boat because more oarsmen, hence power, can be embarked (although this increase has to be balanced against the increase in wetted area and hence skin friction). A beamier boat on the other hand has more transverse stability (Ch. 3) and should also have more capacity. McKee (1983: 81) has defined a *beamy* boat as one with $(L/B) \leqslant 2.6$; a *narrow* boat has $(L/B) \geqslant 3.75$.

If we compare two boats of the same displacement (∇) and draft (T), but one longer and hence narrower than the other, the narrower boat will have less transverse stability, and the greater (L/B) coefficient indicates that this narrow boat will be more directionally stable and thus more difficult to manoeuvre. On the other hand, if beam (B) and draft (T) are kept constant and length increased in proportion to displacement, transverse static stability will only be affected if the position of the C of G alters.

Depth and draft

Depth (D) of the boat amidships will generally determine the *freeboard* (F) and the draft (T) at any one displacement (Fig. 3.1). In a boat with a continuous deck at the level of the top of the boat's side amidships, freeboard is a measure of reserve buoyancy. Providing water *does not come over the sides* this is also true in an open boat. This reserve of buoyancy determines the maximum angle to which a boat can be heeled and still have a righting moment; once the top of the sides (deck edge) is awash resistance to heel begins to diminish. Buoyancy reserve is also important for longitudinal stability when a boat is pitching; increased buoyancy as the ends are increasingly immersed, produces a longitudinal righting moment. Sheer forward (that is an increasing freeboard due to a longitudinal upcurve in the boat's sides) thrusts water away in a seaway and reduces the risk of water coming inboard amidships. Sheer aft can similarly protect a boat in a quartering sea.

With constant displacement a boat of deeper draft will have less transverse stability (Ch. 3). The effect of varying draft in relation to the other dimensions may be summarised in two coefficients:

(L/T) A higher value of this coefficient indicates that the boat will be more manoeuvrable as the relatively light draft will produce less lateral resistance to turning. However, this could be a disadvantage under sail as light draft means less resistance to leeway (but see Ch. 8 and 13 for discussion on methods of resisting leeway).

(B/T) Boats with higher values will generally be 'volume dominated' i.e. good for bulky, low density loads. They will have good transverse stability providing that the loaded boat's centre of gravity (G) is not too high.

SCALE EFFECTS

Varying linear dimensions may have a disproportionate effect on other properties due to an application of the *Square-Cube* law (Gordon, 1978: 192/3). If the length and diameter of a mast are doubled its weight is multiplied by $2^3 = 8$; whereas the area of contact with the mast supporting system is multiplied by $2^2 = 4$. The stress is thus doubled, with the resulting requirement to strengthen the structure or increase the bearing area by a factor of 2.

SPEED POTENTIAL

Speed through the water may be improved by increasing the propulsive force (i.e. the driving force under sail) or by reducing the aerodynamic drag and the hydrodynamic resistance. Hull and rig must be matched, since the structure must be capable of withstanding the loadings imposed by the sail operating at maximum efficiency. In rough sea conditions it may be impracticable to achieve the maximum attainable speed because of pounding and torsional stresses imposed on the hull. Means of increasing the driving force under sail are discussed in Chapter 12, here we consider the means of reducing resistance.

Hydrodynamic resistance

Neglecting resistance caused by waves and spray, total hull resistance to motion may be broken down into the following elements (Rawson and Tupper, 1976/7: 353), some of which interact:

(a) skin friction
(b) viscous pressure (eddy)
(c) appendage
(d) air
(e) wave making

(b)–(e): Residual resistance

(a) *Skin friction* varies directly as the wetted or immersed surface area of the boat, and the smoothness or otherwise of that surface. In general it varies with V^2 (Marchaj, 1964: fig. 155) and is the largest component of resistance from zero up to speeds equivalent to a $(V)/(\sqrt{L})$ ratio of 0.7 (point A on Fig. 11.2)

(b) *Viscous pressure* is caused by anything which disturbs the velocity of water flow. Thus, abrupt changes of form such as a sharp turn at the bilge or a blunt bow increase this type of resistance. An area of negative pressure is created by a blunt stern and the resulting reverse flow and turbulence lead to increased resistance.

(c) *Appendage* resistance is due to (removable) external underwater fittings such as stabilisers on a logboat (McGrail, 1978a: 123).

(d) *Air* – this is aerodynamic drag.

(e) *Wavemaking*. This is the most significant element of residual resistance and is due to the resistance of two trains of surface waves which are

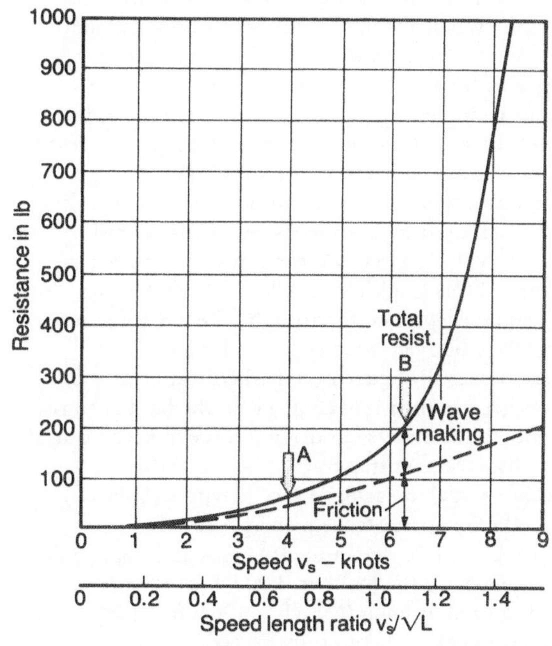

Figure 11.2 Wavemaking resistance and total resistance plotted against speed and speed/length ratio for a modern yacht of waterline length 45.5 ft (13.9 m). After Marchaj, 1964: fig. 155.

produced by, and move with, the boat: the bow produces a wave pattern with a crest at the bow; the stern, a weaker pattern beginning with a trough (Fig. 11.3). Both wave patterns absorb energy which would otherwise be propulsive. The *divergent* elements of these two wave systems do not combine and their influence on total resistance is relatively slight; the *transverse* elements, however, can combine and these have a significant effect on total wavemaking resistance (Marchaj, 1964: 248). The combined resistance to motion will vary depending on whether these two patterns are in or out of phase. It can be shown that the wavelength (λ) of these transverse waves = $2\pi V^2/g = 0.557\ V^2$ (λ in ft V in knots). Thus V (knots) = $1.34\ \sqrt{L}$(ft). Length of wave becomes longer the faster the boat moves. From this relationship it follows that:

$$\text{when } \lambda = \frac{(L)}{(5)} \quad \frac{(V)}{(\sqrt{L})} = 0.6$$

$$\text{when } \lambda = \frac{(L)}{(2)} \quad \frac{(V)}{(\sqrt{L})} = 0.95$$

$$\text{when } \lambda = (L) \quad (V)/(\sqrt{L}) = 1.34$$

At $(V)/(\sqrt{L}) \simeq 0.6$ wavemaking resistance begins to have an effect and by $(V)/(\sqrt{L}) \simeq 1.1$ (point B on Fig. 11.2) this resistance is increasing in proportion to V^3 or even V^6 and can be of greater value than skin friction. When λ = L (i.e. the boat's bow and stern waves are in phase with a wavelength equal to the length of the boat) then wavemaking resistance begins to increase markedly. In this condition, at a $(V)/(\sqrt{L})$ = *c.* 1.34, the boat has a large wave crest at the bow and another at the stern with the trough amidships; this distribution of buoyancy causes the bow to rise. At higher speeds, and consequently greater $(V)/(\sqrt{L})$, the wavelength will increase; the trough of this single wave will move aft and the stern will squat down into the trough where displacement and hence buoyancy are reduced (Fig. 11.3). Resistance is thus increased markedly at $(V)/(\sqrt{L})$ ratios greater than 1.34 (Fig. 11.2); the boat requires an enormous increase in driving force to break out of this 'displacement trap'.

Boats which do achieve speeds where $(V)/(\sqrt{L}) > 1.40$ may appear to be planing. At $(V)/(\sqrt{L}) \simeq 1.70$ dynamic lift begins and boats will semi-plane; by $(V)/(\sqrt{L}) = 3.20$ a boat is truly planing or skimming (Marchaj, 1964: fig. 158). Displacement boats can only exceed $(V)/(\sqrt{L}) = 1.40$ in ideal conditions or with excessive use of mechanical power. Modern yachts generally operate around 0.90, while cargo ships seldom exceed 0.50 (Marchaj, 1964: 254). Speeds appropriate to these values of $(V)/(\sqrt{L})$ are given for varying lengths of boat in Table 11.1.

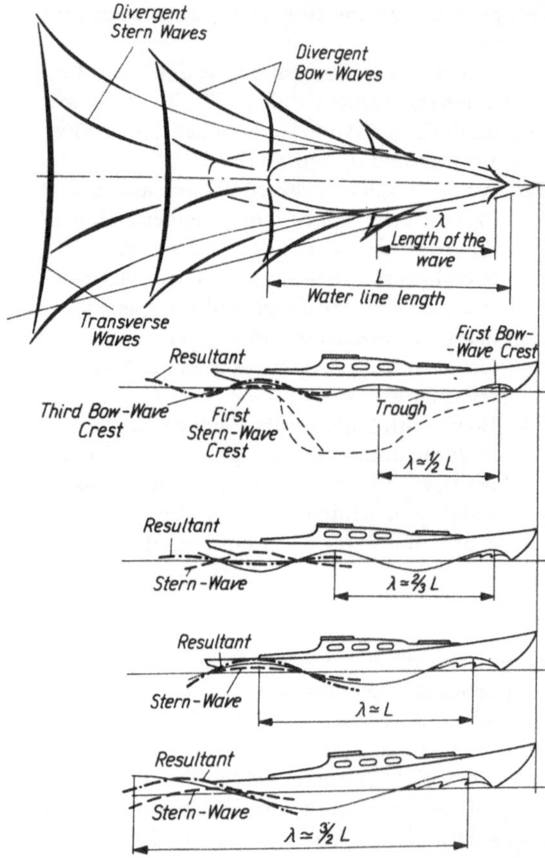

Figure 11.3 Diagram to illustrate wavemaking resistance and its relationship to wave length and speed of boat. After Marchaj, 1964: fig. 153.

SPEED ESTIMATES

Speed assessments are difficult to make from the lines of a boat even when the propulsive power is accurately known. Naval architects of today prefer to have practical confirmation of calculated assessments, due to the theoretical unpredictability of the formation of vortices with consequent energy loss (Coates, 1977: 225–6). Reasonable estimates may not be available until a small-scale model has been tank tested or even until after first of class sea trials. There are, however, certain readily calculated coefficients which, at least for twentieth-century forms of boat and ship, give an indication of speed potential.

The *prismatic coefficient* $(\nabla)/(A.L)$ may be used in the range of speeds equivalent to $(V)/(\sqrt{L}) = 0.6$ to

Table 11.1 Theoretical speeds achievable by boats of different waterline length

Waterline length (ft)	16	25	36	49	64	81	100
Waterline length (m)	4.88	7.62	10.97	14.94	19.51	24.69	30.48
Speed (kn) with $V/\sqrt{L} = 0.50$	2	2.5	3	3.5	4	4.5	5
Speed (kn) with $V/\sqrt{L} = 0.90$	3.6	4.5	5.4	6.3	7.2	8.1	9
Speed (kn) with $V/\sqrt{L} = 1.40$	5.6	7.0	8.4	9.8	11.2	12.5	14

1.1 to compare the wave-making resistance of different hulls. A low value, say 0.55 to 0.53, is given by a fine form away from the station of maximum section (in planked boats generally near amidships) and indicates low resistance and hence a potentially fast boat.

At the lower speed ranges three other coefficients may also be used to compare speed potential:

Slenderness Coefficient	$= (L)/(B)$.	High value (> 5) indicates good speed potential.
Midships Coefficient	$= (A)/(B \times T)$.	Low value (< 0.85) indicates good speed potential.
Block Coefficient	$= (\nabla)/(BLT)$.	Low value (< 0.65) indicates good speed potential.

The 'displacement trap' effect begins at $(V)/(\sqrt{L}) = c.\ 1.1$ and wave-making resistance begins to predominate, and the prismatic coefficient becomes unreliable as a guide to total resistance (McGrail and Corlett, 1977). The volumetric coefficient $= (\nabla)/(L^3)$ is more useful in this higher speed range. It has been shown that with values of $(\nabla)/(L^3)$ less than 2×10^{-3} excessive squat and wave-making will not occur and such a boat may escape the displacement trap without the massive increase in driving force which other forms require. This applies to boats under oars as well as sail.

The *faering* (Gokstad 3) a ninth-century, small-oared boat (Fig. 8.17) has a theoretical maximum speed, estimated from waterline length, of 5.9 kn. Her low block and prismatic coefficients (Table 11.2) suggest that speed/length ratios $(V)/(\sqrt{L}) > 1.1$ might be achievable, and her very low volumetric coefficient suggests that her underwater form is such that she could exceed the theoretical maximum speed for displacement craft, by 'semi-planing'. In sea trials of a replica of this *faering* a cruising speed of 6 kn and a sprint speed of 7.4 kn were in fact achieved (McKee, 1974: 21–3) equivalent to a $(V)/(\sqrt{L})$ of *c* 1.7. For other examples of semi-planing see Binns (1980: 163) and Christensen and Morrison (1976).

Theoretical maximum speeds and four indicators of speed potential for ten vessels are tabulated in Table 11.2: they are ordered chronologically from *c.* 1500 bc to *c.* AD 1000.

Volumetric coefficient. Hjortspring and Gokstad 1 and 3 have coefficients $< 2 \times 10^{-3}$ which indicate that they are designs of high speed potential (McGrail and Corlett, 1977).

Block coefficients. These are all low compared with the values expected of twentieth-century designs, Graveney being the only one approaching the value for a modern yacht (*c.* 0.50 – Rawson and Tupper, 1976/7: 13). As with the prismatic coefficient some of this difference is due to the 'double-ended' nature of the ancient boats, whereas modern designs have a blunter stern. Another difference is that the length of modern boats is measured between 'perpendiculars' whereas for ancient finds the full waterline length is used. The low values in Table 11.2 are characteristic of fine forms. Hjortspring is fuller at the ends than might be expected.

Prismatic coefficient. The range here is nearer that expected today for a fine form (*c.* 0.55). Hjortspring and Gokstad 1 are relatively fuller at the ends.

Slenderness coefficient. Skuldelev 1 and 3 and Graveney are relatively broader than twentieth-century designs (range 5 to 8 – Rawson and Tupper 1976/7: 572). The ranking here is similar to that for the volumetric coefficient.

Inspection of Table 11.2 suggests that these coefficients may not be directly relevant to the assessment of the speed potential of the Brigg and Ferriby boats which are of a radically different form from the others in the table. These two have therefore been omitted from Table 11.3 and the boats there are ordered by the value of their volumetric coefficient – the lowest numbers and therefore the highest speed potential at the top. The slenderness coefficient gives a similar ordering, as does the block coefficient, with the exception of Hjortspring. This boat and Gokstad 1 are in anomalous positions when assessed by the prismatic coefficient, otherwise the boats fall into a similar pattern when ordered by this coefficient. On this very limited test of these speed assessments it does seem that they can be validly applied to medieval boat forms to give a *relative* assessment of speed potential. To date, the

Table 11.2 Theoretical coefficients which indicate the speed potential of certain boats of N.W. Europe

Boat	*Lwl (m)*	*Bwl (m)*	*Draft (m)*	*Displacement volume (m^3)*	*Maximum* speed (kn)*	*L/Bwl*†	*Volumetric‡ coeff.*	*Block§ coeff.*	*Prismatic¶ coeff.*	*References‖*
Ferriby 1	14	2.5	0.40	10.50	9.5	5.60	3.8×10^{-3}	0.75	0.94	Wright, 1976
Brigg 2	12.3	2.3	0.46	12.50	8.8	5.30	6.7×10^{-3}	0.96	0.97	Coates, 1981
Hjortspring	14.7	1.5	0.31	2.64	9.7	9.80	0.83×10^{-3}	0.36	0.69	Rosenberg, 1937
Sutton Hoo 2	19	3.7	0.71	18.00	11.0	5.14	2.8×10^{-3}	0.36	0.56	Bruce-Mitford, 1975
Gokstad 1	22.5	4.2	0.91	17.96	12.0	5.36	1.58×10^{-3}	0.21	0.60	McGrail & Corlett, 1977
Gokstad 3	5.5	1.0	0.26	0.31	5.9	5.37	1.88×10^{-3}	0.21	0.42	McGrail & Corlett, 1977
Graveney**	10.1	3.6	0.65	11.00	8.0	2.77	10.7×10^{-3}	0.46	0.59	Corlett, 1978
Skuldelev 3	11.68	2.9	0.82	7.60	8.7	4.00	4.77×10^{-3}	0.27	0.50	Crumlin-Pedersen (p.c.)
Skudelev 1	15.6	4.5	1.27	34.00	10.0	3.45	8.96×10^{-3}	0.38	0.54	Crumlin-Pedersen (p.c.)

* $1.4 \sqrt{Lwl}$ (ft). Max. theoretical speed achievable by a displacement boat of this waterline length. Whether a boat can achieve such a speed depends upon (a) propulsive power available (b) waterline shape (c) wind and sea conditions.
† Slenderness coeff. Higher values suggest greater speed potential.
‡ Displacement volume/$(Lwl)^3$. Values $< 2 \times 10^{-3}$ suggest a hull with high speed potential
§ Displacement volume/Lwl Bwl T. Low values suggest speed potential. 1.00 = box form
¶ Displacement volume/section area Lwl. Low values suggest speed potential. 1.00 = consistent transverse section.
‖ Dimensions and displacements were taken from the References, except Ferriby's displacement which was calculated by Coates and Sutton Hoo's by Corlett.
** Coefficients were calculated from data in Figs 10.1.1 and 10.2.1 in the Reference and are different from those in Corlett's text.

Table 11.3 Speed potential coefficients – data from Table 11.2 rearranged

Boat	*Volumetric coeff.*	*Block coeff.*	*Prismatic coeff.*	*Slenderness coeff.*
Hjortspring	0.83×10^{-3}	0.36	0.69	9.8
Gokstad 1	1.58×10^{-3}	0.21	0.60	5.4
Gokstad 3	1.88×10^{-3}	0.21	0.42	5.4
Sutton Hoo	2.8×10^{-3}	0.36	0.56	5.1
Skuldelev 3	4.77×10^{-3}	0.27	0.50	4.0
Skuldelev 1	8.96×10^{-3}	0.38	0.54	3.5
Graveney	10.70×10^{-3}	0.46	0.59	2.8

publication of the Greenwich *faering* trial is alone in giving comparisons of speed assessment coefficients and sea trial achievements for an ancient boat design. If more such calibrations can be done more confidence would be generated in the use of these coefficients to forecast likely speeds.

Potential high speed can only become actual high speed if the appropriate propulsion power is available. Here again the reconstructor can be on shaky grounds, especially when attempting to determine the size of sail (Ch. 12). If the outfit of oars can be established then an approximation of the rowing power can be made (see Ch. 12).

CARGO CAPACITY AND TONNAGE

Cargo capacity depends on the internal dimensions of the boat, the space occupied by crew and equipment, and the reserve of buoyancy (relative to the buoyancy at light displacement). This capacity may be increased by reducing the weight of the boat or the number of crewmen (hence increasing the reserve of buoyancy) and by varying the boat's transverse section; thus, for fixed beam and depth a rectangular transverse section will hold more than a rounded one. The actual payload that may be carried will depend on the bulk density of the cargo and on displacement/draft/stability calculations. Payload may be described in two ways: *tonnage* related to the volume of the cargo (a medieval concept) and *deadweight* (related to the weight of cargo (a recent concept). Deadweight (W) is the difference between *light displacement* (weight of water displaced by the boat and equipment but without cargo) and *loaded displacement* (weight of water displaced by the boat and equipment with cargo). It is sometimes useful to distinguish between *gross deadweight* which includes the weight of crew as well as cargo and *net deadweight* which is the weight of the cargo. So that comparisons may be made between boats (both ancient and modern) it is advisable to estimate carrying power of hypothetical reconstructions in terms of deadweight derived from hydrostatic curves by stability calculations rather than using assessments of tonnage such as Builders Old Measurement (B.O.M.)

$$= \frac{(L - 3/5B) \times B \times \frac{1}{2}B,}{94}$$

as advocated by some authors. In the case where only length and breadth are known and B.O.M. or similar formulae are used, the results should be treated with caution as the B.O.M. formula is only applicable to bluff-bowed, full-bodied vessels. A useful tool for making comparisons between boats is the Deadweight Coefficient = (W)/(Δ): this measures the ability of a boat to carry cargo, in particular cargo of high density.

In medieval times the problem of equating a weight with a volume in terms of a standard unit of cargo (such as a cask of wine, which could readily be measured and counted) had to be solved to give an indication of a vessel's capacity for costing and taxation purposes. Different units were established in various parts of Europe (*tun, tonneau, toneladas, botta, last*) with different relations between weight and volume. Lane (1964) has shown that medieval statements of a ship's 'tuns burthen' may best be translated into twentieth-century measures by the following conversions:

Cargo weight 1 tun ≏ 2240 lb ≏ 1 tonne
1 tun ≏ 1.6 botta
1 tun ≏ 0.5 last
Cargo volume
1 tun ≏ 60 ft^3

Morken (1983) generally agrees with these conversions but believes that 2 botta = 1 tun.

Tuns burthen is thus the number of standard units (e.g. a cask of wine weighing 2240 lb and occupying a space of 60 ft^3) which a boat could carry. The space for cargo being of finite size, more cargo by weight could be carried of materials with density greater than that of wine in a cask (*c.* 1) and less of materials with lower bulk densities, subject in both cases to stability considerations (Ch. 3).

ESTIMATES OF CARGO CAPACITY

The assessment of cargo capacity depends directly on an accurate knowledge of the boat's underwater shape, which in turn depends on which waterline is selected; boats can float at a range of drafts (Ch. 3). How can the operational draft, and hence the waterline and freeboard, of an ancient boat be established? Ethnographers have seldom recorded boat operational parameters in any detail; the most that could be deduced from a search of the literature about logboats, for example, was that on inland waterways they were frequently operated with remarkably little freeboard (McGrail, 1978a). There are, however, other methods of tackling this problem:

1. Modes of operational use may be defined for specific classes of boat involved in certain work: for example it has been postulated that logboats on inland waters may be used in one of four roles:
   (a) draft restricted to 30 cm (minimum depth of water)
   (b) at a standard freeboard of 15 cm (safety consideration)
   (c) minimum freeboard as a function of transverse stability (upper edge of sides awash at 10° heel)
   (d) maximum number of crew there is space for, providing freeboard is not reduced below (1c).
2. Logboat roles (1c and d) may be adapted to suit larger boats and ships in one of two ways:
   (a) Man-carrying vessels. Estimate the crew, equipment and victuals and calculate weight, and see whether the resulting freeboard is 'safe' by some criteria. Different estimates of an average man's weight can lead to different estimates of draft and, in small boats, can lead to different appreciations of the boat's transverse stability.
   (b) Cargo-carrying vessels. The above procedure may be extended to cargo boats but unless a representative sample of cargo was excavated with the find, there may be problems in establishing a typical cargo; this would make mean bulk density estimates uncertain and thus volumes and weights unreliable.
3. Logboat role (lb) may also be applied to larger boats. The most promising development here is based on one of the medieval Icelandic laws in the Grågås Codex (Morcken, 1980: 178): a cargo vessel is fully loaded when $\frac{2}{5}$ of the total depth is freeboard.

Table 11.4 gives estimates of gross deadweight which in this context is taken to be the weight of the crew and their equipment plus cargo and victuals, everything except the weight of the boat and her equipment (light displacement). It should be

Table 11.4 Data on cargo-carrying ability of selected boats and ships from N.W. Europe

Boat	*Date (approx)*	*Shape*				*Freeboard (m)*	*Draft (m)*	*Deadweight*† *(tonne)*	*Dwt. Coeff.*	*Reference*
		*Overall** *dimensions (m)*	L/B	L/D	B/D	F	(*T*)	(*W*)	(*W*)/(Δ)	
Ferriby 1	1500 bc	15.4 × 2.6 × 0.7	5.9	22.0	3.7	0.36 0.26	0.30 0.40	3.0 5.5	0.54 0.52	Coates (p.c.)
Brigg 2	650 bc	12.3 × 2.3 × 0.34 0.55	5.3	10.0 20.5	6.8 4.2	0.09 0.09	0.25 0.46	1.54 7.16	0.23 0.57	Coates, 1981: 267–8
Hjortspring	300 BC	18.0 × 1.92 × 0.7	9.4	25.7	2.7	0.39	0.31	2.11	0.80	Rosenberg, 1937
Sutton Hoo 2	AD 600	27.0 × 4.3 × 1.4	6.3	19.3	3.1	0.62	0.71	12.75	0.71	Corlett (p.c.)
Gokstad 1	AD 850	23.2 × 5.2 × 2.0	4.5	11.6	2.6	0.79	0.91	9.5	0.53	Christensen, 1975
Graveney	AD 900	13.5 × 4.0 × 1.0	3.4	13.5	4.0	0.35	0.65	6 or 7	0.55 0.64	Corlett, 1978
Skuldelev 3	AD 1000	13.3 × 3.3 × 1.4	4.0	9.5	2.4	0.54	0.82	4.6	0.61	Crumlin-Pedersen (p.c.)
Skuldelev 1	AD 1000	16.5 × 4.6 × 2.1‡	3.6	7.9	2.2	0.85	1.27‡	24.0‡	0.71	Crumlin-Pedersen (p.c.)
Bremen cog§	AD 1380	23.5 × 7.5 × 5.3	3.5	5.7	1.4	—	—	130.0	—	Ellmers, 1972: 257
Bergen§	14th century	26.0 × 8.5 × 4.2	3.1	6.2	2.0	—	—	180.0	—	Ellmers, 1972: 257

* These are given in the order: overall length; maximum beam; depth from sheer to keel at position of maximum beam. Overall dimensions are some guide to shape and relative cargo space, but they can be misleading: see, for example, the Graveney and Hjortspring data which include the overhanging bow and stern and protruding extensions. The ratios (L/B), (L/D) and (B/D) are similarly only general guides to shape. McKee, (1983: 81) has proposed the following definitions:
L/B ⩽ 2.6 = beamy B/D ⩽ 2.0 = deep
L/B ⩾ 3.75 = narrow B/D ⩾ 3.0 = shallow

† The weight of crew and equipment and cargo and victuals.

‡ Crumlin-Pedersen (1984: 32) gives length 16.3 m, breadth 4.5 m carrying 15 to 20 tonne with a draft of 1.5 m.

§ The Bremen Cog and the Bergen ship – partial data are included for comparison. Elsewhere Ellmers has estimated the Bremen Cog deadweight as 100 tonne at a draft of 1.80 m whilst Fliedner has estimated 130 tonne.

emphasised that this table includes most, if not all, the information currently available about the cargo capacity performance of medieval and prehistoric boat finds in N.W. Europe. The bases for the assessments in Table 11.4 are:

Ferriby 1	A reasoned estimate of safe freeboard in (a) average weather, (b) fair weather.
Brigg 2	Reconstructions with two different heights of side were used. Freeboard was calculated to be safe under stated conditions of heel.
Hjortspring	Load taken to be 24 men (1730 kg) and equipment.
Sutton Hoo 2	Load taken to be 60 men (3000 kg) and equipment.
Gokstad 1	Load taken to be 70 men (5500 kg) and equipment.
Graveney	This appears to have been based on a choice of 0.58 m as the 'designed draft' and an experienced eye identified the deepest draft as 0.65 m.
Skuldelev 1 and 3	The Grågås rule was used to establish safe freeboard.
Bremen cog Bergen	Method not stated.

A number of comments may be made about the data in Table 11.4. Firstly, estimates of gross deadweight have been obtained by different methods with different assumptions and thus they are not truly comparable. For example, in some cases it has not been established that the vessel so loaded would have the necessary transverse stability. Secondly, differing parameters have been used: for example, the weight of an average man varies from 50 kg (Sutton Hoo 2) to 80 kg (Gokstad 1). Others have used 60 kg (Brigg), 72 kg (Hjortspring) and, elsewhere 70 kg (Greek trireme, Foley *et al* 1982). Thirdly, in view of the uncertain validity of some of the reconstructions analysed, it is to be regretted that in only two cases has more than one variant been assessed. Taking into account the circumstances of most maritime archaeological projects, a probabilistic rather than a deterministic approach to performance assessment is appropriate.

The ten boats in Table 11.4 span *c.* 3000 years and thus it is unlikely that this is a representative sample. Any conclusions drawn from this data must therefore be tentative. Bearing this in mind, we may postulate a general increase in cargo carrying ability with time. This increase was not achieved by any significant increase in length (Sutton Hoo 2 was longer overall than the Bremen cog), but the deadweight coefficient and the three shape coefficients indicate that forms became fuller, and beam and depth of hull were significantly increased relative to length.

CARGO BULK DENSITY

Estimates of deadweight made after safe freeboard or operating draft have been defined need to be examined in the light of the bulk density of likely cargo. For example, the deadweights suggested for Brigg 2 were investigated on two counts:

(a) to see whether there was space for the equivalent number of animals (sheep or cattle of a standard size and weight) and crew; and
(b) to ensure the boat still had adequate stability under heel and with leaky seams letting in bilge water.

When the nature of the cargo is unspecified as in the Skuldelev calculations then the mean bulk density of a load which would just fill the available cargo space may be calculated and evaluated as practicable or not in the light of any known trading patterns. For Skuldelev 1 this bulk density is 0.60 which is not unreasonable, but that for Skuldelev 3 is only 0.38, when even turf has a bulk density of 0.48 (corn is *c.* 0.68 and stone averages 2.5) and it may be that not all cargo space theoretically available in this boat could be so used.

CARGO PER CREW-MEMBER

A useful comparative index which may give some idea of relative costs of transport is weight of cargo per crew–member. Unger (1980) has attempted this type of calculation for certain early medieval boats but it is doubtful whether the data he uses is of sufficient quality to bear the inferences placed on them. The validity of such inferences depends not only on accurate estimates of deadweight (tuns burthen) but also on estimated crew sizes. On the basis of experience with replicas, Crumlin-Pedersen considers that a crew of five would be adequate for Skuldelev 3 and six for Skuldelev 1. The cargo per crew index then becomes 0.9 tonnes for boat 3, and 4 tonnes for boat 1. Platt (1973, 71) has noted from fourteenth-century documents that for vessels of 60 to 240 tuns burthen this index varied from 2.86 to 4, whilst Friel (1983: fig. 3) gives values in the range 3.33 to 8.33.

SPECIALISED CARGO VESSELS

Some of the finds listed in Table 11.4 are evidently regarded as specialised cargo boats by their excavators, others as ferries, warships or general-purpose vessels. Again, like is not being compared with like if we interpret the table simplistically. Unless there are significant cargo remains clearly associated with a boat find, as, for example the whetstones found in the ninth/tenth-century AD Klåstad boat (Christensen and Leiro, 1976), it may be difficult to decide a boat's function. An analysis of 15 medieval boat finds which were claimed by their excavators to be cargo carriers suggests that the following criteria may have been used:

(a) A hold located amidships – recognised by the absence of (upper) beams or by the presence of partitions as in the Kollerup cog (Crumlin-Pedersen, 1981).
(b) Rowing stations only at the ends.
(c) Dunnage present – as in the Klåstad (Crumlin-Pedersen, 1981) and Sjøvollen (Christensen, 1968) finds.
(d) $L/B \leqslant 5$.
(e) $(L/D) \leqslant 10$.
(f) Relatively massive frames in the midships section.
(g) Closely-spaced frames (Table 11.5).

When a hold has been identified a simple ratio of length of hold to length of boat gives some measure of the relative importance given to cargo carrying. It is not a measure of relative space which is more accurately estimated by the deadweight coefficient.

Table 11.5 Inter-frame spacing on certain medieval vessels

Boat	Date (approx) AD	Frame spacing (cm)	Reference
Nydam 2	400	100 to 110	Åkerlund, 1963
Sutton Hoo 2	600	91	Evans and Bruce-Mitford, 1975: 369
Kvalsund 1	650	130	Shetelig and Johannessen, 1929
Kvalsund 2	650	105	
Oseberg	800	100 to 110	Brøgger *et al.*, 1917
Gokstad 1	850	90 to 105	Nicolaysen, 1882
Tune	850	100	Ellmers, 1972; 256–7
Graveney	900	52	McKee, 1978
Hedeby 1	1000	84	Crumlin-Pedersen, 1984b
Hasnaes 2	1000	80 to 83	Crumlin-Pedersen, 1981
Skuldelev 1	1000	92	Olsen and Crumlin-Pedersen, 1967
2	1000	*c.*71	
3	1000	93	
5	1000	*c.* 91	
6	1000	*c* 94	
Falsterbo	1100	50	Crumlin-Pedersen, 1981
Galtaback	12th century	48 to 63	Ellmers, 1972; 256–7
Schleswig	12th century	38 to 49	Crumlin-Pedersen, 1981
Lynaes	1150	72	Crumlin-Pedersen, 1983a
Vordingborg	1200	50	Crumlin-Pedersen, 1981
Sjøvoll	1200	35 to 60	Christensen, 1968
Ellingå	1200	54	Crumlin-Pedersen, 1983a
Kyholm	13th century	69	Crumlin-Pedersen, 1983a
Kalmar 1	13th century	48	Crumlin-Pedersen, 1983a
Halmstad	14th century	65	Crumlin-Pedersen, 1981
Sørenga	1400	65	Crumlin-Pedersen, 1981

Note: McKee (1983; fig. 113) gives *c.* 91 cm as the optimum spacing between oarsmen, one behind the other, and the minimum spacing as *c.* 76 cm, when oars are used in the sit/pull mode. This data may be used, in conjunction with other evidence, in the interpretation of fragmentary remains.

Table 11.6 gives this hold index for certain medieval vessels and suggests that there was some increase between the tenth/eleventh century and the twelfth/thirteenth century.

LOGBOAT PERFORMANCE

In general there is little direct evidence of the precise way prehistoric and medieval logboats were used. However, theoretical calculations (McGrail, 1978a: 131–161) based on the hypothetical reconstruction of 24 of the logboats from southern Britain give an indication of performance in relation to size and suggest what general types of load they were best suited to carry. From freeboard, draft and stability calculations these boats may be divided into four groups as shown in Table 11.7.

It must be stressed that this classification is relative and based on the analysis of only 24 logboats. Thus, although we may say that Brigg was twice as good at carrying high density loads as the Banks logboat per unit volume of log (McGrail, 1978a: 320, 348), we do not yet know how this compares with the general capabilities of ancient logboats, nor do we know the absolute values of these measures of performance – more logboats need to be analysed and practical trials undertaken. The representative cargoes used in the calculations behind the data in Table 11.7 consisted of stone, grain and turf thus giving a spread of bulk density from 2500 kg/m^3 to 435 kg/m^3. Stability requirements mean that where a load of turf (very light) is given, the same weight of grain (medium) or stone (heavy) could alternatively be carried; and where grain is given, stone could be carried. Loads given in terms of stone are for logboats which in the particular configuration would only be stable with this weight of dense cargo (virtually ballast). This is because loads of low bulk density (turf, hay, etc.) have a higher centre of gravity than loads of greater bulk density: thus low bulk density loads have to be carried in logboats with good stability if full use is to be made of the space and buoyancy available (Ch. 3).

Logboats classified as all round performers in Table 11.7 can carry a full load of cargo of a wide range of specific density and can carry a relatively great number of men for their size. High-density cargo carriers, on the other hand, can generally carry a full load only if they are ballasted. The bulky, low-density cargo carriers were evidently 'designed' to make good use of available space to carry voluminous loads such as turf or hay: they could also carry loads of greater bulk density, but in this role and in personnel carrying these boats were for their size, relatively inefficient. The personnel carriers carried a good number of men in relation to their size; they were not so relatively good in the cargo-carrying roles.

Table 11.6 Length of hold index for selected medieval vessels

Boat	Date (approx) AD	LOA (L) (m)	Length of hold (L_h) (m)	Lh/L	References
Graveney	900	13.5	5.2(?)	0.385	McKee, 1978
Skuldelev 3	1000	13.3	3.7	0.278	Crumlin-Pedersen, 1983a
Skuldelev 1	1000	16.3	5.5	0.338	Crumlin-Pedersen, 1983a
Lynaes 1	1150	25.0	14.0	0.560	Crumlin-Pedersen, 1983a
Ellingå	1200	14.5	7.0	0.483	Crumlin-Pedersen, 1983a
Kyholm	13th century	13.0	6.2	0.477	Crumlin-Pedersen, 1983a
Kalmar 1	13th century	11.1	5.3	0.477	Crumlin-Pedersen, 1983a
Køllerup	13th century	20.0	8.0	0.400	Crumlin-Pedersen, 1983a

Note: Apart from Graveney and Køllerup these vessels were built generally in the Norse tradition.

Table 11.7 Load capabilities of selected logboats from southern Britain. Source: McGrail, 1978a

Boat	L × B × D (m)	Loads			Classification
		Maximum men (stand/kneel)	Standard freeboard	Minimum freeboard	
Baddiley Mere	5.49 × 0.92 × 0.61	1/6	1 man 1282 kg grain	1 man 1532 kg stones	All-round performance
Banks	5.03 × 1.22 × 0.61	1/7	1 man 920 kg turf	1 man 1120 kg turf	All-round performance
Hylton 2	3.51 × 0.82 × 0.59	1/4	1 man 543 kg grain	1 man 733 kg stone	All-round performance
Poole	10.01 × 1.52 × 0.50	2/16	4 men 1448 kg turf	4 men 1723 kg turf	All-round performance
Walton	5.7 × 0.96 × 0.60	1/8	1 man 1372 kg grain	1 man 1712 kg grain	All-round performance
Brigg	14.78 × 1.37 × 1	2/26	5 men 10391 kg stone	5 men 10891 kg stone	High-density cargo carrier
Kew	4.42 × 0.79 × 0.71	1/2	1 man 1203 kg stone	1 man 1493 kg stone	High-density cargo carrier
Hulton Abbey	2.25 × 1.24 × 0.60	1/2	1 man 384 kg turf	1 man 509 kg turf	Bulky, low-density cargo carrier
Irlam	2.90 × 0.85 × 0.36	1/1	1 man 25 kg turf	1 man 180 kg turf	Bulky, low-density cargo carrier
Oakmere	3.6 × 0.79 × 0.35	1/2	1 man 62 kg turf	1 man 262 kg turf	Bulky, low-density cargo carrier
Warrington 2	3.76 × 0.86 × 0.38	1/3	1 man 118 kg turf	1 man 293 kg turf	Bulky, low-density cargo carrier
Clifton 1	8.55 × 0.76 × 0.36	1/5	2 men 361 kg turf	2 men 871 kg grain	Personnel carrier
Clifton 2	9.25 × 0.76 × 0.38	1/6	2 men 533 kg turf	2 men 1013 kg grain	Personnel carrier
Hardham 1	3.99 × 0.84 × 0.45	1/4	1 man 299 kg turf	1 man 549 kg grain	Personnel carrier
Hardham 2	3.88 × 0.84 × 0.47	1/3	1 man 363 kg turf	1 man 613 kg grain	Personnel carrier

12 Propulsion, steering and anchoring

PROPULSION

Boats may be propelled in four main ways:

(a) By tidal stream or river current – water power. The main problem here is control of direction and this is considered in the section on *Steering*.
(b) By pole, paddle or oar – muscle power.
(c) By tow – manpower, animal power or another boat.
(d) By sail – wind power.

In some of these methods, propulsion may be combined with steering (see below); in others, separate steering arrangements are required. Two or more means of propulsion may be available in the same boat (the Broighter boat model of the first century BC, has mast, poles and oars) and may be used separately or in combination, for example: sail may be combined with paddle (Haddon and Hornell, 1938: 47, fig. 33; Phillips-Burt, 1971: 34) or with oar (Tilley and Fenwick, 1980: 152–3); towing with poling (Beaudouin, 1970: 90); river current and ebb tide downstream with pole or tow or flood tide upstream. Working the tides leaves no artifactual evidence; pole or paddle may leave no evidence on the hull except possibly in the form of walkways or thwarts; towing and propulsion by oar and sail, on the other hand do require specific fittings onboard which may survive to be excavated.

POLING

Propulsion by pushing against a pole in the river or seabed is known as *punting* if the man stands at the stern or at the bows where it may be combined with guiding the boat clear of obstacles (Beaudouin, 1970: 90); or *quanting* or *setting*, if he walks from forward to aft along a walkway – more than one man may quant a boat. This shallow water method of propulsion is known worldwide: Melanesia (Lewis, 1932: 145), S. America (Lothrop, 1932: 232) and Denmark (Rasmussen, 1953: 24), to quote but three examples. Length of pole is related to size of boat and expected depth of water: a 3.5 m pole would be suitable for a small boat and *c.* 6 m for larger ones, anything longer becoming unwieldy. Such poles are frequently multi-purpose, having subsidiary uses as sail spar, depth gauge or even as a temporary mooring post (McKee, 1983: 133). Combined pole-paddles have been used in recent times in Estonia (Manninen, 1927: 6) and pole-oars in Germany, the latter have a forked wooden or iron foot to prevent them sinking too far into the mud (Fig. 12.1). Similar metal terminals of open rectangle or Y-shape have been recovered from Roman period contexts (Ellmers, 1978: 10; 1974b: 137, 141; 1969: 92, figs 5, 8, 9). The Broighter boat model of the first century BC has three poles with Y-shaped ends (Fig. 12.2). Y-shaped and single-prong metal terminals have also been found in Iron Age contexts at La Tène, from Lake Biel near Möringen and from Manching (Ellmers, 1969: 95, fig. 11), and at the

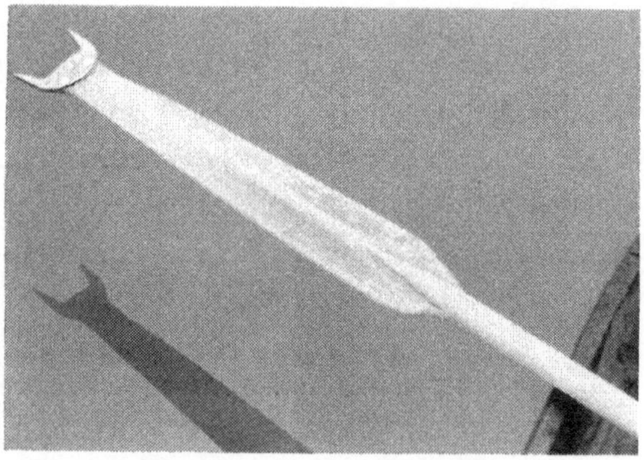

Figure 12.1 A pole-oar as used on the waterways of the S. Baltic coastal region. (Photo: Wolfgang Rudolph.)

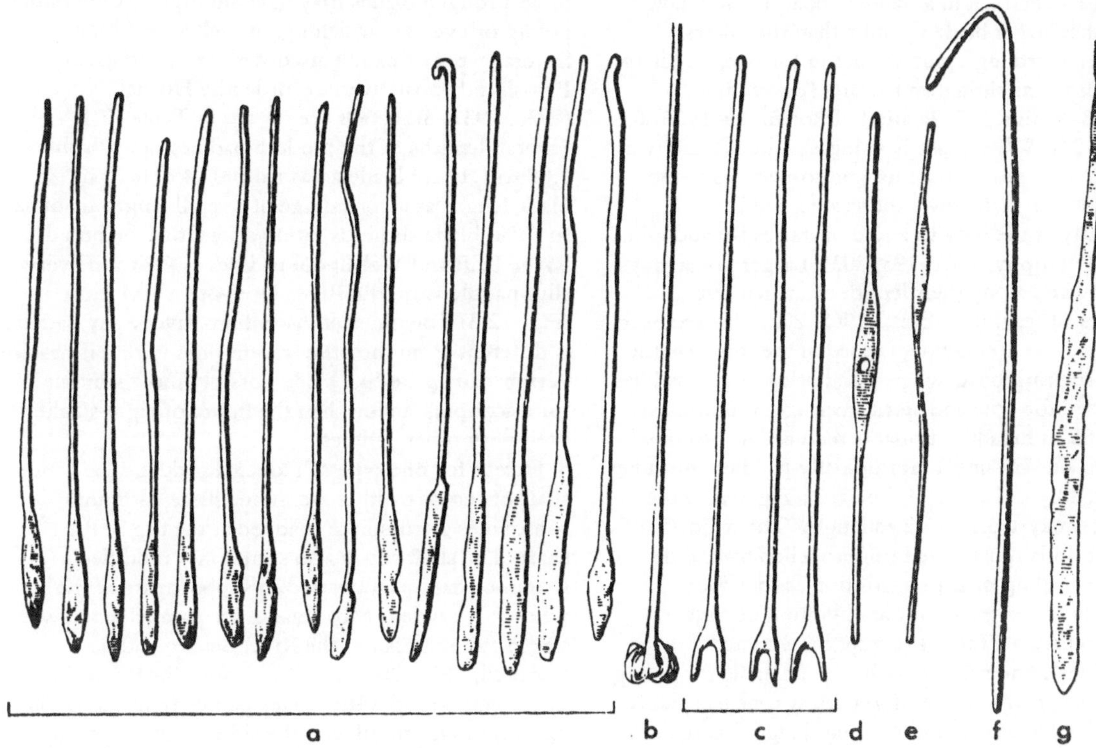

Figure 12.2 Model oars (a), grapnel (b), poles (c), yard (d), spar (e), mast (f) and steering oar (g) found with the Broighter boat model of the first century BC. (Drawing: National Museum of Ireland.)

Roman period site of Pommeroeul, Hainaut, Belgium (de Boe and Humbert, 1977: fig. 48).

Artifacts of these general forms continued in use into recent times on the River Havel at Potsdam near Berlin and in the River Trent near Nottingham where eighteenth- or nineteenth-century Y-shaped pole terminals were found which had been fastened by nails or rivets to ash poles (*stoer* or set poles) and rectangular-shaped double-pronged terminals which had been fastened to willow quant-oars (oar stoer or scull stoer). Typical sizes of the Y-shaped terminals are 13 cm prongs some 12 cm apart, and 16 cm prongs, 17 cm apart. Oar stoer terminals have 8 cm long base with 8 cm prongs 18 cm apart, whilst another had 24 cm prongs 25 cm apart (C. Salisbury, p.c.). A modern combined boat hook-quant pole was also found in the River Trent, and boat hooks with one pointed and one curved terminal are known from Roman age contexts (Ellmers, 1974: 137, 141; de Weerd and Haalebos, 1973: 390, fig. 6; de Boe and Hubert, 1977: fig. 49). These latter may also have doubled as quant or punt poles.

Pommeroeul boat 4, a 'barge' of the first/second century AD is the only early boat find with definite evidence for a poling walkway (de Boe and Hubert, 1977: 30, figs 33–5) but other river boats with less remains surviving may also have had them. It would have been possible to have arranged the cargo in the Ferriby boats and Brigg 2 so that they could have been quanted from the bottom of the boat. Other boats, as the Broighter model and scenes from the Bayeux tapestry demonstrate, normally propelled by oar or sail, may have been poled in special circumstances. A boat with more than *c.* 2 m of freeboard is unlikely to have been poled.

PROPULSION BY PADDLE

There is worldwide evidence for boat paddles with a variety of blade shapes, the length of shaft varying with mode of operation which may be standing, sitting or kneeling (Joass, 1880–1: 180; Lewis, 1932: 41; Lothrop, 1932: 232; Longstaff, 1930: 260; de Laguna, 1963: 222). An unusual mode of use has been noted by Brindley (1926, 1927) who saw paddlers on Lake Inle in the Shan States standing on one leg with the other around the paddle and achieving remarkable speeds. Where there is a specialist steersman in a paddled boat he may have a somewhat larger bladed paddle than the others.

In recent ethnographic studies in Europe, paddlers are recorded as sitting on thwarts (Eskeröd, 1956: 63–4, figs 4, 5, 7a and b; Itkonen, 1941: 55–6, figs 23, 24). When there is a single paddler this thwart is sometimes near amidships and sometimes at the stern. In two-man crews the second paddler is frequently at the bow to avoid obstacles (Beaudouin, 1970: 90; Guppy, 1961: 195, 303). Larger boats may have thwarts along their length on which two paddlers sit (e.g. de Laguna, 1963: 222): this seems to have been the arrangements used in the third-century BC Hjortspring boat, where in addition larger paddles were used for bow and stern steering. A minimum of *c.* 0.85 m breadth of boat is required for paddlers to work side by side; longitudinally paddlers must be separated by *c.* 1 m or, if they are staggered, three paddlers may work in a 2 m length. The majority of logboat finds must have been propelled by paddle from a kneeling or sitting position as they are generally too narrow for oars (without outriggers) and too unstable for sail (except in restricted conditions), and some may be too unstable for poling where the crew's centre of gravity was unacceptably high. The prehistoric Hjortspring, Brigg 2 and Ferriby boats were propelled by paddles, and medieval plank boats primarily propelled by oar may also have been paddled in special circumstances – a paddle was found with the oared Årby boat of the ninth/tenth century (Arbman, 1940).

Paddles are generally gripped in both hands, the lower one providing power, the upper preventing the blade from twisting during the stroke, but single-handed side use with the end of the shaft pressed against the upper arm is also known (Hornell, 1936: fig. 1). Paddles may be seen in use on many Scandinavian rock carvings generally dated Bronze or Iron Age (Ellmers, 1975: fig. 1, 5). Many paddle finds (mostly undated) are recorded in the archaeological literature; it must be remembered, however, that paddle-shaped objects may also be used in the field or dairy. Lerche (1977) has suggested that artifacts with a blade at each end, comparable with modern double paddles, and found in prehistoric contexts, are double digging sticks or spades.

Modern boat paddles vary widely in length and shape; for example, some blades are almost aerofoil shape in plan and equal in length to the shaft, whilst others have much shorter blades with a pointed end or even with three prongs like a fish spear, leister or fizgig (Bidault, 1945: 34, 35, 182). These three-pronged blades may have multiple uses possibly poling or even spear fishing, as well as paddling. Several types of paddle used with recent Irish and British hide boats were recorded by Hornell (1936: 1937–8): details are set out in Table 12.1. Overall lengths of the modern paddles range from 1.88 to 1.17 m; blade areas from 0.04 to 0.10 m^2: blade length as a percentage of overall length from 32 to 59%; blade slimness ratio varies from the broad River Teifi and Welshpool paddles of 4:1 to the very slim paddle from the River Dee with a 10:1 ratio (Fig. 12.3) The nineteenth-century River Spey paddle is different from the other recent ones in that it has a very broad spade-like blade possibly more suitable for a scooping action than the figure of eight stroke used with other paddles.

Except for one type of Donegal paddle, blade and shaft of recent paddles are in one piece, although sometimes a crosspiece is added at the top of the shaft. The shaft top is also sometimes modified in shape so that it can be used to support part of the boat when it is carried. Most blades are parallel-sided with sloping shoulders but the River Severn paddle (Hornell, 1936: 294, fig. 9b) is more curved at the upper end, and the River Dee paddle (Hornell, 1936: fig. 9e) has no shoulders, the sides diverging from the bottom of the shaft (*cf.* the Nydam 2 oars, below). Most blades are plano-convex in section, but the Donegal type of paddle is concave/convex, and the Towy (Hornell, 1936: 31, fig. 2) and the Dee (Hornell, 1936: 294, fig. 9e) are biconvex.

Details of five early paddles are also noted in Table 12.1. The paddle from the River Tay, now in Dundee Museum, has a plano-convex blade *c.* 2.5 cm thick, which is parallel-sided with well-defined shoulders (Fig. 12.4). There is a splayed hand grip at the top of the shaft. An unusual feature is a *c.* 2.5 cm diameter hole through the blade about one-third down its length from where it blends into the shaft; this may have been used to secure the paddle to the boat when not in use. This paddle may be compared in general proportions with the Nydam 2 rudder or steering oar which has a similar hole (see below).

The Dutch paddle from the Assendelver polder has a similar shaped blade to that of the River Tay paddle, but is slimmer. The Essex paddle has a much slimmer blade tapering to a point.

In addition to steering paddles, 16 propulsion paddles of a variety of blade shapes, were found with the Hjortspring boat (Fig. 12.5) but only two were complete – these represent two different types (Rosenberg, 1937: fig. 62b and 62c and 63b). The paddle figured in 62b has a triangular grip at the top;

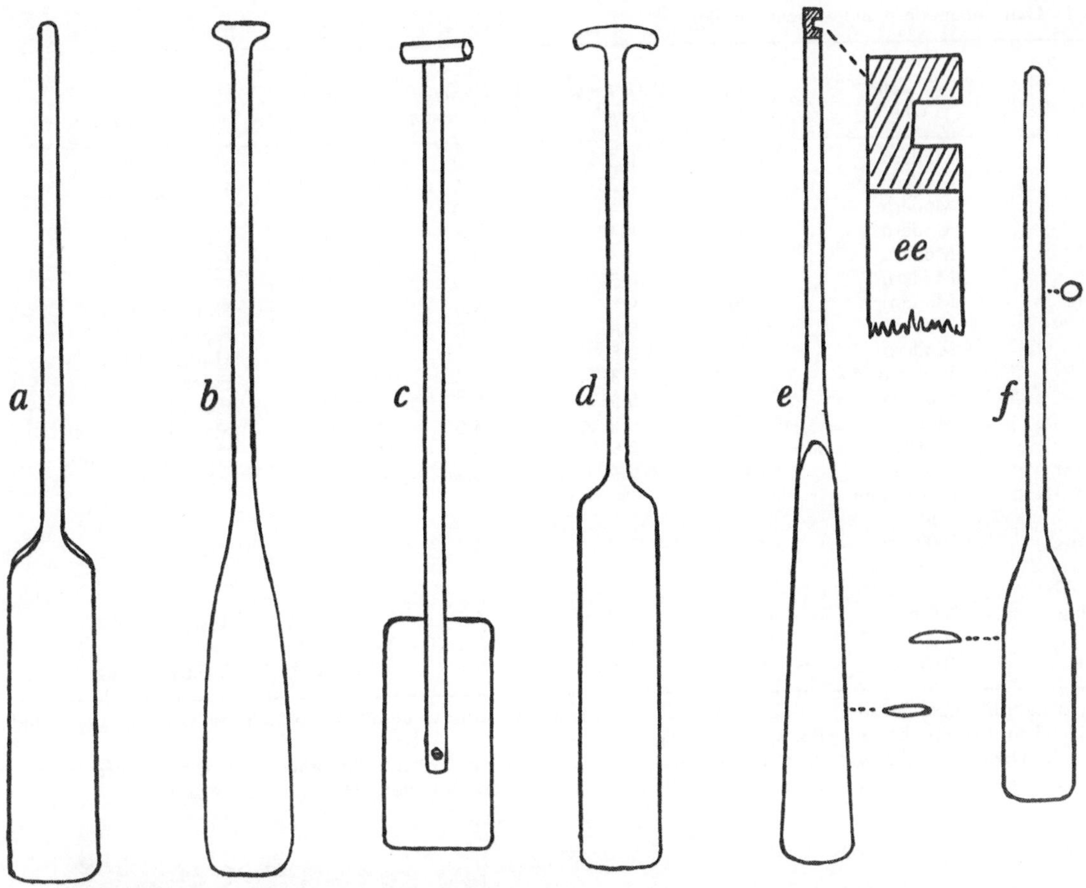

Figure 12.3 Types of twentieth-century coracle paddles:
(a) Wye and Usk
(b) River Severn (older)
(c) River Severn (recent)
(d) Shrewsbury
(e) Bangor on Dee – this paddle is 5 ft (1.5m) long.
(f) Welshpool
After Hornell, 1936: fig. 9.

the blade is biconvex in section with pronounced shoulders. The paddle in Fig. 62c is bigger than the other; the blade is biconcave in section and has curved shoulders. Despite these differences, the two types are similar in relative length of blade (one-third of paddle length) and in slimness of blade (5:1, generally broader than modern hide boat blades).

The Årby paddle has a knob at the end of the shaft, and a blade with a 'scoop' cross section (Arbman, 1940: 61, fig. 21).

OARED PROPULSION

Oars are levers of a second order with a fulcrum on some part of the boat, either the sides or the stern and (rarely) the bow. Many and varied are the types of oar, oar pivot and oar use, and a comprehensive analysis of the subject has long been needed. McGrail and Farrell (1979) have listed 12 variables, each with several states, which may be used to describe methods of oar use; to this list must be added important concepts such as rowing geometry, and hydrodynamic, biomechanic and physiological considerations. Analysis of known rowing actions using even 12 variables is inappropriate here and discussion will be limited to consideration of number of oars used, posture of oarsman, leverage action, nature and position of pivot, and details of oar and of rowing geometry.

Table 12.1 Data for modern and ancient paddles

Paddle	Date (approx)	Length (m)	Blade length/ Length	Blade length/ Blade breadth	Blade area (m^2)	Reference
Teifi	Modern	1.30	0.32	4.0	4×10^{-2}	Hornell, 1936
Towi	Modern	1.88	0.36	7.5	8×10^{-2}	Hornell, 1936
Taf	Modern	1.52	0.40	6.0	6×10^{-2}	Hornell, 1936
Cleddau	Modern	1.30	0.47	7.4	5×10^{-2}	Hornell, 1936
Wye and Usk	Modern	1.52	0.33	4.0	6×10^{-2}	Hornell, 1936
Severn	Modern	1.52	0.50	6.0	10×10^{-2}	Hornell, 1936
Welshpool	Modern	1.30	0.35	4.0	5×10^{-2}	Hornell, 1936
Shrewsbury	Modern	1.50	0.47	7.0	7×10^{-2}	Hornell, 1936
Dee	Modern	1.52	0.50	10.0	6×10^{-2}	Hornell, 1936
Spey	19th century	?	0.36	1.6	12×10^{-2}	Fenton, 1972
Donegal	Modern	1.45	0.55	7.5	8×10^{-2}	Hornell, 1937–38
Boyne	Modern	1.17	0.59	4.5	11×10^{-2}	Hornell, 1937–38
R. Tay	*c.* 1420 AD	1.22	0.58	3.0	15×10^{-2}	Dundee Museum
Assendelver Polder (AOA 797)	Late 1st millennium BC	1.64	0.41	5.2	—	Veryan Heal
Canewdon Essex (AOA 899)	1000 bc	2.08	0.29	4.3	—	Veryan Heal
Hjortspring	300 BC	(1.44	0.30	5.5	3.2×10^{-2})	Rosenberg, 1937
		(1.59	0.30	5.1	3.8×10^{-2}	
Årby	AD 900	1.51	0.40	5.6	3.7×10^{-2}	Arbman, 1940

Notes: 1. Blade (thickness/ breadth) ratios are difficult to measure but seem to be *c.* 0.10 to 0.25 for modern paddles and 0.13 to 0.16 for the two Hjortspring paddles.
2. The Donegal curach was seagoing and measured 2.54 × 1.07 × 0.61 m. The other modern boats were river craft between 1.37 and 1.83 m in length; 0.91 to 1.37 m in breadth and 0.33 to 0.41 m in depth.

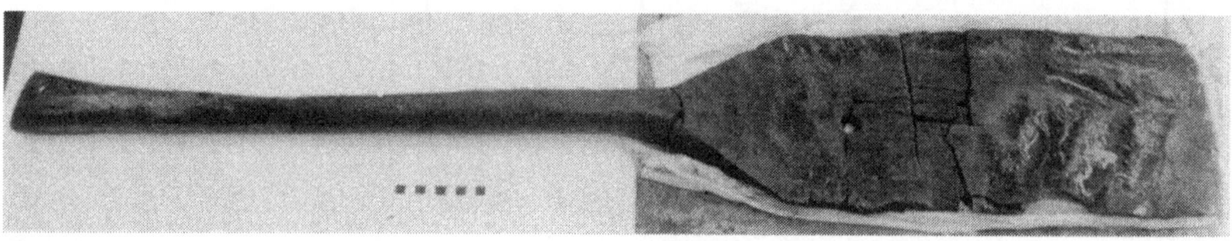

Figure 12.4 A paddle found in the River Tay. Scale centimetric. (Photo: Dundee Art Galleries and Museums.)

The following definitions are used:

Rowing: general term for working pivoted oars or sweeps (large oars generally worked by more than one man).

Pulling: the oar handle is pulled towards the oarsman on the power stroke.

Pushing: the oar handle is pushed away from the oarsman on the power stroke.

Sculling: the use of an oar over the stern with alternate push and pull power strokes (see *Föreningen Allmoge Båtar*, 1971: 12 for diagram).

Oarsmen may use one oar or two for pulling or pushing but only one for sculling. There is a breadth of boat, which depends partly on oar angle used (vertical angle between water surface and oar shaft at mid-stroke), beyond which two oars cannot be used by one man. Conversely, there is a breadth (which again varies with rowing style) below which two oars can only be used on outriggers, as in Japan (Folkard, 1870: 240) or on Lake Ostrova, Albania where the oars are pivoted on the ends of projecting thwarts (Traegar, 1904). In large boats, boats such as the Portuguese *xavega* (*saveiro*) (Johnstone, 1980: fig. 8.8) there may be two or more men to each oar. When oarsmen have one oar they may be double-banked with two men at the same station (e.g. sitting on same thwart) or they may be at alternate stations. In this latter case oarsmen will generally be stationed on

Figure 12.5 Paddles and paddle fragments found with the Hjortspring boat. Rosenberg, 1937 (by permission of the Danish National Museum).

the opposite side of the boat to their pivot, to gain greater leverage; the gearing is thereby reduced, permitting the oarsmen to pull for longer at a stretch. McKee (1983: fig. 112) has suggested that the normal spacing between oarsmen in the sit/pull mode is 3 ft (91 cm) with 2 ft 6 in (72 cm) the minimum. In a boat powered solely by oars the oarsmen are generally disposed regularly along the length except at the ends, with equal numbers on either side or with a difference of one oarsman. In boats which have a cargo space amidships oars may be stationed towards the ends; there may be unequal numbers at bow and stern to avoid interference with the steersman. Oarsmen may stand or sit, although other postures are known such as that recorded by Greenhill (1966: 30) and Casson (1971: 18) where the oarsman moves from a standing to a sitting position during each stroke.

The majority of rowing actions (the main power action, not the recovery and not when backwatering) may be described by using the several states of three variables: numbers of oars, posture and leverage action (Table 12.2). Thus in Fig. 12.6 one man is using the combination two oars/sit/push whilst the other is two oars/sit/pull; in Fig. 12.7 the action is of one oar/stand/push. By this method major traditions may be identified; other variables such as oar angle and rowing geometry must be considered if more detailed analysis is required. Unusual propulsion combinations, such as that used in Shaohsing China (Worcester, 1955) where the crewman push-rows one oar with his feet while using a paddle with his hands, are probably beyond analytical description at this stage. Of the 12 theoretical combinations of the seven states shown in Table 12.2, two are impossible – two oar/sit or stand/sculling which by definition are one oar activities. Otherwise ethnographic examples are known of all these modes, (McGrail and Farrell, 1979: table 2) but it is found that those which include stand/push and sit/pull are generally preferred, probably because they are more efficient.

Table 12.2 Major rowing variables

Variables	*No. of oars used by oarsman*	*Posture*	*Leverage action*
States	One oar Two oars	Sit Stand	Push Pull Stern-sculling

Note: Stern-sculling is the use of a single oar in a pivot on a transom or near the sternpost.

Figure 12.6 Å *heuger* on the Oder lagoon with push and pull rowing. (Photo: Wolfgang Rudolph.)

Rowing geometry

The dimensions of the human body, affect the relationship between oar length, *gearing* (ratio of length of oar outboard to inboard) and angle, and they place limits on the range of values for the horizontal and vertical distances between thwart, feet and pivot (Fig. 12.8). These relationships have been investigated by McKee (1974: 18, 32; 1983: 135–9) for sit/pull modes of rowing with a range of oar angles between 17° and 25°. This data may be used when evaluating the rowing potential of a boat find or evaluating ethnographic or iconographic evidence (McGrail and Farrell, 1979). It must be borne in mind, however, that sit-pull is only used in two of the 10 possible modes of rowing; that oar angles of up to 40° are known to be used even in the sit/pull mode (McGrail, 1978a: 72); and that the rowing geometry may be complicated by styles which allow the oarsman's hands to be crossed (as in the twentieth-century curach) and by a two-oared oarsman sitting or standing over to one side of the boat (see, for example, McGrail, 1978a: fig. 170 where the oarsman is using the sit/pull mode with crossed hands and oars of unequal length).

Angle of sweep may vary from 80° to 48°, while rates can vary up to 35 or even 50 strokes/minute (Alexander, 1927: 230). Today, oar leverage ratios (gearing) are generally between 3 and 4; a racing eight, 2.42, a whaler 3.85 (Alexander, 1927: 230), a Kerry curach, 4.02. Using a gearing of 3 it can be shown that for one man to use two oars in the sit/pull mode without crossing his hands the length of each oar should be slightly less than twice the breadth of the boat at that station (Fig. 12.9). One man sit/pulling one oar may use an oar almost three times the breadth of the boat. For a crew using different lengths of oars at different stations it is essential that the gearing is kept the same so that they may keep in time.

For sea rowing where breaking stroke can easily lead to loss of way and swamping, blades of simple form (symmetrical in long section) and relatively narrow are used. With such a blade (rather than the spoon-shaped blade of a racing oar) the thrust can be varied relatively easily to match changing conditions on water entry. The short, deep stroke appropriate to rough water is more efficient and minimises the risk of 'catching a crab'.

McKee (1974: 32) has recorded key measurements of the rowing geometry for pull/ sit rowing in a

Figure 12.7 Stand/push rowing a *Kahn* near Rostock. (Photo: Wolfgang Rudolph.)

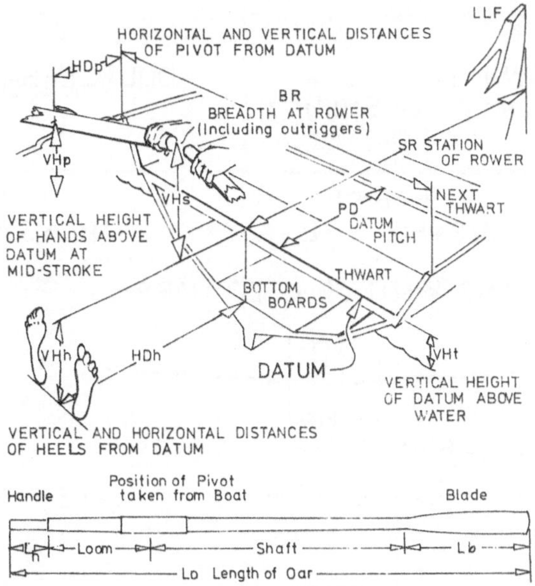

Figure 12.8 McKee's rowing geometry terminology. After McKee, 1974 (copyright NMM Greenwich).

range of British boats for a man of average stature 5 ft 8 in. The range of values he found for the vertical and horizontal distances between the pivot, the seat, and the oarsman's heels, permit the oarsman to use longer or shorter strokes, different oar angles (17–25° from the horizontal) and to lower his hands on the recovery stroke to clear the waves in difficult sea states. The rowing geometry for other forms of rowing (e.g. push/stand) has not yet been documented. This needs to be done, and boats from other countries need to be investigated fully for, although these ranges of value are tied to human parameters, they are also determined by the style of rowing.

McKee's data is listed in Table 12.3 together with data from other sources for comparison. More information than is shown in this table may become available for the Skuldelev finds, as the post-excavation research on these boats progresses. Meanwhile it can be seen that although the evidence for Skuldelev 1 is compatible with oarsmen at bow and stern pulling single oars and sitting on crossbeams at 11th strake level, the possibility of stand/push mode may be considered. Olsen and Crumlin-Pedersen (1967) have indeed noted that the surviving evidence for Skuldelev 6 may only be compatible with stand/push rowing. The Skuldelev 3 evidence is more complete and appears to be compatible with sit/pull rowing from crossbeams at the ends.

Examples of using this method to evaluate evidence on logboats may be found in McGrail (1978a: 63, 75–7). In the case of the logboat from Vaaler Moor (Åkerlund, 1963: 118–120) it has been argued that this boat was paddled from thwarts placed almost level with the top of the sides (Paret, 1930: 106). However, rowing geometry analysis suggests that rowing was possible from thwarts set down *c.* 20 cm below the top of the sides, with the oarsman pulling two oars at every station if oar angles of 40° were allowed at the ends, or 8 out of 10 stations if oar angles remain within McKee's range of 17 to 25°.

Oar pivots

Oar pivots provide a fulcrum for the oar on the power stroke and a restraint during the recovery stroke, and allow the oarsman to use a reverse stroke (backwater) when manoeuvring, and often to leave his oar unattended safely trailing. The former existence of oar pivots may be deduced, even though precise evidence has not survived, from the presence of reinforced or thicker planking which transmits the loads locally imposed by rowing to other parts of the structure (Alexander, 1927). A range of oar pivots is

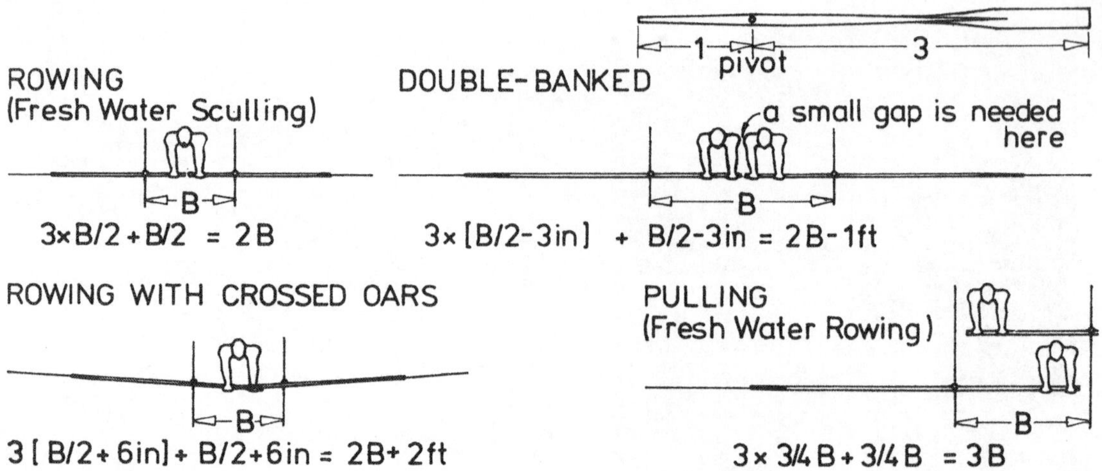

Figure 12.9 Oar lengths in open boats, with a gearing of 1:3. After McKee, 1983: fig. 109 (copyright NMM Greenwich).

Table 12.3 Rowing geometry data

	HD_p	VH_p	VH_h	$VH_p + VH_h$	*Source*
McKee's range	28 to 41	18 to 31	27 to 64	48 to 85	McKee, 1974: 18, 32
Danish lifeboat	29	22	36 to 41	58 to 63	Olsen and Crumlin-Pedersen, 1967: 141
Sixern	40	21	64	85	McKee, 1974: 32
færing replica	33	20	19	39	McGrail, 1974
Skuldelev 1	—	20?	40 to 50?	60 to 70	Olsen and Crumlin-Pedersen, 1967
3	*c.* 40	*c.* 24	26	*c.* 50	
5	27 to 54 average 42	26	—	—	
6	—	80	—	—	
Årby	⩽ 45	—	—	28?	Arbman, 1940
Kvalsund 1	52 to 95	—	52		Shetelig and Johannessen, 1929
Kvalsund 2	25	28	58 to 68	86 to 96?	
Gokstad 1	31	—	—	40	Nicolaysen, 1886
Dingle Curach	37 to 40	18 to 19	18 to 20	36 to 39	Kentley measurements

Notes: 1. McKee surveyed seven twentieth-century boats. His standard oarsman was 1.72 m in height who rowed in the pull/sit mode with oar angles 17°–25°.

2. HD_p = datum to pivot (horizontal)
VH_p = datum to pivot (vertical)
VH_h = datum to heels (vertical)
} Datum = after edge of thwart

3. Blank entries are due to incomplete survival of boat finds.

discussed below, some of them are at present only known ethnographically and doubtless there are other forms not listed here. Oar pivots may be used in more than one mode of rowing – see for example Fig. 12.6.

1. ***Crutch.*** A U-shaped, swivelling crutch, proud of the top strake is shown in use in Fig. 12.7.

With this form may be included the oarfork of the Venetian gondola (Rubin de Cervin, 1956: 210, fig. 1) which is carved from a wooden block, is set about 0.30 m above the top strake and is non-swivelling, although its special form allows several different angles of use.

2. ***Oarlock or Rowlock.*** The U-shaped opening worked in the top strake, or in a separate board

('saxboard') fastened to the strake, has been recently used in Royal Navy cutters (*AMS*, 1, 1964: 236) in Swedish churchboats (Eskeröd, 1973: 55, 139), and on the thirteenth/fourteenth-century AD Kentmere boat 1 excavated by Wilson (1966) see Fig. 6.15.

3. ***Oarports.*** Round or rectangular holes cut through a strake were used in nineteenth-century crabbers on the English east coast (Marsh, 1970: 150–2), in the Viking Age Gokstad and Oseberg vessels, and are depicted on the Bayeux tapestry. Planking with such oarports has been excavated from the twelfth-century site at Wood Quay Dublin; one port had a diameter of 13 cm plus a 4 cm slit so that it could take an oarblade up to 17 cm in breadth. The Gokstad 1 oarports have lockable shutters inboard and have a diameter of *c.* 12 cm, the slit here goes upwards and astern from each port and would allow a blade of up to 16 cm to pass through from inboard to outboard. See also the thirteenth-century illustration published by Fenwick (1978: fig. 8.25–1A).

   The oarports in Skuldelev boat 1 are at the ends in the top strake and are semi-circular in shape, the lower straight edge being on top of an outboard stringer. Skuldelev 3 has two types of port: forward they are rectangular *c.* 9–10 cm sided with a 12–13 cm diagonal; aft they are trapeze-shaped 12 cm in height, lower edge 11 cm, upper 4 cm – the diagonal here measures 13 cm. An oarblade found nearby was 12 cm broad and 1 cm thick at the edges and so would just have passed through these ports. Skuldelev 5 also has two types of port one of 9–10 cm diameter in a re-used plank, the other rectangular 11 × 9–10 cm with a diagonal of *c.* 12 cm.

   The first evidence for oarports in northern and western Europe seems to be on the small gold model of the fifth century BC from Dürrnberg near Hallein (Fig. 12.10) (Ellmers, 1978: fig. 1). This model has ports near each end of the boat both on the same side, which might cast doubt on the accuracy of the model were it not that a similar asymmetric arrangement is found on the boats of Lake Scutari and Lake Ochrida in Albania (Traegar, 1904; Tozer, 1869). The oars with this Dürrnberg model appear disproportionate in size especially in breadth, but Ellmers (1978: 10–11) considers they were probably used in the stand/push mode.

4. ***Two tholes.*** Two wooden pins protruding vertically from the top strake can be seen in use in Fig. 12.6. This form of pivot was widespread in the nineteenth and early twentieth century:

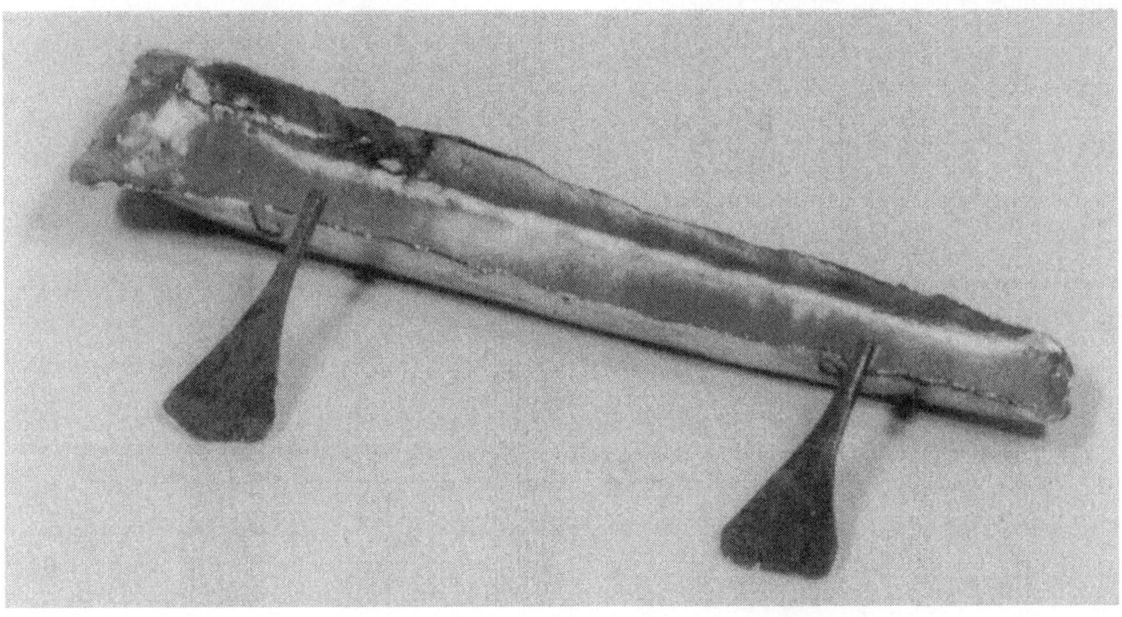

Figure 12.10 A 6.5 cm gold model of the fifth century BC from Dürrnberg with two oars. The model is now in the Kelten museum, Hallein. (Photo: Deutsches Schiffahrtsmuseum.)

the Norwegian nordlandsboat, Newquay gigs (March, 1970: 224–5) and the *chaland* of the River Adour, France (Beaudouin, 1970: 83, fig. 9) for example, and at the stern of the late 18th-century Bayonne chaland (Beaudouin, 1970: pl. 3).

5. ***Single thole and 'bull'.*** A pin vertically protruding from the top strake passes through a hole in the oar loom or in a 'bull', a wooden extension fastened to the oar (Fig. 10.8). This arrangement prevents the oar working inboard – a natural tendency and a major loss of efficiency. This form is used in the Irish curach (Hornell, 1937: 150–3, 174), the Portuguese *xavega* (Johnstone and Tilley, 1976: 18, fig. 3), the *barco de dornas* of the River Miño, Portugal (Fernández, 1966: 2), and some Swedish churchboats (Eskeröd, 1973: 139, fig. 83). It has been suggested by Seaby that a group of excavated logboats from the north of Ireland may have had this form of pivot, as may the logboat from Loch. Dowalton, Wigtonshire, Scotland (McGrail, 1978a: Ch. 3). However, for want of surviving evidence, it cannot be known whether these conjectural pivots were used with a 'bull' rather than the grommet described in the next section.

6. ***Single thole and grommet.*** The loom of the oar passes through a grommet fastened either direct to the thole or to an adjacent hole. Hook-shaped tholes with leather grommets are referred to in Greek classical literature (Morrison and Williams, 1968: 52), but the precise method of use is unclear. Vertical tholes with grommets are used on the Maltese dghaisa which is primarily propelled in the mode two oar/stand/push but also one oar/sit/pull (Tilley, 1969, pl. 20 and 21) here the oarsmen evidently push or pull against the grommet attached to the thole.

   A different method is used on several boats from N.W. Europe: the Shetland sixern (Marsh, 1970: 58), the Norwegian Oselver boat (McGrail, 1974: fig. 9), several Swedish churchboats – see for example Eskeröd (1973: 82, fig. 58, 94, fig. 65) and oared boats of the Norse tradition. The normal rowing mode in these boats is one or two oars/sit/pull and the oar is pivoted against the thole which is hooked to accommodate the oar (Figs 8.17 and 8.43). The grommet aft of the thole assists oar control during the recovery stage of the stroke; only when using the oars to backwater would the grommet be used as a pivot in this style of rowing. Single tholes shaped to take an oar aft of them can be seen on the Sutton Hoo impression (Evans and Bruce-Mitford, 1975: 403, fig. 308) and are also found on the small boats in the Gokstad Viking Age find (Fig. 8.17) where a grommet is also fitted. Similar shaped tholes have been found on the fragmentary fourth-century AD Halsnøy find (Shetelig, 1903), the Nydam boat 2 (Åkerlund, 1963), the sixth/seventh-century Hasnaes 1 boat (Crumlin-Pedersen, 1981) and the ninth/tenth-century Årby boat (Arbman, 1940).

7. ***Grommet.*** A simple grommet through a hole in or near the top strake can be used as a pivot. *Doni* boats of the Reddi are propelled on the River Godaveri, India in the mode one oar/sit/pull (von Fürer-Haimendorf, 1975: 1162) in this manner, and oarsmen of twentieth-century logboats on the Mondsee, Austria also used a grommet but in the mode of one oar/stand/push (Salemke, 1972: 5). The oars on the 1st century BC boat model from Broighter, Co. Derry, Ireland (Fig 10.9 and 12.2) pass through loops of gold wire near the sheer line and although this might be no more than a model builder's convention, these loops may represent simple grommet pivots (Farrell and Penny, 1975). The oarsmen in the stern of the boat on Blusus's tombstone (Fig. 12.32) may be using this minimal evidence form of pivot.

Oars

Table 12.4 lists details available of certain oars ethnographic, iconographic and excavated. The *faering* (Gokstad 3) is *c.* 1.30 m in breadth at both oar stations, giving an oar length of *c.* 2.60 m for two oar/sit/pull mode with a 3:1 gearing. In fact the oar said to be from this boat is 3.26 m long which would have given a gearing of 4:1. This oar may therefore belong to one of the larger boats and in the trials of the Greenwich *faering* replica oars of 2.51 m were used.

The Årby boat has a beam of *c.* 1 m at the tholes giving an ideal oar length of *c.* 2 m and the oar figured by Arbman (1940, fig. 21) is indeed of that length. The blade is lenticular in plan and appears to have a 'scoop' cross section: it is a relatively broad blade (4:1) but as a fraction of oar length it is the shortest of those analysed. This oar's small blade area ($1 \times 10^{-2} m^2$) corresponds to the size of the boat (3.95

Table 12.4 Oar data

Boat	*Loa (m)*	*Blade length (m)*	*Blade B. max. (m)*	*Blade length/ Loa*	*Blade length/ B. max.*	*Blade area ($m^2 \times 10^{-2}$).*	*Handle*	*Blade section*	*Source*
Dingle curach	3.23	0.93	0.06	0.29	15.5	5.4	Yes	—	Kentley
Donegal curach	3.20	0.97	0.08	0.30	12.7	7.3	Yes	Bi-convex	Hornell, 1946
Nydam 2	2.94	1.30	0.15	0.45	8.5	15.0	No	Bi-convex	Åkerlund, 1965
	3.42	1.52	0.13	0.45	12.0	16.0	No	Bi-convex	
Kvalsund 1	2.01+	0.79	0.13	—	6.08	7.1	—	Bi-convex	Shetelig and
Kvalsund 2	3.33 (e)	0.81	0.13	0.24	6.23	8.1		Bi-convex	Johannessen, 1929
	1.51+	0.75	0.16	—	4.69	8.0	—	Bi-convex	
Årby	2.04	0.24	0.06	0.12	4.00	1.0	Yes	Concave	Arbman, 1940
Gokstad 3 (faering)	3.26	0.85	0.10	0.26	8.5	4.3	Yes	Bi-convex	Nicolaysen, 1882
Gokstad 1	5.30	0.96	0.12	0.18	8.0	6.7	Yes	Bi-convex	Nicolaysen, 1882
Oseberg	3.72	0.86	0.17	0.23	5.06	10.0	Yes	Bi-convex	Brogger *et al.* 1917
Broighter	—	—	—	0.29	7.2	—	—	—	—
	—	—	—	0.37	9.0	—	—	—	—
Blussus	—	—	—	0.29	3.0	—	—	—	Ellmers, 1969
				0.33	2.0	—	—	—	
McKee's		—	—	0.25–0.33	slim	11.2–20.4	Yes	—	McKee, 1983: 3–4; 1974
seaboat	*c.* 4.5					16.7–22.3	Yes		
beachboat	*c.* 4.5								
openwater	*c.* 3.0	—	—	0.20	broader	9.29–13.9	Yes	—	
river	*c.* 3.0					4.78–9.3	Yes		

Notes: 1. McKee's boats are bigger than all but Gokstad 1, Oseberg, Nydam 2 and Kvalsund 2 and the data for them may thus not be directly applicable to the smaller boats.
2. In a seaway relatively slim blades of simple form are used today so that thrust can be varied easily to match changing conditions.

× 1.04 × 0.33 m). Two oar blades fully immersed would give a total underwater area of 0.02 m^2 which may be compared with an area of 0.17 m^2 for the *faering's* oars (two oarsmen each with two oars).

Kvalsund 1, as reconstructed by Johannessen, has two oar stations where the beam measurements are 1.44 m suggesting an oar length of *c.* 2.88 m: the fragmentary oar is 2.01 m in length; the blade is generally aerofoil shape in plan with the maximum breadth near the rounded end. Kvalsund 2 is more conjectural in reconstruction: there are ten stations for oars and boat breadths vary from 2.20 to 3.20 m. The oar which was recovered with only minor damage to the handle is estimated to have originally been 3.33 m long, and oars up to 4.40 m could have been used at all stations with a gearing of 3:1, and the oarsmen pulling single oars double-banked on the thwarts. This data may have been used by Johannessen when evolving his reconstruction. The blades of the two oars associated with this boat have a projecting tongue, comparable with the Oseberg oars. The Kvalsund oars are near the top end of the known range for blade area 7.1 to 8.1 × $10^{-2}m^2$ being similar to the Donegal curach and slightly bigger than Gokstad 1; the blades are of lowish length/breadth ratio (4.7 to 6.2).

Engelhardt deduced that there were fourteen oar stations each side of Nydam 2 but recent investigators believe there were fifteen. The oars were said to range in length from 3 to 3.5 m and the two illustrated by Engelhardt (1865) measure 2.94 and 3.42 m, although Shetelig (1930) gives 3.02 and 3.52 m; they appear to have no specific handle. They have biconvex blade cross sections and in plan the blades as illustrated seem almost to be an extension of the loom, without shoulders. These are slim blades, yet because their length is great, both absolutely (1.30 to 1.52 m) and as a propoıtion of oar length (0.45) their areas (15 × 10^{-2} and 16 × 10^{-2} m^2) are the greatest in the survey. A larger blade area makes a given gearing feel much stiffer for biomechanical reasons and it increases the efficiency of the oar (Watson, p.c.). Large blades can be difficult to handle in a seaway but the slimness of the Nydam blades means that the thrust can readily be varied to match changing conditions.

There are difficulties with the rowing geometry of this boat (Åkerlund, 1963) which a 0.45 blade/oar ratio compounds. It may be that, as with the rudder (see below) these oars were an interim, experimental form or the logic behind their design and use may not yet be understood.

The Oseberg vessel had fifteen oar stations on each side; no evidence for crossbeams at an appropriate level for thwarts was found and so it is presumed that the oarsmen sat on portable chests, two manning single oars at each station (Brøgger and Shetelig, 1971: 96) as the oarports are too close to the floorboards (0.40 m) to be practicable for stand/push rowing. The pine oars with this find varied in length from 3.70 to 4.0 m and could have been used at all stations. The blades in plan have a protruding tongue. With an area of $10 \times 10^{-2}m^2$ and a slimness ratio of *c.* 5:1 these oars are evidently about average for this size of craft.

Gokstad 1 had sixteen oar stations and thirty-two pine oars between 5.30 and 5.85 m in length. The oarsmen are presumed to have used chests as in Oseberg for sit/pull rowing. The greater depth of hull and therefore presumably greater freeboard, of Gokstad 1, meant that longer oars were necessary to achieve similar oar angle and gearing. The Gokstad oar blades are slimmer than those of Oseberg and of less area and are a more regular lens shape in plan. With full complement of oars Gokstad would have had 2.14 m^2 of oar blades immersed compared with 3.00 m^2 of Oseberg. However, such simple comparisons do not convey the whole story: the propulsive power available would also depend on the gearing used and the ability of the crew to work heavy, though balanced oars at a steady rate for long periods.

Rowing power

An oarsman in a rowing eight (sit/pull) can develop over 1 hp (550 ft/lbs/sec; 750 Watts) for a short time (Alexander, 1927: 232), but it is generally held that $\frac{1}{3}$ hp (250 Watts) is sustainable for say 10–20 min (Corlett, 1978: 305; Foley, 1982: 315; Landells, 1978: 166). The efficiency of oars in transmitting power from hands to water has been variously estimated between 70 and 22% (Landells, 1978: 166; Corlett, 1978: 305; Alexander, 1927: 232); the balance of opinion favours something of the order of 33 $\frac{1}{3}$%. Thus, for sustained rowing, nine men are required to generate each useful hp (750 W). Considerably more effort is required to propel a boat into a head sea and wind; additionally spray and rain can soon reduce morale. Rowing an open boat in such conditions is only practicable with regular changes of oarsmen.

TOWING

Boats may be towed by other boats (Casson, 1971: fig. 193), by men (Johnstone, 1980: fig. 12.3) or by beasts (Wilson, 1972: 9). Men may tow by swimming and holding onto the boat or from the river bed (Beaudouin, 1970: 90), see Fig. 5.5, but more generally it is done from the bank where it is known as *tracking*. The pull is thus at an angle (which will vary with the length of tow) to the desired direction and, in general, will tend to bring the boat into the bank unless countered by steering. However, the proximity of the bank upsets the pressure field around a boat (Rawson and Tupper, 1976: 511) resulting in a lateral force which may be thought of as a cushion of water holding the boat off the bank. This lateral force generally produces a yawing moment and it is found empirically that this offsets the tendency of a tow line to bring the vessel into the bank if the tow is fastened on the centreline between 20% and 40% of the length from the bow – the boat will then be towed parallel to the near bank (McKee, 1983: fig. 104) with minimum use of the steering device.

The tow line may be fastened to a bollard on the centre line at this point, 30% ± 10% from the bow, or, in small boats it may be made fast on the side to a position which gives the same angle of pull. On the other hand where the boat has a high bow or there is deck cargo, or obstructions along the water's edge, or other boats to be overtaken, then a point of attachment on a mast has to be used (Fig. 12.11). Towing masts and their fittings may be difficult to differentiate from masts for sails, and, indeed dual purpose masts are known. They are, however, generally of less scantlings than might be expected for a sailing mast for the size of boat, with less or no, standing rigging, and stepped in a less elaborate step and supporting timbers – see, for example, the *galupe* and the *chaland* of the River Adour, S.W. France (Beaudouin, 1970: figs 20, 23, pl. 1-H). Beaudouin's illustration of the *galupe* (Fig. 12.11) shows the widespread practice of passing the tow line through a block on the mast to a strong point aft, the height of the block being variable to match towing conditions. These River Adour towing masts are 24 to 31% from the bow. A towing mast some 26% from the bow is depicted on a Gallo-Romano relief from Cabrieres d'Aygues, now in the Musee Calvet Avignon (Johnstone, 1980: fig. 12.3) where three towlines lead from the mast head. The boat on the Blussus's stone (Ellmers, 1978: fig. 5) has a towing mast at *c.* 40% from the bow (Fig. 12.32).

Figure 12.11 Towing and poling *galupes* on the River Adour, France After Beaudouin, 1970: fig. 23 (courtesy Musée de la Batellerie, Conflans–Sainte–Honorine).

Figure 12.12 The fourteenth-century seal of Kiel. Compare sail with that in Fig. 12.16. (Photo: NMM Greenwich.)

A maststep and evidence for standing, and even running, rigging do not therefore necessarily mean that a boat find was propelled by sail: if the step is at *c.* 30% from the bow and the boat is of barge-like form an alternative hypothesis of towing mast should be considered – see discussion on Romano-Celtic boats below. A square sail on a mast so far forward of amidships, with no second mast with balancing sail would restrict use to following winds, as in winds from other directions the Centre of Effort (CE) of the sails will be forward of the Centre of Lateral Resistance (CLR) of the hull, and the resulting couple would give the boat excessive, and possibly dangerous lee helm. On the other hand, a fore and aft sail such as sprit, lateen or lug, on a mast in such a forward position could have a CE much closer to CLR. The Humber sloop and keel have similar hulls, but the sloop has fore and aft rig on a mast stepped some 30% from the bow, while the keel has square rig on a mast at 40% (Finch, 1976: 81, 83). From iconographic evidence the sprit was in use in the Mediterranean from the second century BC, and the lateen from second century AD (Casson, 1971: 244) but neither are known in northern Europe until well into the medieval period (Moore, 1925: 87, 147). However the lug may have been used on Romano-Celtic boats (see p. 234). Furthermore, it can be difficult to differentiate a lug from a square sail in iconographic representations, see for example the 13th century seal of Kiel (Fig. 12.12) and of Hastings, and the possibility of early use should not be ruled out.

Details of the mast steps in boats of the Romano-Celtic tradition (McGrail, 1981a: 22–4) are given in Table 12.5. The Zwammerdam, Bevaix and Blackfriars steps are at positions within the optimum range for a towing maststep, but which might also

Table 12.5 Details of mast steps in selected Romano-Celtic boats

	Distance from bow / Loa	Step L × Th × D (cm)	Notes	Reference
Zwammerdam 2	*c.* 0.25	11 × 7 × ?	step is in a longitudinal timber.	de Weerd, 1978
3	0.25	—	extended logboat	
4	0.25	—	—	
6	0.25	—	—	
Bevaix	0.29	13 × 15 × 10	with a slope lead-in on aft side	Arnold, 1975
Blackfriars 1	0.30	25.4 × 33.7 × 12.7	—	Marsden, 1966
Bruges	—	14.5 × 20 × 7	had an associated mast 9.30 × *c.* 0.16 m	Marsden, 1976
	—	14.5 × 20 × 4	in re-used timber	

Notes: L = longitudinal measurement
Th = thwartship measurement (siding)
D = depth (moulding)

have been used for a square sail in a following wind. The size of the step and the scantlings of its keelson suggest that Zwammerdam 2 did sail and this may also have been the case for boats 4 (which had a 'large' step 'with additional features') and 6 (which had a substantial mast beam as well as a step). Logboat Zwammerdam 3 may also have had a small sail as washstrakes give her flared sections with increased beam, hence stability. The primary means of propulsion for the Zwammerdam and Bevaix boats would seem to have been river current and ebb tide downstream, with the flood tide, towing from the bank and possibly poling or rowing upstream; an auxiliary sail possibly being used on either leg when the wind was fair. The Druten boat and other barge-like river craft of this Romano-Celtic tradition may have been similarly propelled.

On the other hand, Blackfriars 1 was a seagoing vessel and her mast was probably for a sail which may well have been a lug.

PROPULSION BY SAIL

The wind may be harnessed to propel a boat by merely holding or propping up a leafy bush as early settlers noted was done by the Micmac Indians (Adney and Chapelle, 1964: 65) and in mid-nineteenth century Tongareva (Haddon and Hornell, 1938: 45); or by temporarily fastening a blanket, a skin or a piece of bark to a stick or paddle – other methods used by N. American Indians (Waugh, 1919: 30). Such *ad hoc* methods require no permanent fittings on the boat and thus such usage cannot be archaeologically demonstrated. These methods are only of use, however, in light, following winds and when there is little sea-induced motion on the boat. In stronger winds, and in choppy seas and when it is required to make progress across or against the wind, more elaborate fittings are required to hold the mast or sail-supporting spar firmly, and to disperse the wind and sea-induced stresses through the boat's structure. The requirements here are a socket or pivot point to take the heel of the mast, usually on the bottom of the boat, and further support higher up the mast – either rigging lines from mast to the boat's structure or a mast-beam or thwart at about the level of the top of the boat's sides. Boats are however known without standing rigging: in India (Folkard, 1870: 32, 268, 299) and Britain (Finch, 1976: 70, 90). Even where there is rigging it may be of the simplest kind, leaving little trace on the boat, stays and shrouds being made fast to the thwarts or to fittings at the bow and stern which have another, structural role.

Thrust on the sail is proportional to area (*ceteris paribus*) and with inadequate rigging sail area may have to be limited. The ideal sail should set well, resist distortion and be easily controllable so that adjustments can readily be made to match changing wind and sea conditions. The minimum running

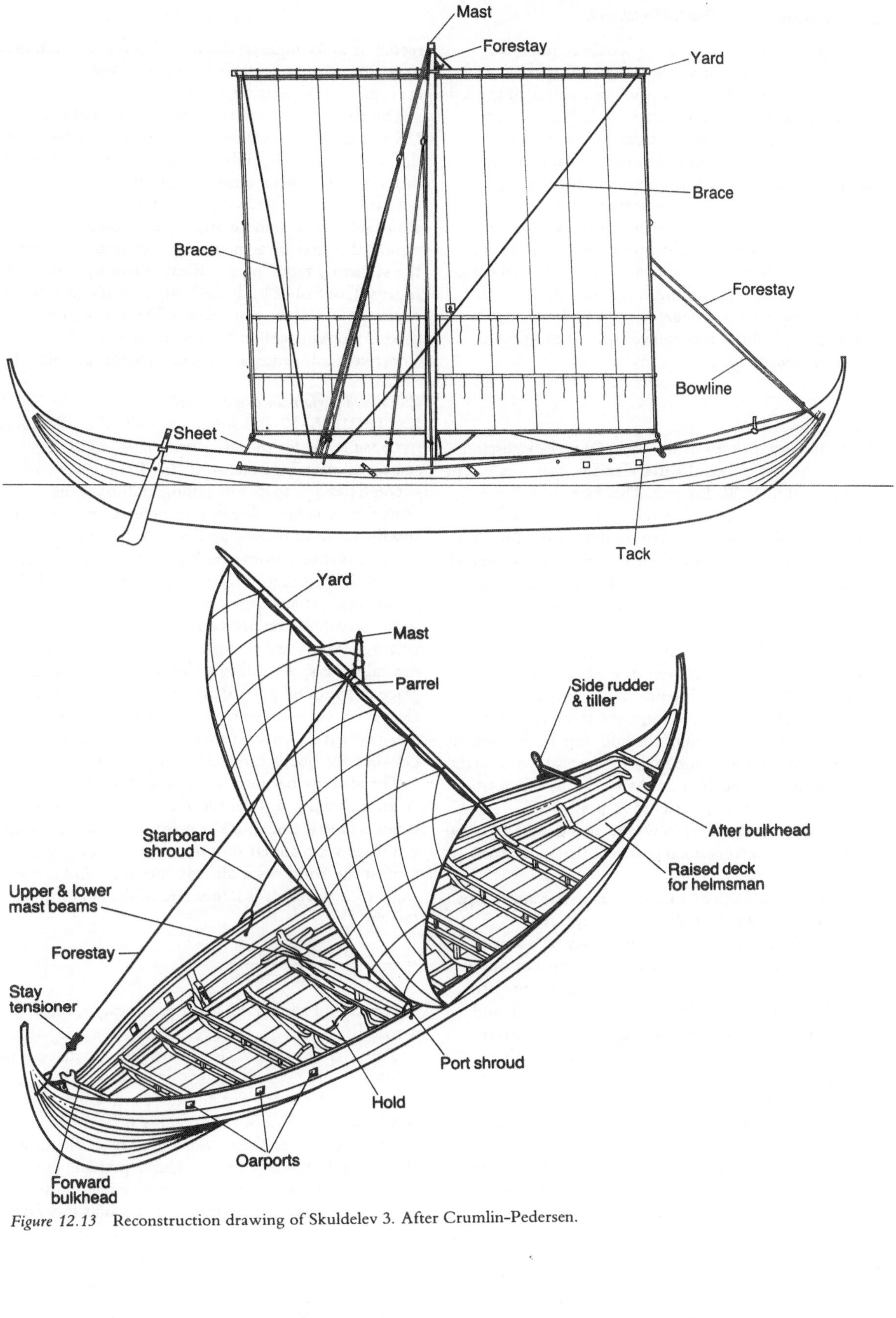

Figure 12.13 Reconstruction drawing of Skuldelev 3. After Crumlin-Pedersen.

rigging (Fig. 12.13) such a sail would require is a halyard to hoist and lower yard and sail (partly or fully) and ropes (*sheets*) to the lower corners of the sail to adjust the sail to the relative wind. However, for windward sailing to be practicable *braces* to each yardarm would be required to give the appropriate adjustments; and a means of holding the leading edge of the sail (*luff* or *forward leach*) firmly into wind would be advantageous (*bowlines, beating spars*). *Lifts* to further support the yard, and a *parrel* to hold it to the mast are further refinements. The ability to match the area of sail to prevailing conditions is essential. This can be done to a degree by partly lowering the yard; better control is obtained by the use of *reefs*, *brails* or *bonnets*.

The square sail

Until the last century of the period under discussion, the single, loose-footed **square sail** was the main type of rig used in N.W. Europe. This form of sail was rectangular in shape and was bent to a yard which was generally square (at right-angles) to the fore and aft line of the boat or ship but which might be braced round through a horizontal arc of some 4–5 points (45–56 ¼°) either side of the bow. Heeling and turning moments are imposed by the wind acting through the Centre of Effort of a square sail (somewhere near the geometric centre) and the hydrostatic force which tends to resist this is considered to act at the Centre of Lateral Resistance (CLR), about which point the hull may be considered to pivot around both the midship vertical axis and the horizontal fore and aft axis. As with CE, the position of CLR continuously varies in response to environmental changes; however, it is generally taken to be near the geometrical centre of the underwater body.

With a single square sail on a mast near amidships CE and CLR will be only slightly separated longitudinally and thus there will only be a small couple tending to turn the boat either into or out of the wind. Slight weather helm is generally preferred i.e. the boat has a tendency to turn into wind, and this may be obtained by having CE slightly aft of CLR, in the condition of sailing balance. In these circumstances a rudder can generate a force opposing leeway (see section on Steering).

For stable, predictable performance it is desirable that CE remains in the same longitudinal position for all relative angles to the wind no matter how much of the sail is reefed. In practice the square sail does not remain symetrically disposed about the mast as it is reefed or as its angle to the wind is varied, but moves over a limited range due to stretching and asymmetrical distortion and bellying.

The couple resulting from the vertical separation of CE and CLR causes the boat to heel away away from the wind until an equilibrium is reached with the new distribution of displacement. For safety reasons it is desirable that the vertical position of CE should be adjustable quickly and easily in response to changing winds: this may be achieved by reefing or unreefing the sail or by reducing its effective area by some other means. Control of both the longitudinal and vertical positions of CE may be achieved by the use of a second (topsail) sail on a single mast – such improvements came late in the medieval period.

Windward performance of a square sail

Kay (1971: 15–17) has suggested that (after allowance has been made for differing waterline length i.e. standardisation by a factor of $(1/\sqrt{L})$) increases in average passage speeds of sailing craft over the centuries have been due not to increased maximum speeds but to improved ability to sail to windward, an improvement from say, 7 points to 3½ points off the wind. This sort of improvement has the added advantage that less time need be spent hard on a wind with consequent reduced stress in the hull and rigging. Windward performance is also important in coastal waters; the ability to leave or make a landing place or to round a headland may depend on how close a boat can sail to the wind and how little is her *leeway* (drift downwind measured as an angle between the boat's fore and aft line and the wake).

The analogy between a semi-rigid aerofoil such as an aircraft wing and the set of a flexible, variable sail is not a precise one, but the reason why an aeroplane can fly is very similar to why a boat can be sailed against the wind – not directly into the wind but at an angle to it. Camber is a measure of the curvature of a sail and

$$= \frac{\text{bow of sail section}}{\text{chord of sail section}}$$

A curved sail adopts the form of a (near) vertical aerofoil.

The relative air flow (RAF) around this shape is the resultant of the wind's velocity and the boat's velocity. For an range of angles of incidence (between RAF and the chord of the sail) this airflow is accelerated around the leeward (downwind) side of the foil and thus an area of negative pressure occurs on the leeward side of the sail, with a region of positive pressure to windward (Bernouli's theorem).

The resultant force (RF) acts to leeward (in the horizontal plane) at right angles to the sail; it may be further resolved into Drag in the downwind direction and Lift at right angles to the RAF. This lift may be further resolved into a useful driving force (D) and an undesirable force (H) which tends both to (heel) and to set the boat to leeward (leeway).

The driving force (D) is considered to act at the CE of the sail (see above) and its magnitude and direction are determined by five main factors:

1. Dynamic pressure of the relative wind (RAF)
2. Sail area
3. Sail shape
4. Sail material and cut
5. Angle of incidence to RAF (α) and camber.

Factors 2, 3 and 4 are determined by the sailmaker; 5 is controlled to a large extend by the seaman and, in conjunction with the design parameters of hull and rigging, determines the efficiency with which (1) is converted into useful driving force.

1. ***Dynamic pressure of the relative wind.*** The stronger the wind the greater the dynamic pressure of the relative wind and the greater the useful driving force (D): but also (but not necessarily by the same proportion) the greater the force (H) which causes the boat to increase leeway and to heel further to leeward until it is in equilibrium with a new distribution of displacement. There are limits to the loading that sail and rigging can sustain; there is a limit beyond which a boat cannot safely be heeled; and leeway should be minimised as it effectively increases the boat's angle to the wind and hence reduces progress to windward. The rigging stresses are greatest when the boat is close-hauled and, as the dynamic pressure varies as V^2, transmitted loads can be very great. Overstress can be prevented by adjusting the sail's camber by means of sheets and bowlines or by letting the boat's head pay off from the wind and if necessary shortening sail and thus lowering CE. If heel is allowed to exceed a certain angle water will be taken on board leading in extreme conditions to swamping, or, if G gets above B because of excess bilge water, she will capsize. The measures taken to deal with excessive rigging stresses will also limit the tendency to heel; alternatively it may be possible to shift ballast or crew to the windward side. At the design stage good transverse stability and greater freeboard will increase the boat's tolerance to the heeling force. The use of a low aspect ▭ rather than a high aspect □ sail will mean a lower CE and hence a reduced heeling moment, but also reduced efficiency (see below).

   Insertion of horizontal battens in a sail allows the sail to bulge between these supports so that as wind pressure increases the radius of curvature of these bulges diminishes, with the result that tension in the sail and rigging stresses remain roughly constant (Gordon, 1978: 124–6). A similar effect is obtained by making a sail from individual squares or lengths of cloth and allowing each section to bulge and reduce stress. However, too sloppy a sail will be aerodynamically inefficient and there may be handling problems.

2. ***Sail area.*** The Driving Force depends on the area of sail presented to the wind. There is, however, a limit to the size of sail that can be conveniently handled and controlled. The Square-Cube Law applies; for example if a replica hull is built to $\frac{2}{3}$ linear dimensions of the original, the displacement volume will be reduced to *c.* ($\frac{8}{27}$) of the original but a sail of $\frac{2}{3}$ linear dimensions would be some ($\frac{4}{9}$) area of the original and the replica would be over-canvassed. A more appropriate sail would have an area of ($\frac{8}{27}$) the original. Additionally, if linear dimensions are doubled, heeling moment due to sail area will be multiplied by 4 but the righting moment (proportional to volume of displacement) will be multiplied by 8. Thus bigger boats are relatively more stable than small ones; that is, stiffer. The sail area/displacement coefficient = $As/\sqrt[3]{\nabla^2}$ is some measure of power available in relation to bulk to be moved through the water and this may be a guide to relative sailing potential. Table 12.6 shows that, for three Viking Age boats where sail area has been estimated, the values of this coefficient range between 18.0 and 11.9. These are lower than the range for modern U.S. yachts of 10 to 20 m in length – the range there being 12.4 to 22.8 (Gillmer 1979: 169) – but these two groups of boats are disparate when sail rig and hull form are considered. Nevertheless as more evidence accumulates, especially from rigorously evaluated replica trials, this coefficient should prove useful in assessing the *relative* performance of ancient boats of similar hull form.

3. ***Sail shape.*** Aspect ratio (AR) gives a measure of

Table 12.6 Sail area coefficients for selected medieval boats

Boat	Displacement (m^3)	Overall length (m^2)	Sail area (m^2)	Sail area coeff.*	Reference for sail area
Skuldelev 3	7.60	13.3	46	11.9	Crumlin-Pedersen (p.c.)
Gokstad 1	17.96	23.2	70	10.4	Christensen, 1975: 266
Skuldelev 1	34.00	16.5	84	8.0	Crumlin-Pedersen (p.c.)

* Sail area coeff. $= \dfrac{\textit{Sail area}}{(\text{Displacement volume})^{2/3}}$; see Gillmer (1979)

See also Bruce No. $= \dfrac{\sqrt{A\ (\text{ft})}}{\sqrt[3]{\Delta\ (\text{lb})}}$ Doran, 1981: 63

a sail's proportions. Generally this is Height^2/Area which for a rectangular sail simplifies to Height/Breadth. A sail's efficiency is reduced by air 'leakage' around the sail from windward to leeward, especially over the head and under the foot; this creates turbulence and increases drag. This leakage is less in a □ sail of high aspect ratio than in a ▭ sail of low AR (Kay, 1971: 78) and optimum camber is easier to achieve. Gillmer (1979: 176) has estimated that an efficient square sail should have an AR of 1.5 or even 2 to sail close to the wind; for fore and aft sails this optimum AR is *c.* 4.5 (Kay, 1971: 99 and fig. 52). Such a high AR (*taunt*) sail may certainly point higher (be used closer to the wind) (Binns, 1980: 90) but it imposes greater stress on the hull and rigging, and may not be practicable; the twist imposed on the yard and sail makes braces essential and a wider shroud base advisable. In addition CE will be higher which increases the heeling moment. Experimental work has shown that the optimum AR for reaching (wind on or near the beam) is 1 i.e. a square-shaped fore and aft sail, whilst a square sail of low aspect ratio ($\frac{1}{3}$) also gave good results on a broad reach (Kay, 1971: 99).

4. ***Sail material and cut.*** The ideal material for sails would have a low porosity, be stable without stretch or distortion, and have little frictional resistance to airflow along its surface. Any dimensional changes mean that the CE will not be constant, leading to sail balance, heeling and steering problems. Flax, long used for sailcloth, distorts easily and the cloth creeps (Gordon, 1978: 147, 253); the latter may be partly cured by pre-stretching the sail, but if the tensions on a sail are at an angle to the weft and warp threads the sail distorts and in time will have a permanently baggy set. Flax fibres swell in thickness with moisture and wetting sails is one way of decreasing porosity and hence increasing effective driving force (Gordon, 1977: 143).

5. ***Angle of incidence to relative airflow and camber.*** Lift (and drag) from an aerofoil increases with increasing incidence angle (angle of attack = angle of sail chord relative to apparent wind) up to a theoretical *c.* 15° (depending on camber). Up to this point the useful driving force (D) also increases. Increasing angles of incidence lead to breakaway of the airflow from the leeward side of the sail beginning at the trailing edge. The aerofoil stalls: that is there is a sharp drop in lift whilst drag continues to increase. Thus to achieve maximum driving force when close-hauled, the sail must be kept as near this critical angle as fluctuating wind and sea will allow, and the aerofoil shape of the sail maintained by tension on the sheets, braces and beating spars or bowlines. This can lead to excessive stress in the rigging and hull, and it thus may pay in the long run not to drive a boat close to the wind in the condition for maximum lift (and hence maximum driving force) but to ease the sheets and allow the boat's head to pay off slightly to leeward; thus sailing 'full and by', or even 'by and large', rather than 'by the wind'.

The onset of stall may be delayed in a two-masted boat by the combination of two sails one aft of the other. The restriction imposed on the relative airflow (apparent wind) where the two sails are closest improves the airflow and causes the air to accelerate around the leeward side of the after-sail (Bernoulli effect) and thereby increase the lift for any particular angle of attack.

Much of the foregoing discussion was theoretical, as the art of handling a single square sail was almost lost in N.W. Europe and is now having to be relearned. Practice and theory, however, suggest that the single square sail is more efficient at running and reaching than beating to windward; it is a better driver than lateen. In order to sail close to the wind the plane of the sail has to be held at the optimum angle to the relative airflow; this is achieved by swinging the yard and sail closer to the wind and by bringing the camber of the sail to the optimum angle. The angle through which a square-rigged yard and sail can be rotated about the mast is restricted to *c.* 4 to 5 points either side of dead ahead by the forestay and by the lee shrouds. The smallest camber obtainable is determined initially by the cut of the sail and then by fine adjustment in use. The stronger the wind the smaller the camber required for efficient sailing (*cf.* a laminar flow aircraft wing). The optimum position for maximum camber is ⅓ to ½ along the chord from the leading edge (luff).

The Lateen sail

The **lateen sail** (Fig. 12.14) is a triangular sail bent to a long yard which is generally in the fore and aft line of the boat. The yard which may be ⅓ longer than the boat generally consists of two spars fished together at the centre which allows the yardarms to whip more freely and thus ease the setting of the sail. The mast is stepped forward of amidships and is generally raked forward since no forestay is rigged as it would impede operation of the sail. The yard is suspended on the leeward side of the mast with its forward end bowsed down so that it is set obliquely with a high peak to the sail; vangs are sometimes secured to this upper end of the yard to control its position and impose tension to the leading edge of the sail. In general the lateen has a better aerodynamic performance than the square sail, and thus is more weatherly. However, the lateen is more difficult to handle – Columbus had the *Nina* re-rigged from lateen to square for oceanic voyages 'to obtain more tranquility and less danger': quoted by Phillips-Burt (1971: 78). This is because, when changing tack, the yard is swung around the masthead almost vertically so that it can be reset to leeward although it may be used to windward of the mast for short tacks. In addition, when running with the wind from astern, there is a danger that CE may well get ahead of CLR, causing steering problems and, as there is a downward component of the propulsive force in a sail, the boat may dig in her head and plunge.

The 'Arab' lateen (one with a short luff, a *settee*) was in use in the Mediterranean from the second century AD and the pure triangular form from the fourth century (Casson, 1971: 244). A similar form seems to have been in use in S. American waters in pre-Columbian times (McGrail, 1981a: 76).

The Spritsail

The **spritsail** is a loose-footed fore and aft sail of four-sided shape, spread by a sprit (spar) from peak to tack where it is held to the lower mast in a rope *snotter* (Fig. 12.15). The mast is stepped forward of amidships to bring CE very slightly forward of CLR. This simple rig gives weatherly performance and manoeuvrability and is the easiest and quickest rig to set and drop when a medium to small sail is used; a large sail requires complex reefing (McKee, 1983: 143).

The spritsail was in use in the Mediterranean from at least the second century BC (Casson, 1971: 244)

Figure 12.14 A triangular lateen sail.

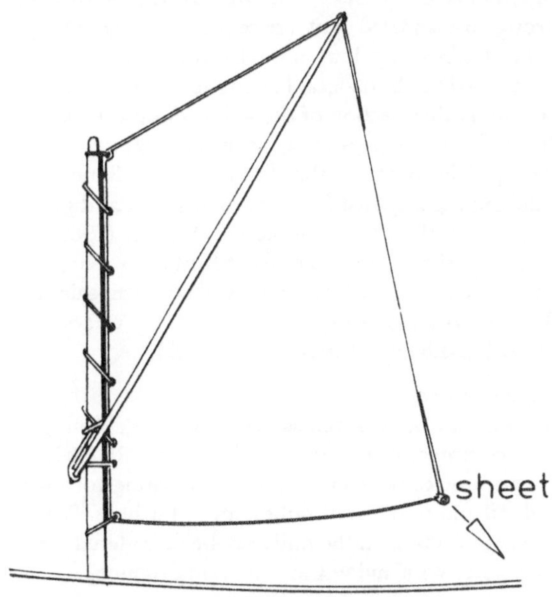

Figure 12.15 A spritsail. After McKee, 1983

Figure 12.16 A lugsail

but its use in N.W. Europe is not documented before the fifteenth century (Moore, 1925: 147).

The Lugsail

The *lugsail* is a loose-footed fore and aft sail set on a lug (yard) which is suspended from the mast so that *c.* $\frac{1}{3}$ to $\frac{1}{4}$ of the sail area is forward of the mast (Fig. 12.16). When changing tacks the yard and sail have to be lowered partly (dipped) to allow the leading end of the lug and the sail tack to be passed abaft the mast to the new lee side. However, when only $\frac{1}{4}$ sail is forward, it is practicable to leave the sail to windward and use it as a standing lug. If the tack is bowsed well down, the sail is raked to counteract the movement aft of CE when reefed. The lugsail draws well and works well to windward but a strong co-ordinated crew are required to change tack.

The Chinese version, in use from the twelfth century AD has horizontal battens and this permits each individual section of the sail to assume a curve, thus reducing stresses in sail, mast and rigging. Multiple sheets are another feature of the Chinese rig, reducing the imposed forces in the rigging and giving better control of the set of the sail. Ellmers (1969) has suggested that representations of Romano-Celtic boats of the second/third century AD on a mosaic at Bad Kreuznach and a gravestone from Junkerath depict lugsails with battens (Fig. 12.25).

Sailing stresses

A mast may be regarded as a pole or strut carrying axial compressive forces (Gordon, 1978: 226, 280), the purpose of which is to provide a seating for yard and sail and to transmit sail forces to the hull. The stresses imposed on the hull may be considered for two cases, with and without standing rigging.

Without stays or shrouds. The mast must be held at its heel in a fitting such as a mast step locked by wedges; it may also be supported a little higher up at crossbeam level. The mast then acts as a cantilever and bends away from the driving force. A relatively slender mast will be more resilient and transmit less loading to the hull. The leverage on the cantilever multiplies the effect at the heel where the vertical and horizontal components of an almost-point loading must be distributed to the keel and shell by such timbers as knees, keelsons and frames. At cross-beam level the forces may be transmitted by beam, knees and longitudinal beam-shelf/risings/stringers to the shell, a stout upper strake is especially useful in this respect.

Without standing rigging the cantilever mast is generally limited to a relatively small sail area and to running and reaching rather than beating when stresses are greatest.

Mast with standing rigging. Such a mast is a strut supported at the heel as in the unstayed case, but also near the mast head by stays running in the fore and aft directions (the after one may duplicate as a halyard) and possibly by shrouds to the sides of the boat. The mast cannot now deflect so much, although slack rigging will allow some bending. The thrust downwards of the mast equals the tension in the rigging; the horizontal forces at the heel of the mast are less and thus sails of greater area may be carried, and the extra loading due to close wind sailing may now be acceptable. Stress has to be distributed not only away from the mast step and crossbeam regions but also from the points of attachment of the rigging to the hull: often stempost and sternposts for the stays; and strengthened top strake for the shrouds, although appropriate crossbeams may be used. The shrouds need to have as wide a base as possible to reduce tension; thus a lead outboard of the top strake is desirable, but if this is insufficiently strong they may be taken to the junction of the crossbeam and sides – whichever method is used the point loading must be distributed to other members of the structure. Edge-fastened planking, especially if it is radially cleft (Ch. 4) will more readily take shrouds as it can take stress tangential to the plank face.

The combined effect of thrust down on the mast heel and tension up at the sides tends to force the hull apart; hence the requirement for transverse strength in the frames and crossbeams.

The tension loads in the rigging are best diffused into several ropes as weight is thereby saved (Gordon, 1978: 306–7, 309–10), thus there is less stress when multiple stays and especially shrouds are used. Operational considerations, extra windage and maintenance may make this complexity impracticable, but where it can be used there is much less chance of over-stressing and failure of rigging. On the other hand, it is more economical to support a heavy compressive load than a light one, thus from this point of view one mast is better than several for the same sail area (Gordon, 1978: 309–10). There are advantages to multimast rigs, however: the CE is lower, the boat may be brought closer to the wind, sailing balance can be more easily achieved, sail area can be more readily varied in response to changing conditions, and handling two small sails is easier than one large one. Thus the structural and operational advantages must be balanced against one another; the advantages of two masts probably predominate when operations in more difficult weather are to be undertaken.

SAIL IN N.W. EUROPE

None of the earliest surviving boats – Cheops of the mid-third millennium BC and the four Dahshur boats of *c*. 2000 BC – were propelled by sail. Insufficient remains survived of the C. Gelidonya sea-going wreck of *c*. 1200 BC, to determine whether she had sail but it seems likely. The earliest artifactual evidence is the mast step in the fourth-century BC Kyrenia boat (Katsev, 1972: 50–2). In northern Europe, the earliest plank boat finds – Ferriby 1 to 3 of mid-second millennium BC, the Brigg 'raft' of seventh century BC, Hjortspring of *c*. 350–300 BC, and the roughly contemporary boat from Ljubljana were all boats for inland or inshore waters and did not have evidence for sail. The earliest finds in N.W. Europe with mast steps are in fact some of the Romano-Celtic boats (de Boe and Hubert, 1977; de Weerd and Haalebos, 1973; Marsden, 1976), although some of these may have been for towing rather than for sailing masts; and it is not until the Oseberg vessel of AD. *c*. 800 that we have indisputable, albeit incomplete, artifactual evidence for sail.

The iconographic evidence is, however, more informative about the early and widespread use of sail, although again the evidence is far from complete and is sometimes ambiguous. A clay model of *c*. 3400 BC from Eridu in S. Mesopotamia (Bass, 1972: 12) and a painting on an Egyptian late-Gerzean vase of *c*. 3100 BC (Bass, 1972: 13) are the earliest evidence for the use of sail anywhere in the world: the Gerzean pot (Johnstone, 1980: fig. 7.10) evidently shows a pole mast and rectangular sail stepped towards one end of the boat – this would limit the boat to a following wind. Earlier representations from Egypt appear to have a palm branch in the bows which could be decorative but may be for propulsion as such means are known from recent times (Folkard, 1870: 247; Waugh, 1919: 30); others (see Johnstone, 1980: p. 76–7 and fig. 7.11) evidently have skin shields set up like banners on a pole which may also indicate use of sail.

By *c*. 2000 BC mast, rigging and rectangular sail are shown on ships carved on Minoan jewels (Casson, 1971: 33) and the use of the square sail is reasonably well documented from this time on in the Mediterranean. The spritsail (Casson, 1971: 244) was in use from at least the second century BC; the 'Arab' lateen (that is, a 'settee' or quadrilateral sail with a short luff) from the second century AD and the triangular lateen from the fourth century AD (Casson, 1971: 244). Unless she had a second mast further aft (for which Casson (1980) has noted iconographic examples from the fifth century BC) the fourth-century BC Kyrenia ship with her mast step *c*. ⅓ away from the bows could have had a square sail with sailing generally limited to before the wind; or a fore-and-aft sail – e.g. a sprit or lateen which would then be earlier than the evidence from iconography.

It has been claimed that some of the early Scandinavian rock carvings (see Farrell, 1977; van der Porten, 1978; Farrell, 1979; Schovsbo, 1980 – for examples) show mast and sail. The majority of these carvings are difficult to date and equally difficult to interpret: some may depict the use of trees and boughs for temporary propulsion by the wind, but the question is probably best left open for the lack of firm evidence.

Pytheas's voyage in *c*. 325 BC (Hawkes, 1975: 44) from the western Mediterranean to northern Europe would have made the sail known there, if not already known. The only examples of indigenous sail in northern waters during the prehistoric period are the Broighter boat model of the first century BC (Fig. 10.9) with mast and yard, and Caesar's description of the boats of the Veneti (*B.G.* 3.13). However, the fourth-century late-Roman poem *Ora Maritima* has a passage suggesting that the Celts were able to sail between France, Britain and Ireland in the sixth century BC. (Hawkes, 1977). In northern Europe, evidence for sailing fittings is mainly from two groups: the Romano-Celtic boats of the third-fourth centuries AD; and the boats of the Norse tradition and subsequently the cog.

Masts

Masts have seldom been excavated, although the lower portions have been found with certain finds, e.g. the Viking Age Gokstad and Oseberg finds. The mast from the Bruges boat of AD 180 ± 80 (HAR-472) was recorded by sketch measurements (Jonckheere, 1903), but only small fragments now survive (Marsden, 1976: fig. 7). The original length was said to be 9.30 m with maximum diameter of *c.* 0.16 m, the upper 2 m is of rounded cross-section, the lower probably square (Marsden, 1976: 31). Holes through tenth/eleventh century mast partners (see below) from Århus and Hadsund indicate that the mast associated with them had diameters of *c.* 20 and 25 cm (Crumlin-Pedersen, 1972b).

Attempts have been made to deduce length of mast from known rules of thumb such as:

(a) the traditional Vestland Norwegian rule (Andersen, 1975: 90–6) that the mast from heel to halliard hole should equal the girth of the boat at that station;
(b) the fifteenth century Italian rule (Anderson, 1925b) that the mast should be four times the ship's breadth or 1.6 times the keel length. This ms. also gives the girth of the mast at deck level as $\frac{1}{14}$ its length.

Such rules only apply directly to boats and ships built in a certain tradition and within a limited size range: thus a boat half the size of another of similar design and rig does not need a mast half as long but somewhat shorter, due to the Square-Cube law (see above).

Christensen (1975: 266) has used a different approach to estimate the length of the Gokstad mast. He noted that handling a mast probably weighing over 800 lb would have been hazardous if it projected outboard during raising and lowering (although Binns, 1980: 124 does not believe this is a seriously limiting factor). Christensen therefore concluded that the Gokstad mast would have been just long enough to lie forward of the cross beams in the stern when lowered: this gives a mast length of *c.* 10 m. Application of the nineteenth/twentieth-century Norwegian rule suggests something approaching 12 m. However, this rule applies to a sail of high aspect ratio and it is not certain that Viking Age craft had such taunt sails (see below).

Howard (1979: 29) has applied the Italian rule designed for multimasted ships to the keel length of the *Grace Dieu* of 1418 – 125–130 ft (38–40 m). The result is to suggest a main mast of 180 ft (55 m) length with a 14 ft (4.27 m) girth whereas a contemporary account noted that the mast was 200 ft (61 m) × 22 ft (6.7 m); although the rule's forecast of length appears close to reality it is not so accurate for girth, which emphasises the difficulty in attempting to extrapolate rules outside their range of validity. Masts of the size of *Grace Dieu*'s mainmast and doubtless smaller ones, were composite made from several timbers in length and breadth, as can be deduced from the many representations of wooldings, ropes around the masts.

In a survey of the evidence for Viking Age rigging Christensen (1979) has concluded that masts were generally of round cross-section but with the foot squared off to fit in the mast step. He considers that, although the mast generally tapered slightly upwards, near the top there was a swelling around the halliard hole to strengthen this hole and to provide a seating for standing rigging.

Mast steps

Masts are generally stepped in a socket (step) in a keelson (longitudinal timber on top of the keel) as in the Norse tradition; or in a transverse frame as in several of the Romano-Celtic boats (Fig. 12.17), in the E. Baltic variant of Viking Age tradition, and in certain boats of the cog tradition, e.g. Kollerup, Denmark and N5 from the Dutch polders (Crumlin-Pedersen, 1983b, Reinders, forthcoming). However, certain Dutch medieval finds have a mast step in a 'chock' offset to the sides – possibly these are for towing masts (Reinders, forthcoming).

In the keelson mast-step boats the mast is further supported at lower beam level by a mast partner or mast beam. In the Skuldelev boats keelsons are notched to fit over the frames, to which they were fastened by knees and *snelles* (a timber combining the functions of two keelson knees and a stanchion for the lower crossbeams (Crumlin-Pedersen, 1984b) Keelsons in Skuldelev 1, 2, 3 and probably 5 had a vertically projecting fork some 45 to 60 cm high, just forward of their mast steps, which supported the mast beam and provided a forward seating for the mast when under sail. Oseberg has a similar projection some 80 cm high (Crumlin-Pedersen, 1972b: fig. 6). Other keelsons with such projections include the tenth/eleventh-century Hasnaes 2 (Fig. 12.18) and Schleswig (Crumlin-Pedersen, 1972b: figs 4 and 5) finds. The main function of these keelsons is to distribute pressure from the heel of the mast to the keel and the frames – the ratio of keelson to keel length is some measure of the stress dispersal (Table 12.7). The keelson of Skuldelev boat 2 is not only the largest known (absolutely and proportionally) but also has the greatest sided and moulded dimensions and thus must have considerably reinforced the longitudinal structure of

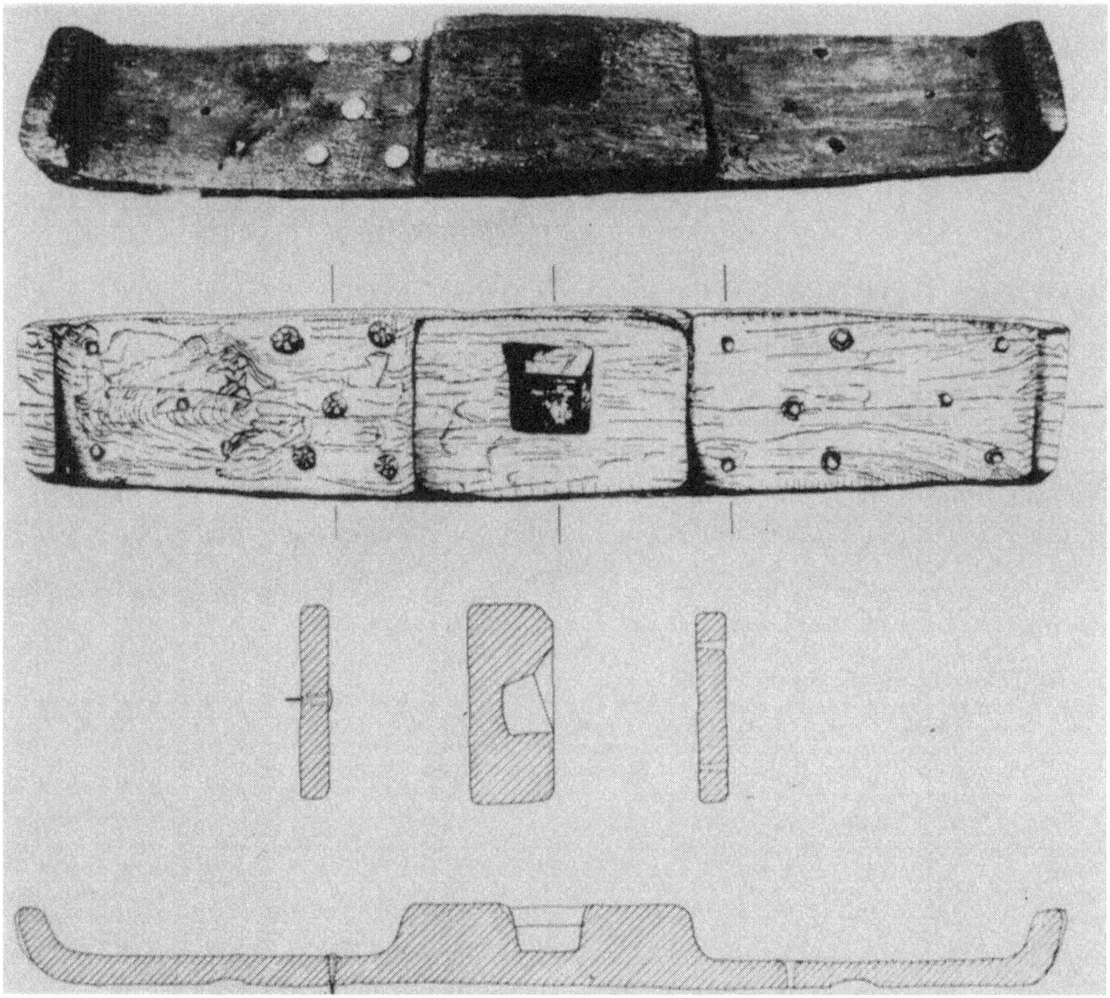

Figure 12.17 The mast step from the Bevaix boat. After Egloff, 1974 (courtesy Musée Cantonal d'Archéologie, Neuchâtel).

this great longship (Olsen and Crumlin-Pedersen, 1967).

A keelson fragment from Wood Quay (WQT 54) made from half an oak log with the pith side uppermost, had a groove leading into the step, similar to that on Hasnaes 2 (Fig. 12.18); this may have been closed by a chock when the mast was stepped.

The oak frame of the Romano-Celtic Bruges boat which housed the mast step is of inverted T-section away from the central position so that it combines strength with minimal weight. The mast step (where the frame is 40 cm broad and some 13.5 cm deep) is now 14.5 cm long × 20 cm across × 12 cm deep (Marsden, 1976: fig. 10); allowance for shrinkage since excavation indicates the length may have been up to 17 cm (in fore and aft line of the boat). A second timber in this boat appears to have formerly been a mast frame with a now-broken step of similar breadth to the other one but of shallower depth (*c.* 4 cm).

In the Norse tradition masts were stepped (near) amidships, an appropriate position for a single square sail. Six Dutch cogs of the fourteenth century, on the other hand, have their mast step further forward, between 24 and 34% from the bow (Reinders, forthcoming) and the steps in two Danish cogs Kollerup and Kolding (Crumlin-Pedersen, 1979: fig. 2.13) are at 29% and 34%. This raises the question as to whether these vessels had a square sail limited in use to a near-following wind or a fore and aft sail (sprit or lug). The Bremen cog of 1380 and the Vejby cog of 1350 have mast steps at *c.* 43%, a similar position to those in Norse vessels.

Figure 12.18 Hasnaes boat 2. Keelson with mast step. (Photo: Ole Crumlin-Pedersen.)

Table 12.7 Details of selected medieval keelsons

Boat	*Keelson length (m)*	*Frames spanned*	*Keelson/keel*	*Keelson cross-section near amidships b × d (cm)*	*Keelson cross-section near ends b × d (cm)*	*Step l × b × d (cm)*	*Reference*
Oseberg	1.65	2	0.07	23 × 12	—	—	Brogger *et al.* 1917
Gokstad 1	3.60	4	0.17	30 × 40	18 × 35	27 × 26 × 20	Nicolaysen, 1882
Hasnaes 2	*c.* 2.70	4	—	30 × 38	17 × 23	17 × 15 × 19	Crumlin-Pedersen 1972b: fig. 4
Skuldelev 1	5.1	6 or 7	0.42	36 × 25	7 × 18	16 × 17 × 10	
Skuldelev 3	3.7	3 or 5	0.39–0.41	25 × 30	6 × 14	10 × 11 × 12.5	
Skuldelev 2	*c.* 13.3	19/20	*c.* 0.50	38 × 30	11 × 9–11	18 × 18 × 8	Olsen and Crumlin Pedersen, 1967
Skuldelev 5	*c.* 3.7	5	*c.* 0.25	26 × 25	5 × 12	—	
Skuldelev 6	1.32	—	0.16	17 × 21	—	7 × 8 × 6	
WQT 54 (frag)	1.02+	—	—	32 × 23	—	20 × 12 × 10	—
Schleswig	*c.* 2.70	6	—	30 × 25	10 × 20	12 × × 10 × 10	Crumlin-Pedersen, 1972b: fig. 5
Bremen cog	11	—	0.72	—	—	—	
OZ 43	8	—	0.60	—	—	30 × 25	
M107	1.8	—	0.13	—	—	11 × 11	
N5	—	—	0.03	—	—	12 × 7	Reinders, forthcoming
NZ42	2.2	—	0.18	—	—	15 × 12	
Q75	*c.* 1.0		0.10	—	—	—	
NOPA 57	+10.0	—	—	—	—	65 × 25	

Mast partners and mast beams

Heavy mast partners, with openings aft, further supported the Gokstad, Oseberg and Tune masts at *c.* 90 cm–1 m above the keel, that of Gokstad spanning six crossbeams (4.8 m long); Oseberg *c.* 3.15 m long and spanning four crossbeams and Tune 3.91 m long and spanning five crossbeams. The Oseberg mast partner is let into the beam just forward of the mast, otherwise it rests on the crossbeams, being fastened only at its ends to the underlying beam by one treenail – this mast partner had split, either during drying after manufacture, or during use, and had been repaired in antiquity by two iron bands. The Tune and Gokstad partners are notched to take the underlying cross-beams thus restraining movement; in addition the Gokstad partner has had wood removed from two longitudinal grooves underneath, thereby reducing weight whilst maintaining strength. The mast was held in position by a removable chock in a horizontal groove in the partner, aft of the vertical mast hole. Thus in these three ships the mast was supported generally over its lowest 1 m, thereby transmitting horizontal forces otherwise concentrated in the mast step, to the keel and frames and thus out to the shell of the boat. A mast partner from Århus Bay (Crumlin-Pedersen, 1972b: fig. 9B) is similar to the ones described above, being 3.4 m long and spanning five beams (Fig. 12.19). On the other hand one from Hadsund (1972b: fig. 3), 3.70 m long and spanning five beams, is hollowed along its length at one end, not forked as are all others so far found. Crumlin-Pedersen (1972b) has pointed out that such a partner could be fitted so that the mast would be lowered towards, and raised from, the bow, as seen on the Bayeux tapestry and on two late-twelfth century drawings (Fenwick, 1978: fig. 8.25) rather than aft as with the Gokstad and Oseberg arrangements.

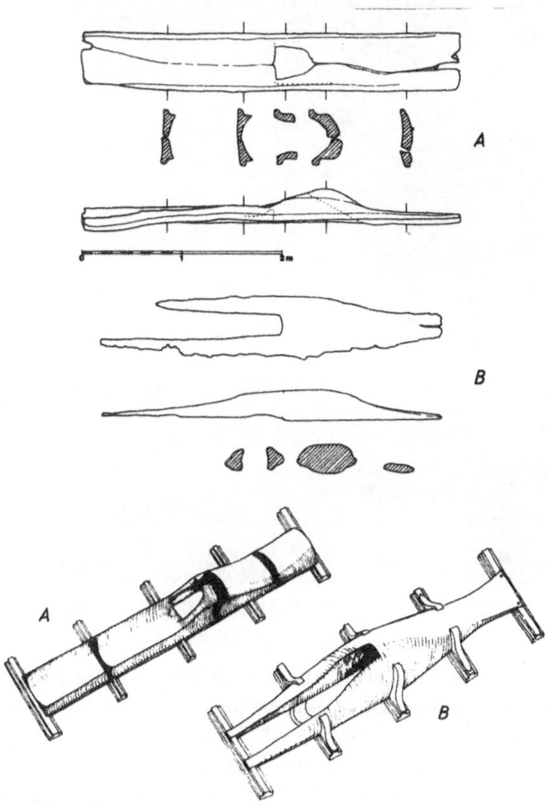

Figure 12.19 Mast partners:
A Hadsund
B Århus Bay After Crumlin-Pedersen, 1972b.

Transverse timbers have been used rather than longitudinal partners to support masts on other boats. (Fig. 12.20). The Broighter boat model has a vertical hole through the central thwart or crossbeam (Fig. 10.9) and this solution is also seen in the Bruges boat (Marsden, 1976: fig. 4.7). The corresponding elements in the five Skuldelev boats have either not survived or are fragmentary, but Ole Crumlin-Pedersen has deduced that, in general, the mast took against the cross-beam forward of it, probably let into in for additional support and possibly even lashed to it. Crumlin-Pedersen also conjectures that these masts may have been further braced by longitudinal timbers on either side of the mast fastened to the uppermost crossbeams.

Standing rigging

Apart from iconographic evidence, little has survived of the standing rigging required to support a mast and this meagre evidence is mainly from the Norse tradition: iron rings through the topstrake of the Ladby boat (Thorvildsen, 1957: fig. 22) and osier rings with two cleats each side projecting outboard of the topstrake some 5–6 cm, within $\frac{1}{2}$ to 1 m of the midship station of Skuldelev 3 (Olsen and Crumlin-Pedersen, 1967: fig. 43) – these are best explained as shroud rings although Binns (1980: 155) believes they may be fairleads for either sheets or halliards. Taking the representational evidence as seen on the Bayeux tapestry and other sources (e.g. the 1199 Dunwich and the 1230 Pevensey seals) see Fig. 12.21 – it seems that the masts of ninth- to twelfth-century vessels were supported by a forestay to the stem or a forward beam (also used when stepping and lowering the mast), shrouds to the top strakes or to a convenient crossbeam, and the halyard (see below) may have doubled as a back stay to brace the mast from an after-thwart or beam. Knotted rope and toggles have survived from certain Viking Age finds (see, for example, Christensen, 1979: fig. 11.5) as has a deadeye (a rigging 'block' without sheaves): their precise use onboard is uncertain. A replica of the deadeye from Gokstad and toggles were successfully used to tension shrouds on the recent *Krampmacken* replica (Roberts, 1984). Four other wooden objects (*jomfru*) from the Gokstad find (Christensen, 1979: fig. 11.6) have also been interpreted as rigging tensioners. Two methods of use have been devised:

Figure 12.20 Mast beam of Zwammerdam boat 4. (Photo: courtesy the Universiteit van Amsterdam, A. E. van Giffen Instituut voor Prae-en Protohistorie (copyright holders).)

Landström (1961: 62) suggests they were used in pairs one fastened to the ship's side, one to the rigging with a further line between them used as a purchase, whilst Crumlin-Pedersen (1977: 30) has proposed that a single *jomfru* should be fastened to a strong point by a lanyard or strop through its lower hole; the arms of the *jomfru* are then used as cleats to which two lines are belayed after they have been used to tension the rigging line by purchase against a toggle (or an eye) on its end. Such use may be seen in Crumlin-Pedersen's reconstruction drawing (Fig. 12.13) of Skuldelev 3 (1967: fig. 44); and they also been used in certain replicas e.g. in *Odin's Raven*, the 2/3 replica of Gokstad (see the cover of *Viking Voyages* by A.

Figure 12.21 Seal of Dunwich of 1199 depicting a sailing vessel with stays and shrouds. (Photo: NMM Greenwich.)

Binns, 1980, where one is used to set the backstay).

Early Medieval cordage is generally thought to have been made from lime bast or possibly from hemp (Crumlin-Pedersen, 1977: 30). In other cases where special strength was required ropes were made from whale, walrus or seal hides cut spirally (*svardreip*) (Crumlin-Pedersen, 1977: 30). The whale and seal ropes described by Ohthere to King Alfred (Lund, 1984: 20) were 60 ells long (15 fathoms). Sandahl (1958: 93) has noted that in the account of the building of the galley *la Phelipe* at Lynn in 1336–7 *russhewale* i.e walrus hide is listed as cordage material (see also Ch. 4).

It is generally concluded that boats (and ships) which relied (almost entirely) on sail for propulsion, as for example the cargo vessels Skuldelev 1 and 3, would seldom, if ever, have their masts unstepped. On the other hand, war craft such as the Viking longships and Skuldelev 2 and 5 which had a sail but used oars in unfavourable winds, would have had the yard and sail sent down and the mast unstepped to reduce resistance before any serious rowing was undertaken. This is borne out to a degree by the fact that, in proportion to their size, Skuldelev 1 and 3 have much deeper mast steps than other Viking Age craft (Table 12.7) and thus masts would be proportionally more difficult to step and unstep.

Spars not in use are seen on some medieval seals supported in a two pronged fitting known as a *myke* (Anderson, 1926: 82). Other mykes are shown unused (e.g. fifteenth century trader drawn by 'WA' see Nance, 1911: 66). The myke is also sometimes seen with the backstay fastened to it or supporting coils of rope or an anchor. Anderson suggests that from its shape and position it was originally used to support the mast when lowered, as may be seen on a third-century AD mosaic from Sousse of a Mediterranean boat (Casson, 1971: fig. 191). Two re-used mykes have recently been excavated from twelfth/thirteenth-century levels at Wood Quay Dublin. They are both broken at the base but measure 1.90 m (WQT 51) and 1.47 m (WQT. 58) in height and have been made from natural oak crooks (Fig. 12.22). These mykes would readily have supported several spars or one large one of up to 2.4 m girth (WQT. 51) or 2.7 m (WQT. 58); such a single mast would lie in size between that of Gokstad (0.94 m) and that of *Grace Dieu* (6.7 m). Using the Norse (length/girth) factor of, say, 10, the Italian one of 14 and the *Grace Dieu* one of 9, masts of this girth would be between *c.* 22 m and 38 m in length, and would probably have been composite. If of pine (*Pinus sylvestris*) they would have weighed 4 to 6 tonnes and 5 to 9 tonnes depending on their length.

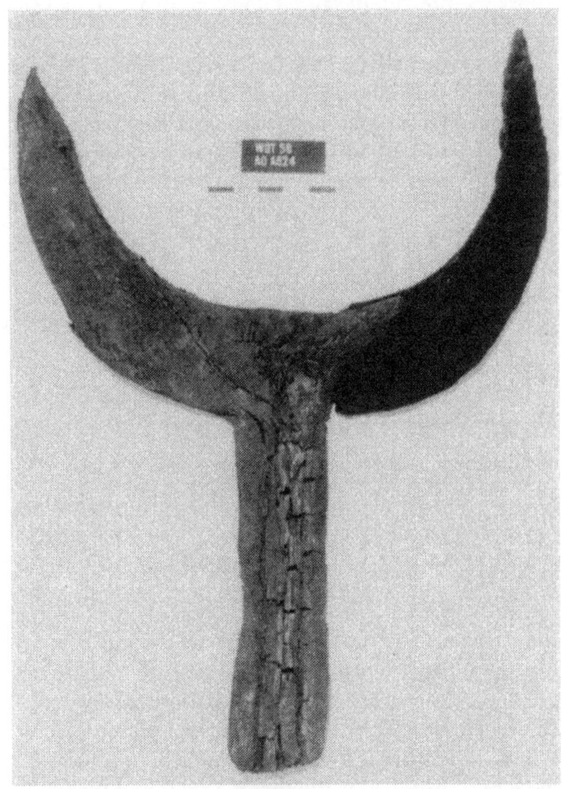

Figure 12.22 A *myke*, a crutch for masts and spars, from the late twelfth-century at Wood Quay, Dublin. Scale has 5 cm divisions. (Photo: NMM Greenwich.)

Yards

Examples of yards are scarce: there is one with the Broighter model, and complete and fragmentary yards were found with the Gokstad small boats (Nicolaysen, 1882: 39, pl. IV.12) and in medieval contexts from Bergen and Tønsberg (Christensen, 1979: 189). With the Gokstad boats there were three spars which Nicolaysen deduced to be yards, of these the one illustrated by Nicolaysen measures *c.* 3.63 m in length with a diameter of *c.* 6 cm at the centre, tapering to *c.* 4 cm at the arms. About 5 cm from each end is a hole of 2–3 cm diameter where a cringle at the corners of the head of the sail would have been fastened by a rope earing. Such a sail would thus have been *c.* 3.5 m in breadth at the head.

The yard with the Broighter model (Fig. 12.2d) now measures 5.5 cm which would be 5.8 cm if it had formerly been symmetrical about the hole for the mast. This compares with a mast length of 8.6 cm, giving a ratio of mast to yard of 1.48 or 1.56 to 1. The yard length may also be compared with the maximum breadth of the model (7.6 cm) as the necessity for sheeting the corners of the foot of the sail to the boat (especially when running before wind) limits the breadth of saii, and hence yard, which can conveniently be used.

Christensen (1975: 266) and Binns (1980: 124) have noted that the outer two of the three uprights with crosspieces 6 ft proud of the Gokstad decking are 35½ ft apart (*c.* 10.8 m). A spar fragment found near these supports measured *c.* 20 ft (6 m) and may have been half a broken yard. Another spar 41 ft long (12.5 m) found with Oseberg may be a whole yard and this would have fitted the three Gokstad uprights with a 1 m overlap at each end. It thus seems possible that a yard of *c.* 12 m may have been used in Gokstad and Oseberg, giving a sail breadth of say 11 m. This may be compared with their beam measurement at this station of 5.2 m and 5.1 m respectively.

Halliards and braces. Yards on single-masted, square-rigged north European boats of the nineteenth and early twentieth centuries were hoisted by and suspended from a halliard (halyard) rope, sometimes with the assistance of a purchase, and the set of the yard was controlled by braces running from the yard arms to strong points aft on the hull (see for example, Faerøvik and Christensen, 1979). Halliards are depicted on the early medieval Gotland stones (Christensen, 1979: 190) and later representations. Dumb holes through which the halliard was rove have been found near the top of a mast and three model masts from Bergen; an equally simple system was used by nineteenth-century Norwegian fishermen who well tallowed the rope to reduce friction through the hole. On the seventeenth-century curach depicted by Capt. Phillips (Fig. 10.2) there is a fork near the top of the mast to take the halliard. Braces are not depicted in Viking Age representations, but are visible on seals of the thirteenth century (e.g. Dunwich (Fig. 12.21). Sandahl (1982) has noted that *bras* occurs in a twelfth century text; and *yerdrope* in 1294–5, thus braces may have been in use from the mid-twelfth century. Control of the yard is vital when beating to windward, braces being especially important; lack of evidence from the Viking Age must place in question their ability to beat. Binns (1980: 20) has noted that a post-Viking law specified that the crew stationed amidships bailed out but also 'watched the yard' which he interprets as tending the braces, and braces were fitted to the *Odin's Raven* replica (1980: 155).

Parrels. The fitting or traveller which holds the yard to the mast is known as a *parrel*, or *rakke*. The one found with the Oseberg find (Brøgger, 1917: fig. 121) is a grown crook with holes through the ends (Fig. 12.23). Similar ones have been excavated from medieval contexts in Bergen, Tønsberg and Oslo (Christensen, 1979: 189), and from early thirteenth-century contexts at Wood Quay (Fig. 12.24): nineteenth-century Norwegian parrels are very similar (Faerøyvik and Christensen, 1979: 18). Such crooks fitted against the afterside of the mast and were lashed to the yard which was before the mast. In gusty conditions it is important that the yard and sail can be brought down quickly and easily, thus rope downhauls were fitted to nineteenth-century parrels (Faerøyvik and Christensen, 1979: 19); medieval parrels may also have had them. Binns (1980: 161–2) found that the *rakke* on *Odin's Raven* seized on the mast: he therefore lashed only one end of the *rakke* to the yard; the rope from the other end was led over the yard and then down between yard and sail to the foot of the mast where it could be slackened thus allowing the rakke to open away from the mast, and the yard and sail to be lowered without difficulty.

The better-preserved Wood Quay parrel (WQT. 356) is fashioned from an oak crook and has clenched nails at each end to prevent or to repair a split. The crook has been worked to a near-rectangular cross-section of *c.* 10 × 10 cm with some sapwood left on. The holes near each end are approximately 6 × 4 cm and, although there has been some shrinkage across the timber, they must always have been elliptical. This parrel would have fitted a mast of *c.* 40–50 cm diameter (1.26–1.57 m girth):

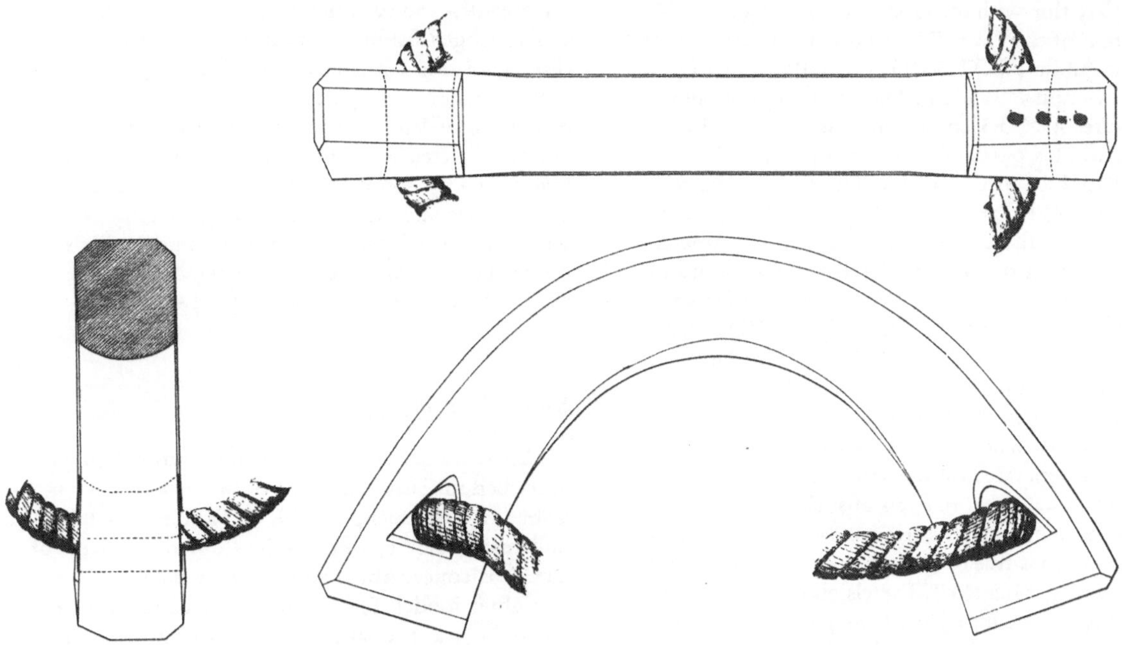

Figure 12.23 A parrel from the Oseberg find. After Brøgger *et al.*, 1917 (courtesy the University Museum of National Antiquities, Oslo, Norway (copyright holders)).

Figure 12.24 A parrel from Wood Quay, Dublin. Scale has 5 cm divisions. (Photo: NMM Greenwich.)

compare this with the Gokstad mast of girth 0.94 m and that of the *Grace Dieu* of 6.7 m. A second parrel from Dublin (WQT 360) is generally similar but of cross-section 6 × 8 cm. The surviving hole now measures 4 × 3.5 cm and may have originally been circular. This parrel would have fitted a *c.* 30–40 cm (0.94 to 1.26 m girth) mast. The Oseberg parrel (Fig. 12.24) has near triangular-shaped holes and would have fitted a mast of *c.* 22–25 cm diameter (69–79 cm) girth, and the Oseberg mast is said to have been *c.* 63 cm girth (Christensen, 1975: 266).

Another timber from Wood Quay (WQT 220) may be a parrel rib, lath or slat. Later medieval and post-medieval parrels are known to have been composite ones (Anderson, 1955: drawing 10) consisting of wooden beads or trucks threaded on a rope and separated by wooden spacers or ribs – there could be two or more rows of such trucks; (for example one recovered from the sixteenth-century *Mary Rose* has five rows each with six trucks (Mary Rose Trust, 81a: 2437). Parrels on nineteenth-century Norwegian square-rigged fishing boats were essentially similar but of only two or three rows and with fewer trucks (Faerøvik and Christensen 1979: 19). The Wood Quay timber measures *c.* 69 cm × 10 cm and is 2 cm thick, the four holes are *c.* 18 cm apart and *c.* 2.8 cm in diameter, so they would have taken a *c.* 8 cm rope (*c.* 3 in). The galley *la Phelipe* built at Lynn in 1336 had walrus hide cords for her *rakke* (parrel) ropes (Tinniswood, 1949: 308): possibly the Wood Quay parrel had been similarly fitted.

Lifts. Standing lifts running from the mast head to each yard arm take some of the weight of the yard, but they are not compatible with the use of the halliard to quickly lower and raise the yard and sail – the working area of the sail must then be adjusted by other means – see below. Emanuale (1977) has deduced that some Egyptian ships of the second millennium BC, e.g. those depicting Queen Hatshepsut's fleet at Punt, had a different type of standing lift which took the weight of the yard when lowered to a position near deck level. On the other hand the *c.* 1500 BC ships depicted on the Thera frescoes (Emanuele, 1977: fig. 1) appear to have running lifts which not only support the yard but assist in raising and lowering it. Classical merchantmen appear to have had standing lifts (e.g. Casson, 1971: figs 144 and 149 but classical warships none at all.

Caesar's ability to cut down the sails of the Veneti boats by cutting through their *antemna* (*B.G.* III.14) may indicate that the Veneti used no lifts. Nineteenth- and twentieth-century Norwegian square-rigged fishing boats also had no lifts (Faerøyvik and Christensen, 1979).

It is thus not unexpected to find no lifts on depictions of Viking Age craft or on the medieval vessels depicted on seals. The earliest use of the term *lyfte* traced by Sondahl (1958) is 1294–6. As ships grew in size and yards became bigger and heavier, lifts were introduced: where yards had to be readily sent down such lifts were running ones (McGowan, 1981: 12 – a figure of a *Kraeck* of *c.* 1470).

Sails and running rigging

Celtic. Caesar (*B.G.* III.13) and Strabo (4.195) described the seagoing ships of the first-century BC Veneti, a Celtic people of N.W. France, which had sails made of hide, whereas the Roman sails were of flax. It is conceivable that the boat which the Broighter model (Fig. 10.9) represents had a leather sail and it was probably a square sail, the aspect ratio of which may be estimated from the model as *c.* 0.60 – a low A.R. sail.

Ellmers (1969: pl. 16; 1975: fig. 8; 1978: fig. 3) has suggested that a boat on a *c.* AD 250 Oceanus mosaic from Bad Kreuznach, Rheinland-Pfatz and another on a gravestone fragment from Jünkerath of the second/third century AD, represent Celtic river craft used on the Rhine, Saar and Moselle. These two boats have an unusual form of rectangular sail strengthened by three horizontal battens (Fig. 12.25) comparable with a Chinese lugsail but possibly made of leather. The aspect ratios here are relatively high: 1.4 and 1.2 to 1. The mast is stepped *c.* $\frac{1}{3}$ from the bow on the mosaic boat and so the small sail depicted may be a square sail only used with a following wind or it may have been a lugsail which would have allowed the boat to sail closer to the wind. Blackfriars 1, with a mast step also *c.* $\frac{1}{3}$ from the bow, may also have had a lugsail.

Ships on the two British coins of the early first century AD (Fig. 12.26) attributed to Cunobelin (Muckelroy *et al.*, 1978), have a mast near amidships (44.5%); and the yard/mast combination indicates sail AR of 0.70 and 0.48, similar to that of the Broighter model.

Medieval sails. The earliest depictions are on Gotland stones dated to the sixth/eighth century AD (Lindquist, 1941). These have been interpreted in different ways – what may be deduced seems to be:

(a) the sails depicted are generally low aspect ratio

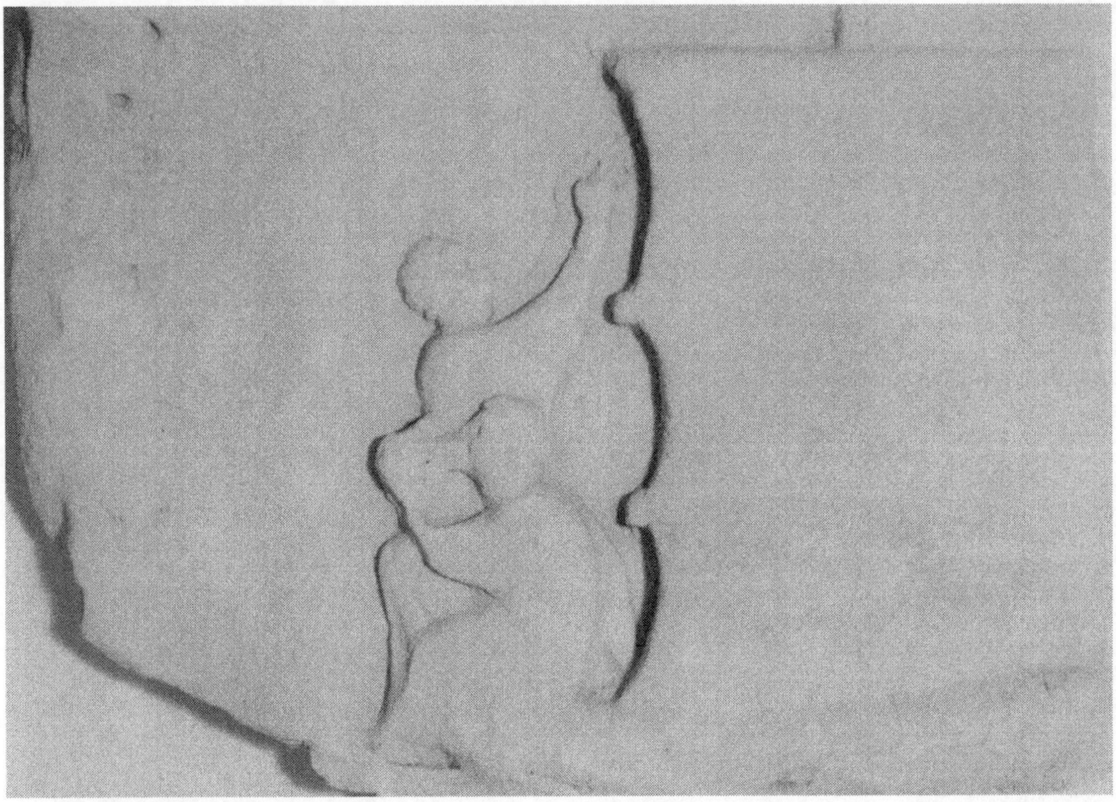

Figure 12.25 Carving of a sail on a second/third-century gravestone from Junkerath. (Photo: Deutsches Schiffahrtsmuseum.)

e.g. the one on stone Stenkyrka Smiss 1 (Fig. 12.27) has an AR = *c.* 4/8 = 0.50.

(b) they are generally rectangular in shape and were used as square sails.

(c) at least one appears to show a boom or lower yard.

(d) they were controlled by a complex of ropes.

The earliest artifactual evidence for sail is the Oseberg vessel of *c.* AD 800. The Askekarr trading vessel with sail, formerly thought to have an early ninth-century date, is now thought to be mid-tenth or even eleventh century (Westerdahl, 1982). From a consideration of length of mast and yard (see above) Christensen has deduced that Gokstad 1 probably had a *c.* 10 × 7 m sail of low aspect ratio 0.70. Sails of such proportions have been used on one recent replica of a Viking Age vessel *Imme Gram*, but other replicas have had sails of aspect ratios > 1, probably influenced by the known advantages of such a sail shape (see above), and by the fact that nineteenth-century Norwegian square-rig boats had similar sails (Faerøvyik and Christensen, 1980). Broad, low aspect, square-sails are not favoured by many twentieth century naval architects, nevertheless, they seem to predominate in representations of early Mediterranean vessels: e.g. of 37 vessels with square sails illustrated by Casson (1971), only 10 have sails with aspect ratio ⩾ 1 (Figs 19, 120, 137 (six), 142, 169 and possibly 143). And apart from the two depictions of Celtic sails described above the vast majority, if not all N.W. European illustrations up to *c.* 1300 show low aspect ratio sails, a possible exception being the ship of the 1293 Sluys seal which may have AR = 1.09. Strong arguments must be advanced therefore before high aspect ratio sails can be postulated for early-Medieval northern Europe.

Recently Andersen (1976) has systematically measured key aspects of traditional Norwegian square-rigged boats to determine whether there are any regularities; for example, between length of yard or breadth of sail and the position of fairleads for sail-controlling ropes. Andersen's aim was to use this

Figure 12.26 Early first-century AD coin of Cunobelin found at Canterbury. (Photo: Canterbury Archaeological Trust.)

data to interprete holes and fittings found on excavated boats of the Nordic tradition. It might then prove possible to estimate the breadth of sail with some confidence. However, this nineteenth/twentieth-century Norwegian data is applicable to high aspect ratio rigs and thus it is doubtful whether length of mast or length of sail can be deduced for low AR rigs.

Medieval documents indicate that wool was the usual material for Viking Age and later sails (Christensen, 1979: 190), or possibly flax, and it may be that cloth sail fragments were excavated at Gokstad, but they were identified as a tent (Nicolaysen, 1882: 37) and no longer survive.

Woollen sails were used in the Thames estuary in the late-fifteenth century (Woodruff 1875: 33, quoted by Fenwick, 1978: 251) and nineteenth-century Norwegian woollen sails survive today (Christensen, 1979: 190).

Sail control

To achieve efficient and safe sailing both the set of the sail and its area have to be readily controllable. The yard, to which the sail is laced by *robands* (known iconographically from the the eleventh century – see the Bayeaux tapestry, and mentioned as *robendes* in 1312–3) sets the shape of the head of the sail. The foot on the other hand appears to be loose on Viking

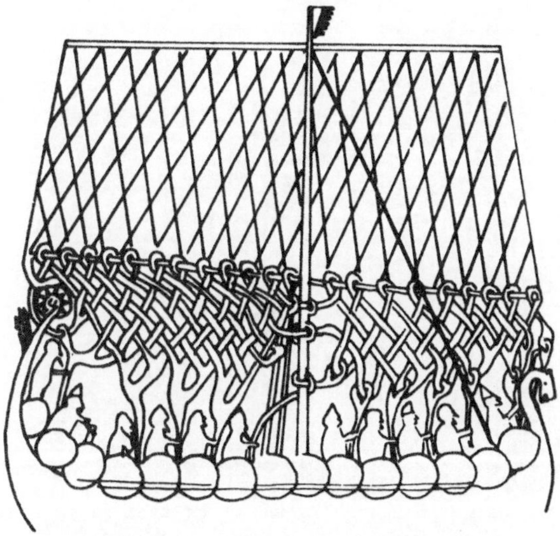

Figure 12.27 Carving on an eighth-century gravestone Stenkyrka Smiss 1, Gotland. After Christensen, 1979.

Age and later representations controlled by lines to the two corners: tack to the weather corner; sheet to the lee corner. However, one of the two sails identified as Celtic by Ellmers (1969) appears to have a boom or lower yard at its foot, as does at least one of the sixth/eighth-century representations on the Gotland stones (Christensen, 1979: fig. 11.3). Such a boom could well have advantages, which outweighed operational complications, with a broad, low aspect ratio sail which could be difficult to control by sheets alone when running.

On nineteenth-century Norwegian fishing boats the tack was led through a hole in the topstrake forward and fastened there, either direct or to a belaying pin (Christensen, 1979: 190). A similar lead may have been used in Oseberg and in Skuldelev boat 3 where there are holes in comparable positions, as there are on a thirteenth century model boat from Dublin (Fig. 12.28). Oseberg also has cleats forward (Christensen, 1979: 190) which may also have been used for the tack. Skuldelev 1 and 5 have holes aft through the top strake or an inter-frame timber, and Skuldelev 3 has two cleats inboard near amidships.

Oseberg has cleats aft as has Gokstad 1 where there are three on each quarter: these could have been used for the sheets or other running rigging. Bone objects (Christensen, 1979: fig. 11.8) which may have been used as fairleads (altering the run of a rope to a more convenient direction for working) were found at Gokstad and are known from other medieval contexts.

The complexity of control lines seen below the sails on some of the Gotland stones (Christensen, 1979) (Fig. 12.27) are difficult to interprete, although 30 years ago Åkerlund (1956) suggested they had been used to shorten sail by brailing – a method used in the Classical world; see Casson's diagram (1971: fig. 188). In a recent experiment Nylen has interpreted these lines as multiple sheets which could be adjusted along the length of the foot thereby trimming the sail and controlling airflow; in use they appear to be effective (Roberts, 1984). Binns (1980: opp. p. 24) used an analogous multiple sheet system in the replica *Odin's Raven*. This *priare* (or *prier*) was used in nineteenth-century Norwegian boats

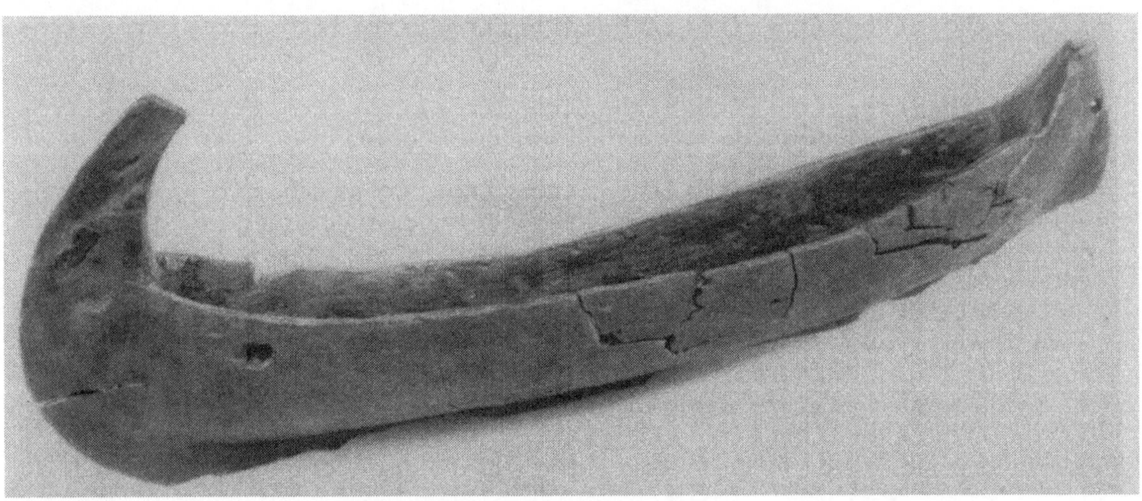

Figure 12.28 A 36 cm wooden model from the thirteenth-century site at Winetavern Street, Dublin. (Photo: National Museum of Ireland.)

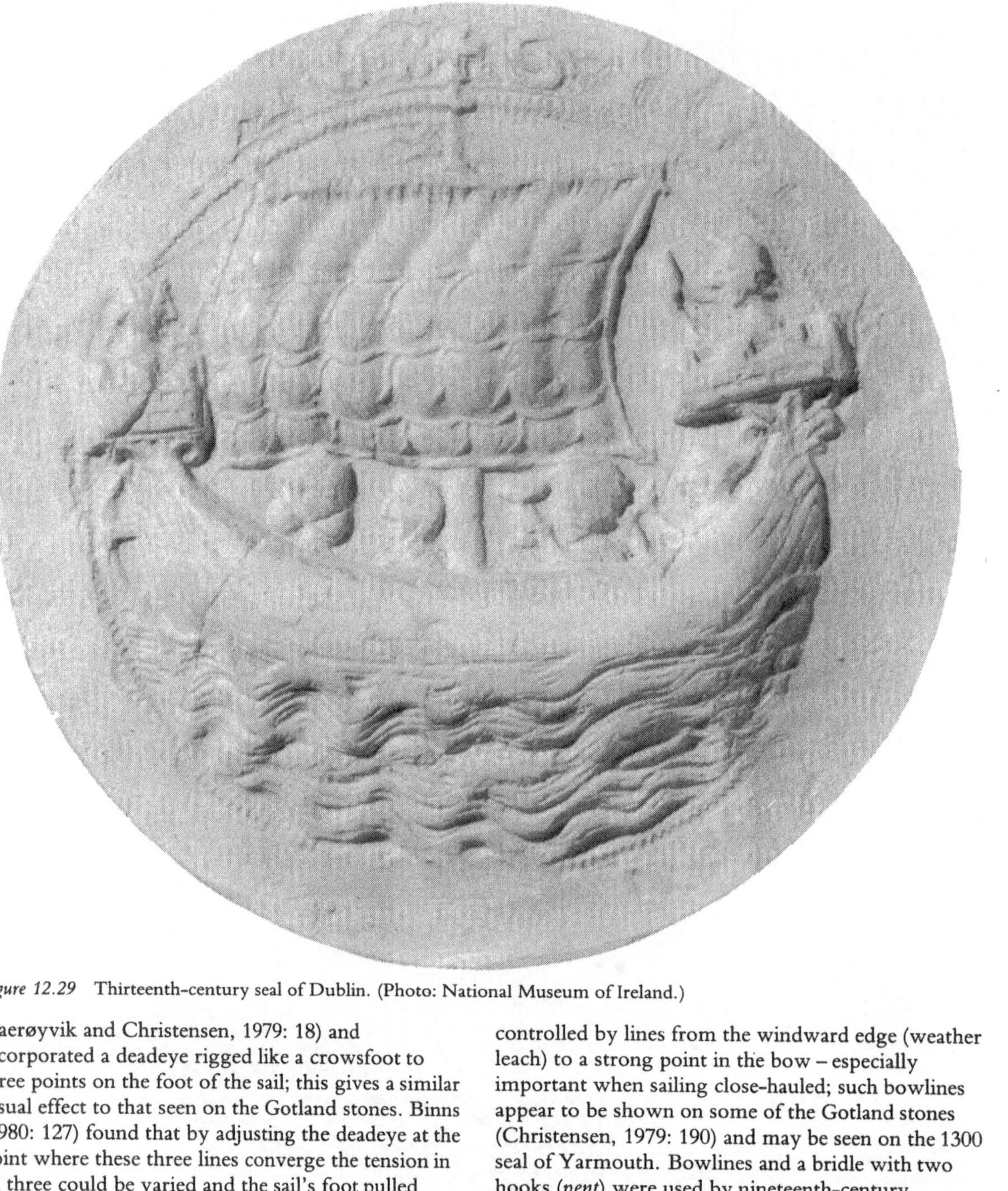

Figure 12.29 Thirteenth-century seal of Dublin. (Photo: National Museum of Ireland.)

(Faerøyvik and Christensen, 1979: 18) and incorporated a deadeye rigged like a crowsfoot to three points on the foot of the sail; this gives a similar visual effect to that seen on the Gotland stones. Binns (1980: 127) found that by adjusting the deadeye at the point where these three lines converge the tension in all three could be varied and the sail's foot pulled down and to windward thereby ensuring optimum airflow. The terms *prial* and *prial-rope* are known from 1399 (Sandahl, 1982: 71) and they appear to refer to this type of sail control line.

In addition to sheet and tack (seen on the Dublin seal of 1297) (Fig. 12.29) the set of the sail may be controlled by lines from the windward edge (weather leach) to a strong point in the bow – especially important when sailing close-hauled; such bowlines appear to be shown on some of the Gotland stones (Christensen, 1979: 190) and may be seen on the 1300 seal of Yarmouth. Bowlines and a bridle with two hooks (*pent*) were used by nineteenth-century square-rigged Norwegian boats (Lees, 1977); other N. Sea boats used a *diva*, a three-part tackle, to hold the lower part of the weather leach vertical and into wind (Binns, 1980: 160–2) – evidence for the crucial importance of bowlines and similar lines in performance to windward.

Skuldelev 1 (Olsen and Crumlin-Pedersen, 1967: fig. 22) and Gokstad 1 (Christensen, 1979: fig. 11.2) both have blocks of wood treenailed to the sides forward of the mast; each block has indentations angled forward, three in Skuldelev 1 and two in Gokstad. These probably received the heel of a tacking spar (*beitass* in old Norse; *lof* in Middle English, known from *c.* 1205). This spar may have been extended over the bow to give a better lead to the bowline and thus improve windward performance; alternatively, the spar may have been used to give a direct push forward on the weather leach and thus keep it taut as was done in the early twentieth-century Cornish luggers of Beer Head (Moore, 1925: 255). As these luggers braced the heel of their spars (*vargord*) against the mast, no evidence for such use need survive. It is also possible that two medieval *beitiass* or *lufs* were used to bear out both corners of the foot of the sail when running before the wind (Crumlin-Pedersen, 1977: 32).

The beitass or lof thus seems principally to have been brought into use when the boat was beating to windward and a bowline had to be used. The permanent successor to this temporary spar was the bowsprit stepped in a chock on or near the centre line of the vessel and extending ahead of her. A bowsprit may be seen on thirteenth-century seals from Yarmouth and Poole and was first recorded as *bousprete* in 1294–6 (Sandahl, 1958).

The area of usable sail may be varied by: lower/raise yard; reef/unreef; brail/unbrail; replace mainsail by stormsail and *vice versa*; add/remove bonnets; increase/decrease number of sails (in multi-masted or multi-sail ships).

To reef a sail, ropes protruding from both sides at reefpoints lying along a line parallel to the sail's foot are tied around the bunched-up or rolled sail thereby reducing the area. Further sets of reef points higher up the sail enable more reefs to be taken in. One of the earliest representations of reef points is on the thirteenth-century seal of Dublin (Fig. 12.29): this sail has three rows nearer the foot than the head. Later representations show reef points nearer the head; others show reefs all over the sail (Bergen, 1299; Rye 1400). In these latter cases a reef could be taken in without having to shift tack and sheet. The term *reef* has not been noticed in documents until the mid-fourteenth century when it appears as *riff* or *riffe*. Buntlines, from foot of mast to yard, may have been used to gather up the sail before reefing and are possibly visible on the thirteenth-century seal of Ipswich.

Brails are ropes running from leach to mast and down to deck, they are used to bring one leach nearer the mast, thereby reducing the effective area. They have not been noted in representations in Northern Europe up to the fourteenth century, but are known from the Classical world (Casson, 1971: 234). The word *brail* is, however, noted from 1409–11 in N.W. Europe and *liche* which may have had a similar purpose (Sandahl, 1958) from 1336–7.

The replacement of the mainsail by a smaller sail (stormsail) has been documented from fifth-century BC evidence in the Mediterranean (Casson, 1971: 236), but it is not known in northern and western Europe until fourteenth-century inventories.

A bonnet is an additional strip of canvas laced to the foot of the mainsail; they are first mentioned in northern seas in 1351–2 as *bonett*, evidently being first used in the Mediterranean region in the early fourteenth century (Sandahl, 1958: 23).

Although top masts are not mentioned as a separate item until 1485 (*toppe maste*), topsails (*topseil*) were mentioned in 1390. It may be, therefore, that the first topsails were set on the same mast as the mainsail. A second mast, known in the Mediterranean from early times – sixth century BC (Casson, 1980) – does not appear to have been used in N.W. Europe until the early fifteenth century (Friel, 1983). The development of the multimasted ship is outside the scope of this book.

STEERING

There are two aspects to steering: the maintenance of a given heading, relative to the wind or some other datum, and the alteration of course to a new heading. Both aspects may be undertaken at the same time and by the same means as certain forms of propulsion: thus the poler and the paddler, facing forward to propel the boat, may also steer by using an asymmetric action. Oarsmen who face forward can also readily steer: for example in the stand/push mode (Fig. 12.7), or when using an oar or sweep (large oar) over the stern as a combined propulsion/steering device – a method which reached its apogee in the Chinese *yoloh* or 'self-feathering propellor' as Needham (1971: 622) described it. It is also possible, to combine steering with propulsion when oarsmen face astern as in the sit/pull mode but for safety in confined or crowded waters the oarsmen need to be conned by a forward facing helmsman or passenger. On the other hand, when a boat is towed, uses the tidal stream, or is under sail, independent steering arrangements are neccessary and special control devices are fitted by means of which hydrodynamic lift is used to turn the boat. In addition to steering, all such devices, to a greater or lesser degree, may act to oppose leeway. Conversely, by adjustment and fine tuning, devices with the primary task of reducing leeway may be used to steer a boat or raft.

These steering devices may be grouped under four headings:

1. *Paddle* – freely held and used with a two handed grip.
2. *Guares* (or *firrer*) – a blade with a handle at the top, mounted parallel to the fore and aft line of the boat and constrained so that it can be moved only in the vertical direction, thereby varying its immersed area (Fig. 5.6) and thus its turning effect.
3. *Steering oar* or sweep – pivoted at one point, which may be on the stern or stem, or in the quarter or bow, so that it can be moved in an oblique plane generally nearer the horizontal than the vertical, thus varying its turning effect; it may have a tiller in line with the blade of the oar.
4. *Rudder* – a shaped blade of (near) aerofoil cross-section with a stock above it to which is attached a tiller. It is mounted on the hull either at the stern (*median* rudder) or on the quarter (*side* rudder) at two or more points (or in a sleeve) which are disposed nearer the vertical than the horizontal, and which become the axis about which the rudder may be rotated in the (near) horizontal plane, thus varying its turning effect. A median rudder has a tiller generally in the fore and aft line of the boat; the side rudder's tiller is (near) athwartship. Side rudders may be used singly or in pairs. Paired side rudders minimise any reduction in immersed rudder area when a boat heels, although this effect is not significant for a single rudder up to angles of heel of *c.* 15°. If paired rudders are linked they double the turning power. The median rudder's advantages over the side rudder are that a slightly greater turning moment is available and it is independent of heel.

STEERING ACTION

As with any (sub) aerofoil shape immersed in a fluid a steering device will generate lift and drag whenever it has a suitable angle of incidence to the relative flow. At a neutral position – determined by experience – this lift (in the horizontal plane) will oppose and just equal any turning moment due to relative position of CE and CLR. The horizontal lift from a single side rudder or *firrer* should also balance the turning moment created by its own asymmetric position off the centre line). When the device is moved either way from that position (vertically in the case of the *firrer*) an effective turning moment is applied to the boat.

When under sail, drift to leeward (leeway) induces a relative water flow over a steering device in the neutral position for steering: thus a horizontal lift force is generated which opposes the leeway inducing it. When beating to windward a steering device in the fore-and-aft line of the boat may have zero or even negative incidence to the relative water flow but a positive angle of incidence will result if slight weather helm can be maintained (Marchaj, 1964: 350). Weather helm, which may be desirable for safety reasons, may be designed into a boat or may be obtained by adjusting the sail area so that CE moves slightly aft of CLR (McKee, 1983: 141–2).

When using only the tidal stream for propulsion a boat has no motion relative to the water and thus pure steering devices are ineffectual. Steerage way may be achieved by using a combined propulsion/steering device or by slowing the boat slightly to produce relative motion using a form of sea anchor or by dragging a length of cable across the seabed – the rudder or steering oar will then 'bite'. In the latter case the steering device will act in the reverse sense from when the boat is being propelled by normal means.

A boat under sail may be steered by an adjustment of the sail either to hold a course (but easier with a steering device) or to alter course through the wind as when tacking, or when achieving the same alteration but by wearing with the stern passing through the wind. Once the sail has been set to give the required course the steering device is used for fine tuning in response to varying wind, and to counteract any tendency for the boat to come up into the wind by the use of slight weather helm.

The power of a steering device to turn a boat is dependent on:

- cross-sectional shape
- aspect ratio (immersed depth/immersed breadth)
- immersed area
- waterflow (speed)^2
- any turbulence in the waterflow
- angle of incidence (angle of device to direction of waterflow)
- directional stability of the boat
- density of water

This force increases with increased angles (helm) applied to the device until at angles *c.* 35–45° it stalls that is, due to waterflow breakaway and increased turbulence, the turning power (lift) is markedly reduced (Rawson and Tupper, 1976: 490). A rudder with $\frac{1}{3}$ to $\frac{1}{4}$ of its immersed area forward of the (vertical) axis through its pivots will be dynamically balanced and requires less force to turn it. Modern (median) rudders are given a cross-section designed to maximise lift (power to turn) and minimise drag over a range of likely operating speeds; such a shape is generally an aerofoil section with the point of maximum thickness nearer the leading edge; this reduces turbulence especially at the trailing edge. That the Greeks were aware of the general properties of such a shape and of its suitability for rudders may be inferred from the word used for rudder: *pterugion*, with the root *pteros* = wing (Thurneyssen, 1980: 4).

Boats are generally long and narrow and the greater their (L/B) and (L/T) ratios the more their directional stability and the greater the force required to turn them. They turn about their centre of flotation (centre of waterplane area), and thus the further away from this point (i.e. nearer the stern) the turning force is applied the less this force needs to be. The ideal steering device therefore is shaped and mounted so that:

(a) minimum force is required to move it.
(b) it has a neutral position to which it will return if unattended.

As turning force depends (*inter alia*) on relative water velocity steering devices designed for higher speeds may not be so efficient at lower speeds. Thus the rudder of a boat which is to be rowed as well as sailed may be a compromise shape.

Steering devices, especially paddles, may be used at bow as well as at or near the stern – see for example Ellmers (1975: fig. 7) but the steersman's appreciation of the general problems is clearer from the stern.

PADDLES

Steering by paddle will leave no detectable features or wear upon the hull. In ethnographic contexts steering paddles are frequently bigger than the paddles used for propulsion. Such differentiation may be seen on some of the rock carvings in Scandinavia, for example on the fifth-century AD Gotland stone depicted by Ellmers (1975: fig. 7); and paddles with much longer and broader blades than those of the remaining paddles were found at each end of the *c.* 300 BC Hjortspring boat (Rosenberg, 1937: 88, figs 61, 63a). Steering paddles have to be of sufficient length to permit the steersman to hold the blade in the water, there is thus a height of boat side beyond which paddles cannot be used.

GUARES

The turning force due to a *guares* is varied not by increasing its angle of attack, but by increasing its immersed area as the blade is raised or lowered, as used in the sailing rafts of India, China and S. America. The *firrer*, a paddle with T-shaped handle and broad blade, used on the *kahn* small sailing boats of Steinhuder Meer, north of Hanover (Ellmers, 1979) is used in a similar manner. No excavated evidence for such a steering method has been recognised to date, but Ellmers (1979) has interpreted steering devices depicted on thirteenth-century seals from Lübeck, Neustadt near Lübeck and Niewport, Belgium as *firrer*. No hull fittings are shown on these seals as would be needed if a *firrer* were not to turn or be swept under a boat of any size; in fact the steering device seen on the AD 1280 Lübeck seal (Ellmers, 1979: fig. 1.5) could equally be interpreted as a freely-held steering paddle.

Leeboards of near aerofoil section have been used on flat-bottomed sailing craft in recent times, suspended from the leeside near amidships and adjustable in immersed area; they act on the same principle as a *guares* but their main task is to counteract leeway. Leeboards are thought to have been first used in N.W. Europe in the late sixteenth-century Netherlands (Reinders, 1983: 337; Prins, 1970: 349–53). However, Ellmers and Pirling (1972: 46) have described a leeboard dredged from the Rhine at Krefeld in 1972 in the vicinity of a 15 m barge thought to be from the thirteenth or fourteenth century; details of the board have not been published and it remains undated. Crumlin-Pedersen (1981: 36) has described a large wooden panel recovered from Egernsund, Denmark in 1966–7 as 'similar to the leeboards of Dutch vessels' and this has been recently dated to 1960 ± 75BP = *c.* 10bc (K-2514): further publication is awaited.

STEERING OARS

Steering oars pivoted on or near the stern – in crutch, rowlock or single or double thole pins (Fig. 12.30) – have been used in recent times, inland and at sea, in northern and western Europe: see, for example Beaudouin (1970: pl. 2). They are also seen on medieval documents: see, for example, the mid-fifteenth century Hastings Ms. (Anderson, 1963: pl. 7). Large ones (sweeps) are generally operated by moving the blade in the direction it is desired to turn; thus they work mainly by increasing drag on that side. Small ones on the other hand may be rotated as well as displaced to one side and thus gain turning force (lift) from a variable angle of attack. Large steering oars are generally prevented from rotating, by being wedged between two tholes or pivoted on a single thole to prevent them turning to a feathered position with the blade parallel to the water surface when they would be ineffective. For boats with a wide range of draft and freeboard the steering oar has to be used at two different angles and operated from two different positions, corresponding to deep and light draft (Fig. 12.30): there may be two corresponding wear marks or grooves on the oar's loom. Elaborate fittings are not required for this type of steering device and they are used in S. American boats pushed through a hole in the stern (Hornell, 1928: 131; Lothrop, 1932: 233) or held by a simple lashing (Folkard, 1870: 268).

A 10 m steering oar with a 6.67 × 0.40 m blade (Table 12.8) and a loom notched in two places for different angles of use, has been recovered from Lake Neuchatel near Bevaix, Switzerland (Ellmers, 1977/8): it is dated to 220 ± 110 ad (Fig. 12.31). Such an oar may have been described by Tacitus (*Annals*, II.6) who wrote that German river boats had one at each end (or they may have been large steering paddles). Lehmann (1978a: 264, 262, fig. 5) has suggested that the Roman Age 'barge' from Druten, Netherlands may have had a steering oar pivoted between two vertical timbers set into a raised deck near one end.

The first-century AD monument to Blussus (Fig. 12.32) now in the museum at Mainz, shows a flat-bottomed boat propelled by two, possibly four oarsmen, with steering by bow paddle and by what Ellmers (1969: 86) has interpreted as a steering oar pivoted in a rowlock on the stern with a near vertical tiller operated by a seated steersman. The blade of this steering oar is depicted well above the presumed waterline of the boat and Sleeswyk (1982) has suggested that rather than a steering oar this is a side rudder with a horizontal tiller, depicted in a raised position. On the other hand the length of steering oar depicted may have been determined by the space available to the sculptor, and if the device were used as a steering oar pivoted on a single thole, as in the

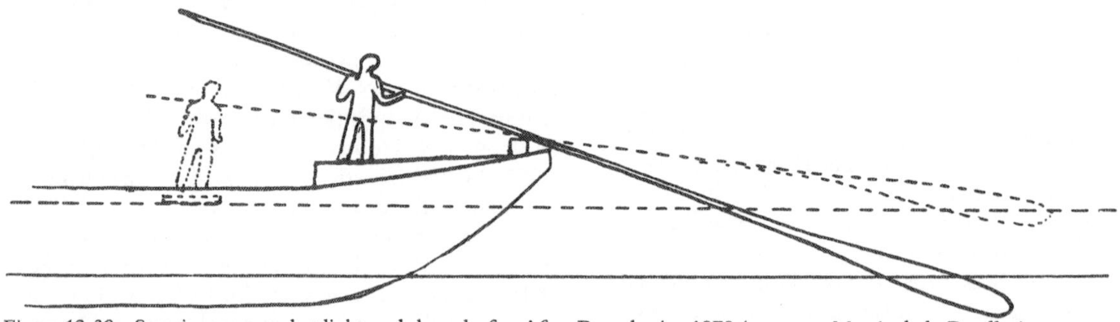

Figure 12.30 Steering oar used at light and deep drafts. After Beaudouin, 1970 (courtesy Musée de la Batellerie, Conflans-Sainte-Honorine).

Table 12.8 Details of early steering oars or sweeps

Site	*Date*	*Overall length* (*m*) L_{so}	*Blade* length l	*Blade* breadth b*	l/L_{so}	l/b*	*Reference*
Broighter model†	1st century BC	7.9	?	?	0.65	14.57	—
Zwammerdam	1st/2nd century AD	5.15	2.35	1.24	0.46	1.90	de Weerd, 1978: 17–20, fig. 24
Newstead‡	1st century AD	1.65	0.70	0.14	0.42	5.00	Marsden, 1963
Bruges	180 ± 80 ad	c. 4.45	2.00	0.70	0.45	2.86	Marsden, 1976: fig. 4
Bevaix	220 ± 110 ad	10.00	6.67	0.40	0.66	16.67	Ellmers, 1977/8

Notes: * Zwammerdam was measured *in situ* before shrinkage, but Newstead, Bruges and Bevaix after shrinkage which would probably have reduced the breadth of blade significantly.

† Bench spacing in the Broighter model suggests a scale of 1:100 and measurements given here are based on that assumption.

‡ The Newstead steering oar may be a paddle – see for example the River Tay paddle in Table 12.1.

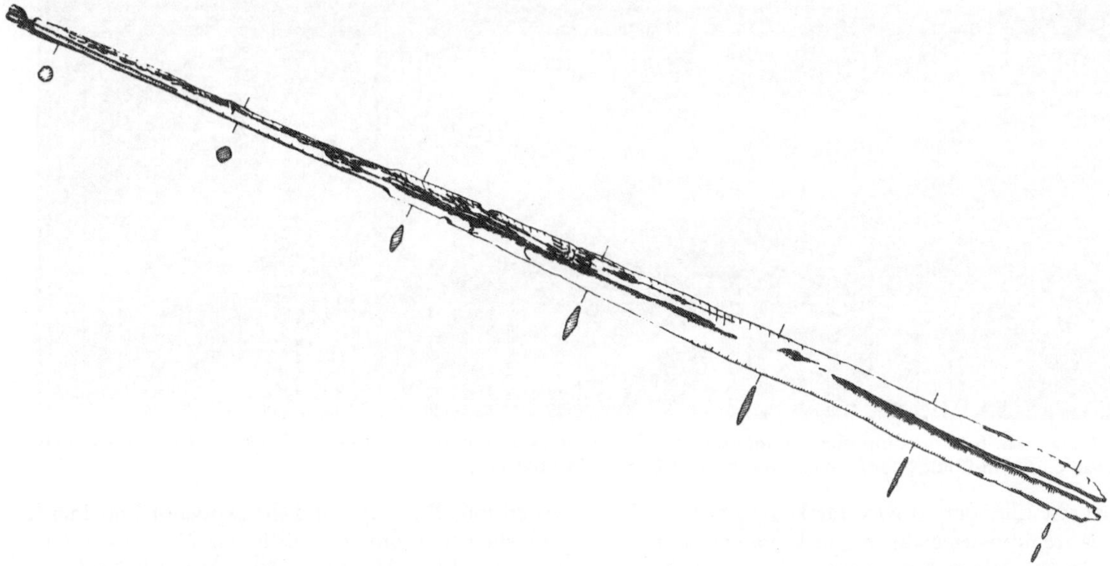

Figure 12.31 A 10 m steering oar from L. Neuchatel, dated to second/third centuries AD. After Egloff, 1974 (courtesy Musée Cantonal d'Archéologie, Neuchâtel).

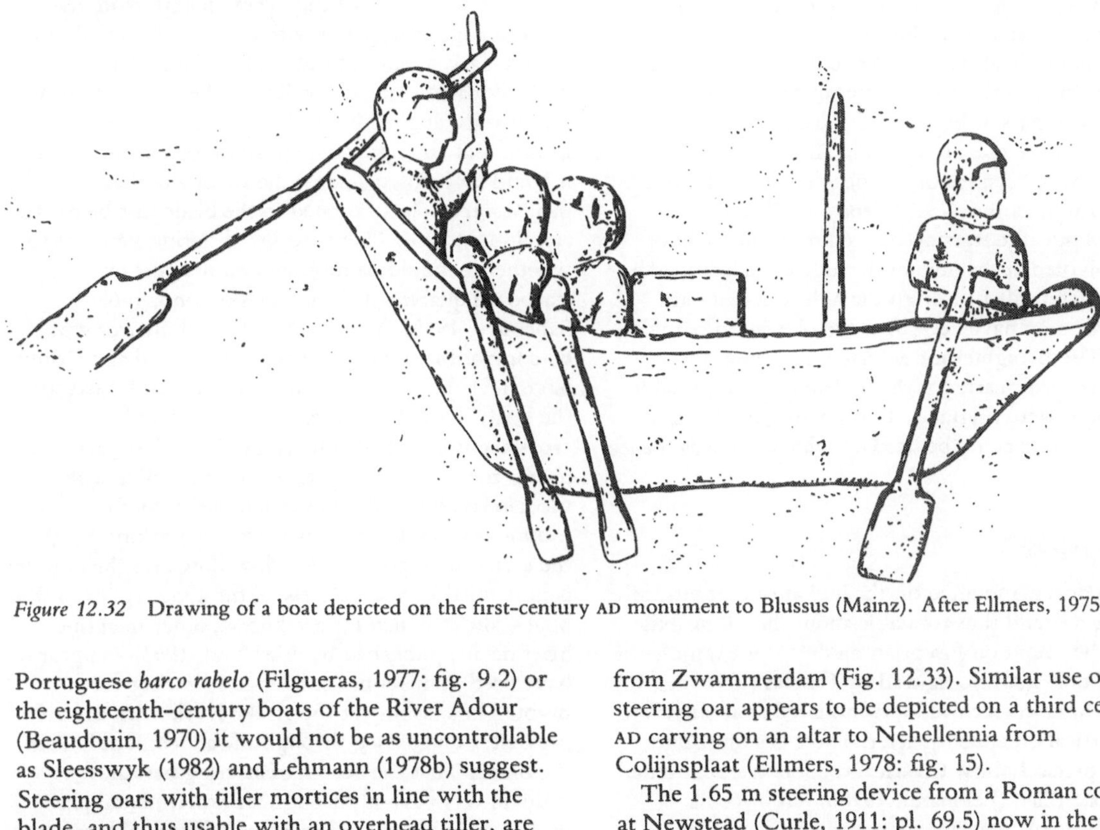

Figure 12.32 Drawing of a boat depicted on the first-century AD monument to Blussus (Mainz). After Ellmers, 1975.

Portuguese *barco rabelo* (Filgueras, 1977: fig. 9.2) or the eighteenth-century boats of the River Adour (Beaudouin, 1970) it would not be as uncontrollable as Sleesswyk (1982) and Lehmann (1978b) suggest. Steering oars with tiller mortices in line with the blade, and thus usable with an overhead tiller, are known: from the Lake Nemi Roman Age barges (Lehmann, 1978b: 97); and the 5.15 m steering oar with a 2.35 × 1.24 m blade (made from three pieces) from Zwammerdam (Fig. 12.33). Similar use of a steering oar appears to be depicted on a third century AD carving on an altar to Nehellennia from Colijnsplaat (Ellmers, 1978: fig. 15).

The 1.65 m steering device from a Roman context at Newstead (Curle, 1911: pl. 69.5) now in the Museum of National Antiquities, Edinburgh, has a tiller hole at right angles to the blade; it may therefore have been a steering oar for use on a small boat with

Figure 12.33 The 5.15 m composite steering oar from Zwammerdam. (Photo: courtesy the Universiteit van Amsterdam, A. E. van Giffen Instituut voor Prae-en Protohistorie (copyright holders).)

a transverse tiller but no wear marks or other features are now detectable on the loom and so its method of pivoting is not known; or it may have been a paddle. One of the objects with the first century BC Broighter sailing boat model (Fig. 12.2) is generally considered to be a steering oar and this also has a tiller hole at right angles to the blade. At present it is attached to the port quarter of the boat by a small ring which passes through this hole and a corresponding one on the hull. The model and its fittings have been restored since they were discovered in 1896 and, as Farrell and Penny (1975: 23–4) have pointed out, the steering oar could originally have been pivoted at the hole in the hull and controlled by a thwartship tiller through the hole at its inboard end; this may have been how the Newstead oar was used. The Broighter model has a matching hole on the starboard quarter of the hull and so it is possible that there were originally two steering oars or that the single one could be used on whichever was the lee side.

SIDE RUDDERS

Side rudders, mounted on the hull at two points, or in a sleeve, and thus rotatable about their long axis, are to be found on Egyptian models, for example the model of *c.* 2000 BC figured by Casson (1971: fig. 10) which has two such rudders mounted at about 45° to the vertical on either quarter. There is some doubt about precisely how Classical rudders were mounted and controlled (Lehmann, 1978b) but it seems probable that they also were paired side rudders.

The earliest example from N.W. Europe may have been that found with the fourth-century AD Nydam boat 2, but only a copy of a copy now survives (Åkerlund, 1965: 257) and the excavator Engelhardt published two somewhat different drawings (1865: pl. II.9 and fig. 24). This 3.30 m long artifact has a broad, paddle-like blade, an attachment hole which is relatively low (39% of the overall length from the bottom edge, compared with the 60–69% of all other known early medieval rudders), and an integral tiller with two arms, one in the line of the stock, the other one at right angles. It does not seem possible to mount it at an appropriate position on the hull where it could be pivoted against the small wooden projection or cleat fastened to the blade just below the attachment hole. There are also differing views as to whether it should be mounted on the port or starboard quarter of the boat (Åkerlund, 1965; Salisbury, 1965): use on the starboard quarter would be compatible with the tiller as now fitted and would have the rubbing or pivoting cleat inboard, next to the boat; it would also result in the aerofoil cross-section of the blade having its widest part nearer the leading edge, as is recent practice with median rudders. Whichever side the rudder is mounted when the tiller is placed at working height the attachment point is very low down on the hull. It is noteworthy that the crew of the 1933 replica of this boat – about which little is known other than the brief details published by Åkerlund (1963) – appear to have used their replica rudder as a steering oar pivoted in a lashing at top strake level, with a loose preventer line passed through the attachment hole in the blade. That the design of steering oars and rudders was passing through an experimental phase in this period may explain some of the problems encountered in discussions of the Nydam boat's steering.

Details of twelve other side rudders are given in

Table 12.9 Details of certain medieval side rudders

Site	Overall length (m)	Length below lower pivot (m)	Length below tiller hole (m)	Blade*: Breadth c. ⅓ from foot (m)	Blade*: Mean breadth (m)	Blade*: Thickness c. 1/3 from foot (m)	Blade*: Area below lower pivot (m^2)	l_1/L_R	l_2/L_R	L^*/b_2	l_2^*/b_2	b_1^*/th	References
	L_R	l_1	l_2	b_1	b_2	*th*	*A*						
Nydam 2†	3.30	1.28	2.98	0.48	0.42	0.07	0.54	0.39	0.90	7.9	3.04	6.86	Åkerlund, 1965: fig. 1
Kvalsund 2	1.27	0.87	1.17	0.12	0.12	0.02	0.10	0.69	0.92	7.6	7.56	6.00	Bruce-Mitford, 1975: fig. 315
Oseberg	3.28	2.00	3.00	0.40	0.38	0.06	0.75	0.61	0.92	8.8	5.34	5.00	Brøgger, 1917: pl. xix
Gokstad 1	3.35	2.00	3.00	0.50	0.50	?	0.99	0.60	0.90	6.8	4.04	—	Nicolaysen, 1882: pl. II, fig. 4
Gokstad 2‡	1.30	(0.85 (0.98	1.15	0.30	0.25	0.03	0.21 0.24	(0.65 (0.75	0.88	5.3	(3.48 (3.92	9.38	Nicolaysen, 1882: pl. V, fig. 2
Gokstad 3	1.01	0.64	0.90	0.23	0.20	0.02	0.13	0.63	0.89	5.2	3.28	9.58	Nicolaysen, 1882: pl. V, f. 3
Gokstad 4	1.30	0.82	1.17	0.18	0.18	0.02	0.14	0.63	0.90	7.4	4.68	7.50	Nicolaysen, 1882: pl. V, f. 4
Tune	1.92	1.20	1.75	0.22	0.26	0.04	0.31	0.63	0.91	7.4	4.62	5.50	Nicolaysen, 1882: pl. II, f. 3
Southwold	3.91	2.26	2.90	0.32*	0.32*	0.07*	0.86	0.58	0.74	12.2	7.06	4.12	G. Hutchinson
Vorsä§	2.80	1.78	(2.50 (2.70	0.31	0.28	0.04	0.49	0.64	(0.89 (0.96	10.2	6.48	7.75	Crumlin-Pedersen 1966: fig. 2
Östra Aros§	1.36	0.88	(1.12 (1.19	0.17	0.13	0.03	0.11	0.65	(0.82 (0.87	10.5	6.76	6.80	Crumlin-Pedersen 1960: fig. 4
Rebaek	4.10	2.48	3.37	0.70	0.66	?	1.65	0.60	0.82	6.2	3.74	—	Sølver, 1946: fig. 1
Jungshoved	3.62	2.30	3.00	0.45	0.41	?	0.55	0.64	0.83	8.8	5.61	—	Crumlin-Pedersen, 1966: fig. 3

* The breadth and thickness measurements for Southwold include a shrinkage allowance. In other cases reports are seldom clear whether measurements were taken before or after shrinkage. If after shrinkage and no shrinkage allowance has been made the stated breadths and thicknesses may be considerably less than when rudder was in use (see Ch. 3). The rudder will also have shrunk in length but relatively insignificantly.

† The Nydam rudder may be a steering oar.

‡ Gokstad 2 rudder has two lower pivot-holes. A. E. Christensen believes the higher of these two relates to former use of this timber.

§ Vorså and Östra Aros rudders have alternative tiller positions.

Table 12.9 – see also Fig. 12.34; the ones from Southwold, Kvalsund 2, Oseberg, Tune and the four from Gokstad are dated; the remainder are thought to be of the Viking period. Three of the four Gokstad rudders and the one from Oseberg are associated with boats on which the rudder mountings survived, and rudder and mountings have been matched. The rudder mountings have not survived on the other dated boats. Notwithstanding this, it is sometimes possible to deduce at which frame the side rudder was

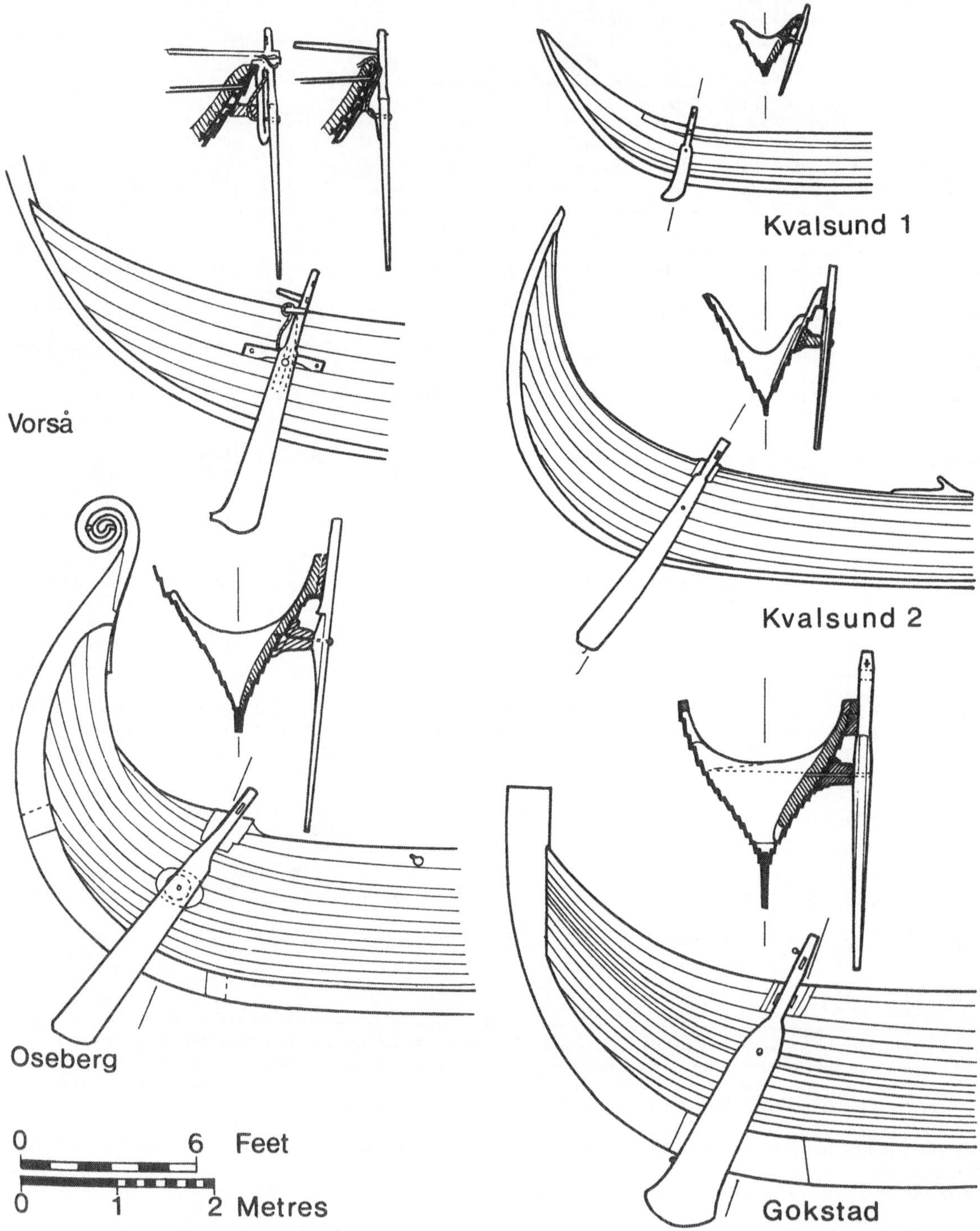

Figure 12.34 Five early medieval side rudders. After Evans and Bruce-Mitford, 1975 (by courtesy of the Trustees of the British Museum). Note that the rudder depicted on Kvalsund 1 is a copy of that found with the Gokstad *faering*.

hung, by the relatively greater scantlings – see Crumlin-Pedersen's work on the Skuldelev finds.

In the Oseberg/Gokstad group the rudders are all mounted on the starboard quarter with a withy rope passing through the attachment hole in the upper part of the blade through a boss about which the rudder pivots and is made fast to the rudder frame inboard. The withy rope takes the weight of the rudder yet is sufficiently flexible to allow it to be rotated. The rudder stock is held in against the top strake by a quick-release leather thong which when cast off, allows the rudder to be rotated in the fore-and-aft vertical plane to a position appropriate to shallow water. The rudder is controlled by a thwartship tiller which is some 8–12% of the overall length below the stock head. The wear pattern and other evidence on these rudders indicates that they were all hung so that the flatter side of their sub-aerofoil cross-section was next to the boat, with the broadest part of this cross section nearer the trailing edge. This also results in the projecting portion (heel) of the bottom of the blade being aft, as is frequently represented in medieval illustrations, for example on the Bayeux tapestry (Fig. 12.35). These rudders are hung at an angle aft of the vertical, varying from *c.* 24° (Gokstad 1) to 39° (Oseberg). Boats on six seals from the thirteenth and fourteenth centuries have rudders hung at 30° (Winchelsea) to 45° (Lübeck) aft of the vertical. The Winchelsea craft also seems to have a sponson on the starboard quarter to protect the rudder when alongside. Rudders also lean in towards the hull some 6° from the vertical for Oseberg, 3° Gokstad 1, and 20° Gokstad 3 (faering).

The analysis used in the compilation of Table 12.9 can be used with unattached rudders as well as those associated with a specific boat. Thus position of upper mounting is not assessed as this requires knowledge of the boat (in absence of wear marks on stock); and underwater area of rudder is not directly assessed as this would involve estimates of waterline and hydrostatic calculations. An assessment of blade underwater cross-sections is not possible with information at present available as published sections are not all at the same relative position: breadth/thickness ratios at non-comparable positions vary from *c.* 5 (Oseberg) to 9.58 (Gokstad 4).

Relative position of lower mounting (withy hole) (Column 9 in Table 12.9). This is expressed as a fraction of overall length from the bottom of the blade. All are within the range 0.60 to 0.69 except for:

Southwold: 0.58 – lowish
Nydam 2: 0.39 – relatively very low
Gokstad 2: 0.75 – higher than average (but see Note ‡)

Relative position of tiller hole(s). (Column 10). Expressed as a fraction of overall length from bottom of blade. All are within the range 0.87 to 0.92 except:

Rebaek	0.82	low
Southwold	0.74	

Figure 12.35 Scene from the Bayeux tapestry (Photo: Photographie Giraudon (courtesy the Musée de l'Evêché, Bayeux, and with the particular permission of the town of Bayeux).)

Slimness coefficient of rudder. (Column 11). Assessed as overall length of rudder compared with mean breadth of blade. Apart from Southwold which with a coefficient of 12.2 is slim, coefficients vary between 10.5 for the 1.36 m rudder from Östra Aros to 5.2 for the Gokstad 3 rudder. There are no obvious correlations and the significance of this coefficient may only become apparent when further finds are dated and provenanced, or after sea trials.

Aspect ratio of blade below withy hole. (Column 12). This is taken as an approximation to the aspect ratio of the wetted part of the rudder. It is estimated by comparing length of blade below withy hole with median breadth. These ratios vary from 7.56 (Kvalsund 2 with a relatively narrow blade) to 3.04 (Nydam 2 with a relatively broad blade).

Area of blade below withy hole. (Column 8). This is taken as an approximation to the wetted area of the blade. It is estimated from length and median breadth of blade below withy hole. This area varies from 0.10^2 (Kvalsund 2) to 1.65 m^2 (Rebaek) and there is some correlation as might be expected with size of boat as crudely estimated by (L × Bmax × Depth) where this is known.

Data from Table 12.9 for rudders associated with known vessels are tabulated in Table 12.10. Four indexes have been calculated to investigate possible correlations between rudder and vessel dimensions:

Column 7: Blade area is compared with the slimness coefficient of the vessel. As slim vessels are more directionally stable it might be expected that their rudder blade areas would be greater. If this is correct then Oseberg and Gokstad 1 have relatively large area rudders.

Column 8: There should be some correlation between depth of hull and the length of the rudder above the withy hole (near waterline). If this is correct the rudder on Nydam 2 is out of proportion – this tends to support the hypothesis of it being used at a large angle to the vertical i.e. a steering oar.

Column 9: If there is a correlation between the overall length of rudder and depth of hull, then again Nydam 2's rudder is too long.

Column 10: There may be a correlation between displacement volume (for which LBD is taken as approximate) and blade area. Gokstad 3 (faering) appears to have a relatively larger blade area than the larger boats; this may be due to the non-linearity of this proposed relationship. The Tune rudder appears to be relatively small in blade area, but as the size of this vessel is not known accurately and the rudder cannot be matched to the remains (Brøgger and Shetelig, 1971: 107) this result may not be significant.

With only five rudders evaluated the deduced data from Table 12.10 must be treated with caution. When it is possible to assess more examples, more reliable conclusions may be drawn.

Rudders seen on certain reconstruction drawings of boat finds are copied from elsewhere: for example the Gokstad *faering*'s rudder has been used on Shetelig and Johannassen's (1929) drawing of Kvalsund 1 (Fig. 12.34). Although a large proportion of Kvalsund 2's rudder was excavated and recorded (Shetelig and Johannassen, 1929), Johannassen's reconstruction drawing of the hull stretches the

Table 12.10 Side rudder data from known vessels

	Boat			*Rudder*					
Site	*LOA (m)* L	*Bmax (m)* B	*Depth (amidships) (m)* D	*Area below withy hole* A	*Length above hole* l_3	$A/(\frac{L}{B})^2$	l^3/D	$L_R^{\star}/D$	$\frac{\sqrt[3]{LBD}}{\sqrt{A}}$
Nydam 2	23.70	3.75	1.20	0.54	2.02	1.35×10^{-2}	1.68	2.75	6.49
Oseberg	21.58	5.10	1.58	0.75	1.28	4.19×10^{-2}	0.81	2.08	6.41
Gokstad 1	23.24	5.20	2.02	0.99	1.35	4.95×10^{-2}	0.67	1.66	6.31
Gokstad 3	6.51	1.38	0.49	0.13	0.37	0.58×10^{-2}	0.76	2.06	4.56
Tune	*c.* 20.00	4.35	1.20	0.31	0.72	1.47×10^{-2}	0.60	1.60	8.41

Sources – as in Table 12.9
★ L_R = overall length of rudder.

fragmented evidence for this find to the limit, and thus it seems best not to use the method of attachment shown as independent evidence.

We cannot be certain how typical is the evidence from the Gokstad/Oseberg group of side rudders, and it would therefore be wrong to assume, for example, that all medieval rudders were mounted to starboard or that they all have broad, trailing edges. That such rudders were generally mounted on the starboard quarter may be deduced from the word itself (*stearboard*), and we see many illustrations of this for example on the Bayeux tapestry. Side rudders are however occasionally shown on the port side (Fenwick, 1978: fig. 8.10b) and the thirteenth-century seals of Folkestone, Rye, Lubeck and the fourteenth-century seals of Dover and Hastings. Crumlin-Pedersen (1966: 258–61) states that the fragmented and distorted rudder from Jungshoved (Fig. 12.36) was probably a port rudder principally because of wear marks near the lower suspension point. This orientation gives the rudder a heel pointing astern, the flatter side of the blade's cross-section is next to the boat, and the blunter end is aft. Having the flatter side of the rudder next to the boat has the theoretical advantage that the sub-aerofoil can then generate a turning force opposite in direction, if not in size, to the turning force created by the asymmetric drag due to the side rudder. This rudder is notable for the short distance of 70 cm between its lower suspension hole (for a withy?) and the tiller hole, which may be compared with over 1 m on the Gokstad 1 and Oseberg rudders which are *c.* 20 to 30 cm shorter than the surviving length of the Jungshoved rudder. These holes on the latter are *c.* 60% and *c.* 80% of the overall length from the preserved bottom edge whereas corresponding values for Gokstad 1 and Oseberg are 60–61% and 90–92%. Thus it is the tiller hole which is relatively low which may mean that the associated boat had low freeboard but more likely that, as Crumlin-Pedersen suggested, the tiller came inboard through the planking: thus the upper suspension point (leather thong) may have been above the tiller.

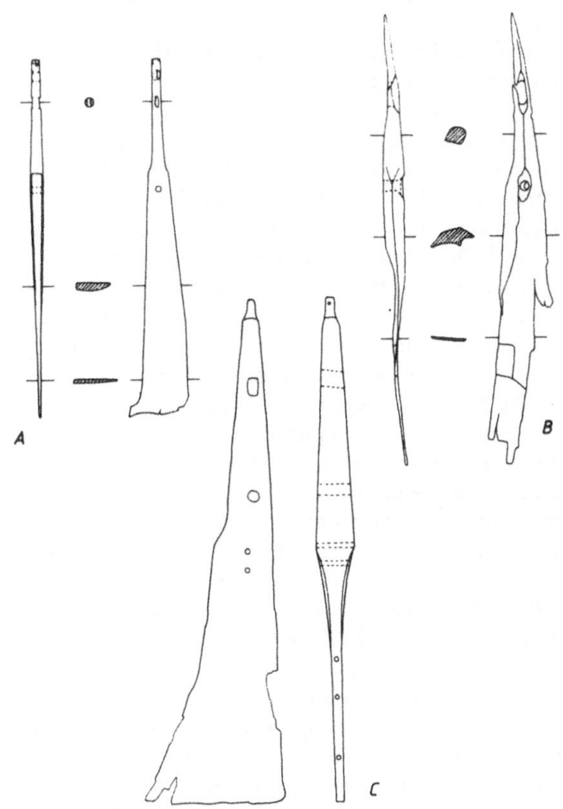

Figure 12.36 Side rudders from A. Vorsa; B. Jungshoved; C. Rebaek. After Crumlin-Pedersen, 1981.

The side rudder found in Östra Aros, Uppsala, in 1934 (Åkerlund, 1965: fig. 4) has two tiller holes at 82% and 87%, the axis of the upper hole being oblique to the rudder blade rather than at right angles as in the lower tiller hole. Crumlin-Pedersen has convincingly demonstrated that, where there are two tiller holes, the upper is for use when the rudder is pivoted about its lower suspension point (withy) to bring the blade above the level of the keel in shallow water. This rotation lowers the stock head, the upper tiller position is then more convenient to the steersman. This argument led Åkerlund (1965: 258) to suggest that this Östra Aros rudder was also a port rudder as a tiller in the upper hole would, after rotation to raise the blade, be in roughly the same position relative to the steersman as it was in the lower tiller hole with the rudder suspended in its usual near-vertical position.

The Vorså rudder found in 1958 in the Kattegat (Crumlin-Pedersen, 1966: 251–8; 1981: fig. 12A) also has two tiller holes (at 80% and 96%), the upper one again being angled 25° away from the thwartship position. Wear marks and the heel, however, indicate that this rudder was used to starboard; in this orientation the flatter side of its cross-section is next to the boat and its thicker edge is aft; and a tiller in the upper hole becomes angled aft and thus more convenient to the steersman when the stock is rotated forward for shallow water operations. The ability to rotate these side rudders into an alternative position which two tiller holes implies suggests that the Vorså and Östra Aros rudders probably projected well below the keel, in comparison with say the Oseberg rudder which does not have this feature and only projects some 20 cm and the Gokstad 3 rudder which projects some 8 cm. It may be that the Vorsa and Östra Aros boats had to be used in shallow waters (up rivers?) for a considerable portion of their regular voyages, not just on the approach to a shelving beach landing place.

In discussion with Åkerlund about the Nydam rudder, Salisbury (1965) highlighted the requirement for a rudder to be self-centring – the alternative is that if it were unattended it would be forced hard over. A further requirement, slightly different in concept, is that a rudder should be balanced forward and aft of its pivotal axis so that the resultant water pressure on the two areas would neither strongly resist the efforts of the steersman to turn the rudder nor would it excessively assist him so that the rate of rotation became uncontrollable but, rather, that a slight positive force was necessary on the tiller. The relationship between these two areas forward and aft of the pivotal axis to achieve this aim has to be established empirically. Tank experiments indicate that a rectangular plate rudder will be so balanced up to stall angles of 35° to 45° of rotation if *c.* 40% of its area is forward of the turning axis: such a rudder should also be self-centring in a waterflow (Crumlin-Pedersen, 1966: 256). Of the rudders listed in Table 12.9. Rebaek and Gokstad 1 to 4 have this relationship of areas; Östra Aros and Tune have approximately equal areas, whilst Oseberg, Vorså and Kvalsund 2 have more than half their underwater blade area forward of the turning axis and thus in theory should be both unbalanced and not self-centring. As all these rudders have a blunt trailing edge in the orientation that has been deduced for them, turbulence is generated aft of the turning axis. Crumlin-Pedersen (1966: 256) has suggested that a rudder area greater than the theoretical 40% would thus be required forward of the turning axis to counterbalance this effect. This theory might account for the area relationships of those with equal areas and those with a greater area forward, but then one would have to assume that turbulence generated near the trailing edges of the four Gokstad rudders and that from Rebaek was negligible or slight. It is impossible to come to a firm conclusion on the evidence at present available; it is noteworthy however that neither this theory nor any other explains the theoretical advantage in having a sharp leading edge to these medieval rudders in contradistinction to the modern rudder's blunt leading edge which minimises turbulence.

When a copy of the Vorsa side rudder was used on a replica of the Ladby boat, it was indeed found to be non self-centring, having a greater area forward of the turning axis than aft, and needed the constant attention of the steersman – this was further accentuated by the tendency of the replica as rigged to bear away from the wind. Crumlin-Pedersen (1966: 256–8) does not specifically say that the rudder was balanced, however, but that it was 'effective' and that the 'slightest turn of the helm is enough to adjust the course'. This rudder replica was not in fact authentic in that it had a thicker cross-section than the original and nylon rather than withy rope was used at the lower suspension point.

Andersen (1895) found that the side rudder he used on his 1893 replica of Gokstad 1 performed 'satisfactorily in every way' but, as Binns (1980: 123, 158) pointed out, the steersman used a relieving tackle to assist him control the tiller and he had the assistance of an anachronistic triangular jibsail on the forestay. In addition, Andersen elsewhere makes it clear that under certain conditions use of the rudder could not prevent this replica turning up into the wind.

Binns (1980), who is very aware of the difficulties of relating replica performance to authentic statements about medieval performance, found that a bow oar was frequently needed on the ⅔ replica Gokstad (*Odin's Raven*) to help them through the wind. This relative inefficiency may be a true reflection of Viking sailing ability; on the other hand it may have been due to incorrect shaping of the replica rudder or the boss on which it pivots or incorrect tensioning on the rope through this lower attachment point, all of which were also possibly incorrect on the Greenwich *faering* replica (McGrail, 1977a: 251–4). McKee (1974: 26) has suggested that, since this Greenwich replica's steering gear was only effective at small angles of rudder, this may have been 'a significant obstacle to the development' in Viking times 'of a sailing boat that could change tacks', but such conclusions may not be fairly drawn from this aspect of the Greenwich experiment.

Binns noted during voyages in the ⅔ replica (1980: 157–8) that the 'inner surface of rudder was often starved of water and more or less stalled even at small rudder angles, creating a large drag and turbulence', thus reducing turning force. In addition, he found that the waterflow on the quarter was often vertically upwards rather than parallel to the waterlines and this was particularly marked on the starboard tack when the hull was being pressed away from the rudder and when the rudder itself was less immersed due to the boat's heel to port. Such matters call for further investigation, theoretically and by model building at first and then, where justified, by practical experiment afloat. Such trials could be extended to test replica rudders installed on a standard hull: in this way, although absolute assessments of rudder effectiveness might not be attainable, relative performance could be. Rudders could also be tested in 'unorthodox' positions such as on the port side or with their blunt edge leading. In this way we may get closer to learning about the virtues and drawbacks of medieval side rudders. In this connection it is of interest to note the experimental work undertaken by Slettebo (1979) from which he deduced that a replica of the Nydam 2 rudder or steering oar was best used blunt edge forward and thus should be mounted on the starboard side.

MEDIAN RUDDERS

Several town seals of the thirteenth and fourteenth centuries (for example Elbing, 1242, Wismar, 1256, Poole, 1325) show vessels with centreline rudders; these may best be described as *ships* as they are clearly decked and no longer open boats. Median rudders are also depicted on the Tournai fonts at Winchester (Fig. 8.18) and Zedelgem, Bruges, Belgium (Fenwick, 1978: fig. 8.12b), which are conventionally dated to the late-twelfth century. Nevertheless early fourteenth-century seals of Faversham (Fenwick, 1978: fig. 8.32c) and Winchelsea (Fenwick, 1978: fig. 8.25.4) show evidently fair-sized vessels with side rudders which, if not an anachronism, may indicate that during the twelfth/fourteenth century both side and median rudders were used. It may be that, at first, median rudders were used on the curved hull form of vessel (*hulc*) depicted on the late twelfth-century font and on the *cog*, rather than the post-Viking development of the Norse tradition (*keel*).

Anderson (1928: 239) noted that in surviving accounts of the building of late thirteenth-century galleys, terms are used which can be interpreted as references to both forms of rudder; thus the Southampton galley appears to have had one side and one median rudder, while the Newcastle galley had two side and one median. The galley built at Lyme had a median rudder, which with all its gear cost 40 shillings, and a side rudder costing 13 shillings (Sandahl, 1951: 121): thus relative costs may have reinforced resistance to change to a new form of steering. Why these royal galleys should have both forms, if this is how the passages should be translated, is not clear. Anderson (1976: 47) considered that the side rudder was an auxiliary to the main median rudder. Another possibility (hinted at by Roberts (1984)) is that these 'side rudders' may in fact have been leeboards for use when the galleys were under sail.

There are no early excavated examples of median rudders or their fittings in N.W. Europe corresponding to the iconographic evidence, but by *c.* 1300 we have evidence for their attachments from two Danish sites (Vigsø and the Kolding cog) (Crumlin-Pedersen 1981) and on the Q75 wreck from the north-eastern polder, Zuider Zee (van der Heide, 1974: 421; Reinders, 1979: 38). The earliest representation of a cog with a stern rudder is on the Elbing seal of 1242 (Ellmers, 1980: 27). The Bremen cog of *c.* 1380 has a well-developed median rudder steering system, the essentials of which remained virtually unchanged to the end of the medieval period. Subsequent developments are described by McGowan (1981: 40–3).

ANCHORS AND OTHER MOORING DEVICES

Boats may be fastened to the sea or river bed in several ways, some with greater holding powers than others:

(a) mooring post
(b) mooring spike
(c) stone sinkers with holes
(d) hook-shaped anchors

MOORING POSTS

Boats may be made fast to posts (temporarily or permanently) driven into the seabed in shallow water – such use may be seen on the Bayeux tapestry. The only equipment required in the boat is a painter (rope or similar material) secured to a strong point near one end. Where a boat has a hole through her bottom within a watertight compartment (Rudolph, 1974: fig. 32) or a hole above the waterline through an overhanging end (Ellmers, 1979: fig. 1.3; van Nouhoys, 1951: 17–18) or below the waterline through a solid end (McGrail, 1978a: 69) portable mooring posts (a secondary use for propulsion poles) may be used. The undated logboat excavated in 1920 from Terbregge near Rotterdam had such a hole and an ash pole was found lying in the clay nearby (van Nouhoys, 1928). Other small boats which were probably moored in this way (pole-setting) include boat 1 from Skaggered, Götenberg, Sweden (Olsen and Sjöberg, 1971: 48, fig. 2), boat 2 from Holme Pierrepont, near Nottingham, England (McGrail, 1978a: fig. 21), and two logboats from near Göttingen (Ellmers, 1973a: 53–4); and possibly those from Ballinclemsig, Co. Kerry (O'Connell, 1941: 81), Baddiley Mere (McGrail, 1978a, fig. 53) and Zwammerdam boats 1 and 3 (de Weerd and Haalebos, 1973: 392, 395: fig. 12) (Fig. 6.14).

MOORING SPIKES

Spikes on the end of a painter of rope or chain (Fig. 12.37) may also be used to hold fast a boat, usually on the margins of a river or lake (McKee, 1983: 108–9, figs 84, 86). The earliest mooring spike so far recorded appears to be that from the sixteenth/seventeenth century small boat excavated from Lake Caldecotte, Milton Keynes and now under examination at the National Maritime Museum, Greenwich.

STONE SINKERS WITH HOLES

Stones with a single hole for a line have been, and are, used in many parts of the world to moor boats (van Nouhoys, 1951: 21; Frost, 1963, 1973, 1979, 1982). Holding power here is entirely dependent on the friction created between heavy stone and sea bed, thus these devices are more useful on a stony or soft mud, rather than a sandy, bottom. The addition of a stick lashed at right angles to the stone as in South America (Edwards, 1965: 75), or sticks protruding through further holes through the stone, as in Portugal (Filgueiras, 1977: fig. 9.19) and in many other places (Frost, 1963: 4; 1973: 407), increases holding power as these projections dig into the sea bed or jam between rocks.

As simple, as well as complex, forms continue in use today, stone sinkers used as anchors are generally impossible to date unless they are found in a datable context or have an inscription. Finds from the Mediterranean region are as early as the second millennium BC. Both one-holed and three-holed types are known and they range in weight from less than

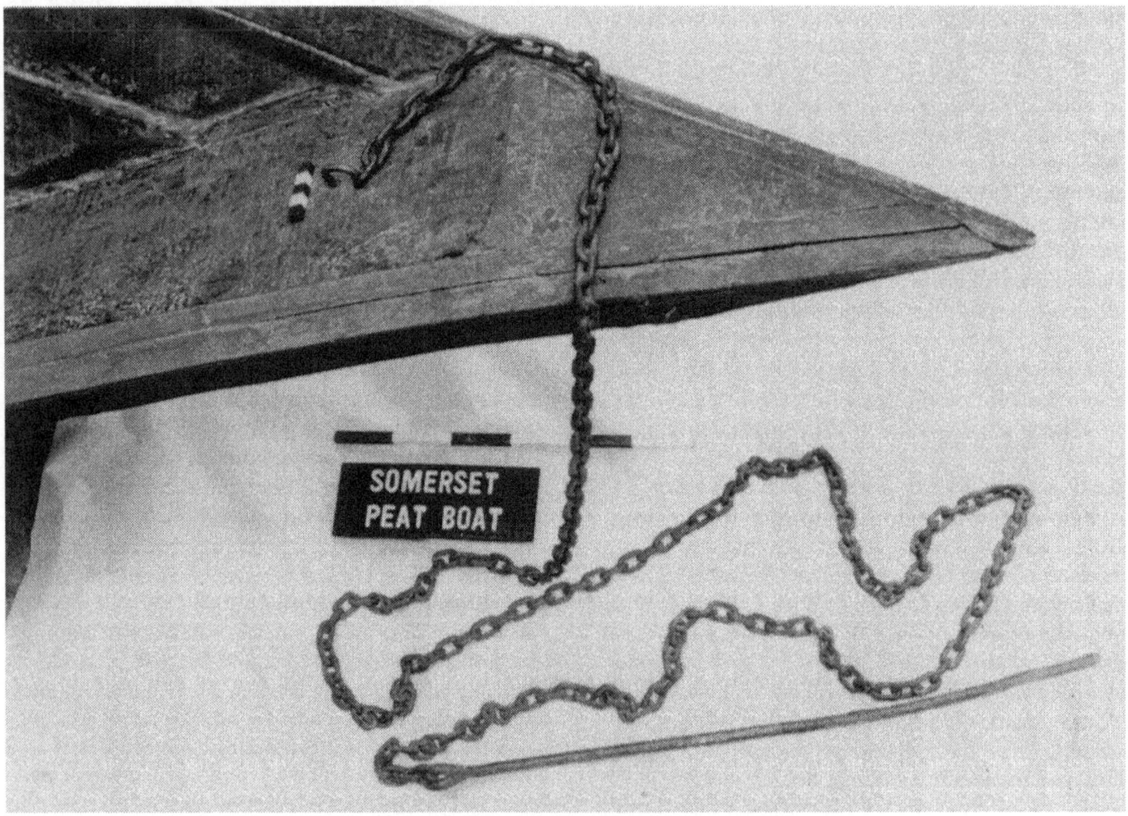

Figure 12.37 Mooring chain and spike of the Somerset peat boat. (Photo: NMM Greenwich.)

20 kg to over 700 kg. Frost (1973: 405) has argued that those too heavy for one man to handle (over 50 kg) almost certainly predate the introduction of metal and wooden anchors. Not all holed stones are used as anchors for they may be fishing net weights, doorpost sockets, tethering points, torch holders and counter weights for olive presses (Frost, 1973: 404); and they are used today in exposed parts of western Ireland and Scotland to anchor thatch ropes and nets. Particular care is therefore necessary in identifying holed stones recovered from land contexts.

Holding power is maximised when the pull on the stone sinker, or any other form of anchor, is horizontal. Such a direction of pull is achieved in modern times by using a metal chain sufficiently long for a great proportion to lie on the sea bed resulting not only in a horizontal pull on the anchor but also increased friction on the sea bed. The ratio of length of cable to depth of water (= *scope*) is usually in the range 3 to 5 for a secure hold. An effect analogous to that of a metal chain could theoretically be obtained by weighting organic ropes with small pierced stones (Frost, 1982: 263–4) – but to date such use has not been recorded either ethnographically or in underwater finds.

HOOK-SHAPED ANCHORS

Composite artifacts with one or more hook-shaped protrusions of wood or antler ballasted by stone are also used in many parts of the world as anchors (van Nouhoys, 1951: figs 1 to 6; Frost, 1963), holding power being dependent on the grip of the hook where it digs into the seabed. The addition of a crosspiece (stock) at right angles to the hook (or the line between two hooks) tends to turn the anchor into an alignment which facilitates the digging-in of at least one hook. Use of a stone or lead stock instead of a wooden one increases the chances that the wooden hooked anchor will lie in good contact with the sea bed and that there will be a near horizontal pull when the boat is riding to her anchor.

In the Classical world hooked anchors of the following types are known (Boon, 1977: 11–13):

(a) wood with stone or lead stock
(b) wood with removable lead stock
(c) iron with removable iron, wood, stone or lead stock.

Anchors with stocks are difficult to stow onboard, hence the emergence of the removable stock. The woodwork of these anchors seldom survives, nevertheless examples of all these types have been recovered, lead stocks being frequent, and additionally there is iconographic and documentary evidence.

In northern and western Europe one lead stock of Mediterranean pattern has been found in Porth Felen, a cove in the Llŷn Peninsular, N. Wales, otherwise knowledge of anchors comes from the Broighter boat model of the first century BC (Farrell and Penny, 1975), the late iron age (first half of first century AD?) hill fort of Bulbury, Dorset (Cunliffe, 1972) and three Viking Age boat burial sites, Ladby, Gokstad and Oseberg (Sølver, 1946; Nicolaysen, 1882; Brøgger *et al.*, 1917). There is also some iconographic evidence from the time of the Bayeux tapesty onwards.

Among the equipment found with the Broighter boat model (Fig. 12.2) was what Evans (1897: 391) called a 'grappling iron'. This is a *c.* 8 cm pole with a group of four hooks or prongs (a grapnel) at one end. Grappling irons on the end of a rope were used in recent times to hold ships together especially when boarding; grapnels on the other hand are small 4-pronged anchors which do not have or require a stock. Grapnel use is widespread: see the thirteenth-century AD Arab ms. al Harriri's *maqamat* (Johnstone, 1980: fig. 13.10) and in modern Portugal (Fernandez, 1933: fig. 8b), France (Beaudouin, 1970: fig. 14) (Fig. 8.35) and the Netherlands (van Nouhoys, 1951: fig. 8). Although the Broighter object has a much longer shank in relation to the size of its four hooks, than modern examples, it seems much more likely to have represented a grapnel (anchor) than a grappling iron. A chain or rope cable would have been attached to the end of the shank.

The iron anchor from Bulbury, Dorset (Fig. 12.38) was part of a hoard found near the centre of the hill fort in 1881 (Cunnington, 1884), some 2 to 3 miles from Poole Harbour up the River Shelford; it is now in the Dorset County Museum, Dorchester. This anchor measures 1.44 m overall, the shank having a rectangular cross-section *c.* 5 × 2 cm (Cunliffe, 1972). The arms, which are also generally rectangular in section, taper to points and are set at an angle to each other, so that in profile they resemble an arrow-head (Type A – Kapitän, 1984); there are no shaped flukes as were occasionally used in late Imperial times (Casson, 1971: 253) and as are found on many recent anchors. At the crown, between the arms, is a circular hole 2 cm in diameter; this was probably for attachment to the ship when onboard, and, when in use, for the attachment of a buoyed rope, which would not only mark the spot, should be anchor be slipped or the cable cut, but could also be used to assist in dislodging the embedded arm when breaking the anchor out of the

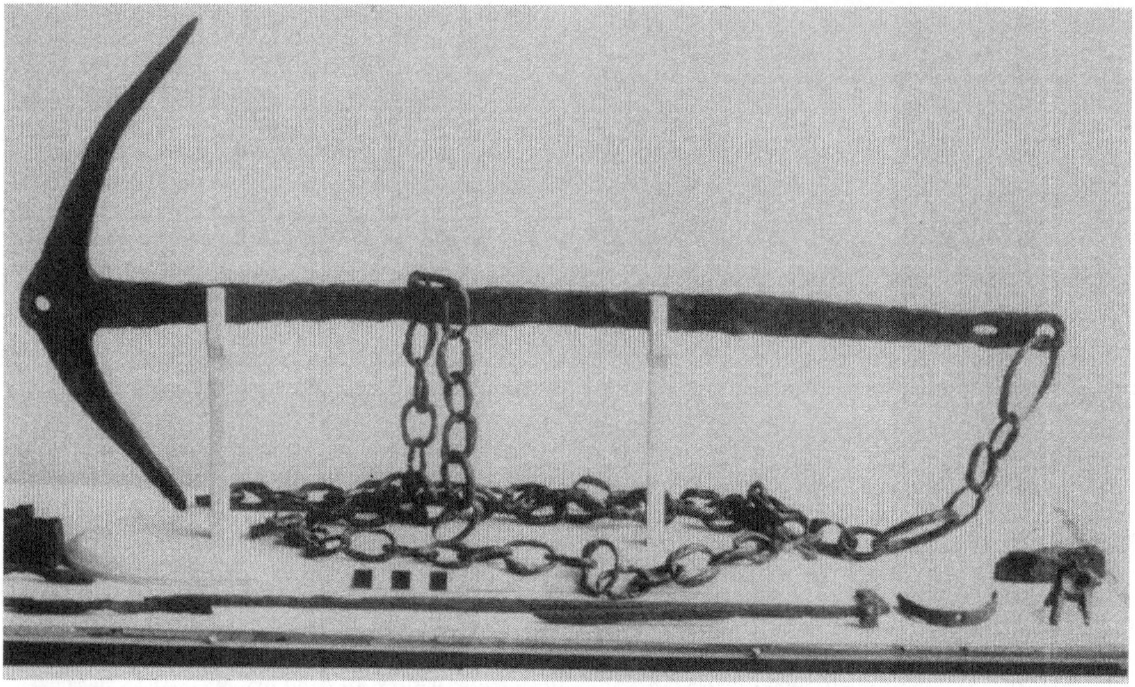

Figure 12.38 The mid first-century AD anchor and chain from Bulbury, Dorset. (Photo: Dorset Natural History and Archaeological Society, Dorset County Museum, Dorchester, Dorset.)

ground. At the other end of the shank is a similar hole for the anchor ring and a rectangular hole, *c.* 3.5 × 1.5 cm, for the stock, which by analogy with a similar anchor from Pompeii, was probably of iron (Cunliffe, 1972. 300). A chain with a reconstructed length of some 6.5 m was found with the Bulbury anchor. Of the 114 chain links, 113 were elliptical, each *c.* 7.5 cm long; the fourth link was similar to the anchor ring, *c.* 13.5 cm in diameter. This chain would have been the forerunner at the end of an anchor cable of some organic material. Such a chain forerunner assists in maintaining a horizontal pull on the embedded anchor, and is more resistant to wear on the sea bed than a cable of cordage. Anchor chains were used in the Mediterranean from at least the fourth century BC but their use appears to have been exceptional (Casson, 1971: 252), so much so that Caesar (*B.G.* III. 13) remarked on the fact that the anchors of the Veneti had iron chains rather than the rope with which he was familiar.

Cunliffe (1972: 300) argued from the Bulbury and Veneti evidence that iron anchors may have had a pre-Roman ancestry in western Europe and his argument may be reinforced by the identification of a grapnel with the Broighter boat model. It seems likely that the Veneti of the mid-first century BC had iron anchors as well as chains, and the Bulbury anchor of the mid-first century AD (Cunliffe, 1972: 304–5) is undoubtedly similar in form to southern European examples. On the other hand, the fact that the Bulbury anchor's stock was flawed with large slag crevices and that Cunliffe considers the whole hoard to have been waste material collected for re-forging, may mean that this anchor may not have been manufactured locally. Further well-dated early examples of iron anchors and cables may throw light on this question of origins.

An iron anchor and cable of chain and rope were found in the Ladby boat grave (Sølver, 1946), an iron anchor and two wooden stocks in Oseberg (Brøgger *et al.*, 1917: pl. XXVI) and an anchor and stock in the Gokstad grave (Nicolaysen, 1882: pl. IV, fig. 17), however the Gokstad anchor disintegrated before it could be recorded.

The *Oseberg* anchor (of *c.* AD 800) measures *c.* 1.02 m in length and weighed *c.* 9.8 kg. The shank, of rectangular cross-section (*c.* 4.5 × 2 cm), had a circular hole through each end, the anchor ring being *c.* 12 cm in diameter, and the ring for the buoy rope *c.* 7 cm. (Fig. 12.39). The arms which are also roughly rectangular in section taper to blunt points without flukes: they are more curved in profile (Kapitän B) than the angular arrow-head shape of the Bulbury arms. There is no hole in the shank for a

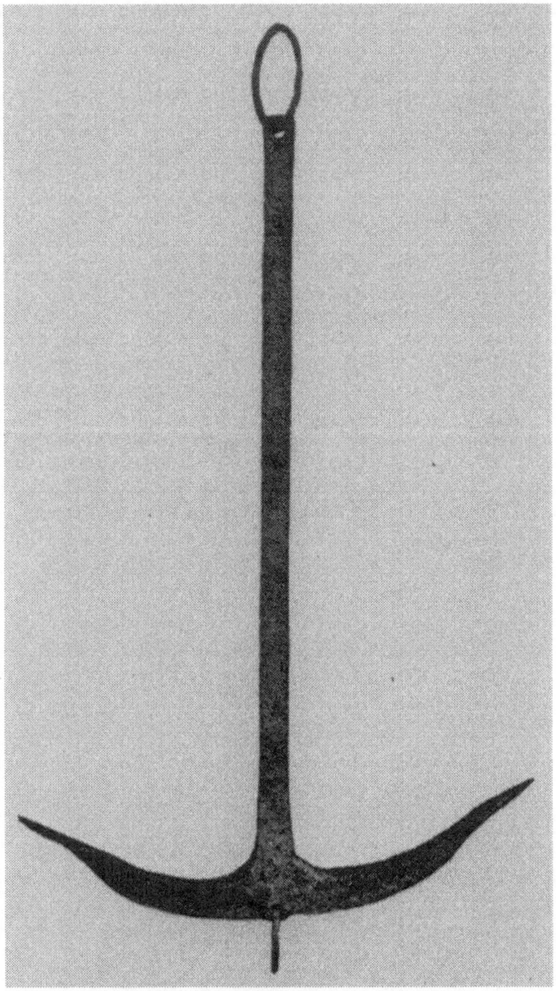

Figure 12.39 The late eighth-century AD anchor from Oseberg. (Photo: courtesy the University Museum of National Antiquities, Oslo, Norway (copyright holders).)

stock and the method of attachment may have been by binding. The two Oseberg stocks are relatively long (Table 12.11) and may be unconnected with this anchor.

The *Ladby* anchor (of *c.* AD 900) of 1.36 m is generally similar in form to the Oseberg anchor, and also has an anchor ring at the top of the stock and a buoy rope ring at the crown. A technical advance is that the Ladby anchor's arms end in incipient flukes, a shaping of the ends to increase holding power. Solver (1946) estimated that the chain forerunner which was fastened to the Ladby anchor by an iron hook moused with twisted bronze wire, had 50 links totalling *c.* 10 m in length; to this had been attached an unknown length of bast rope which was probably three-stranded, righthand lay. Examination carried out by Høeg in 1944 revealed that the chain links were of steel very similar to Swedish forged steel but that each link had been welded using soft Danish bog or lake ore iron (Sølver, 1946: 120–2). The Ladby stock did not survive but by analogy with others could have been 0.60 to 1.10 m in length.

The Gokstad stock (Sølver, 1946: fig. 11) is some 2.75 m in length. The shank of its (unrecorded) anchor seems to have fitted into a rectangular 3-sided hole in the centre of the stock, and was then clamped in place by a wooden locking piece inset in the shank and fastened to it by treenails (Nicolaysen, 1882: fig. 5).

The anchor being carried ashore from one of William's ships on the Bayeux tapestry (Fig. 12.35) seems to be generally similar in form and size to the Ladby anchor and has prominent flukes; no stock is visible, but there appears to be both a ring at the crown, and an anchor ring to which a cable of unknown material is fastened.

Engelhardt (1866) mentioned that an iron anchor was found in 1864 by German excavators in the same bog from which he had excavated Nydam boat 2 the previous year. Ilkjaer and Lønstrop (1981/2) have found a further reference to this anchor in the diary of the clerk of a nearby parish. The drawing in the diary shows an anchor with angularly set arms with large flukes; at the other end of the stock there appears to be an anchor ring. The accuracy of this drawing may be questioned, and in any case such a find cannot be dated with any certainty.

Anchors are shown on several town seals in the medieval period from the thirteenth century – see for example that of Portsmouth where the anchor has both an anchor ring and a ring at the crown. Cork was used for anchor buoys in the early fifteenth century (Rose, 1982: 134) and possibly earlier. Late Medieval ships carried several anchors including *grapnels* for boats; a *kedge* for shifting berth; *stream* anchors for general use; and, for use in strong streams, *bower* anchors of which the best one was the *sheet* anchor (Howard, 1979: 22–3).

FITTINGS ASSOCIATED WITH ANCHORS

Small anchors (up to say 50 kg) may be thrown or lowered overboard by one or two men; except possibly for wooden fairleads through which the cable is passed, no fittings are required, the inboard end of the cable being made fast by taking a turn around a convenient timber. In certain cicumstances

Table 12.11 Details of selected anchors and stocks

Site or Anchor	*Date (approx)*	*LOA (L) (m)*	Bill to bill (*B*) (*m*)	Profile of arms (*Kapitän, 1984*)	Angle between arms	Stock (*S*) (*m*)	L/B	L/S	Reference
L. Nemi wood	1st century AD	5.55	2.60	A	49°	2.35	2.13	2.36	Speciale, 1934
L. Nemi iron	1st century AD	3.60	1.75	B	99°	2.92	2.06	1.23	
Bulbury	1st century AD	1.44	0.78	A	108°	—	1.85	—	Cunliffe, 1972
Oseberg	AD 800	1.02	0.65	B	125°	—	1.57	—	Brøgger *et al.* 1917
Oseberg	AD 800	—	—	—	—	2.28	—	—	
Oseberg	AD 800	—	—	—	—	1.44 (+)	—	—	
Gokstad	AD 850	—	—	—	—	2.75	—	—	Nicolaysen, 1882
Ladby	AD 1000	1.36	0.84	B	132°	—	1.62	—	Sølver, 1946
Holigost	early 15th century	4.85	3.35	—	—	—	1.45	—	Howard, 1979: 22–3
Long shank	late 18th century	—	—	A	110°	—	1.61	0.94	Upham, 1983
Admiralty	early 20th century	—	—	B	90°	—	1.46	1.00	*Man. Seamanship* 1. (1926)

Note: 1. The Oseberg anchor was said to weight 9.8 kg and the Ladby anchor 27.95 kg after excavation: it is difficult to make an allowance for weight loss by corrosion.

the anchor may be lowered into a small boat for letting go some distance away, or taken ashore by hand as may be seen on the Bayeux tapestry. This size of anchor may also be weighed and stored on deck when not in use.

As boats increased in size to become ships, anchors became correspondingly bigger. By the medieval period these sizeable anchors were secured outboard against the bow – see for example the AD 1297 seal of Dublin (Fig. 12.29). The precise method of securing anchors in this position is not known, but anchors with fixed stocks are difficult to secure – in the Mediterranean this early led to anchors with removable and movable stocks (Kapitän, 1984) and these may have been used in medieval N.W. Europe but the evidence is lacking. With such an anchor the stock could be turned into the same plane as the shank making tight stowage possible when at sea. When anchoring became a possibility or was intended the stock would be returned to its position at right angles to shank and arms, and the anchor *catted* i.e. suspended from some convenient timber or block ready for letting go. (Fig. 12.40).

Such anchors became correspondingly more difficult to weigh by hand and windlasses (known in the Classical world – see Casson, 1971: 251–2,

Figure 12.40 The seal of New Shoreham of *c.* AD 1295. The inscription links the depicted ship with the former name of the town, Hulkesmouth. (Photo: NMM Greenwich.)

332–3) came into use not only for weighing anchor, and probably riding to anchor, but also for hoisting and lowering the yard (Marcus, 1980: 102) which had also increased in size and weight. The drum or barrel of a windlass rotates about a horizontal axis and is turned by hand spikes or levers. Weighing anchor by a windlass mounted well aft is depicted on the thirteenth-century seal of Winchelsea (Fig. 12.41). A windlass was found in a similar position in thirteenth-century Kalmar boat 1 (Åkerlund, 1951) and the twelfth/thirteenth-century Ellingå wreck has a base aft for a windlass (Crumlin-Pedersen, 1981).

The Bremen cog of *c.* 1380 has a windlass forward and aft there is a *capstan* the barrel of which is turned about a vertical axis by bars.

Hawseholes are depicted from the fifteenth century: these are holes in the upper decking in the bows through which the anchor cable passes out through the ship's sides, thereby replacing the method of leading the cable over the side through a fairlead.

Figure 12.41 The thirteenth-century seal of Winchelsea. (Photo: NMM Greenwich.)

13 Seamanship

Seamanship may be defined as 'the art and science of working a vessel so that a voyage is conducted safely and efficiently'. This term includes navigation and pilotage which are discussed in the following chapter; here the discussion is limited to such topics as making ground to windward, working the tides and operating from landing places and harbours. The archaeological evidence for such activities is at present almost non-existent, apart from that for medieval harbours, but, as with certain other matters discussed in this book, knowledge of what to look for may lead to identification.

ENVIRONMENTAL CHANGES

It might be thought that the prehistoric seaman was combating the same forces and tackling essentially the same problems as post-medieval man. In a *general* sense this is true, but *details* not generalities are of importance to a seaman operating in a particular environment; thus the shape and character of the coastline, the depth of water and nature of the sea-bed, the position of shoals, reefs, banks and bars, the tidal regime and the weather sequences including the predominant wind are all of immediate concern to him. How different from present conditions these were in former times is difficult to quantify (McGrail, 1983: 303). Yet, before the problems facing the prehistoric and medieval seaman can be understood, an attempt must be made to reconstruct the ancient maritime environment or at least to deduce the order of variation from twentieth-century conditions.

Sea-level changes

Sea-level investigations have recently been pursued worldwide, see, for example, the *Proceedings of the Geologists' Association* for 1982, which contains reports on sea-level movements during the last 15,000 years by a British group taking part in an international 8-year project. Where detailed work has been undertaken it has proved possible to make theoretical reconstructions of earlier coastlines and river courses with varying degrees of confidence: see, for example, Devoy (1982: 67, fig. 2) for S.E. Britain in general, and Cunliffe (1980: figs 18 and 19) for the Romney Marsh region in particular; Thoen (1978: maps 1–8) and Ryckaert (1980: 76, figs 1–4) for Belgium and the Netherlands; and Ambrosiani (1981: 1982) for aspects of such work in the region of Lake Mälar, Sweden.

In Scandinavia, there has been a marked rise of the land in post-glacial times, whereas in the southern North Sea region there has been a general rise in mean sea-level, but at a diminishing rate, from *c.* 7000 BC to the present day, (Devoy, 1982; Heyworth and Kidson, 1982; Thoen, 1978; Ryckaert, 1980) with occasional regressions. The rate of this rise varies regionally, and much further research is undoubtedly required, but Heyworth and Kidson's curve for MHWST applicable to the W. Channel, Bristol Channel and Cardigan Bay region may be quoted as an example – see Table 13.1.

In conjunction with other factors, mean sea-level determines at any one time the general form of the coastline, the presence or absence of shoals, sands, reefs and tidal races; river gradients and hence rates of erosion and deposition, and the presence or absence of spits and bars; and it indirectly influences local tidal effects. The precise effects caused by changing sea-level vary at each particular time and place. It is, however, possible to make three general observations.

Firstly, Although MHWST in the southern N. Sea and Channel region has been lower than today's level during the past 5000 years, it has generally been

Table 13.1 Sea level changes in S.W. Britain

Date	*Height of MHWST (approx)* *(m)*	*Change* *(m)*
7000 bc	−31	—
6000	−17	+14
5000	−11	+6
4000	−7	+4
3000	−5	+2
2000	−3	+2
1000	−2	+1
ad/bc	−1	+1
1000 ad	−0.4	+0.6
1950 ad	0	+0.4

Source: Heyworth & Kidson, 1982: fig. 2
Notes: 1. Ordnance Datum is at *c.* −5 m.
2. MHWST = mean high water spring tides

within today's tidal range (average *c.* 6 m). Before *c.* 3000 BC, on the other hand, with lower sea levels more land would have been exposed; around 7000 BC for example, MHWST would have been *c.* 26 m below Ordnance Datum i.e. somewhere in the region of the present day 10–15 fathom line.

Secondly, in former times there would generally have been less silting of estuaries and if spits and bars across their entrances existed they would not have been so prominent as today. However, the long-term effects of increasing silting and the deposition of eroded material by coastline currents have not been quantified and much detailed work remains to be done.

Thirdly, from *c.* 3000 bc onwards the tidal regime would probably have been generally as it is today. Before that date however differences in topography, fetch and water depth would have resulted in different tidal patterns (Devoy, 1982: 85).

Man-induced changes:

In addition to Man's influence from early times on the rate of water run-off and thus on river flows and deposition rates, drainage schemes and other engineering works in the past 500 years have considerably affected the appearance of much of the coastline, especially harbour sites. For example, until the seventeenth century Liverpool was confined to a peninsula jutting into the Mersey estuary (Nicolson, 1981: 5); St Malo was an island in the estuary of the River Rance until the late nineteenth century, and connected to the mainland by a causeway, whilst nearby Alet was a peninsula until 1965 (Belin, 1764: chart 44; Purdy, 1842: 219; Langouët, 1977: 38).

WEATHER PATTERNS

Weather expected and weather experienced are most important factors in seafaring, but ancient weather patterns are difficult to define except in the most general terms. Lamb (1977: 372–4, 384–5) has outlined some of the variations in the post-glacial European climate:

1. up to 6000 BC. A temperature rise with winters generally milder and summers warmer than today.
2. 6000 to 3500 BC. A 'climatic optimum' with mild winters; the humidity was greater than before or after. The wind was generally westerly.
3. 3500 to 1000/500 BC. A generally warm settled regime with some serious interruptions and fluctuations of temperature and humidity at *c.* 200 years intervals.
4. 1000/500 BC to AD 100. A decisive shift to a colder, wetter climate in N.W. Europe, the most marked change being from *c.* 1000 to 700 BC, so that by the second half of the first millennium BC the weather would have been comparable with today. The winds in this period were generally N.W./N. in summer and W. in winter.
5. AD 100 to AD 400. Some recovery of warmth and a tendency to be drier.
6. AD 400 to AD 800. Reversion to colder, wetter weather.
7. AD 800 to AD 1300 Improvement culminating in a warmer epoch in twelfth and thirteenth centuries.
8. AD 1300 onwards. With some significant exceptions (e.g. the 'little Ice Age' of the seventeenth century) the climate is thought to have been much the same as it is today within the level of deduction that can at present be made.

The best that can be done in any investigation into weather effects in former times is probably to assume that back to *c.* 1000 BC the climate was generally as it is now; future work by palaeometeorologists may enable more specific and quantitative assessments to be made.

Seasonal sailing

Some slight confirmation of the assumption of a generally similar climate in the recent past may be seen in the fact that thirteenth-century Scandinavians avoided putting to sea between November and March (Marcus, 1980: 101). The early English poem 'The Seafarer' indicates that the prudent Anglo-Saxons also did not put to sea until after the first Cuckoo (line 62) – pointed out to me by Alan Binns. This winter season was also avoided in the Classical Mediterranean (Casson, 1971: 270). Off the coasts of N.W. Europe today gales of force 7 and above are eight times more frequent in winter than in summer, and rough seas may be expected on average every fourth day in winter compared with every twelfth in summer, and the sea, and especially the air, temperatures are significantly lower in winter. Such low temperatures combined with exposure to wind and wetness from rain or sea water (unavoidable in an open boat) soon make a man numb. Cloud cover is greater in winter than in summer and the whole area may be overcast for several successive days. Rain falls more often and lasts longer in winter with a probability of 66% compared with 33%, and freezing rain and snow can be encountered. There were thus good reasons why winter seafaring would only be undertaken in exceptional circumstances in the prehistoric and early medieval periods.

In the late Medieval period, however, there are examples of planned winter voyages, some of which ended in shipwreck (Platt, 1973: 126; Rose, 1982: 35, 48, 51). In the fourteenth century it became important for English merchants to import wine from the Bordeaux region as early in the season as possible. Thus in 1384 the wine fleet left Southampton in September and returned there in the following January and February (Platt, 1973: 70, 74, 129). In 1440 coastal voyages were undertaken in November (Platt, 1973: 154–5); these were probably similar to nineteenth-century coastal voyages under sail which consisted of broken passages with many long waits for briefs periods of settled weather (Greenhill, 1978: 201–4).

Weather at sea

The weather today off the coasts of N.W. Europe is variable and actual weather experienced can differ considerably from seasonal means. Forecasting the weather for short periods ahead without instruments is practicable, especially the sequences of a depression with its high cloud forewarning out to 200 nm, often preceded by a change in swell pattern. Although timing the passage of associated fronts may be difficult, simple observations of the clouds, wind, sea and airmass, and the behaviour of seabirds provide useful indications of future weather (Reids *Nautical Almanac* 1983: 714–5; Slade and Greenhill, 1974: 24) and were well within the capabilities of early man.

The Icelandic sagas have many references to weather signs (Marcus, 1980: 101) indicating their importance to the thirteenth-century mariner. That Viking Age seamen needed, and indeed had, a detailed knowledge of weather sequences has been emphasised by Binns (1961; 1980: 37–9) who has shown, for example, that the ninth-century Norwegian Ohthere chose the only weather sequence which would give him the right winds to take him from a point on the N.W. Norwegian coast around N. Cape into the White Sea (Lund, 1984: 18–19). Burland (1950) has also emphasised the medieval Scandinavians' ability to choose the weather for passages across the N. Atlantic to Iceland, Greenland and America, and for the return voyage.

WINDS AND THEIR USE

Today the predominant winds and the resultant swell off N.W. Europe are from between S.W. and N.W. throughout the year, although with travelling depressions moving N.E. the mariner can experience winds from S. through W. to N., or even from the east, depending on his position relative to the depression's centre. Similar conditions were probably experienced in general, as far back as 1000 BC.

The wind's strength and direction are modified by local topographical effects. For example, because of the effects of land, W. and S.W. gale force winds are less severe and last half as long in the southern North Sea and eastern Channel than in the western Channel and the Bay of Biscay; conversely, the N. Sea is more exposed to N. and N.E. winds. In the lee of headlands, there will be an area of light winds, then an area of turbulent winds and seas. Furthermore, land and sea breezes, caused by unequal heating and cooling modify the general wind pattern, with a tendency for there to be sea breezes in the afternoon and land breezes in the forenoon, especially in settled summer conditions.

Early twentieth-century decked fishing boats would not put to sea in winds above Force 6, it therefore seems a realistic assumption that the open boats of the early-medieval and prehistoric periods would be limited to Force 5 winds (fresh breezes of *c.* 20 kn) and below. If unexpectedly exposed to stronger winds (there is a *c.* 20% probability of such winds during the summer months in the Channel, the North Sea and the Irish Sea) it seems likely that a hide boat could ride out the weather and cope with the associated seas better than a planked boat, because of greater seakeeping qualities.

CLOSENESS TO THE WIND

In making progress to windward the aim is to achieve

that heading and speed which, when combined with leeway, gives a resultant velocity which makes most ground to windward in a given time. This angle between the boat's heading (course) and the true wind will not be the closest the boat can physically get to the wind and, in general, it is found that boats make best progress to windward when 1 point ($11\frac{1}{4}$°) or so off this minimum angle.

To achieve this optimum condition the driving force must be maximised at the same time as leeway is minimised (Fig. 13.1), consistent with not exceeding the safe angle of heel and not overstressing hull, rigging and crew. In other words the qualities of hull, rig and crew have to be matched. Driving force may be maximised by setting the sail to maintain the optimum aerofoil shape for the prevailing conditions, with a firm leading edge. Leeway may be reduced at the 'design' stage by giving the boat an underwater form which in the relative waterflow will generate horizontal lift to oppose leeway – this aspect is discussed in more detail in Chapter 8 (keels) and Chapter 12 (sails). Lateral resistance of the hull may also be increased by the incorporation of sharp projections such as a *gripe* at the bow or a *skeg* at the stern (McKee, 1983: 141). When underway leeway may be further reduced by the steering device and by leeboards (Ch. 12).

The distinction between the heading closest to the wind and the heading on which the boat makes best progress to windward (Vmg is maximum) is not always made in the literature. Published estimates of 'closest to wind' headings are given in Table 13.2. Sailors are frequently optimistic in their assessment of how close to the wind they can sail and have been known to give this heading relative to the *apparent* wind (to which they are indeed closer). Scientifically controlled sailing trials of authentic replicas are needed to provide trustworthy data and the information in Table 13.2 must therefore be approached with some scepticism. Binn's estimate has the virtue that it was based on fixes and therefore was relative to the true wind; and Haasam's simulation result agrees with it. Thus the evidence at present available suggests that Viking Age ships would make

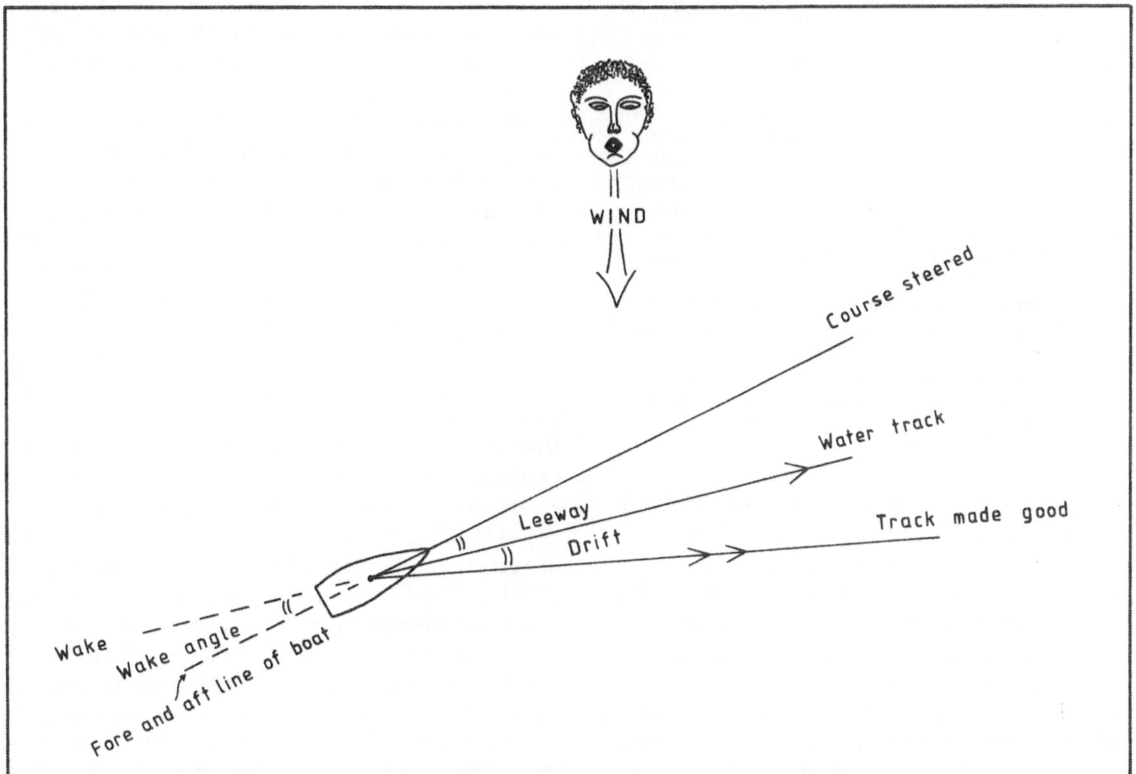

Figure 13.1 Diagram, not to scale, showing leeway as an angular displacement from course steered, and drift due to current and tidal stream (here acting in the same direction as the wind) as an angular displacement from the track through the water. The combined effect of leeway and drift gives the track made good over the sea-bed.

Table 13.2 Estimates of windward ability

Vessel	Angle between track & true wind* (points)	Leeway angle (points)	Reference	Notes
Classical	7	(included ?)	Casson, 1971: 274.	Estimate
Norse	6	1	Haasum, 1974: 116–7.	Based on simulation
Square sails in general	5	1	Roberts, 1984: 129, 130, 138.	Small-scale model and square sail on dinghy
Norse	10	(included ?)	Haasum, 1974: 118.	1893 Replica *Viking*
Norse	7	(included ?)	Haasum, 1974: 118.	1963 Replica *Hugin*
Norse	6	1	Binns, 1980: 165, 168.	1979 Replica *Odin's Raven*
Norse	5	(not included ?)	Crumlin – Pedersen, 1984a: 32–3.	Replicas and modern boats with square sail

Notes: 1. All vessels had a single square sail (except replica *Viking*).
* Binns, Crumlin–Pedersen and Haasum state that they are considering the heading on which greatest speed could be made to windward. Crumlin–Pedersen estimates this speed to be 1.5 to 2 kn, whilst Haasum suggests this is 2 kn.

the best progress to windward when 6 pts (67½°) off the wind with a leeway of *c.* 1 pt (11¼°) giving a combined effect of 7 pts off the wind. In these conditions, discounting current and tidal stream, their speed made good to windward would be *c.* 2 kn. What effective speed to windward individual boats of early N.W. Europe could achieve would have depended on their rig, on the prevailing wind and sea conditions, and on the abilities of their crew. Sailing hard on a wind can exhaust the crew and overstrain fittings and rigging, thus would be done only when essential and for short periods.

Severin's experimental hide boat *Brendan* (1978: 290) sailed to within 6 pts of the wind with a leeway of 2 pts (22½°) and thus made nothing to windward in these conditions. Although this boat and her rig (of two square sails) cannot be taken as an authentic representation of a medieval hide boat, it seems not unlikely that this performance may be representative of large square-rigged hide boats.

AVERAGE SPEEDS ON PASSAGE UNDER SAIL

Whether a boat under sail will make a fast passage depends not only on her speed potential and on the propulsive power available but also on wind velocity relative to the required track. In general, boats sail fastest with a wind on the beam or just aft of it, and make best progress to windward when one point (11 1/4°) or so off the closest heading they can make relative to the wind: see, for example the polar diagrams in Doran (1978: 21), Sandison (1954: 51) and Marchaj (1964: 294–6). Polar diagrams for sailing replicas would help in the assessment of windward performance and thus in estimating average transit speeds, but none has yet been published.

When sailing with a fair wind high average speeds may be achieved. On the other hand, a boat whose angle to the wind for optimum windward performance is *c.* 6 points will have 50% further to travel when tacking to make an upwind destination than she would if she could sail a direct track with a fair wind. In such foul wind conditions overall average speeds will be lower.

Table 13.3 gives details of documented sea passages during the Classical and Medieval periods, and also includes the achievements of certain replicas of Viking Age vessels. Not counting the second leg of Ohthere's voyage to the White Sea (in which the length of day sailed is in doubt) there are thirty voyages said to be in fair winds. These have a range of average speeds from 3.42 to 7 kn, with a mean of 5.18 kn. There are twelve voyages said to be in foul winds (not counting the 1893 replica *Viking* which was unauthentic in rig) with a range of 1.54 to 3.33 kn and a mean of 2.22 kn. Of the remaining fifteen voyages in which it is unknown whether the wind was generally fair or foul, that of the reed raft is discounted as being of a different type of craft, and the remainder fall either within the range for fair weather voyages (seven with a mean of 5.13 kn) or

Table 13.3 Classical and Medieval sea passage details and those of certain replicas

*Vessel**	*From*	*To*	*Distance (nm)*	*Days*	*Average speed (kn)*	*Wind*	*Reference*
Indian reed raft†	R. Ganges	Ceylon	900	20	1.88	—	Pliny *NH* 6 XXIV: 82
Classical†	R. Ganges	Ceylon	900	7	5.36	Fair	Pliny *NH* 6 XXIV: 82
Classical†	—	—	130	1	5.41	Fair	Herodotus IV: 86.
Classical	Ostia	Africa	270	2	5.63	Fair	Casson, 1871: 283.
Classical	Messina	Alexandria	830	6	5.76	Fair	Casson, 1971: 283.
Classical	Ostia	Gibraltar	935	7	5.57	Fair	Casson, 1971: 283.
Classical	Ostia	Hispania–Citerior	510	4	5.31	Fair	Casson, 1971: 283.
Classical	Ostia	Provincia–Narbone -nsis	380	3	5.28	Fair	Casson, 1971: 283.
Classical	Messina	Alexandria	830	7	4.94	Fair	Casson, 1971: 283.
Classical	Puteoli	Alexandria	1000	9	4.63	Fair	Casson, 1971: 283.
Classical	Corinth	Puteoli	670	4.5	6.20	Fair	Casson, 1971: 284
Classical	Abdera	R. Danube	500	4	5.21	Fair	Casson, 1971: 284
Classical	Rhegium	Puteoli	175	1.5	4.86	Fair	Casson, 1971: 284
Classical	Carthage	Gibraltar	820	7	4.88	Fair	Casson, 1971: 284
Classical	Syrtes	Alexandria	700	6.5	4.49	Fair	Casson, 1971: 284
Classical	Alexandria	Ephesus	475	4.5	4.40	Fair	Casson, 1971: 284
Classical	Carthage	Syracuse	260	2.5	4.33	Fair	Casson, 1971: 284
Classical	Plycus	Alexandria	450	4.5	4.17	Fair	Casson, 1971: 284
Classical	Puteoli	Tauromenium	205	2.5	3.42	Fair	Casson, 1971: 284
Classical†	Ibiza	Gibraltar	400	3	5.56	Fair	Casson, 1971: 287
Classical†	Epidamnus	Rome	600	4.5	5.56	Fair	Casson, 1971: 287
Classical†	Crete	Egypt	310	3 or 4	4.31) 3.23)	Fair	Casson, 1971: 287
Classical†	Rhodes	Alexandria	325	3.5	3.87	Fair	Casson, 1971: 287
Classical†	Sea of Azov	Rhodes	880	9.5	3.86	Fair	Casson, 1971: 287
Classical	Cyrene	Crete	160	2	3.33	Foul	Casson, 1971: 289
Classical	Ascalon	Thessalonica	800	13	2.56	Foul	Casson, 1971: 289
Classical	Rhodes	Gaza	410	7	2.44	Foul	Casson, 1971: 289
Classical	Alexandria	Marseilles	1500	30	2.08	Foul	Casson, 1971: 189
Classical	Puteoli	Ostia	120	2.5	2.00	Foul	Casson, 1971: 289
Classical	Gaza	Byzantium	855	20	1.78	Foul	Casson, 1971: 289
Classical	Rhodes	Byzantium	445	10	1.85	Foul	Casson, 1971: 289
Classical	Caesarea	Rhodes	400	10	1.67	Foul	Casson, 1971: 289
Classical	Alexandria	Cyprus	250	6.5	1.60	Foul	Casson, 1971: 289
Classical	Sidon	Chelidonia	350	9.5	1.54	Foul	Casson, 1971: 289
Medieval	Hedeby	Truso	410	7	2.46	Foul	Ellmers, 1972: 250
Medieval‡	Kaupang	Hedeby	400	5	3.33	Foul	Ellmers, 1972: 250
Medieval	Venice	Byzantium	1310	24	2.29	?	Ellmers, 1972: 250 – (in Mediterranean)
Medieval	Schonen	Birka	350	5	2.92	?	Ellmers, 1972: 250
Medieval	Wolling	Novgorod	850	14	2.50	?	Ellmers, 1972: 250
Medieval	Birka	Novgorod	550	5	4.58	?	Ellmers, 1972: 250
Medieval	Alborg	Viken	130	1	5.42	?	Ellmers, 1972: 250
Medieval	Viken	Trondheim	820	5	6.83	?	Ellmers, 1972: 250
Medieval	Helgoland	England	270	3	3.75	?	Ellmers, 1972: 250
Medieval	Ribe	England	330	3	4.58	?	Ellmers, 1972: 250
Medieval	Ribe	Bruges	325	2	6.79	?	Ellmers, 1972: 250
Medieval §	Hedeby	Birka	580	20	2.42 (1.81)	?	Ellmers, 1972: 250
Medieval §	Hålgoland	Kaupang	950	30+	*c.* 2.67 (2.00)	?	Ellmers, 1972: 250
	Denmark	Novgorod	1100	30	3.08 (2.31)	?	Ellmers, 1972: 250
Medieval §	Alborg	Iceland	950	30	2.67 (2.00)	?	Ellmers, 1972: 250
Medieval §	Hålgoland	*c* N. Cape	260	6	3.58 (2.69)	?	Lund, 1984: 18–19

Table 13.3 (con't)

*Vessel**	*From*	*To*	*Distance (nm)*	*Days*	*Average speed (kn)*	*Wind*	*Reference*
Medieval¶	N. Cape	White Sea	330	4	6.88 (5.16)	Fair	Lund, 1984: 18–19
13th century							
Saga†	C. Stad	Iceland	525	3.5	6.25	Fair	Binns, 1980: 86
Saga†	Iceland	Ireland	690	5	5.75	Fair	Binns, 1980: 86
Saga†	Kvalsund	Lindesnes	150	1	6.25	Fair	Binns, 1980: 86
Saga†	Bergen	Trondheim	286	2	5.96	Fair	Binns, 1980: 86
Saga†	Baffinland	Herjolfsnes	590	4	6.15	Fair	Binns, 1980: 86
1894 Replica[11] *Viking*	Bergen	St. Johns	2122	26.5	3.34	Foul	Morcken, 1970b: 347
1979 Replica *Odin's Raven*	On passage	Norway to Shetland	?	?	5	Fair	Binns, 1980: 185
Various replicas	—	—	—	—	6–8	Fair	Crumlin-Pedersen, 1984: 33

Notes: *. All are single square-sail planked vessels except the Indian reed raft and replica *Viking* (two sails). All voyages are considered to be continuous sailing day and night except those marked (§) and (¶).

†. These are not records of actual voyages but measures of distance using a standard 'day's sail'.

‡. Crumlin-Pedersen, 1984: 33 considers this was a day-only voyage. However, in the text Ohthere's voyage from Kaupang to Hedeby is described in different terms from that of his voyage Hålgoland to Kaupang which was said to be day only.

§. These are voyages on which there is thought to have been no night sailing. Average speeds are for 12 h sailing and (in parentheses) 16 h sailing.

¶. The first 2 legs of Ohthere's exploratory voyage into the White Sea. Binns (1980: 37) has pointed out that there is a favourable current of 2 kn on the 2nd leg.

[11] Replica *Viking* had an unauthentic rig, including a foresail.

that for foul wind voyages (seven with a mean of 2.65 kn).

This group of voyages is not ideal for analysis in that the early ones are known because of chance survival of documentation and thus may be unrepresentative, and the later ones were undertaken in replicas some of which may not have been fully authentic and in most cases the trials were not undertaken under controlled conditions. Nevertheless the estimates marked (†) in Table 13.3 were undertaken at a speed which contemporary people considered to be standard in generally fair weather conditions, and thus may be considered representative. A tentative conclusion therefore, from this admittedly non-ideal data, is that, in general, ancient vessels from the Classical Mediterranean and Early Medieval N.W. Europe on open sea voyages in fair weather conditions could average *c.* 5 kn. In unfavourable conditions, of foul winds or when hampered by passages through archipelagoes, overall speeds could average 2 to 2½ kn. The latter figure is supported by Crumlin-Pedersen's assertion (1984a: 32) that replicas of Viking Age vessels and related modern boats with a single square sail can make 1.5 to 2 kn to windward.

CURRENTS, TIDES AND TIDAL STREAMS

The predominant winds in the regions are between S.W. and N.W. throughout the year and this generates a slight surface current which in the English Channel for example, averages 6 nm/day in a N.E. direction. In the Bay of Biscay, on the other hand, the inshore current tends to be parallel to the coast in a S.E. direction from Ushant.

Superimposed on these horizontal movements of the sea caused by predominant winds are the regular ebbs and flows of the tidal stream generated by astronomical forces and modified locally by topography and by weather. In mid-Channel the set of this stream is N.E./S.W. at speeds of 1–3 kn, depending primarily on the phases of the moon; the streams are weaker in the Bay away from the entrance to the Channel. During each tidal cycle the mid-Channel stream flows N.E. for *c.* 6 1/4 h, then S.W. for *c.* 6 1/4 h (Fig. 13.2) resulting in a theoretical net horizontal movement of zero in a 12½ h period. Cross-Channel boats, however, will generally experience some net displacement when they traverse areas with differing streams.

The northern part of the Bay of Biscay and the

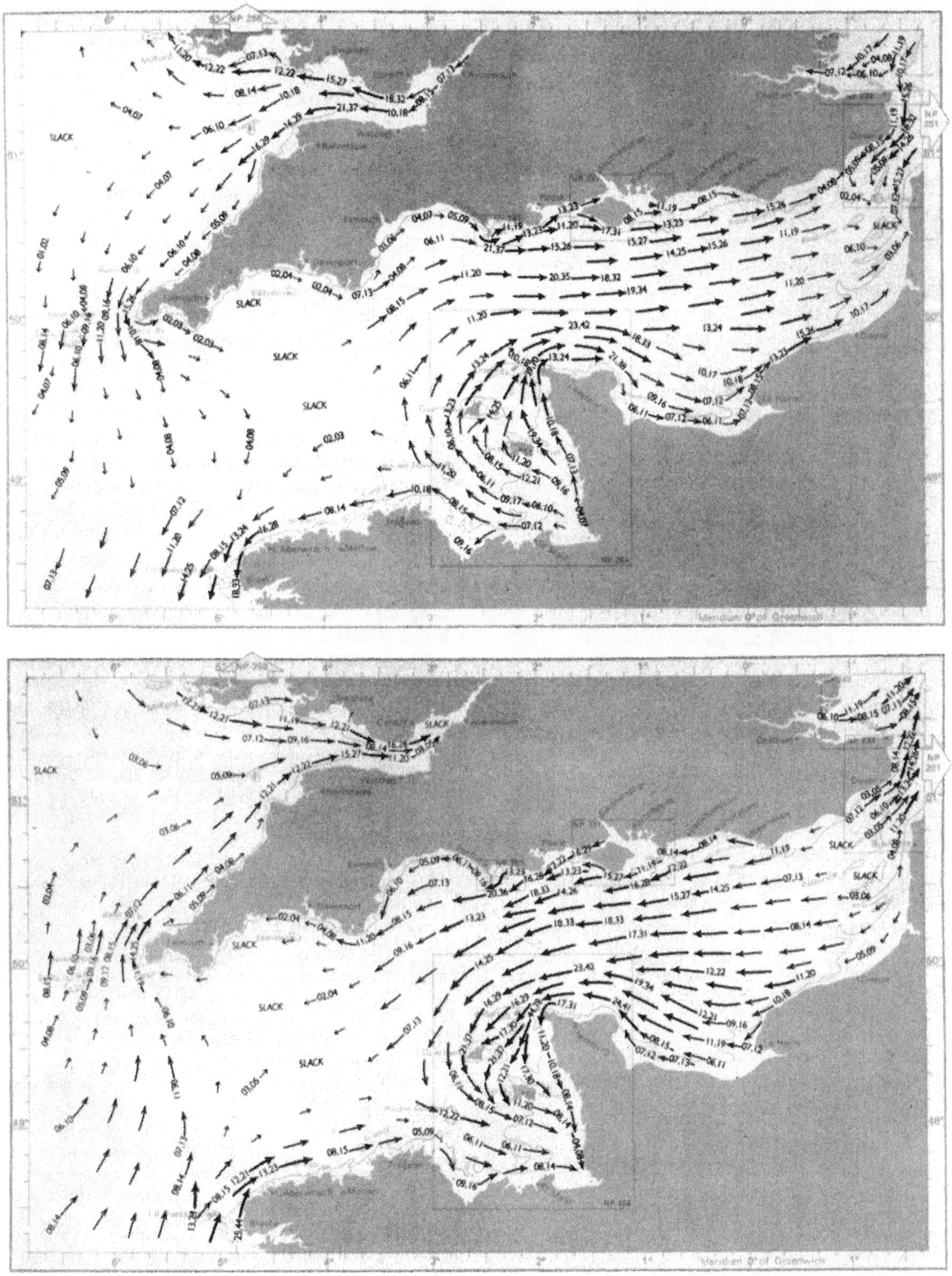

Figure 13.2 The charts show tidal streams in the Channel and adjacent waters at *c.* 3 h before (upper) and 3 h after (lower) datum – in this case the time of high water at Dover. The arrows indicate direction of flow and are graded for strength of stream, so that a thick arrow means a strong stream. (Produced from portions of *Tidal Stream Atlas* (1973) with the sanction of the Controller, H. M. Stationery Office and of the Hydrographer of the Navy.)

English Channel, towards the edge of the Continental shelf, are to leeward of the predominant SW wind, in the path of Atlantic depressions, and downstream of the North Atlantic current. The Bay and the Channel absorb much of this energy and, in particular, the wedge shaped Channel narrowing from the west restricts the movement of a large body of water, and thus funnels and intensifies these natural elements, and strong tidal streams of complex pattern are generated. Thus passages within this region can present a wide variety of problems to the seaman, both inshore and in more open waters. See, for example, Caesar's account of his cross-Channel passages in 55 and 54 BC (*B.G.* 3 to 5).

In the open water between Ushant and Scilly the set of the tidal stream traces out an ellipse, so that for about half the cycle it is generally N. and S. rather than E. and W. There is a similar circulation inshore in the Bay, with a S.E./N.W. alignment. The stream-pattern is modified generally in inshore waters where sets are at an angle to the mid-Channel alignments, and the times of flow-reversal (slack water) are altered. The rates of tidal stream also vary inshore, and speeds up to 10 kn can be experienced in narrow channels such as that to the east of Alderney.

There are also difficult passages due to funnelling effects around the Isle of Anglesey, off N. Wales, in the N. Channel between Ireland and Scotland, and in the Channel Islands generally and in the Orkney archipelago. The Pentland Firth between Caithness and Orkney can be especially dangerous. The sailing directions say: 'The extreme violence of the race, especially with W. or N.W. gales can hardly be exaggerated . . . Only personal experience can make anyone realise what a seething cauldron of tide-rips, whirlpools, and short choppy seas the firth can become . . .' (*N. Sea Pilot* 2). Strong tidal flows may also be found around promontaries and headlands such as Portland Bill, in inlets such as the Bristol Channel and Solway Firth, where the tide resonates, and off the estuaries of the N.W. flowing rivers between Rhine and Elbe.

In the Baltic there is little, if any, astronomically-generated tide, but when winds prevail from one quarter for several consecutive days, and where there is a reasonable fetch, sea levels can alter by up to 1 m.

WORKING THE TIDES

Favourable tidal streams with more than a certain depth of water are always desirable and often essential when leaving or approaching landing-places (Greenhill, 1978: 199–205; Eglinton, 1982: 1–29): thus the times of flow-reversal and of high and low water (not necessarily the same) need to be known with some precision. Local knowledge is also required to estimate the variations that persistent strong winds impose on the tidal regime, and to identify contra-flows in estuaries so that progress may be made against an unfavourable stream.

The sequence of high and low water at landing-places and generally in inshore water is linked to the tidal stream-pattern and to the age and bearing of the moon, the chief arbiter of tides (Waters, 1978: 31); in the pre-literate era this complex relationship had to be learned by observation and by rote for each landing-place. The tidal range (the vertical distance from high water to low water) at individual landing-places and the height of tides at times between high and low water had also to be learned and used to determine the depth of water in the approach channel which, together with the slope of the beach, influences the decision to beach or to moor a boat, and also determines how and when she will be loaded or unloaded.

Around Britain the tidal range can vary from *c.* 14 m at Springs, 6 m at Neaps in the River Severn Estuary/Bristol Channel to *c.* 2 m at Springs, 0.6 m at Neaps off the Dorset/Hampshire coast. In this latter region between Selsey Bill and St. Alban's Head, there is an unusual tide pattern with double high water. At Poole, for example, there is very little range of tide at neaps and only *c.* 2 m at springs, and the tide stands around high water for over 3½ hours; thus there is effectively HW for 14 hours out of 24, a distinct advantage in boat operations and cargo handling. The tidal range is least off Swanage where it is *c.* 1.7 m. Off N.W. France the tidal range is 6 to 11 m with a maximum of *c.* 12 m in the Baie du M. St Michelle. On the E. coast of Ireland the maximum range is *c.* 3.6 m with 4.5 m on the W. At Reykjavik the range can be up to 5 m, whilst at Narvik on the W. coast of Norway the maximum range is *c.* 3.5 m and up to 4 m off the N. coast of USSR.

On coastal passages especially when rounding headlands or traversing races it is necessary to work the tides when the wind does not allow sufficient headway to be made against them. During foul streams the vessel may be anchored or may lie hove-to to await fair streams (Eglinton, 1982: 12–13; Greenhill, 1978: 200). It may only be possible to attempt such difficult inshore passages in daylight unless there is a moonlight night (McGrail, 1983: 323).

LANDING PLACES

Away from technically-advanced and economically-complex societies, twentieth-century boatmen use informal landing sites (boat places). Examples may be quoted from Britain, the Indian Ocean, Poland, Germany, California and Ecuador (McGrail, 1983: 34, 36). For seagoing vessels, beaches in natural harbours or estuaries, protected from the predominant wind and swell are preferred but landing places on open beaches approachable only through surf are also known. Boats of suitable form and structure are operated direct from such sites; after beaching they may be manhandled further inland on portable rollers (Laures, 1983; Finch, 1976: 85) or carried on oars. In tidal waters boats may be allowed to take the ground on an ebb (falling) tide; or they may be held off shore at anchor or fastened to mooring posts or stones (Ellmers, 1972).

Beach operations are not limited to small boats: the coasting ketch *Charlotte*, operating from the beaches of S.W. England in the early years of this century (Fig. 13.3), had a deadweight (approximate cargo capacity) of 80 tons; and Eglinton (1982) has described his early life in vessels of a similar size

Figure 13.3 The coasting ketch *Charlotte* discharging cargo on the beach at St. Ives, Cornwall c. 1908. (Photo: Gillis Collection.)

operating from mud landing places on the N. Somerset coast.

By analogy we would expect to find similar informality in ancient Europe – outside the influences of Classical Civilisation – until sometime in the Medieval period when a relatively high state of technology was achieved, with economic and commercial complexity. The archaeological and documentary evidence available to date suggests that this was indeed the case, notwithstanding the argument advanced by Needham and Longley (1980 and 1981) that the ninth/eighth-century BC settlement site near the River Thames at Runnymede had a timber wharf and jetty; the main wooden structure seems best interpreted as a protective revetment and the outlying stakes and withies as a fish weir or possibly a hard (McGrail, 1985c). The three Ferriby boats of the mid-second millennium bc (Wright, 1976) and the Brigg 'raft' of the mid-first millennium (McGrail, 1981b) were all operated from the foreshore of the Humber estuary. And the other boats of N.W. Europe known to have been used in the prehistoric period – logboats and hide boats – could readily have been operated from such boat places or in the case of hide boats from protected coastal sites.

The Old English place-name elements 'hythe' (hȳ ð) and 'staith' (staeð) were originally applied to such informal sites (see for example Field, 1980) and thus we know that early medieval landing places existed for example at Chelsea, Erith, Lambeth, Putney and Stepney on the R. Thames; Greenhithe and Riverhead in Kent; Stonehaven in Kincardine; Lakenheath in Suffolk; and Staithes in N. Yorkshire, Stathe in Somerset, and Stafford on the River Sow.

Beach landing places are shown on the eleventh-century Bayeux tapestry with one boat being beached, another coming to anchor in the shallows and others fastened to mooring posts. The ninth/tenth-century Graveney boat (Fenwick, 1978) was excavated from such a landing place on the muddy intertidal margin of a tributary of the River Thames (Fig. 13.4).

Crumlin-Pedersen (1983: table 3) has noted that several medieval wrecks in the Baltic were found near known sites of 'beach markets'. The wrecks from Skuldelev (Olsen and Crumlin-Pedersen, 1967: 108–9, 117–18, 130, 144–5) and Utrecht boat No. 1 (Ellmers, 1972: 138) among others, have wear on their keels or bottom planking consistent with them having taken the ground on sand or gravel beaches. The wear pattern on transverse holes low on the stems of Nydam 2, Graveney and Kentmere 1 indicate that these boats were probably hauled up on the beach by rope. Illustrations from the *Life of St. Alban* dated to *c.* ad 1240 and now in Trinity College,

Figure 13.4 A reconstruction model of the ninth/tenth-century Graveney boat on a hard of light timbers on the margins of a tributary of the R. Thames. (Photo: NMM Greenwich.)

Dublin (Fenwick, 1978: fig. 8.25) show beach operations, including the off-loading of horses. Lighters at Southampton are known to have used beach hards (Platt, 1973: 76, 83) in that century, and fifteen vessels on the strand there were burned by the men of Winchelsea on 30 September 1321 (Platt, 1973: 107). Even as late as the fifteenth century royal ships of some size spent the winter at Greenwich drawn up on *les woses* – mudflats (Rose, 1982: 31); and ships in reserve were treated similarly at Winchelsea and the R. Hamble (Rose, 1982: 54, 86, 91).

The informal landing places described above have little, if any, fixed structures and thus it may be difficult to recognise them archaeologically, except on the few occasions when boats are found beached. There may, however, be indirect evidence and if a waterside site is found to have two or more of the following characteristics of the appropriate period there must be a strong presumption that a regularly-used, early landing place has been located:

1. Slipway or causeway (McGrail, 1983: 36–7).
2. A cleared area on an otherwise stony foreshore, possibly marked by parallel rows of stones at right angles to the shoreline (Rolfson, 1974: 136).
3. A hard (McGrail, 1983: 36) as at Ferriby (McGrail, 1985c), Graveney (Fenwick, 1978: 18), Norwich (Ayres, 1981: 278–9) and by the River Thames at New Fresh Wharf (Marsden, 1979).
4. The remains of isolated mooring posts or stones.
5. Anchors (often difficult to date accurately) or other items of boat's equipment.
6. Remains of beach capstans. See Fig. 13.5 for modern examples.
7. Artifact clusters near the shoreline, the position of which is otherwise inexplicable.

LOCATION OF LANDING PLACES

Inland

Where a relatively fast river joins a slower one deposition frequently creates a gradually shelving beach where boats may easily be put ashore. When such a landing place gave access to a settled hinterland it could well become an inland trading place. This would especially be the case with a site near the head of a tidal estuary (which might also be near the lowest fordable point) where decreasing depths of water could force a change to boats of less draught (or a shift to land transport) and thus the site would become a natural break point for river traffic.

Landing places may also be found where tracks lead to unfordable waterways and ferries have to be used. Medieval sites of this type are known, and prehistoric examples are at N. Ferriby and Brigg in Humberside.

Coastal

Sheltered landing places can generally be found in estuaries, river mouths and bays (see the section below on international maritime trade), but landing places on open coasts, unprotected from wind and swell by offshore island or by the arms of a bay, may be used by fishermen and traders when sites with greater shelter are not available; or by raiders when secrecy is paramount.

Landing places are also needed in uninhabited locations or near temporary and seasonal settlements. Thus, fishermen may operate from a forward base much nearer the fishing grounds than where they live, as recently practised in Shetland (Sandison, 1954). The requirements of defence, revenue collection and pilotage may lead to the establishment of landing places in sheltered positions near a river mouth or the entrance to a bay, fjord or natural harbour. Those who tended coastal beacons or lighthouses also needed landing places which would usually be on sites unattractive for settlement. Landing places may also be found in sheltered places on either side of headlands, especially where there is access to drinking water: coastal vessels would wait here for a fair wind and/or tidal stream so that they could round the point.

International maritime trade

A principal outcome of the trade within the pre-Westernised Indonesian archipelago (McKnight, 1973) was the creation of emporia ('ports of trade' – Polanyi, 1963; 'gateway communities' – Hirth, 1978). Documentary and archaeological evidence indicates a similar situation in early medieval Europe (Hodges, 1982) and, although ceremonial gifts may have been relatively more common in the prehistoric period, maritime trade, regulated to a degree by political authority, undoubtedly took place between the early analogues of emporia on the northern and western seaboard of pre-Roman Europe.

Theoretical analysis suggests that there are three main viewpoints influencing the choice of landing places for such trade: the seaman, the trader and the political authority (two of these roles may be combined) each with different ideas as to what constitutes the ideal landing place (McGrail, 1983: 311–13).

The *seaman's* ideal landing place is sheltered from the predominant wind and swell; it must have a readily identifiable landfall nearby and the approaches

A

B

Figure 13.5 A A beach capstan on the southern shores of the Baltic. (Photo: Wolfgang Rudolph.)
B A beach windlass on the shores of Luonnonmaa, Finland. (Photo: courtesy the National Museum of Finland.)

to it must be defined by landmarks and possibly seamarks; it should be usable almost regardless of wind direction or tidal state; and there should be access to drinking water and material for repairs. These sites are to be found within natural harbours such as estuaries, especially where they have wide, uncluttered approaches with a short fetch (open water) and an open sector away from the predominant wind.

The *trader*, on the other hand, requires a landing place immediately adjacent to the emporium for the regional economy, which may lie well inland. The third party in this international maritime trade, *political authority*, requires a landing place to be near the coast, i.e. as near as possible to the common 'frontier', in a readily defined area where traders and seamen can be segregated, protected and supervised, and tolls imposed (Hodges, 1982: 53–5). Landing places on coastal promontories, peninsulas or small islands are thus suitable from the political viewpoint.

These three somewhat different sets of requirements may best be met by:

(a) the use of landing places on an island, on a well-defined promontory, or at a river confluence within a natural harbour fed by a river system giving access to the regional emporium (if one exists) and to other settlements. For example: a site on Hengistbury Head within Christchurch Harbour, Dorset; sites on the islands of Birka and Helgö in Lake Mälar, Sweden, on Dalkey Is. in Dublin Bay, or at Domburg in the River Scheldt; or on a promontory site between the River Itchen and River Test in Southampton Water, or at Dorstad at the confluence of the River Lek and River Rhine

or

(b) landing places within archipelagoes or in the lee of skerries, with access by water to other coastal sites or to a hinterland. For example, Bergen in W. Norway.

or

(c) landing places at protected sites near the base of an isthmus where river travel and land portage obviate a long or difficult sea passage. For example a site at Hedeby in the Baltic (Hodges, 1982: 78; Sawyer, 1984: 49; Crumlin-Pedersen, 1984b: 241–2) where a portage leads to the River Traene and thus to Hollingstedt on the N. Sea coast; or at Lubeck with a possible portage to the River Elbe near Hamburg. A site on St Michael's

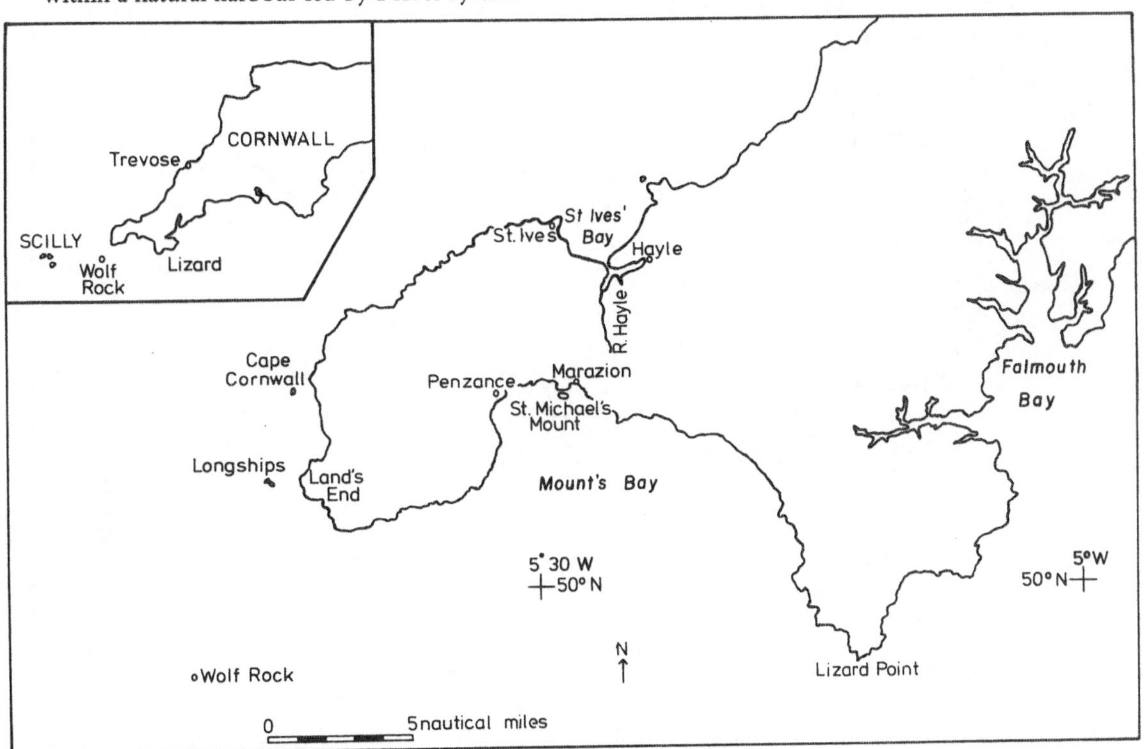

Figure 13.6 The S.W. of Cornwall showing the route of an overland portage from St. Michael's Mount on the S. coast to St. Ives Bay on the N. (Drawing: NMM Greenwich.)

Mount off S.W. Cornwall could be connected by portage to the N. coast of Cornwall near Hayle.

Peacock (1978: 49) has suggested that, in the Roman world, overland portages cost at least 22.6 times as much per mile as transport by sea. If part of the overland journey could be done by river then costs would be reduced to *c.* 5 times that of sea transport, but costs due to multiple handling would be increased. When Peacock's figures are used to compare a coastal voyage from a landfall at the Lizard around Land's End to a position 5 nm north of Hayle (Fig. 13.6), with a journey which includes an overland portage from St Michael's Mount, the latter costs over 2½ times as much as the sea voyage. However, the passage around Land's End can be difficult and may indeed be impossible for several consecutive days until there is a favourable wind shift. Thus portage may not only be more reliable but could even be cheaper if the costs of delays due to foul winds are counted.

The documentary and archaeological evidence for international trade across the English Channel (La Manche) in the late-first millennium BC may be examined in the light of the foregoing analysis (McGrail, 1983). Five main cross-Channel routes (Fig. 13.7) may be suggested, and the location of the settlement sites (emporia) in their terminal areas may be tentatively identified (Table 13.4) but the location of the landing places associated with these international trading settlements is much more

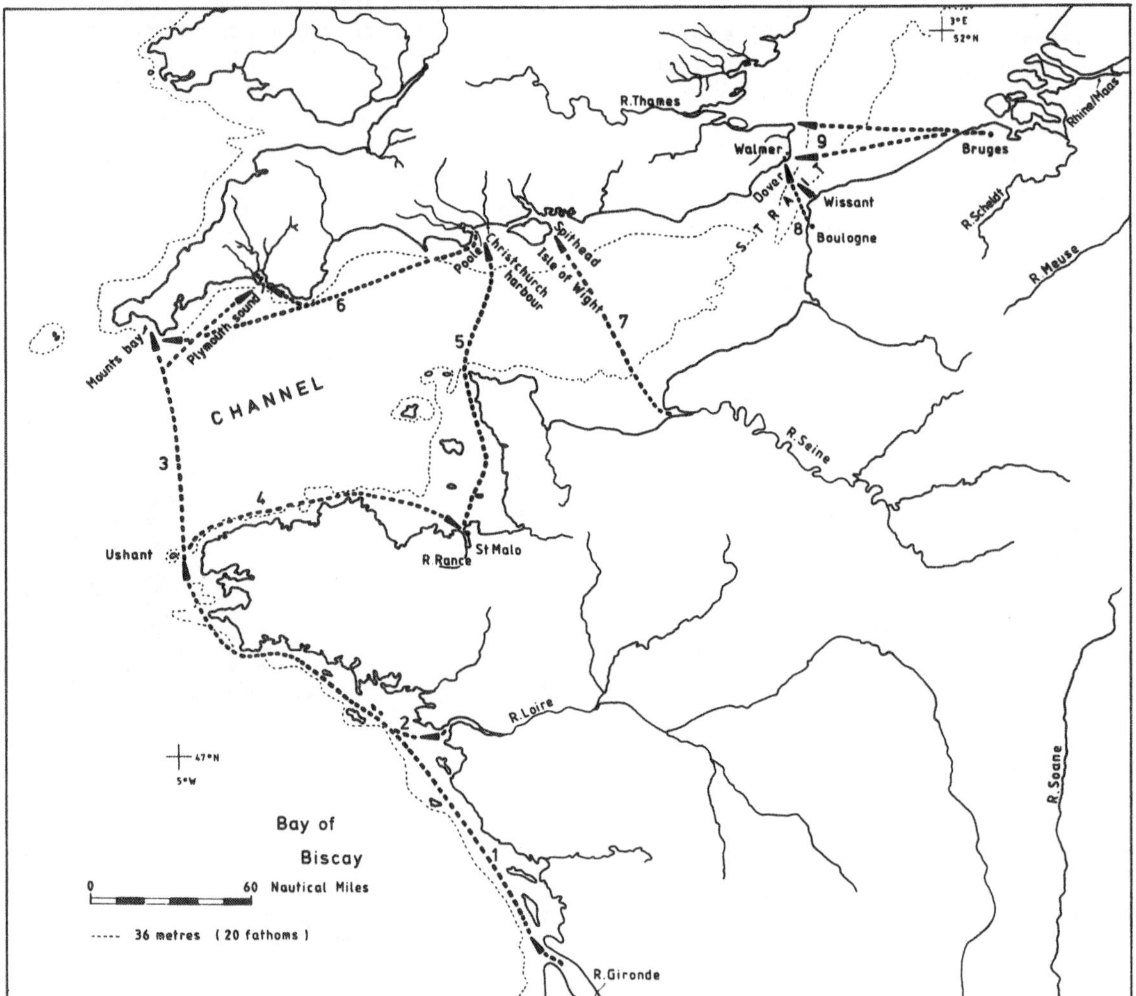

Figure 13.7 Sea routes for trade between S. Europe and Britain during the late first millenium BC. The cross-Channel routes are numbered 3, 5, 7, 8 and 9. (Drawing: NMM Greenwich.)

Table 13.4 Possible cross-Channel routes in the 1st millennium BC*

Route	*Continental*		*British*	
	Terminal area	*Emporia*	*Terminal area*	*Emporia*
Rhine/Thames	Rhine estuary or thereabouts	Domburg? Colijnsplaat? Bruges?	Thames, Stour (Kent) Blackwater and Colne estuaries	Thanet? Canterbury? Colchester?
Strait of Dover	W. Pas de Calais	Wissant?† Boulogne?†	Thames and Stour (Kent) estuaries	Thanet? Canterbury?
Seine/Spithead	Seine estuary	Rouen?	Solent and adjacent waters	Isle of Wight?
Mid-Channel	Rance estuary	Alet/St. Malo?	Christchurch Harbour and Poole Harbour	Hengistbury Head Green Is.? Hamworthy?
West Channel	Loire estuary and Gironde estuary.	Nantes?Bordeaux?	Plymouth Sound and Mounts Bay‡	Mount Batten Mt. St. Michael?

Notes: *. See McGrail, 1983a and 1985c for details
†. Limited access inland by water
‡. Possible portage to Hayle in St Ives Bay, N. Cornwall

Figure 13.8 A conjectural model of a fifteenth-century harbour such as Southampton. (Photo: NMM Greenwich.)

difficult to determine. Nevertheless, with the exception of the W. Pas de Calais region (where Wissant and Boulogne may have been more important militarily than commercially), each terminal area listed in Table 13.4 has stretches of coastline in the vicinity of the emporia with landing places which could have satisfied the seaman, the trader and the political authority. Early trading use of the major rivers in continental N.W. Europe is reasonably well-documented (McGrail, 1983: 324) but to date there is little direct evidence that the middle and upper reaches of the British and Irish rivers were similarly used in this period.

HARBOURS

The critical 'take-off' point when formal harbours with waterfront structures were built in N.W. Europe was reached earlier in some regions than others (McGrail, 1985c). On present evidence the earliest waterfronts were built in the ninth century AD and most economic regions seem to have had at least one such site by the twelfth century AD (Hodges, 1982: 97–8). Deep water berths alongside wharfs and jetties with warehouses and cranes facilitated cargo handling for larger vessels, whilst moles and breakwaters increased their protection from wind and swell in the otherwise open sector, and lighthouses, beacons and artificial leading marks made pilotage simpler (Fig. 13.8).

The 'hegge' dug out of the mud near the River Hamble to make a 'dook' for the *Grace Dieu* and for other early fifteenth-century ships (Rose, 1982: 54–5, 92, 122) may have been the protective wall around a dock, or artificial inlet, into which the ship could be floated from her mud berth on a spring tide. A permanent stone-built dock some 34.50 × 11 × 2.50 m dated to *c.* 1430 has been excavated by Marsden (1973) at Blackfriars near the River Thames.

14 Pilotage and navigation

Navigation, as described here, is that aspect of seamanship dealing with the determination of a vessel's position and the art and science of taking her in safety from one place to another. This term includes *pilotage* which may be defined as the art and science of conducting a vessel within waters restricted by the proximity of land and which are generally shallow. Navigation is now regarded as a science but has not always been so; until relatively recently it was also an *art* – in the sense of the use of personal skills – achievement depended on handed-on experience, local knowledge and detailed observation of natural phenomena (Waters, 1978: 3–38). Much early nineteenth-century navigation in N.W. Europe was of this nature (Binns, 1980: 79–80; Olsen, 1885; Greenhill, 1978: 199–205; Kemp, 1976: 123–4). Polynesian methods of navigation used in the early twentieth century were similar (Lewis, 1972) and may be described as environmental rather than instrumental.

In the late twentieth century we have exceedingly high expectations of accuracy and timekeeping and our expectations of risk, delay and indeed loss are correspondingly low; such was not the case in antiquity. Binns (1980: 83) has pointed out that, if the initial course is well chosen on a voyage from S. Norway to Britain (a frequent occurrence in the ninth to eleventh centuries) an unnoticed error of 90° to the southward would not, in the long run, affect matters much; the voyage would be prolonged but the vessel would still fetch up in British waters. As *Njals saga* put it: 'Many lands there are . . . which we might strike with the weather we have had – the Orkneys or Scotland or Ireland' (quoted by Marcus, 1980: 109).

Modern man is, in addition, generally only a superficial observer of such natural phenomena as the weather sequences, the tides, the movements of the sun, moon and stars, and the habits of birds and fish. By contrast in earlier times survival could depend upon perceptive interpretation of these observations and such a body of orally transmitted knowledge was an important aspect of a culture.

Needham (1971: 555) has identified three stages in the development of navigational techniques:

1. non-instrumental
2. quantitive
3. mathematical

The third stage, mathematical navigation, made little impact until after the medieval period and is thus outside the scope of this book. The change from the first to the second stage proceeded at different rates in northern and in southern Europe, much being accomplished in the Mediterranean in the thirteenth and fourteenth centuries that was not achieved in north-western waters until the fourteenth and fifteenth and even sixteenth centuries. Thus until almost the last century of the period under discussion, seamen of N.W. Europe were, in the main, using non-instrumental techniques. The artifactual evidence is thus very meagre: we have far more evidence for early boatbuilders than we have for early navigators and pilots. However, by identifying the nature of navigational problems, many of which are not specific to region, culture or time, and integrating the meagre documentary evidence from the proto-historical period, it is possible to make reasoned suggestions about early navigational and pilotage techniques by studying how recent pre-industrialised maritime cultures have tackled analogous problems.

It is clear from archaeological evidence that early N.W. European man did indeed use the sea. Artifact distributions show that islands were colonised at early dates: Man, the Orkney archipelago and Gotland from Neolithic times; Jura was reached by 6000 bc at the latest and Orkney by 4000 bc (Dennell,

1983: 125). Ireland appears to have been separated from Britain by the N. Channel for the past 19,000 years (except for periods of winter freezing) and was colonised by sea before *c.* 7000 BC. Britain became separated from the Continent by rising sea levels about 9500 years ago (Devoy, 1982; p.c.) and contacts between the Continent and Britain which were established from at least the early Neolithic had to be by sea. Most, if not all, such voyages could have been undertaken at short crossing points where prominent headlands on both sides were visible from a point in mid-channel. But, if early man had been entirely restricted to the shortest sea crossings and then only on days of good visibility and fair winds (in other words *pilotage*), he would not only have been severely constrained in his crossings of the eastern part of the Channel between France and Britain, but would also not have been able to cross this Channel at all west of the E. Brittany/Dorset route. There is, however, evidence that the Celts could undertake such voyages which took them out of sight of land in the late prehistoric period (Table 13.4) and it seems probable that these *navigational* techniques had been acquired at a much earlier date.

NON-INSTRUMENTAL NAVIGATION

EARLY SCIENTIFIC KNOWLEDGE AND AIDS TO NAVIGATION

The earliest archaeological and documentary evidence comes mainly from the Classical world. Herodotus (2.5.2) was familiar with sounding leads which had a cavity to sample the seabed, and several have been recovered from Classical wreck sites (Casson, 1971: 246) (Fig. 14.1): those in the range 4 to 5 kg suitable for depths down to 25 fathoms (*c.* 50 m) (*Manual of Seamanship* 1 (1915) 121–3) and those of *c.* 13 kg for depths down to 50 fathoms (*c.* 100 m). Similar finds may be expected in due course from N.W. European waters.

Classical authors were familiar with land-based methods of determining time and a form of latitude (Strabo, 1.4.4, 2.1.18, 2.5.8; Pliny *NH* 7 – LX.213, 215) but these methods were scarcely usable at sea because of ship movement. However, stars were used at night for course steering and orientation (Homer, *Odyssey*, 5.271–8; Pliny, *NH* 7 LVI – 209; Strabo 2.5.2) and even to obtain an approximation of change of latitude or movement in a N/S direction by measurement against mast and yard (Lucan, *Bel.Civ*, 8.177–81). Distances were estimated in terms of a standard day's sail (Strabo, 3.2.11, 15.1.14). The influence of the moon's phases on the tides was

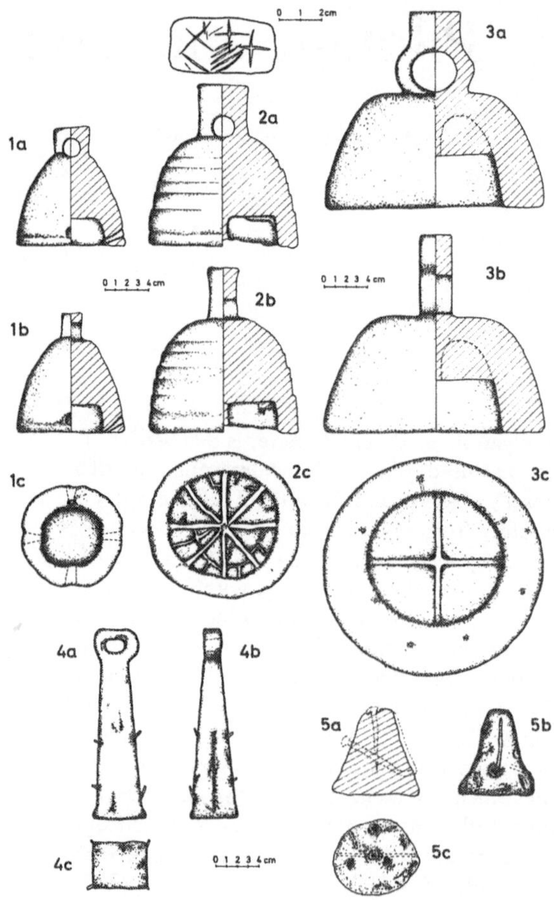

Figure 14.1 Sounding leads from the French Mediterranean coast and dated to the first half of the second century BC (No. 4 and 5) and to the middle of the first century AD (No. 1 and 2). After Fiori and Joncheray, 1973.

known (Strabo, *Geog.* 1.3.11; 3.5.8–9), as were the sea conditions associated with certain weather sequences and the directions from which predominant winds blew were used as reference bearings to produce a 'wind-rose' (Taylor, 1971: 14–20).

In N.W. Europe the alignment of certain megalithic structures in Britain, France and Scandinavia suggests a knowledge of the sun's movements and possibly those of the moon and stars (Heggie, 1981: 222–3; Ellmers, 1981). Woods (1978: 184–8) considers direct application to navigation or tidal predication unlikely, nevertheless such knowledge would have been useful at sea and the maritime members of these Neolithic and Bronze Age societies certainly had to cope with related problems though this may well have been done empirically rather than by theoretically-based methods. The Celtic world had a good working knowledge of astronomy (Caesar, *BG*, 6.14) with particular emphasis on the moon (Piggott, 1974: 104–5): such learning would have had direct application to direction finding at sea and to tidal prediction.

The seaman of N.W. European waters in any age had much to learn as the N. Sea and the waters around Britain and Ireland present problems differing in kind or in degree from other waters (e.g. Baltic or the Mediterranean) – see McGrail (1983: 304–7). Early man would have been unable to undertake successfully the sea voyage he undoubtedly did without the ability to observe and forecast the weather, observe and forecast tidal conditions, make simple astronomical observations, recognise coastal landscapes and retain in his mind a good estimate of his constantly changing position relative to point of departure and destination. Like many illiterate peoples, the Celts (and probably their predecessors) paid great attention to memory training for transmission of learning and culture (Chadwick, 1970: 45–7; Ross, 1970: 125–6) and the oral teaching of the Gauls, for example, included the motion of the stars (Caesar, *BG* 6.14). Stereotyped phrases and rules of thumb about the motions of the heavenly bodies, the ebb and flow of the tides, the weather sequences and the 'directions and distances' between landing places could have been memorised and used empirically. Recent equivalents are known (McGrail, 1983: 319).

The seaman also had to know his boat: her best point of sail for making ground to windward, how close to the wind she would sail when rounding a headland, speeds achieved and leeway experienced in different sea states and loadings: for these characteristics helped him to estimate his position.

In the exposition of possible early navigational techniques which follows, modern units of measurement such as the *knot* and modern concepts such as *leeway* are used (McGrail, 1983: 313). This is not intended to imply that these were used or explicitly understood by the ancient seaman. It seems probable, for example, that *current* (surface movement of water due to predominant wind), *tidal stream* (horizontal ebbs and flows of the sea in a *c.* 6¼ h cycle) and *leeway* (lateral movement of boat downwind) were not thought of separately but rather as one entity which varied in a complex way with the phases of the moon and with the prevailing wind and which could have an effect readily apparent and potentially dangerous in inshore waters but which could also displace a boat on passage. The seaman noted (or deduced) the effects of wind and sea on his boat and took action to allow for them without necessarily having an analytical understanding of their causes.

OUTBOUND PILOTAGE

When preparing for a cross-Channel or coastal voyage the seaman would intuitively compare actual wind, sea and tide, and the conditions to be expected in the next day or so, with those generally prevailing on the intended route at that time of year. With this information he could estimate the course to be steered and the time required for the crossing as deviations from the course generally used and the time usually taken. He would thus form a mental picture – a 'chart' – of his forthcoming passage, which he would revise during the voyage in the light of further observations of natural phenomena (see Slade and Greenhill, 1974: 27 on the planning of early twentieth-century voyages, very little of which was committed to paper). He would probably leave on an ebb (outflowing) tide with a fair wind at a time which would enable him to make a landfall in the vicinity of his destination or at a prominent point on route in daylight, preferably at slack water before a flood tide. With a number of only roughly quantified variables in these calculations such a crossing would be achievable only in settled summer weather but would nevertheless be an ideal to aim at. After leaving the shore channels appropriate to the wind direction would be used, steering by reference to landmarks and visible shoals and reefs. Oars could be used to clear a windbound estuary or natural harbour when offshore winds were known to be favourable.

Once clear of land the pilot would check wind, sea and swell against a known directional relationship such as a transit between two headlands, the bearing of an island, the general trend of the coast, or in relation to a back bearing (in Oceania fires were lit ashore for this – Lewis 1972: 61), and work his way to windward in foul wind conditions. Before losing sight of land he would 'take departure' by estimating his position relative to some prominent landmark – this would be the first datum point on his 'mental chart'. In good visibility high ground can remain in sight for a considerable distance to seaward and Table 14.1 lists typical distances from sea level. In theory another 2.57 nm should be added to these distances to allow for the height of the observer, but as it may be difficult to distinguish a headland (without a beacon or lighthouse) from the horizon the tabled figures are

Table 14.1 Theoretical visibility distances from sea-level

Height of land	Distance (nm)
30 m (100 ft)	11.5
61 m (200 ft)	16.2
152 m (500 ft)	25.7
305 m (1000 ft)	36.3

Source: Inman's Nautical Tables 1920: 12
Notes: 1. An approximation is given by: D(n.m) = 1.1 $\sqrt{H}$ (ft). 2. D(km) = 3.8 $\sqrt{H.}$ (m). 3. In meteorological conditions of refraction high ground may be seen at more than the theoretical distance. In poor visibility the distance is much less.

probably the best estimate of visibility distance in good weather.

With a fair wind the pilot would then set a course which would allow for his estimate of the set of the tidal stream and current and any leeway. With a foul wind the boat would be brought to the heading on which the boat was known to make the most ground to windward, and an estimate made of the displacement caused by current, stream and leeway. With due regard to the likelihood of a shift in such a wind, the number and lengths of tacks required to make the landfall could then be assessed. It would be essential to plan to make the landfall in daylight or bright moonlight. Other things being equal it would also be prudent to aim to make a landfall slightly to windward or uptide of the destination, thus having a less arduous leg for the final approach.

INBOUND PILOTAGE

Early warning of land, the 'expanded' target of David Lewis (1972), may be given by orographic cloud, the flight of seabirds at dawn and dusk, seaweed and jellyfish in the water (Binns, 1980: 79), and other adventious aids (Marcus, 1980, 114–5). The lead and line may be used to obtain warning of shallows and hence the nearness of land and to note the nature of the bottom so that safe channels may be recognised. Reflected wave patterns or the sound of surf breaking in the shallows or even the smell of sheep may indicate the heading on which land lies.
Shore-sighting birds – as used by Noah – are said to have been used in Viking times (*Landnamabok*. Jones, 1964: 118) as they were in the waters off Ceylon in Classical times (Pliny *NH* 6.24.83). Colour changes of the water also give an indication of position as for example when approaching the mouth of a large river. A good landfall is made when the pilot recognises the silhouette and other characteristics of the coast as it becomes visible at a range when the boat is still clear of inshore hazards (Table 14.1). In poor visibility fires may be lit ashore to identify the destination and perhaps act as leading marks. Thomas (1981) has suggested that such a beacon may have been sited on Gt. Ganilly Island, Scilly in the late first-millennium BC: see also Eglinton (1982: 1–2) for a recent example.

When the landfall has been identified and the position of the boat relative to the destination established a safe course is set for the landing place. Oars may be used against an ebb tide or to catch a flood, or the boat may be held off to await favourable conditions. Leading marks especially transits (two objects in line) are of great use when approaching a harbour entrance or an individual landing place. Prehistoric barrows are known to have been used as sailing marks in recent times in Swedish waters (Westerdahl, 1980: 324). Similar use in earlier times seems probable, especially in the light of Beowulf's dying command to his chieftains 'to build upon a headland near the sea a splendid tumulus . . . so that seamen who steer their great ships far over the misty sea may call it Beowulf's Barrow' (38.2794ff – Wright, 1957: 93). Certain medieval churches may have been sited for similar reasons, as suggested by Binns (1980) for the Humber region and by Crumlin-Pedersen (1983) for the southern Baltic – Wismar, Rostock and Stralsund.

Whether the site of any particular barrow visible from sea-level was chosen so that it would be a useful landmark from seaward may now be impossible to determine, but should be borne in mind when surveying a coastal area. A large coastal barrow or group of barrows would certainly assist the pilot to recognise that part of the coast, and a transit bearing, with the edges of two barrows in line for example, might be available as a leading mark to indicate a safe line of approach.

LANDFALL ON AN UNFAMILIAR COAST

Providing the general track followed since leaving the last firm position is known a seaman in sight of unfamiliar land is not lost although he may not know his position until he has obtained more information e.g. investigating the colour and composition of the cliffs, the nature of the beach material etc. After a lengthy storm drift to an unknown shore the seaman can orientate himself and estimate a course to destination or back to point of departure by noting the bearing of the celestial pole (North) or the noon sun (South). By landing or anchoring in sheltered

Figure 14.2 Using a *kamal* to measure the altitude of a heavenly body. The tablet is held at the distance form the observer where it apparently just fills the gap between horizon and star. Knots on the string correspond to the angular altitude of the star. (Photo: T. Severin.)

water and waiting for clear skies he can make more accurate observations and obtain some measure of the distance he had travelled north or south of his departure point by estimating the altitude (angular elevation) of the celestial pole, or of the noon sun if direct vision can be avoided – see e.g. Strabo (2.1.18) possibly against the ship's mast or by using a simple device such as the Arab *Kamal* (Fig. 14.2), a tablet and knotted string (Taylor, 1971: 128–9, fig. XI; Severin, 1982: 92–3) or finger breadths or palms/fist or finger spans at arms length (Taylor, 1971: 49), or a stick calibrated in the sun's apparent radius (Marcus, 1953: 120), or by the use of a *gnomen* or sundial to measure the ratio of index mark to shadow (known to the Classical world – see Strabo 1.4.4 and 2.5.8). In all these cases the seaman would compare his observation with similar known data for destination or departure point. Southward deviation in the northern hemisphere is indicated by decreased pole altitude and increased sun's meridian altitude. Shorter hours of daylight in summer also indicate southward displacement but this would require the use of a water clock (*clepsydra*) known to have been used on land in the Classical world (Pliny *NH* 7-LX 215; Ceasar, *BG*, 5.13). Change of latitude (N/S movement) is also indicated by changes in the azimuth (bearing) of sunrise and sunset which draw nearer to due east or west (i.e. at right angles to the N/S line) with southerly movement. As with the other indicators, such changes would be made evident as a difference from the data known for a standard site, probably the departure point or 'home port'.

New lands were indeed encountered by unplanned storm drifts – see for example the Vikings' 'discovery' of Iceland (Foote and Wilson, 1970: 52), Greenland and America (Taylor, 1971: 74, 79–80). In meteorological conditions of refraction theoretical visibility distances from sea-level (Table 14.1) can be significantly increased and previously unknown lands seen. The flight path of migrating birds can indicate the direction of new lands and this may have been the way eighth-century Irishmen were led to Iceland and other northern islands as recorded by Dicuil (Taylor, 1971: 76). Orographic cloud may be seen over land well out of sight and an abrupt change in a swell pattern may indicate that the boat has entered the 'shadow' of a distant, previously unknown, island which is distorting the otherwise predominant swell.

COASTAL PASSAGES

On coastal passages the seaman would frequently establish his position relative to prominent natural landmarks and seamarks. Should visual contact be lost through poor weather, or when heading across the arms of a bay, or because of having to haul out to seaward to avoid some hazard, he would use his 'mental chart' to record estimated position until he could close the coast and regain his visual bearings. On such a passage it would be necessary to work the tides, possibly including anchoring or lying hove-to during foul streams (Eglinton, 1982: 12–13; Greenhill, 1978: 200) and it could prove necessary to wait for a fair wind before being able to round headlands (Marcus, 1980: 103). Alfred's account of Ohthere's ninth-century voyages along the Norwegian coast contains references to such events (Lund, 1984). It would also probably be necessary to anchor during non-moonlight nights – but see Marcus (1980: 116) on thirteenth-century pilots who could find their way up long Norwegian fjords by starlight.

NAVIGATION

Methods other than pilotage have to be used when out of sight of land. The navigator has then to rely on his assessments of tracks and speeds made good relative to the sea bed or more probably relative to the last known position on his 'mental chart', i.e. his point of departure, or relative to his destination. Courses steered and speeds achieved relative to the water give a DR or dead reckoning position. When an angular allowance is made for leeway and drift an EP or estimated position is obtained which, if the allowances are reasonable, gives the track made good (Fig. 13.1) relative to point of departure.

The early navigators' problem when out of sight of land may be resolved into two parts:

1 how to estimate directions such as courses, leeway and drift; and
2. how to estimate speeds or distances.

Direction from the celestial pole
The moon and the stars appear to move, in relation to the Earth, but the celestial pole or null point (near *Polaris* nowadays) maintains a fixed direction and is readily identifiable without instruments in a clear night sky. This null point was known from at least early Classical times (Homer, *Odyssey* V. 271–6) although the star then nearest the pole was *Kochab* rather than *Polaris* (Taylor, 1971: 9–12, 43): its use as a reference heading for direction in prehistoric N.W. Europe seems likely. The Anglo-Saxon ninth-century AD vocabulary included two phrases for the pole star, *scip steorra* and *lad steorra* which indicate its use at sea, and there are other references in the German and Scandinavian languages which testify to its navigational importance (Ellmers, 1981) and its use is mentioned in a number of Icelandic sagas (Marcus, 1980: 108).

Ellmers (1981) has postulated that in early times the pole star was the only directional aid available to seamen, and was the *sine qua non* in N.W. European waters for navigation out of sight of land, as exemplified by the 'night leap' (*saltu per noctem*) methods used on certain Baltic sea crossings described by the eleventh-century missionary Adam of Bremen (Tschan, 1959). The corollary of this hypothesis is that seamen could only sail by day when in sight of land, that two or more consecutive nights could not be spent at sea and thus radius of action was limited to the distances that opposite coastlines were visible from sea level plus the distance that could be sailed in one night of eight or fewer hours. Ellmers's proposals would also limit navigation out of sight of land to nights when there was good visibility, little cloud cover and a reliable steady wind.

In his arguments Ellmers does not stress the requirements to leave estuary or coast on an ebb tide and to make a landfall in daylight. These constraints would make it difficult, if not impossible, to ensure that time out of sight of land coincided with the hours of darkness which it would have to do if the early seaman had been limited to *Polaris* as the only directional aid. There are indeed well-documented early uses of 'night-leaps' in addition to those mentioned by Adam of Bremen, for example Julius Caesar used night crossings on his voyages from France to Britain in 55 and 54 BC (*BG* IV 21–36; V, 2–23; Strabo 4.3.4; 4.5.2; see also McGrail, 1983: 328–9). But with the limitations postulated by Ellmers – coast visible by day and *Polaris* by night – channel crossings west of the E. Brittany/Normandy to Hampshire/Dorset route could not be attempted as there are distances of up to 80 nm out of sight of land (McGrail, 1983: table 3) which could only be traversed in one night in the April/September sailing season by a boat making good 7 to 13 kn, which would be unusual.

The first documented out-of-sight-of-land voyages

in N.W. European waters for two consecutive nights are those described by the early ninth-century Irish monk Dicuil (Taylor, 1971: 69–70) from Ireland to some of the northern islands in the late eighth century. But before that date the ability to navigate out of sight of land during the day, and at night with cloud cover must have been essential, especially when unusual winds were experienced or when the times of ebb and flow were incompatible with Ellmers's preferred methods. The methods described below can be used in such circumstances.

Direction relative to wind or swell

Wind and sea maintain a fairly constant direction for only a relatively short time in N.W. European waters, unlike such areas as the Indian Ocean where there are seasonal trade winds. However, in northern waters winds from different directions may be recognised by the weather accompanying them; thus a warm wet wind comes from the S.W. and a cold wet wind from the N.W. Direction can be established in relation to the wind blowing at any one time and the Vikings are known to have used such a 'wind rose' (Taylor, 1971: 8).

Swell (the sea motion originated by a strong predominant wind and which persists after the wind has died down) on the other hand gives a more certain indication of direction for longer periods, although less pronounced in summer than winter. In the Bay of Biscay the swell comes from between W. and N.W., and in the Channel generally S.W. to W., although a moderate N.E. swell may sometimes be encountered in the Strait and a N.E. swell is also likely to be encountered in the North Sea.

Navigators in Oceania placed great reliance on swell direction to give them their bearings. Lewis (1972: 90) found, for example, that one navigator was able to steer by swell direction through several wind changes and made a good landfall after covering over 45 nm in overcast conditions. The predominant swell has to be differentiated from wave motion, from any lesser swells, and from wave diffraction-patterns from islands or promontories. When not visually recognisable it can be identified by feel, the navigator lying on deck to analyse the boat's motion, the testicles being the finest sensor (Lewis, 1972: 86–92).

Walton (1974: 10), March (1970: 39) and Binns (1980: 20–1) have noted that in recent times Shetland fishermen were also able to detect and use the underlying S.W. swell ('Mother Wave') in a confused sea. Such ability was probably more widespread in earlier times, but atrophied with the advent of instrumental aids.

Directions from celestial observations

Navigators in Oceania were able to establish their bearings and steer accurate courses by reference to single stars or constellations (Lewis, 1972: 52–4). The obliquity of celestial motion in higher latitudes rules out such precise use in N.W. European waters, except for the Pole Star (as discussed above). In daytime the fixed direction (south) of the sun's zenith (noon) position may be determined with reasonable accuracy when visible, as there is a period of very little change in azimuth and altitude at that time; north may similarly be identified at midnight in high latitudes. At dawn and dusk east and west may be approximately identified, providing the seasonal directions of sunrise and sunset (amplitude) have been established before the voyage relative to a known bearing ashore. On E/W voyages the sun's amplitude would change by less than 1°/day, whilst on N/S passages the change could be *c.* 5°/day. Thus on a five-day voyage the maximum change in the bearing of sunrise and sunset from the known bearing before leaving land would be 25° (*c.* 2 pts). With more difficulty, the navigator may be able to maintain his direction at other times of the day by reference to the changing altitude and azimuth of the sun (a feature of bird navigation): the Oceanic navigator did this by 'delicate judgement' as distinct from scientific measurement.

Leeway and drift allowance

To determine the track actually sailed over the sea-bed and thus to know the bearing at any one time of the point of departure or of the destination, each course steered has to be modified by the angular amount of leeway experienced due to the wind, and by the angular displacement due to current and tidal stream (Fig. 13.1). Leeway is negligible when the wind is from a sector some 4 points either side of dead astern; otherwise it may be estimated by wake angle. Drift allowance depends on the navigator's evaluation of the changing state of the tidal stream since taking departure; an alternative method is to assume that in general these displacements cancel out during each period of 12½ hours, leaving a slight N.E. drift due to current. This latter method is not valid when the boat passes through radically different tidal systems.

Distance estimation

In order to obtain a 'mental fix' of his position, the navigator has to establish how far he has travelled along his track (bearing from departure point) which he has estimated by applying leeway and drift to the course steered. Distance travelled is determined by speed and time; both are difficult to estimate without aids, and it seems unlikely that any of the known Classical devices for time measurement were used at sea. However, at any one season of the year the ratio of daylight to darkness could be deduced from measurements on land, and thus sunrise, noon and sunset could become 'fixed' points in time; and some measure of the passage of time during daylight could be obtained fom the sun's change in azimuth providing it could be noted relative to a 'fixed' bearing such as swell direction. At night an estimate of elapsed time since sunset may be made by noting the position of the circumpolar constellations.

In several maritime societies distances have been expressed in units of a day's sail (Casson, 1971: 281–96; Morken, 1968; Gelsinger, 1970; Taylor, 1971: 50–2, 79, 126); implicit in this is the concept of a 'standard' speed to be expected from the 'usual' sort of boat in fair wind and sea conditions. Where landfalls can be seen from departure points the experience to establish such a standard may soon be acquired; on longer routes average performance could take longer to establish and might not be so accurate.

When documented Viking Age and medieval voyages are analysed the day's sail (*doegr sigling*) appear to vary between c. 30 and 150 nm. Morken (1968) believes the definition of 'day's sail' depended upon whether the boat was under oars or sail and whether the 'day' was one of 12 hours (daylight only) or 24 hours: he postulates the following definitions:

(a) 36 nm rowing at 3 kn for 12 hours
(b) 72 nm sailing at 6 kn for 12 hours
(c) 144 nm sailing at 6 kn for 24 hours

Morken found that most documented early medieval voyages could be fitted into one of these groups.

A similar change in definition was noted by Pliny (*NH* 6 XXIV–82): the distance between the River Ganges and Ceylon was formerly believed to be 20 days' sail – a standard set by reed craft; when the standard was set by Roman vessels the distance became 7 days sail. The 'days sail' was thereby increased from c. 45 to 120 nm.

Estimates of actual speeds achieved may be expressed in relation to this standard speed linked to the concept of a day's sail. Lewis (1972: 119) found that the Oceanic navigators used a multitude of indications including spray turbulence and wind pressure. Another method is to note (when it can be seen) the position of a second bow wave created by the boat; the further aft this is, the faster the boat is moving (Marchaj, 1964: 251, fig. 154). Yet again a Dutchman's log (bubbles or an object thrown overboard) may be timed between bow and stern (Greenhill, 1978: 195); or the time to make a passage between two known coastal marks may be noted: in both cases a standard phrase might be chanted to give a measure of elapsed time (see also Drillis, 1963: 439–440). Such usage is not recorded, but chants are known to have been used to keep Classical (Morrison and Williams, 1968: 196) and Saxon (Colgrove, 1927: 26) oarsmen sychronised. Another timing mechanism might be the pulse or heartbeat. By such means, an experienced seaman learnt what speeds his boat achieved on different courses in various combinations of wind, sea and swell, the speeds being estimated relative to the 'standard' speed.

Estimated position

Thus knowing his track (course steered relative to celestial pole, wind or swell, amended for leeway and drift) and the distance sailed along it (as a relative speed or as fractions of a 'day's sail'), the navigator can 'mark' his estimated position on his mental chart at noon, sunrise, sunset and possibly at each change of tack, as an estimate of his progress since the last EP or as an estimated bearing and distance from departure point and from destination. A traverse board (Fig. 14.3) known to have been available from the sixteenth century, can be similarly used (Waters, 1978: 37; Binns, 1980: 81). As his appreciation developed of how his progress was deviating from the 'norm' for that route the navigator would be able to decide whether to adjust his track (if the wind would allow this) or to adjust his ETA with consequent changes in the tidal streams and depths of water to be expected inshore, or even to change his destination to a more readily accessible one.

NAVIGATIONAL CAPABILITIES UP TO C. ELEVENTH CENTURY AD

Needham (1971: 555) and Taylor (1971: 85) have both quoted a fourth-century AD description of the capabilities of an Indian or Arabian pilot which could equally be applied to the seamen of prehistoric N.W. Europe:

> He knew the course of the stars and could readily orientate himself; he also had a deep knowledge of the value of signs, whether regular, accidental or abnormal, of good and bad weather. He distinguished the regions of the ocean by the fish, by the colour of the water, by the nature of the bottom, by the birds, the mountains and other indications.

In inshore waters the seaman had to be familiar with

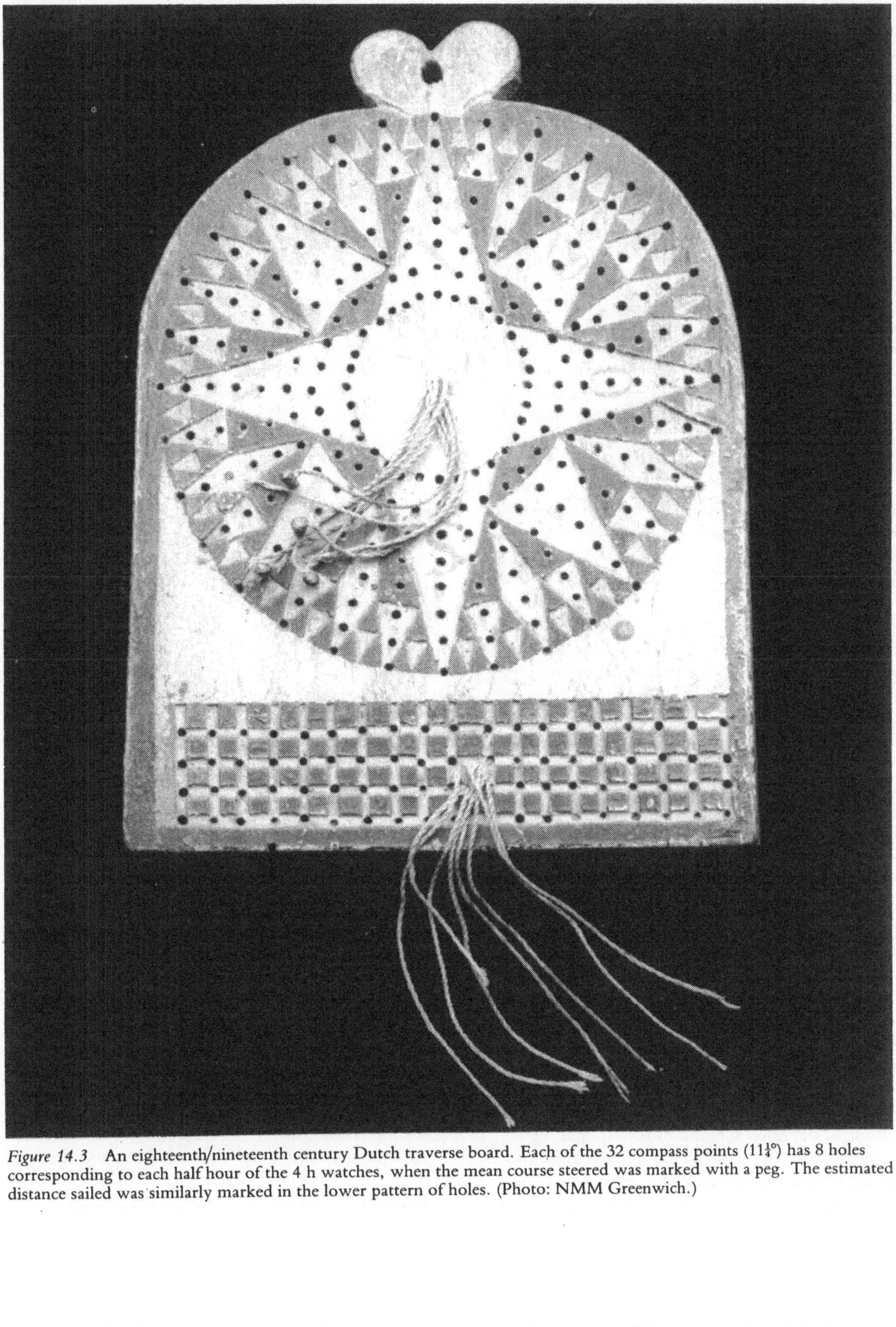

Figure 14.3 An eighteenth/nineteenth century Dutch traverse board. Each of the 32 compass points ($11\frac{1}{4}$°) has 8 holes corresponding to each half hour of the 4 h watches, when the mean course steered was marked with a peg. The estimated distance sailed was similarly marked in the lower pattern of holes. (Photo: NMM Greenwich.)

the details of the coastline, offshore soundings and the nature of the sea bottom, the safe channels to the main landing places, their sailing marks and hazards and the times of high and low water and slack streams in relation to the moon's phases. As a navigator out of sight of land he had to set the best course he could make relative to the wind and keep his 'reckoning' of his movements and adjust them where possible to allow for deviations from the norm so that he made a good landfall. His aim would be at any one moment to be approaching his destination on as constant a bearing as the elements would allow.

Coastal voyages throughout N.W. Europe and the channel crossings between France and Britain and Britain and Ireland were navigable by such means, as were crossings in the Baltic, and archaeological evidence confirms that such voyages were undertaken early in the prehistoric period. North Sea crossings were also theoretically possible but here the early archaeological evidence is not so clear.

Similar pilotage and non-instrumental methods of navigation continued in use into and throughout the first millennium AD. Viking Age voyages from Northern Norway to the White Sea and also to Hedeby in Denmark via Kaupang in Oslo Fjord were recorded by Alfred as told to him by the Norwegian merchant Ohthere. Alfred also recorded Wulfstan's voyage from Hedeby to Truso, near Danzig (Lund, 1984). Ohthere's voyages were coastal ones, proceeding from one known point to the next, recognising, for example, the recommended departure points on the Norwegian coast for Iceland, the Faroes and Shetland, and Britain. Wulfstan, on the other hand, was under sail for seven days and nights and out of sight of land for some of this time. Nevertheless he gave an account of his progress eastward by reference to points of land out of sight, and evidently kept his reckoning on a 'mental chart'.

Sailing directions for certain passages have been recorded in some of the Icelandic sagas which evidently encapsulated inherited wisdom (Marcus, 1980: 112–3, 117). The directions in *Landnámabók* (Foote and Wilson, 1970: 256) for sailing to Greenland have sometimes been taken to indicate that the Norwegian Vikings practised latitude sailing, for departure (Bergen) and destination (Hvar of Greenland) are both c. 60° N (Marcus, 1980: 112). Changes in latitude as represented by changes in the altitude of the noon sun and of *Polaris*, and by changes in the length of the day and the amplitude of the sun would undoubtedly have been noted, but latitude sailing requires accurate measurement and it is doubtful that the necessary instruments were available.

The Landnámabók directions, however, make it clear that three legs were sailed, the first one of which must have been well south of west passing some 40 nm N. of Shetland 'so that you can sight it in clear weather'; the second leg was to a position some 30 nm S. of Faroes 'so that the sea appears half-way up the mountain slopes'; and the third leg passed some 90 nm south of Iceland which would be out of sight but 'so that you may have birds and whales from it'. Thus the Viking seaman was doing as his predecessors had done, proceeding from one landfall or known position to the next. The fact that when Viking Age seamen lost their bearings and were driven towards an unknown shore – as when Gunnbjörn Úlfssom discovered Greenland (Marcus, 1980: 117) – they could not only return home but arrived there with sufficient information for others to repeat their voyage intentionally, testifies to their competence as navigators.

QUANTITIVE NAVIGATION

The quantitive period of navigation may be said to have started in N.W.Europe with the use of the mariner's compass. The first known reference to this was by Alexander Neckham, Abbot of Cirencester. He had lectured in Paris from 1180 to 1186 and thus had experience of sea crossings. In his books *de Utensilibus* and *de Naturis Rerum* he described how a magnetised needle floating in a bowl of water would indicate the four cardinal points. From his description it seems that this compass was used at sea only in foul weather and then only for checking wind direction – the mariner thereby had an improved wind rose but continued to conn his vessel by reference to the natural elements i.e. the wind rather than by reference to a compass course (Waters, 1978: 22). Marcus (1980: 117–8) believes the compass was not in general use in the N. Sea and N. Atlantic until the mid-thirteenth century, but as the instrument became more reliable and as precision was gained more use was made of it.

St. Bede had given a general account of the monthly tidal cycle in AD c. 725 and this was known to the compilers of the thirteenth century Norse books *Konungs Skuggsja* (King's Mirror) and the fourteenth century *Rimbegla* (Taylor, 1971: 81–2, 136; Marcus, 1980: 110). Tide tables giving high water times at London Bridge were compiled by the monks of St. Albans in the thirteenth century

(Waters, 1978: 31) and more useful compilations became available in the Mediterranean region in the fourteenth century (Taylor, 1971: 137). Seamen in N.W. Europe however continued for some time after this date to commit to memory the times of high water at various ports on the days of full and new moon ('full and change') in terms of the bearing of the moon (the establishment of the port). Times of intermediate high water were found by interpolation using a retardation of 45 min (instead of the more precise 48 min) which was conveniently equivalent to one point (Waters, 1978: 31–2).

The fifteenth century saw an influx of navigational aids: the sandglass, used at sea from the thirteenth century in the Mediterranean; traverse tables, which simplified estimation of distance made good to windward; and astrolabes and quadrants for taking the altitude of sun or stars, were introduced to N.W. European waters. Rutters or sailing directions (pilots) for various N.W. European coastal voyages were also printed for the first time early in this century, replacing memorised directions. Some of these earlier oral sailing directions had been incorporated into written work such as the fourth-century AD *Ora Maritima* of Avienus in which he included extracts from a sixth-century BC Periplus (Hawkes, 1977); a first-century AD Periplus included sailing directions for the Red Sea (Huntingford, 1980); and some of the thirteenth-century AD Icelandic sagas included directions for the N. Atlantic (Marcus, 1980: 112). Printed sailing directions were thus based on much older lore, handed-down and improved over centuries.

Rutters made no marked change to the methods of pilotage, however, vessels still sailed from one known landmark to the next one (as they continued to do until relatively recent times) and the course was frequently referred to as the 'caping of the ship' (Waters, 1978: 11). The sounding lead, the only prehistoric aid to pilotage, continued to be an essential aid in the medieval period and later. Indeed great emphasis was placed on its use in the rutters and pilots.

Portolans or charts with rhumb line courses marked on them, used in the Mediterranean from the thirteenth century and in Portugal from the fifteenth, did not begin to be used in N.W. Europe until the sixteenth century. A reliable series of charts for British coastal waters did not become available until the late seventeenth century.

Lighthouses were built at Dover, Boulogne and at La Corunna, N.W. Spain in Roman times, the one at Boulogne being reinstated by Charlemagne in AD 811 (Ellmers, 1981). These were useful as visual aids in daylight as well as at night. Informal fires may also be lit onshore as leading marks. Thomas (1981) has suggested that such a beacon may have been sited on Gt. Ganilly Island, Scilly in the late first millennium bc – see also Eglinton (1982: 1–2) for a recent example. The installation of beacons to identify a place or to mark channels and hazards begin to be mentioned from the twelfth century (Morken, 1970; Crumlin-Pedersen, 1983; Schnall, 1975), but it was not until the sixteenth century that formal bodies such as Trinity House, Deptford, were set up to be responsible for the marking of channels in important estuaries.

The thirteenth-century Norwegian seaman (Taylor, 1971: 81–4) depicted in the *Konungs Skuggsja* was familiar with finding his direction from the stars and could divide the horizon into a wind rose of 8 'points'. He knew the weather sequences to be expected, had a good knowledge of the movements of whales, fish, walrus and seals and sea-ice, and could calculate the tides in relation to the moon's phases. Chaucer's fourteenth-century shipman (Coghill, 1951: 35–6) could also reckon tides, tidal streams, phases of the moon, and his tracks and distances; he knew well the harbours and havens from Gotland to C. Finistere and the creeks of Brittany and Spain. The prehistoric seaman also used these skills as did Julius Caesar or his navigator in the late-first millennium BC. Essentially the same body of pilotage and navigational expertise continued in use in certain parts of N.W. Europe right into the twentieth century (Greenhill, 1978: 199–205).

References

ABBREVIATIONS USED

AA *Acta Archaeologica, Copenhagen.*
Ant. J. *Antiquaries Journal*
Ant. *Antiquity*
BAR *British Archaeological Reports*
CBA *Council for British Archaeology*
IJNA *International Journal of Nautical Archaeology*
MM *Mariner's Mirror*
Med. Arch. *Medieval Archaeology*
PPS *Proceedings of the Prehistoric Society*
PSAS *Proceedings of Society of Antiquaries of Scotland*
UJA *Ulster Journal of Archaeology*

Adney, E. T. and Chapelle, H. I. (1964) *Bark Canoes and Skin Boats of N. America*, Washington.

Åkerlund, H. (1932) *Skeppsfyndet vid Falsterbo.*

Åkerlund, H. (1951) *Fartygsfynden i den forna hammen i Kalmar*, Stockholm.

Åkerlund, H. (1956) Áss och beitiáss, *Unda Maris*, Gotenburg.

Åkerlund, H. (1963) *Nydamskeppen*, Göteborg.

Åkerlund, H. (1965) Nydam ship, *MM* **51**, 255–8.

Albion, R. G. (1952) Timber problems of the Royal Navy 1652–1862, *MM* **38**, 5–10.

Alexander, F. H. (1927) Propulsive efficiency of rowing, *Trans. Inst. Naval Architects* **69**, 228–44.

Allen, D. (1971) Ship on Gaulish coins, *Ant. J.* **51**, 96–9.

Ambrosiani, B. (1981) Changing water levels and settlement in the Mälar district since AD 700, *Florilegium Florinis Dedicatum* **14**, 140–3. Uppsala.

Ambrosiani, B. (1982) Mälarstäderna och landhöjningen, *Bebyggelsehistorisk Tidskrift* **3**, 71–82.

Andersen, E. (1975) Hals og skaut, mast og segl, *Norsk Sjøfartsmuseum Årsheretning*, 47–100.

Andersen, M. (1895) *Vikingfaerden*, Oslo.

Andersen, S. H. (1983) En stenalderbåd fra Tybrind vig, *Antikvariske studier* **6**, 162–72.

Andersen, S. H. (1994) New finds of Mesolithic logboats in Denmark, in Westerdahl, C. (ed.) *Crossroads in Ancient Shipbuilding* 1–10. Oxbow Monograph 40, Oxford.

Anderson, A. O. and M. O. (eds) (1961) *Adomnan's Life of St. Columba.*

Anderson, B. W. (1978) *Living World of the Old Testament* (3rd edn).

Anderson, R. and R. C. (1926) *Sailing Ship.* Bonanza, New York. (republished 1963).

Anderson, R. C. (1925) Italian naval architecture about 1445, *MM* **11**, 135–63.

Anderson, R. C. (1928) English galleys in 1295, *MM* **14**, 220–41.

Anderson, R. C. (1955) *17th-century rigging.*

Anderson, R. C. (1976) *Oared Fighting Ships.* Argus Books.

Arbman, H. (1940) 'Der Årby-fund', *Acta Archaeologica* **11**, 43–102.

Arima, E. Y. (1975) *Contextual Study of the Caribou Eskimo Kayak.* Canadian Ethnology Service Paper No. 25, Ottawa.

Armstrong, E. C. R. (1920) *Catalogue of Irish Gold Ornaments in the Collection of the Royal Irish Academy*, Dublin.

Arnold, B. (1975) Gallo-Roman boat from the Bay of Bevaix, Lake Neuchâtel, Switzerland, *IJNA* **4**, 123–41.

Arnold, B. (1976) La pirogue d'Auvernier Nord 1975 (Bronze Final), *Cahiers d'Archéologie Subaquatique* **5**, 75–84.

Arnold, B. (1977a) Some remarks on caulking in Celtic boat construction and its evolution in areas lying northwest of the Alpine arc, *IJNA* **6**, 293–97.

Arnold, B. (1977b) Le nâcon de Cudrefin, *Cahiers d'archéologie subaquatique* **5** (1976), 105–20.

Arnold, B. (1980) Bevaix NE1917: un monoxyle celte et ses courbes hydrostatiques, *Jahrbuch der Schweizerischen Gesellschaft Für ur-und Frühges-chichte* **63**, 186–98.

Arnold, B. (1982) Architectural woodwork of the late Bronze Age village Auvernier-nord, in McGrail, S. (ed.) (1982), 111–28.

Arnold, B. (1990) Some objections to the link between Roman boats and the Roman foot (*pes monetalis*), *IJNA* **19**, 273–277.

Arnold, B. (1992/3) *Batellerie Gallo-romaine sur le lac du Neuchâtel.* 2 vols. Archéologie Neuchâteloise, 12 and 13, Neuchâtel.

Arnold, B. (1995/6) *Pirogues Monoxyles d'Europe Centrale*, 2 vols, Archéologie Neuchâteloise, 20 and 21, Neuchâtel.

Arunachalam, B. (1996) Traditional sea and sky wisdom in Indian seamen and their practical applications, in Ray and Salles, (1996) 261–282.

Arwidsson, G. and Berg, G. (1983) *Mästermyr Find*, Stockholm.

Ashbee, P. (1966) Fussell's Lodge long barrow, *Archaeologia* **100**, 1–80.

Aubaile-Sallenave, F. (1987) *Bois et Bateaux du Vietnam*, SELAF, Ethnosciences 3, Paris.

Ayres, B. (1981) 'Norwich', *Current Archaeology* **80**, 278–80.

Baillie, M. G. L. (1978) Dating of some ships' timbers from Wood Quay, Dublin, in Fletcher, J. (ed.) *Dendrochronology in Europe*. NMM Archaeological Series 4. BAR (Oxford) S. **51**, 259–62.

Barnwell, E. L. (1875) Caergwrle Cup, *Archaeologia Cambrensis* 4th series **6**, 268–274.

Bartlett, J. and Hobbs, X. (1897) Egyptian burial boat in the Field Museum, Chicago. Typescript in Field Museum, Chicago.

Basch, L. (1972) Ancient wrecks and the archaeology of ships, *IJNA* **1**, 1–58.

Basch, L. (1973) Golo wreck and sidelights on other ancient ships culled from Admiral Paris's *Souvenirs de marine conserves*, *IJNA* **2**, 329–44.

Basch, L. (1981) Carthage and Rome: tenons and mortices, *MM* **67**, 245–50.

Bass, G. (ed.) (1972) *History of Seafaring*, London.

Bass, G. F. (1967) C. Gelidonya: a Bronze Age Shipwreck. *Trans. American Philosophical foc.* NS 57(pt 8).

Bass, G. F. & van Doorninck (1982) *Yassi Ada* **1**. Texas A & M.

Beaudouin, F. (1970) *Les bateaux de l'Adour*, Bayonne.

Bell, M. (1992) Goldcliff, (1992) in Bell, M. (ed.) *Severn Estuary Levels Research Committee Annual Report*, 15–30.

Bellabarba, S. (1993) Ancient methods of designing hulls. *MM* **79**, 274–292.

Bellabarba, S. (1996) Origins of the ancient methods of designing hulls: a hypothesis. *MM* **82**, 259–268.

Bellin, J. N. (1764) *Le Petit Atlas Maritime* 5th volume, Paris.

Best, E. (1925) *Maori Canoe*. Dominion Museum Bulletin 7, Wellington.

Beverly, R. (1722) *History of Virginia* Book 3, 2nd ed.

Bidault, J. (1945) *Pirogues et Pagaies*, Paris.

Binns, A. L. (1961) Ohthere's Northern Voyage, *English and Germanic Studies* **7**, 43–52.

Binns, A. L. (1980) *Viking Voyagers*. Heinemann.

Birdsell, J. H. (1977) Recalibration of a paradigm for the first peopling of Greater Australia, in Allen, J. *et al.* (ed.) *Sunda and Sahul*, 113–68. Academic Press.

Blue, L., Kentley, E., McGrail, S. and Mishra, U. (1997) Patia fishing boat of Orissa: a case study in Ethnoarchaeology. *South Asian Studies* **13**, 212–231.

Blue, L., Kentley, E. and McGrail, S. (forthcoming A.) Ethnograpy, Archaeology and India's maritime past. Proceedings of the 1996 Lyon conference.

Blue, L., Kentley, E. and McGrail, S. (forthcoming B.) Vattai fishing boat and other vessels of Tamil Nadu built by distinctive frame-first methods, *South Asian Studies*, **14**.

Boczar, M. L. (1966) Craft in use at the Rivergate of the Dunajec, *MM* **52**, 211–22.

Bonino, M. (1975) Picene ships of the 7th century BC engraved at Novilara (Pesaro Italy), *IJNA* **4**, 11–20.

Bonino, M. (forthcoming) Sewn boats in Italy, in McGrail and Kentley (eds) *Sewn Plank Boats*. BAR Oxford.

Boon, G. C. 1977. A Greco-Roman anchor stock from N. Wales. *Antiquaries. J.*, 57: 9–30.

Booth, B. (1984) Handlist of maritime radiocarbon dates, *IJNA* **13**, 189–204.

Bowen, R. Le Baron (1952) Primitive watercraft of Arabia. *American Neptune* **12**, 186–221.

Bowen, R. le Baron (1956a) Boats of the Indus civilisation, *MM* **42**, 279–90.

Bowen, R. le Baron (1956b) European sewn boats, *MM* **42**, 256–8.

Brassey, T. A. (ed.) (1905) *Naval Annual*.

Brindley, H. H. (1919/20) Notes on the boats of Siberia, *MM* **5**, 66–72, 101–7, 130–42, 184–7; **6**, 15–18, 187.

Brindley, H. H. (1924) Canoes of British Guiana, *MM* **10**, 124–32, 307–8.

Brindley, H. H. (1926) Shan States dugouts, *MM* **12**, 460.

Brindley, H. H. (1927) Shan States racing canoes, *MM* **13**, 91–2.

Brindley, H. H. (1931) Sailing balsa of L. Titicaca and other reed bundle craft, *MM* **17**, 7–19.

Brodrick, H. (1902) Martin Mere, *Trans. Southport Society of Natural Science* 8th Report, 13th session, 5–18.

Brøgger, A. W., Falk, H. J., Shetelig, H. (1917) *Osebergfundet* **1**. Kristiania.

Brøgger, A. W. and Shetelig, H. (1971) *Viking Ships*.

Bronsted, J. (1938) *Danmarks Oldtid*, vol. 2. Copenhagen.

Bruce-Mitford R. (1970) Ships' figureheads in the Migration period and Early Middle Ages, *Ant.* **44**, 146–48.

Bruce-Mitford, R. (1975) *Sutton Hoo Ship Burial*, vol. 1, British Museum.

Brusić, Z. (1968) Istraživanje Antičke Luke Kod Nina, *Diadora* 4, 203–9 and Plates I-X. Zadar. Jugoslavia.

Brusić, Z. (forthcoming) Liburnian boats – their form and construction in McGrail, S. and Kentley, E. (eds) *Sewn Plank Boats*. BAR Oxford.

Buck, P. H. (1927) *Material culture of the Cook Is.* (Aitutaki) Board of Maori Ethnol. Research, vol 1.

Burland, C. A. (1950) By weather to Wineland, *MM* **36**, 81–5.

Cairo, R. (1972) A note on S. Vietnamese basket boats, *MM* **58**, 135–53.

Callender, G. (1913) Ship in the Jerusalem Window, *MM* **3**, 97–103.

Carrazé, F. (1977) Mediterranean hull types compared 3. Jeanne-Garde B wreck at Porguerolles (France), *IJNA* **6**, 299–303.

Carr-Laughton, L. G. (1957) Clove-board, *MM* **43**, 247–9.

Casson, L. (1963) Sewn boats, *Classical Review* N. S. **13**, 257–9.

Casson, L. (1971) *Ships and Seamanship in the Ancient World.* Princeton Univ. Press.

Casson, L. (1980) Two-masted Greek ships, *IJNA* **9**, 68–9.

Castiglioni, O. C. (1967)Le piroghe preistoriche italiane. Problematica e inventario dei reperti, *Natura* **58**, 5–48.

Cederlund, C. O. (ed.) (1993) *Årby Boat.* Stockholm: Statens Historiska Museum

Chadwick, N. (1970) *The Celts.* Penguin.

Chittick, N. (1980) Sewn boats in the western Indian Ocean, *IJNA* **9**, 297–309.

Christensen, A. E. (1959) Faeringen fra Gokstad, *Viking,* **23** 57–69.

Christensen, A. E. (1968a) Sjøvollen Ship, *Viking* **32**, 131–53.

Christensen, A. E. (1968b) *Boats of the North.* Oslo.

Christensen, A. E. (1972a) Scandinavian ships from earliest times to the Vikings in Bass, G. F. (ed.) 1972, 160–80.

Christensen, A. E. (1972b) Boatbuilding tools and the process of learning in Hasslof, O. *et al.* (ed.) *Ships and Shipyards Sailors and Fishermen,* 235–59. Copenhagen.

Christensen, A. E. (1972c) Boatbuilding tools and the process of learning, in Hasslof, O. (ed.) *Ships and Shipyards, Sailors and Fishermen,* 235–259. Copenhagen.

Christensen, A. E. (1973) Lucien Basch: ancient wrecks and the archaeology of ships. A comment, *IJNA* **2**, 137–45.

Christensen, A. E. (1974) Klåstadskipet, *Tjolling Bygdebok* **1**, 542–6.

Christensen, A. E. (1975) The famous Viking longship in Almgren, B. *The Viking,* 247–82. Nordbook, Gothenburg.

Christensen, A. E. (1977a) Comment in McGrail, S. (ed.) *Sources and Techniques in Boat Archaeology,* 112. NMM Greenwich Archaeological Series 1. BAR (Oxford) S.29.

Christensen, A. E. (1977b) Ancient boatbuilding – a provisional classification in McGrail, S. (ed.) *Sources and Techniques in Boat Archaeology,* 269–280. NMM Greenwich Archaeological Series 1. BAR (Oxford) S.29.

Christensen, A. E. (1979) Viking Age rigging in McGrail, S. (1979) 183–94.

Christensen, A. E. (1982) Viking Age boatbuilding tools in McGrail, S. (ed.) (1982) 327–38.

Christensen, A. E. (1984) Sewn boats in Scandinavia in McGrail, S. (ed.) (1984a) 85–96.

Christensen, A. E. (1985) Boat finds from Bergen. *Bryggen Papers,* **1**, 47–278.

Christensen, A. E. (1996) *Earliest Ships.* Conway, London.

Christensen, A. E. and Leiro, G. (1976) *Klåstadskipet* Vestfoldminne.

Christensen, A. E. and Morrison, I. (1979) Experimental archaeology and boats, *IJNA* **5**, 275–84.

Christensen, C. (1990) Stone Age dugout boats in Denmark, in Robinson, D. E. (ed.) *Experimentation & Reconstruction in Environmental Archaeology*: 119–142. Oxbow Books, Oxford.

Clark, (J), G. (D). (1952) *Prehistoric Europe: the economic basis.* Methuen, London.

Clark, J. G. D. (1954) *Excavations at Star Carr.* CUP.

Clark, J. G. D. (1977) *World Prehistory* 3rd edn. Cambridge.

Clark, G. and Piggott, S. (1976) *Prehistoric Societies.* Penguin.

Clarke, R. *et al.* (1993) Recent work on the R. Hamble wreck near Bursledon, Haupshire. *IJNA* **22**, 21–44.

Close-Brooks, J. (1974–5) An Iron Age date for the Loch Lotus canoe, *PSAS* **106**, 199.

Coates, J. F. (1977) Hypothetical reconstructions and the Naval Architect, in McGrail, S. (ed.) *Sources and Techniques in Boat Archaeology,* 215–26. BAR. S29

Coates, J. F. (1981) Safe carrying capacity of the hypothetical reconstruction in McGrail, S. (ed.) (1981b) 261–9.

Coates, J. F. (1984) An index for flat-bottomed boats, *IJNA* **13**, 248–9.

Coates, J. F. and McGrail, S. (ed.) (1984) *Greek Trireme of the 5th Century* BC. NMM Greenwich.

Coates, J. *et al.* (1995) Experimental boat & ship archaeology: principles & methods, *IJNA* **24**, 293–301.

Cochrane, R. (1902) On Broighter, Limavady, Co. Derry and on the find of gold ornaments there in 1896, *J.Roy.Soc. Antiq. Ireland* **32**, 211–24.

Coghill, N. (1951) *Canterbury Tales.* Harmondsworth.

Coles, J. M. (ed.) (1975–84) *Somerset Levels Papers* 1–10.

Coles, J. (1979) *Experimental Archaeology.* Academic Press.

Coles, J. M. and Harding, A. F. (1979) *Bronze Age in Europe.*

Colgrove, B. (1927) *Life of Bishop Wilfrid.*

Collins, A. E. P. and Seaby, W. A. (1960) Structures and Small Finds discovered at Lough Eskragh, Co. Tyrone, *Ulster Journal of Archaeology* **23**, 25–37.

Cordier, G. (1963) Quelques mots sur les pirogues monoxyles de France, *Bull. de la Soc. Préhistorique Française* **60**, 306–15.

Corlett, E. C. B. (1969) Twin hull ships, *Trans. Roy. Inst. of Naval Architects* **111**, 401–38.

Corlett, E. C. B. (1977) Comment in McGrail, S. (ed.) *Sources and Techniques in Boat Archaeology,* 229.

Corlett, E. C. B. (1978) Appreciation of the lines in Fenwick, V. (ed.) (1978), 303–6.

Crumlin-Pedersen, O. (1960) Sideroret fra Vorså, *KUML,* 106–15.

Crumlin-Pedersen, O. (1966) Two Danish side rudders, *MM* **52**, 251–61.

Crumlin-Pedersen, O. (1967) Parallel to Dunajec Craft? *MM* **53**, 31.

Crumlin-Pedersen, O. (1968a) *Traeskibet,* Lyngby.

Crumlin-Pedersen, O. (1968b) Gredstedbro Ship, *AA* **39** 262–67.

Crumlin-Pedersen, O. (1969) *Das Haithabuschiff* Ausgrabungen in Haithabu, Bericht 3, Neumunster.

Crumlin-Pedersen, O. (1972a) Skin on wood in Hasslof, O. *et al.* (ed.) *Ships and Shipyards, Sailors and Fishermen.* Copenhagen, 208–34.

Crumlin-Pedersen, O. (1972b) Kaellingen og Kløften, *Handels og Søfartsmuseet på Kronberg.* Årbog, 64–80.

Crumlin-Pedersen, O. (1977a) *Traeskibto.* Council of Danish Timber Trade.

Crumlin-Pedersen, O. (1977b) *From Viking Ship to Victory.* NMM Greenwich.

Crumlin-Pedersen, O. (1978) Ships of the Vikings. *The Vikings*, 32–41, Uppsala.

Crumlin-Pedersen, O. (1979) Danish cog-finds in McGrail, S. (ed.) *Medieval Ships and Harbours in Northern Europe.* NMM Archaeological Series 5. BAR (Oxford) S.66, 17–34.

Crumlin-Pedersen, O. (1981) Skibe på Havbunden, *Handels og Søfartsmuseets Årbog*, 28–65.

Crumlin-Pedersen, O. (1983a) Schiffe und Seehandelsrouten im Ostseeraum 1050–1350, *Lübecker Schriften zur Archäologie und Kulturgeschichte* **7**, 229–37.

Crumlin-Pedersen, O. (1983b) *From Viking Ships to Hanseatic Cogs.* Third Paul Johnstone Memorial Lecture. NMM Greenwich.

Crumlin-Pedersen, O. (1984a) Ships, navigation and routes in the reports of Ohthere and Wulfstan in Lund, N. (ed.) 30–42.

Crumlin-Pedersen, O. (1984b) Der Seetransport: die schiffe von Haithabu, *Archäologische und naturwissenscheftliche Untersuchungen* **2**, 241–50, Weinheim.

Crumlin-Pedersen, O. (1984c) Experimental boat archaeology in Denmark in McGrail, S. (ed.) (1984a) 97–121.

Crumlin-Pedersen, O. (1991) Badgrave og gravbade pa Slusegard, *Slusegarde-gravpladsen* **3**, 93–263. (Jysk Arkaeologisk Selskabs Skrifter, 14.3).

Crumlin-Pedersen, O. (1996) Problems of reconstruction and the estimation of performance, in Christensen, A. E. (1996) 110–119.

Crumlin-Pedersen, O. (1997) *Viking Age Ships and Shipbuilding.* Viking Ship Museum, Roskilde and Wikinger Museum, Haithabu.

Crumlin-Pedersen, O. and Vinner, M. (eds) (1986) *Sailing into the Past.* Roskilde.

Cunliffe, B. W. (1972) Late Iron Age metalwork from Bulbury, Dorset. *Ant. J* **52**, 293–308.

Cunliffe, B. W. (1978A) *Iron Age Communities in Britain* 2nd edn. R.K.P.

Cunliffe, B. W. (1978b) *Hengistbury Head.*

Cunliffe, B. W. (1980) Evolution of Romney Marsh: a preliminary statement in Thompson, F. H. (ed) *Archaeology and Coastal Change*, 37–55. Soc. Antiquaries Occasional Paper (New Series) **1**.

Cunnington, E. 1884. On a hoard of bronze, iron and other objects found in Bulbury Camp. Dorset. *Archaeologic* 48: 115–20.

Curle, J. (1911) *Roman frontier post and its people: the fort of Newstead.*

Currie, C. R. J. (1972) Scarf-joints in the North Berkshire and Oxford area, *Oxoniensia* **37**, 177–86.

Darrah, R. (1982) Working unseasoned oak in McGrail, S. (ed.) (1982), 219–29.

Davison, H. R. E. (1967) *Pagan Scandinavia.* Thames and Hudson

Day, R. (1888) [Report] *Proceedings of the Society of Antiquaries* **12**, 65–7.

Dean, M. (forthcoming) A boat recovered from the foreshore at West Mersea in Essex, *IJNA* **14**.

De Boe, G. and Hubert, F. (1977) Une installation portuaire d'epoque Romaine a Pommeroeul, *Archaeologia Belgica* **192**, 5–57.

Dechelette, J. (1924–7) *Manuel d'Archeologie Prehistorique* 4 vols 2nd edn. Paris.

de Graeve, M. C. (1981) *Ships of the Ancient Near East.* Orientalia Lovaniensia, Analecta **7**, Leuven.

de Laat, S. J. (1956) Wooden animal heads of Carolingian times found in the River Scheldt, *AA* **27**, 127–37.

de Laguna, F. (1963) Yakutat Canoes, *Folk* **5**, 219–29. Copenhagen.

de la Prynne, A. (1870) *Diary of Abraham de la Prynne.* Surtees Publication.

Delaney, J. (1976) Fieldwork in south Roscommon in Ó Danochair, C. (ed) *Folk and Farm*, 15–29. Dublin.

de Morgan, J. (1895) *Fouilles a Dahchour*, Vienna.

Denford, G. T. and Farrell, A. W. (1980) Caergwrle bowl, *IJNA* **9**, 183–92

Dennell, R. (1983) *European Economic Prehistory*: a new approach. Academic.

Devoy, R. J. (1982) Analysis of the geological evidence for Holocene sea-level movements in S.E. England, *Proc. Geologists' Assoc.* **93**, 65–90.

de Weerd, M. D. (1978) Ships of the Roman period at Zwammerdam/Nigrum Pullum, Germania Inferior in Taylor, J. du P. and Cleere, H. (ed.) (1978) 15–21.

de Weerd, M. D. (1988a) *Schepen voor Zwammerdam.* Haarlem.

de Weerd, M. (1988b) A landlubbers view of shipbuilding procedures in the Celtic barges of Zwammerdam, in Filgueiras, O. L. (ed.) *Local Boats*, 35–51. Oxford: BAR S438.

de Weerd, M. D. (1990) Barges of the Zwammerdam type and their building procedures, in McGrail, S. (ed.) *Maritime Celts, Frisians and Saxons*: 75–6, CBA Research Report 71, London.

de Weerd, M. D. (1994) Rib insertion in phases, in Westerdahl, (1994), 43–44.

de Weerd, M. D. and Haalebos, J. K. (1973) Schepen voor het Opscheppen, *Spiegel Historiael* **8**, 386–97.

Dick-Read, R. (1964) *Sanamu*. Hart-Davis, London.

Dimbleby, G. (1978) *Plants and Archaeology*. Baker, London.

Doran, E. (1971) Sailing raft as a great tradition in Riley, C. L. *et al.* (ed.) *Man across the Sea*, 115–38. University of Texas.

Doran, E. (1978) Seaworthiness of Sailing Rafts, *Anthropological Journal of Canada* **16**, 17–22.

Doran, E. (1981) *Wangka: Austronesian canoe origins*. Texas A & M Univ. Press.

Dove, C. E. (1971) First British Navy, *Ant.* **45**, 15–20.

Drillis, R. J. (1963) Folk norms and biomechanics, *Human Factors* **5**, 427–41.

Dubois, C. (1976) Remarques sur les quilles des navires Romains, *Rev. Archéologique de Narbonnaise* **9**, 155–75.

Dunphy, W. P. (1979) Bark canoes of N. America, *MM* **65**, 77–82.

Durham, G. (1955) Canoes from cedar logs, *Pacific Northwest Quarterly* **46**, 33–9.

Durham, B. (1960) *Indian canoes of the Northwest coast*. Seattle.

Edlin, H. L. (1970) *Trees, Woods and Man*. Collins.

Edwards, C. R. (1965) *Aboriginal Watercraft on the Pacific Coast of S. America*. Ibero-Americáná 47. Univ. of California.

Edwards, R. (1972) *Aboriginal Bark Canoes of the Murray Valley*. South Australia Museum.

Edwards, D. (1981) *Logboats from Scotland*. Unpublished dissertation University College, Cardiff.

Edye, J. W. (1834) Native vessels of India and Ceylon. *J. of Royal Asiatic Society* **1**, 4–14.

Eglinton, E. (1982) *Last of the sailing coasters*. HMSO.

Egloff, M. (1974) La barque de Bevaix, épave gallo-romaine du lac de Nechâtel, *Helvetia Archaeologica* **19/20**, 82–91.

Elgee, H. W. & F. (1949) An EBA burial in a boat-shaped wooden coffin from N. E. Yorkshire. *PPS* **15**, 87–106.

Ellmers, D. (1969) Keltischer Schiffbau, *Jahrbuch des Romisch-Germanischen Zentralmuseums Mainz* **16**, 73–122.

Ellmers, D. (1972) *Fruhmittelalterliche Handelsschiffahrt in Mittel und Nordeuropa*. Neumünster.

Ellmers, D. (1973) Kultbarken, fähren, fischerboote, vergeschichtliche einbäume in Niedersachsen, *Die Kunde* NS24, 23–62.

Ellmers D. (1974a) Vor-und Frügeschichtliche Schiffahrt am Nordrand der Alpen, *Helvetia Archaeologica* **19/20**, 94–104.

Ellmers, D. (1974b) Nautical archaeology in Germany, *IJNA* **3**, 137–145.

Ellmers, D. (1975) Antriebstechniken Germanischer schiffe im 1. Jahrtausend N. CHR., *Deutsches Schiffahrtsarchiv* **1**, 79–90.

Ellmers, D. (1977/8) Eine rudergabel aus dem Rhein bei Weisbaden. *Fundberichte aus Hessen* **17/18**, 339–46.

Ellmers, D. (1978) Shipping on the Rhine during the Roman period in Taylor J. du P. and Cleere, H. (ed.) (1978) 1–14.

Ellmers, D. (1979) The Cog of Bremen and related boats in McGrail, S. (ed.) (1979), 1–15.

Ellmers, D. (1980) Es began mit der Kogge, *Schuredenspeicher Musem* **21–23**. Stade.

Ellmers, D. (1981) Der nachtsprung an eine hinter dem horizont liegeude gegenküste, *Deutsches Schiffahrtsarchiv* **4**, 154–167.

Ellmers, D. (1984) Earliest evidence for skinboats in late Palaeolithic Europe in McGrail, S. (ed.) (1984a), 41–55.

Ellmers, D. and Pirling, R. (1972) Ein mittelalterliches schiff aus dem Rhein, *Die Heimat* **43**, 45–48.

Engelhardt, C. (1865) *Nydam Mosefund*, Copenhagen.

Emanuele, P. D. (1977) Ancient square rigging with and without lifts, *IJNA* **6**, 181–5.

Engelhardt, C. (1866) *Denmark in the Early Iron Age*. London.

Eskeröd, A. (1956) Early Nordic-Arctic Boats, *Arctica*, 57–87. Studia Ethnographica Upsaliensia XI.

Eskeröd, A. (1970) Båtar från Ekstock till Trålare. Lund.

Eskeröd, A (1973) *Kyrkbåtar och Kyrkbåtsfarder*. Stockholm.

Evans, A. J. (1897) Votive Deposit of Gold Objects, *Archaeologia* **55**, 391–408.

Evans, A. C. and Bruce-Mitford, R. (1975) The Ship in Bruce-Mitford, R. (ed.) *Sutton Hoo Ship-burial* vol. 1, 345–435. British Museum.

Ewe, H. (1972) *Schiffe auf Siegeln*. Rostock.

Faerøyvik, B. and Ø., and Christensen, A. E. (1979) *Inshore craft of Norway*. Conway.

Farmer, R. H. (1972) *Handbook of Hardwoods* 2nd edn. HMSO.

Farrell, A. W. (1977) Mast and sail in Scandinavia in the Bronze Age? *MM* **63**, 190.

Farrell, A. W. (1979) Mast and sail in Scandinavia in the Bronze Age? *MM* **65**, 83.

Farrell, A. W. and Penny, S. (1975) Broighter boat: a reassessment, *Irish Archaeological Research Forum* **2.2**, 15–26.

Fatimi, S. Q. (1996) History of the development of the *kamal*, in Ray and Salles (eds) (1996), 283–292.

Feacham, R. W. (1958–9) A dugout canoe in Cambuskenneth Abbey, *PSAS* **92**, 116–7.

Fenton, A. (1972) Curragh in Scotland, *Scottish Studies* **16**, 61–81.

Fenwick, V. (1978) *Graveney boat* NMM Archaeological Series 3. BAR (Oxford) 53.

Fernandez, J. L. (1966) Vellas artes de pesca no Rio Miño, *Revista de Etnografia* **7**. Museu de Etnografia e Historia Junta Distrital do Porto.

Field, J. (1980) *Place names of Britain and Ireland*. David and Charles.

Filguieras, O. L. (1977) The Xavega boat in McGrail, S. (ed.) *Sources and Techniques in Boat Archaeology*, 77–111. NMM Greenwich Archaeological Series 1. BAR (Oxford) S.29.

Finch, R. (1976) *Sailing craft of the British Isles*. Collins.

Finney, B. (1994) *Voyage of Rediscovery*. University of California Press.

Fiori, P. and Joncheray, J-P. (1973) Mobilier métallique provenant de fouilles sous marines, *Cahiers d'Archeologie subaquatique* **2**, 86–9.

Flemming, N. C. (1980) Structures under water in Muckelroy, K. (ed.) *Archaeology under water*, 164–175. McGraw-Hill.

Flemming, N. C. (1996) Sea level, neotectonics and changes in coastal settlements: threat and response, in Rice, E. E. (ed.) *Sea and History* 23–52. Sutton.

Fliedner, S. and Phil-Weber, B. (1972) *Cog of Bremen* 3rd edn. Bremerhaven.

Foley, V., Soedal, W., Doyle, J. (1982) Trireme displacement estimate, *IJNA* **11**, 305–18.

Folkard, H. C. (1870) *Sailing boat* 4th edn. Reprinted 1973, Wakefield.

Foote, P. C. and Wilson, D. M. (1970) *Viking Achievement*.

Forbes, R. J. (1936) *Bitumen and petroleum in Antiquity*. E. J. Brill. Leiden.

Forssell, H. (forthcoming) Sewn boats in Finland in McGrail S. and Kentley E. (eds) *Sewn Plank Boats*, NMM Greenwich BAR Oxford.

Friel, I. (1983) England and the advent of the 3-masted ship *Int. Congress of Maritime Museums* 4th Conference Proceedings, 130–138. Paris.

Friel, I. (1993) Henry V's *Grace Dieu* and the wreck in the R. Hamble near Bursledon, Hampshire. *IJNA* **22**, 3–19.

Friel, I. (1995) *The Good Ship*. British Museum Press.

Frost, H. (1963) From rope to chain *MM* **49**, 1–20.

Frost, H. (1973a) Anchore, in Blackman, D. J. (ed) *Colston Papers*, **23**, 397–410. London: Butterworth.

Frost, H. (1973b) First season of excavation on the Punic wreck in Sicily, *IJNA* **2**, 33–49.

Frost, H. (1976/1981) Punic Ship. *Notizie Degli Scari* **30** (Supplement).

Frost, H. (1979) Egypt and stone anchors *MM*, **65**, 137–161.

Frost, H. 1982 Birth of stocked anchors and the maximum size of ship. *MM*, 68: 263–278.

Geikie, J. (1879) Discovery of an ancient canoe in the old alluvium of the Tay at Perth, *Scottish Naturalist* **5**, 1–7.

Gelsinger, B. E. (1970) Norse 'Days Sailing', *MM* **56**, 107–9.

Giffard, E. (1993) Expanding oak logboats: is it possible? in Coles, J., Fenwick, V. and Hutchinson, G. (eds) *Spirit of Enquiry*, 52–53. University of Exeter.

Gillespie, J. E. (1874) Notice of a Canoe found in Loch Lotus, Parish of New Abbey, Kirkcudbrightshire, *PSAS* **11**, 21–3.

Gillmer, T. (1979) Capability of the single square sail rig in McGrail, S. (1979) 167–82.

Giot, P-R. and Niort, P-L. (1954) La Pirogue Préhistorique D'Ancenis, *Bulletin Archéologigue*, 285–8.

Goddard, D. (forthcoming) Flying proas of Kiribati in McGrail, S. and Kentley, E. (eds) *Sewn Plank Boats* NMM Greenwich. BAR Oxford.

Gomez, J. (1982) Une pirogue monoxyle néolithique dans le lit de la Charente, *Bull. Soc. Préhistorique Francaise* **79**, 61–3.

Goodman, W. L. (1964) *History of Woodworking Tools*.

Gordon, J. E. (1977) *New Science of Strong Materials* 2nd edn., Penguin.

Gordon, J. E. (1978) *Structures*. Penguin.

Göttlicher, A. (1971) Der Einbaum im Morgenster-Museum, *Heimatbund der Männer vom Morgenstern Bremerhaven Sunderdruck aus Jahrbuch* **52**, 45–50.

Green, J. and Intakosai, V. (1983) Pattaya wreck site excavation, Thailand, *IJNA* **12**, 3–13.

Green, H. S., Smith, A. H. V., Young, B. R., Harrison, R. K. (1980) Caergwrle bowl: its composition, geological source and archaeological significance, *Rep. Inst. Geol. Sci.* **80.1**, 26–30.

Greenhill, B. (1959) Japanese inshore fishing boats, *MM* **45**, 3–13.

Greenhill, B. (1966) *Boats of E. Pakistan*. SNR. Reprinted from *MM* **43** (1957)

Greenhill, B. (1971) *Boats and Boatmen of Pakistan*. David and Charles, Newton Abbot.

Greenhill, B. (1976) *Archaeology of the Boat*. A. and C. Black.

Greenhill, B. (1978) *Merchant Schooners* vol. 1 3rd edn. HMSO.

Greenhill, B. (1995) *Archaeology of Boats and Ships*. Conway, London.

Grierson, P. (1972) *English Linear Measures*. Stenton Lecture 1971. Univ. of Reading.

Grimes, W. F. (1931) Two dugout boats from Wales, *Ant. J.* **11**, 136–4.

Guppy, N. (1961) *Wai Wai*. Penguin.

Haasum, S. (1974) *Vikingatidens segling och navigation*. Theses and Papers in North European Archaeology 4. Stockholm.

Haddon, A. C. and Hornell, J. (1938) *Canoes of Oceania* vol. 3 Terms, General Survey and Conclusions, Honolulu.

Hale, J. R. (1980) Plank-built in the Bronze Age, *Ant.* **54**, 118–27.

Hansen, K. and Madsen, J. S. (1981) *Barkbåde*, Roskilde.

Harriot, T. (1590) *Briefe and True Report of the New Found Land of Virginia*. Republished New York, 1972.

Harrison, R. (1746) Letter to the Earl of Derby dated 24 August 1746, now in Lancashire Record Office (DDK 1640).

Hartmann, A. (1970) *Prahistorische Goldfunde aus Europa*. Berlin.

Hasslöf, O. (1972) Main principles in the technology of shipbuilding in Hasslöf, O. (ed.) *Ships and Shipyards, Sailors and Fishermen*, 27–72. Copenhagen.

Hawkes, C. F. C. (1977a) *Pytheas* 8th J. L. Myres Memorial Lecture, Oxford.

Heal, S. V. E. (1978) The character of wood, *PPS* **44**, 23–31.

Heal, S. V. E. (1981) Tools usable in the construction of the 'raft' in McGrail, S. (ed.) (1981b), 253–60.

Heggie, D. C. (1981) *Megalithic Science*. Thames and Hudson.

Heizer, R. F. (1966) Plank canoes of South and North America, *Kroeber Anthropological Society Papers* **35**, 22–39.

Henwood, (1971–3) Report, *J. Royal Inst. of Cornwall* **4**. LXXXVII.

Hermann, J. (1981) Ein neuer bootsfund im Seehandelsplatz Ralswick auf Rügen, *Ausgrabungen und Funde* **26**, 145–58.

Herteig, A. E. (forthcoming) *Proceedings of 1983 Waterfront Conference at Bergen*.

Heyworth, A. and Kidson, C. (1982) Sealevel changes in southwest England and in Wales, *Proc. Geologists' Assoc.* **93**, 91–111.

Heyerdahl, T. (1978) *Early Man and the Ocean*. Allen and Unwin.

Hillam, J. (1981) Tree ring analysis in McGrail, S. (ed.) (1981b), 103–21.

Hillam, J. (forthcoming) Dendrochronology of the N. Ferriby boats, in McGrail S and Kentley E (eds) *Sewn Plank Boats*. BAR Oxford.

Hirth, K. G. (1978) Inter-regional trade and the formation of prehistoric gateway communities, *American Antiquity*, **43**, 35–45.

Hockmann, O. (1997) Roman river patrols & military logistics on the Rhine & the Danube, in Jorgensen, A. N. and Clausen, B. L. (eds) *Military Aspects of Scandinavian Society*, 239–247. National Museum, Copenhagen.

Hodges, H. (1964) *Artifacts*. Baker. London.

Hodges, R. (1982) *Dark Age Economics*. Duckworth.

Hoekstra, T. J. (1975) Utrecht. *IJNA* **4**, 390–2.

Hollstein, E. (1968) Dendrochronologische untersuchungen an den Domen von Trier und Speyer, *Kunstchronik* **21.6**, 168–81.

Hornell, J. (1928) S. American balanced canoes, *Man* **28**, 129–33.

Hornell, J. (1929) Kru Canoes of Sierra Leone, *MM* **15**, 233–7.

Hornell, J. (1930) Tongue and Groove Seam of the Gujarati Boatbuilders, *MM* **16**, 310–12.

Hornell, J. (1936) British coracles, *MM* **22**, 5–41, 261–304.

Hornell, J. (1937–8) Curraghs of Ireland, *MM* **23**, 74–83; **24**, 5–39, 148–175.

Hornell, J. (1939/40) Frameless boats of the Middle Nile, *MM* **25**, 417–32; **26**, 125–44.

Hornell, J. (1941a) Seagoing *mtepe* and *dau* of the Lamu Archipelago, *MM* **27**, 54–68.

Hornell, J. (1941b) Pelota or hide-balsa of S. America. *MAN* **41**, 1–4.

Hornell, J. (1942) Floats: a study in primitive water transport, *J. Roy. Anthrop. Inst.* **72**, 33–44.

Hornell, J. (1943) Fishing and coastal craft of Ceylon, *MM* **29**, 40–53.

Hornell, J. (1944) Outrigger canoes of Madagascar E. Africa and the Comoro Is., *MM* **30**, 3–18, 170–85.

Hornell, J. (1945) Pearling fleets of South India and Ceylon, *MM* **31**, 214–30.

Hornell, J. (1946) *Water Transport*. Cambridge. (republished 1970, Newton Abbot)

Hornell, J. (1948) Making and spreading of dugout canoes, *MM* **34**, 46–52.

Horridge, G. A. (1978) *Design of Planked Boats of the Moluccas*. Maritime Monograph 38, Greenwich.

Horridge, G. A. (1981) Ancient planked boats of the islands east of Asia in Howse D. (ed.) (1981) *Five Hundred Years of Nautical Science, 1400–1900*, 241–58. NMM Greenwich.

Horridge, G. A. (1982) *Lashed-lug boat of the eastern archipelagoes, the Alcina Ms and the Lomblen Whaling boats*. Maritime Monograph 54, NMM Greenwich.

Hourani, G. F. (1963) *Arab Seafaring*. Beirut. (Khayats Oriental Reprints No. 3) Originally published Princeton, 1951.

Howard, F. (1979) *Sailing ships of war 1400–1860*. Conway Maritime Press.

Hughes, M. K., Milsom, S. J. and Leggett, P. A. (1981) Sapwood estimates in the interpretation of tree-ring dates, *J. Arch. Science* **8**, 381–90.

Humbla, P. (1934) *Båtfyndet vid Äskekårr*. Goteborg.

Humbla, P. and von Post, L. (1937) *Galtabäcksbåten och Tidigt Båtbyggeri i Norden*.

Huntingford, G. W. B. (ed.) (1980) *Periplus of the Erythraean Sea*. Hakluyt Society 2nd series No. 151.

Hurault, J. (1970) *Africains de Guyane*, The Hague.

Hutcheson, A. (1897) Notice of an ancient canoe found in the River Tay, near Errol, *PSAS* **31**, 265–272.

Hutchinson, G. (1994) *Medieval Ships and Shipping*. Leicester University Press, London.

Ilkjaer, J. and Lønstrop, J. (1981–2) Ankeret, *Skalk* 12–15.

Illsley, J. S. and Roberts, O. T. P. (1980) Two further boat finds in the Snowdon Lakes, *IJNA* **9**, 342–8.

Imre, M. and Dénes, P. (1973) *Ahajo Története*. Corvina.

Irwin, G. (1992) *Prehistoric Exploration and Colonisation of the Pacific*. Cambridge University Press, Cambridge.

Itkonen, T. I. (1941) Suomen Ruuher, *Kansatieteellinen Arkisto* **5**, 1.

Izikowitz, K. G. (1957) Canoe of the Soloman Is. and its social role, *MM* **43**, 19–27.

Jane, F. W. (1970) *Structure of Wood* 2nd edn.

Jardine, W. G. and Masters, L. J. 1976–7, A dugout canoe from Catherinefield Farm, Locharbriggs, Dumfries, *Trans. Dumfries and Galloway Nat. Hist. and Antiquarian Society*, **52**, 56–65.

Jay, B. A. (1961) *Timber*. TRADA High Wycombe.

Jenkins, J. G. (1974) *Nets and Coracles*. David & Charles.

Joass, J. M. (1880–1) Note on the curach and ammir in Ross-shire, *PSAS* **15**, 179–80.

Johannessen, F. (1940) Båtene fra Gokstadskibet. *Viking* **4**, 125–130.

Johnstone, P. (1972) Bronze Age Sea Trial, *Ant.* **46**, 269–74.

Johnstone, P. (1973a) Stern first in the Stone Age? *IJNA* **2**, 3–11.

Johnstone, P. (1973b) Review of T. Heyerdahl's *RA expedition in INJA* **2**, 216–7.

Johnstone, P. (1973c) Kalnes Bronze Age boat, *Ant.* **47**, 60–61.

Johnstone, P. (1976) Shipwrights and wheelwrights in the Ancient World in Megaw, J. V. S. (ed.) *To Illustrate the Monuments*, 50–6.

Johnstone, P. (1980) *Seacraft of Prehistory*. R. K. P.

Joffroy, R. (1978) Note sur deux ferrets Mérovingiens de collections du Musée des Antiquites Nationales in Fleury, M. and Périn, P. (ed.) *Problémes de Chronologie relative et absolue concernant les cimetières mérovingiens d'entre Loire et Rhin*. Paris.

Johncheray, J-P. (1977) Mediterranean hull types compared 2. Wreck F from C. Dramont (Var) France, *IJNA* **6**, 3–7.

Jonckheere, E. (1903) *L'Origine de la côte de Flandre et al bateau de Bruges*. Bruges.

Jones, G. (1974) *Norse Atlantic Saga*.

Jones, R. (1976) Tasmania: aquatic machines and offshore islands in Sieveking, G. de G., Longworth, I. H. and Wilson K. E. (eds) *Problems in Economic and Social Archaeology* 235–63. Duckworth.

Jones, P. and Simons, E. N. (1960) *Story of the Saw*. Spear and Jackson.

Kahl, H., Neilsen, I., Terp, F., Veje, P., (1971) *Roar linde 71*.

Kapitän, G., (1984) Ancient anchors – technology and classification, *IJNA* **13**, 33–44.

Katzev, M. L., (1972) Kyrenia ship in Bass, G. F. (ed.) (1972), 50–2.

Kay, H. F. (1971) *Science of yachts, wind and water*. Foulis, Henley on Thames.

Keeley, L. (1980) *Experimental determination of stone tool uses*. Chicago.

Keith, D. H. and Buys, C. J. (1981) New light on Medieval Chinese seagoing ship construction, *IJNA* **10**, 119–32.

Kemp, P. (1976) *Oxford Companion to Ships and the Sea*. OUP.

Kentley, E. (1984) Skin boats and basket boats: a note on the classification of water transport, *IJNA* **13**, 81–3.

Kentley, E. (1985) Some aspects of the Masula surf boat, in McGrail, S. and Kentley, E. (eds) *Sewn Plank Boats*: 303–318. BAR S276, Oxford.

Kentley, E. (1993) Sewn boats of the Indian Ocean: a common tradition? in Coles, J., Fenwick, V. and Hutchinson, G. (eds) *Spirit of Enquiry*, 68–71. University of Exeter.

Kentley, E. (1996) Sewn boats of India's east coast, in Ray and Salles, 1996: 247–260.

Kentley, E. and Gunaratne, R. (1987) Madel Parua – a sewn boat with chine strakes. *IJNA* **16**, 35–48.

Kiedal, K-P. and Schnall, U. (1982) *Die Hanse-kogge von 1380*. Bremerhaven.

Knapp, F. V. (1924) Canoe-building tools of the Tasman Bay Maoris, *Journal of the Polynesian Society* **33**, 103–13.

Knuth, E. (1980) *Umiaq'en fra Peary Land*. Roskilde.

Kolchin, B. A. (1953) *Chernoe metallurgie V. Drevnoi Rusi*. Moscow.

Lahn, W. (1993) *Die Kogge von Bremen* 1, Kabel, Hamburg.

Lamb, H. H. (1977) *Climate* volume 2. Methuen.

Landels, J. G. (1978) *Engineering in the Ancient World*. B.C.A., London.

Landström, B. (1961) *The Ship*. Allen and Unwin.

Lane, F. C. (1964) Tonnages, medieval and modern, *Econ. Hist Review* **17**, 213–33.

Lane-Fox, A. (1875) On early modes of navigation, *J. Roy. Anthropological Inst.* **4**, 399–437.

Languoët, L. 4th century Gallo-Roman site at Alet, in Johnston, D. E. (ed.) *Saxon Shore*, 38–45. CBA Research Report 18.

Laures, F. F. (1983) Comment on 'A trireme displacement estimate', *IJNA* **11**, 305–18.

Layard, A. H. (1849) *Nineveh and its remains*. vol. 2.

Layard, A. H. (1853) *Discoveries in the ruins of Nineveh and Babylon*.

Lees, J. (1977) *Norwegian fembøring*, NMM Occ. Technical Paper No. 7, Greenwich.

Lehmann, L. Th. (1978a) The flat-bottomed Roman boat from Druten, Netherlands, *IJNA* **7**, 259–67.

Lehmann, L. Th. (1978b) Mystery of the Graeco-Roman steering oar, *IJNA* **7**, 95–8.

Leigh, C. (1700) *Natural History of Lancashire, Cheshire and the Peak in Derbyshire; with an account of the British, Phoenician, American, Gr. and Rom. antiquities in those parts*, Oxford.

Leighton, A. C. (1972) *Transport and communication in early medieval Europe* AD 500–1100. Newton Abbot.

Lerche, G. (1977) Double paddle-spades in prehistoric contexts in Denmark. *Tools and tillage* **3.2**, 113–24.

Leslie, R. C. (1892) *The sea boat and how to build, rig and sail her*.

Lethbridge, T. C. (1933) Anglo-Saxon burials at Soham, Cambridgeshire, *Proc. Cambridge Antiquarian Soc.* **33**, 152–63.

Lethbridge, T. C. (1937) *Umiak*. Cambridge.

Letts, H. W. (1895) Ancient canoe found near Loughbrickland, Co. Down, *UJA* 2nd series **1**, 153–4.

Lett, H. W. (1897) Canoe found at Portmore, Co. Antrim, *UJA* 2nd series, **3**, 251–2.

Lewis, A. B. (1932) *Melanesians.* N. H. Museum, Chicago (re-issued 1945).

Lewis, D. (1972) *We the Navigators.* Canberra Nat. University Press.

Lewis, D. (1994) *We, the Navigators.* 2nd edn, University of Hawaii Press, Honolulu.

L'Hour, M. and Veyrat, E. (1989) Mid-15th century clinker boat off the northern coast of France. *IJNA* **18**, 285–298.

L'Hour, M. and Veyrat, E. (1994) French medieval clinker wreck from Aber Wrac'h, in Westerdahl, (1994), 165–180.

Lienau, O. (1934) *Die bootsfunde von Danzig-Ohra.* Westpreusischen Gefchichte No. 17.

Lindqvist, S. (1941–2) *Gotlands Bildsteine* 2 vols, Stockholm.

Lindqvist, S. (1942) Boat models from Roos Carr, *AA* **13**, 235–42.

Liou, B. (1974a) La Lugue B, *IJNA* **3**, 325.

Liou, B. (1974b) *L'epave romaine de l'Anse Gerbal à Port-Vendres.* Paris.

Lipke, P. (1984) *Royal Ship of Cheops.* NMM Greenwich Archaeological Series 9. BAR (Oxford) S. 225.

Litwin, J. (1980) Copper wreck, *IJNA* **9**, 217–25.

Longstaff, F. V. (1930) British Columbian Indian Cedar Dugout Canoes, *MM* **16**, 259–62.

Lothrop, S. K. (1932) Aboriginal navigation off the west coast of S. America, *J. Royal Anthropological Inst.* **62**, 229–56.

Lothrop, S. K. (1952) *Zutugil Dugout Canoes.* Notes on Middle American Archaeology and Ethnology No. 111, published by Carnegie Institution of Washington.

Lucas, A. T. (1963) Dugout canoe in Ireland. *Varbergs Museum Årsbok*, 1963, 57–68, Varberg, Sweden.

Lund, N. (ed.) (1984) *Two Voyagers at the Court of King Alfred.* Sessions, York.

Lundstrüm, P. (1971–2) Klinknaglarnas vittnesbörd, *Sjöhistorisk Årsbok*, 81–8.

Lyon, D. J. (1974) Documentary sources for the archaeological diver, *IJNA*, **3**, 3–19.

MacAlister, R. A. S. (1949) *Archaeology of Ireland*, 2nd edn, London.

McDowell, U. (1983) *Logboats of Ireland.* Unpublished MA dissertation. University College, Dublin.

McGowan, A. (1981) *Tiller and Whipstaff.* HMSO/NMM.

McGrail, S. (1974) *Building and trials of the replica of an ancient boat: the Gokstad faering.* Part 1. NMM Monograph 11.

McGrail, S. (ed) (1977a) *Sources and Techniques in Boat Archaeology.* NMM Greenwich Archaeological Series 1. BAR (Oxford) S.29.

McGrail, S. (1977b) Aspects of Experimental Boat Archaeology in McGrail, S. (ed.) *Sources and Techniques in Boat Archaeology*, 245–258. NMM Archaeological Series 1. BAR (Oxford) S.29.

McGrail S. (1977c) Axe, adze, hoe or slice, *IJNA* **6**, 62–4.

McGrail, S. (1978a) *Logboats of England and Wales*, NMM Archaeological Series 2, BAR (Oxford) 51.

McGrail, S. (1978b) Medieval logboat from Giggleswick Tarn, Yorkshire in Annis, P. (ed.) *Ingrid and other Studies* NMM Greenwich Monograph 36, 25–46.

McGrail, S. (1978c) Dating ancient wooden boats in Fletcher, J, (ed.) *Dendrochronology in Europe.* NMM Archaeological Series No. 4, BAR S51, 239–58.

McGrail, S. (ed.) (1979) *Medieval Ships and Harbours in Northern Europe.* NMM Greenwich Archaeological Series No. 5, BAR (Oxford) S 66.

McGrail, S. (1981a) *Rafts, Boats and Ships.* HMSO.

McGrail, S. (1981b) (ed.) *Brigg 'raft' and her prehistoric environment.* NMM Greenwich Archaeological Series No. 6. BAR (Oxford) 89.

McGrail, S. (1981c) A Medieval logboat from the River Calder at Stanley Ferry, Wakefield, Yorkshire. *Med. Arch.* **25**, 160–4.

McGrail, S. (ed.) (1982) *Woodworking Techniques before 1500* NMM Archaeological Series No. 7. BAR (Oxford) S129

McGrail, S. (1983a) Cross-channel seamanship and navigation in the late 1st millennium BC, *Oxford J. of Archaeology* **2**, 299–337.

McGrail, S. (1983b) Interpretation of archaeological evidence for maritime structures in Annis, P. (ed.) *Sea Studies* 33–46 NMM Greenwich Monograph.

McGrail, S. (ed.) (1984a) *Aspects of Maritime Archaeology and Ethnography.* NMM Greenwich Monograph.

McGrail, S. (1984b) Maritime Archaeology present and future in McGrail, S. (ed.) *Aspects of Maritime Archaeology and* Ethnography, 11–40. NMM Greenwich Monograph.

McGrail, S. (1985a) Towards a classification of water transport. *World Archaeology* **16**, 289–303.

McGrail, S. (1985b) Experimental boat archaeology – some methodological considerations, in Proceedings of the Replica Seminar held in Roskilde, 1984.

McGrail, S. (1985c) Early landing places in Herteig, forthcoming.

McGrail, S. (1985d) Brigg 'raft' – problems in reconstruction and the assessment of performance. In McGrail, S. and Kentley, E. (eds) *Sewn Plank Boats.* BAR (Oxford) S276, 165–194.

McGrail, S. (1987/8) Early boatbuilding techniques in Britain and Ireland – dating technological change. *IJNA* **16**, 343–6; 17:158

McGrail, S. (1989a) Shipment of traded goods and of ballast in Antiquity. *Oxford Journal of Archaeology* **8**, 358–8.

McGrail, S. (1989b) Pilotage & navigation in the time of St Brendan, In: de Courcy Ireland, J. and Sheehy, D.C. (eds) *Atlantic Vision*, 25–35 Boole, Dun Laoghaire.

McGrail, S. (1990) *Maritime Celts. Frisians and Saxons.* CBA Research Report 71.

McGrail, S. (1991) Bronze Age seafaring in the Mediterranean – a view from NW Europe, in Gale, N.H. (ed.) *Bronze Age Trade in the Mediterranean*, 83–91.

Studies in Mediterranean Archaeology XC. Paul Astroms Forlag Jonserad, Sweden.

McGrail, S. (1992) Replicas, reconstructions and floating hypotheses. *IJNA* **21**, 353–5.

McGrail, S. (1993a) *Medieval Boat & Ship Timbers from Dublin*. Royal Irish Academy, Dublin.

McGrail, S. (1993b) Future of the designated wreck site in the R. Hamble, *IJNA* **22**, 45–51.

McGrail, S. (1994) Brigg 'raft': a flat-bottomed boat. *IJNA* **23**, 283–8.

McGrail, S. (1995) Romano-Celtic boats and ships: characteristic features. *IJNA* **24**, 139–145.

McGrail, S. (1996a) Ship: carrier of goods, people and ideas, in Rice, E. E. (ed.) *Sea and History* 67–96. Sutton.

McGrail, S. (1996b) Study of boats with stitched planking, in Ray and Salles (eds) (1996) 225–238.

McGrail, S. (1996c) Navigational techniques in Homer's Odyssey, *Tropis* **4**, 311–320.

McGrail, S. (1997a) The boat fragments, in Nayling, N. and Caseldine, A. (eds) *Excavations at Caldicot*, 210–217 CBA Research Report 108, Gwent.

McGrail, S. (1997b) Early Frame-first methods of building wooden boats & ships, *MM* **83**, 76–80.

McGrail, S. and Corlett, E. (1977) High speed capabilities of ancient boats, *IJNA* **6**, 352–3.

McGrail, S. and Denford, G. (1982) Boatbuilding techniques, technological change and attribute analysis in McGrail, S. (ed.) (1982), 25–72.

McGrail, S. and Farrell, A, 1979, Rowing: aspects of the ethnographic and iconographic evidence, *IJNA* **8**, 155–66.

McGrail, S. and Kentley E. (eds.) (1985) *Sewn Plank Boats*. BAR. S 276, Oxford.

McGrail, S. and Switsur, R. (1979) Medieval logboats of the R. Mersey in McGrail, S. (ed.) (1979), 93–115.

McGruer, E. (nd) *Engineering in Wood and a lesson in boatbuilding*. Scottish Country Industries Development Trust. Edinburgh.

McKee, E. (1972) *Clenched Lap or Clinker*. NMM Greenwich.

McKee, E. (1974) *Building and trials of the replica of an ancient boat: the Gokstad faering, part 2: the sea trials*, NMM Greenwich Monograph 11B.

McKee, E. (1976a) Traditional British boatbuilding methods, *MM* **62**, 3–14.

McKee, E. (1978) Set of drawings of parts of the hull and draughtsman's notes; Reconstructing the hull; Model planks; Drawing the replica. Tentative sequence for building the replica in Fenwick, V. (1978), 49–104, 265–302, 307–10.

McKee, E. (1983) *Working boats of Britain*. Conway Maritime Press.

Macknight, C. C. (1973) Nature of early maritime trade, *World Archaeology* **5**, 198–208.

McKusick, M. B. (1960) *Aboriginal Canoes in the West Indies*. Yale University Publications in Anthropology No. 63, New Haven.

McPherson, N. (1877/8) Notes on antiquities from the Isle of Eigg. *Proc. Soc. Antiquaries of Scotland*, **12**, 594–6.

Magnus, B. (1980) Halsnøybåtens tekstiler, *Arkeo* 1980, 22–5.

Malinowski, B. (1922) *Argonauts of the Western Pacific*. RKP (reprinted 1978).

Mann, L. M. (1933) Some recent discoveries, *Transactions of the Glasgow Archaeological Society* NS. 8, 138–51.

Manninen, I. (1927) Zur ethnologie des einbaumes, *Eurasia Septentrionalis Antiqua* **1**, 4–17.

Manning, W.H. and Saunders, C. (1972) Socketed iron axe from Maids Moreton, Bucks with a note on the type, *Ant. J.* **52**, 276–92.

March, E. J. (1970) *Inshore Craft of Britain*. 2 vols, Newton Abbot.

Marchaj, C. A. (1964) *Sailing Theory and Practice*. Adlard Coles.

Marchaj, C. A. (1979) *Aero-hydrodynamics of Sailing*. Adlard Coles.

Marchaj, C. A. (1996) *Sail Performance*. Adlard Coles, London.

Marcus, G. J. (1953–4) Factors in early Celtic Navigation, *Etudes Celtiques* **6**, 312–27.

Marcus, G. J. (1980) *Conquest of N, Atlantic*. Boydell Press.

Marsden, P. R. V. (1963) Newstead steering oar, *MM* **49**, 224.

Marsden, P. R. V. (1965a) County Hall Ship, *Trans. London & Middlesex Arch Soc.* **21**, 109–17.

Marsden, P. R. V. (1965b) A boat of the Roman period discovered on the site of New Guy's House, Bermondsey, 1958, *Trans. London & Middlesex Arch. Soc.* **21**, 118–131.

Marsden, P. R. V. (1966) *A ship from the Roman period from Blackfriars in the city of London*. Guildhall Museum.

Marsden, P. R. V. (1967) *Roman ship from Blackfriars, London. London*: Guildhall Museum.

Marsden, P. R. V. (1972) Blackfriars wreck 3 *IJNA* **1**, 130–2.

Marsden, P. R. V. (1973) A 15th-century dock at Blackfriars London, *IJNA* **2**, 192.

Marsden, P. R. V. (1976) A boat of the Roman period found at Bruges, Belgium in 1899 and related finds, *IJNA* **5**, 23–55.

Marsden, P. R. V. (1979) Medieval ships of London in McGrail, S. (ed.) (1979), 83–92.

Marsden, P. (1994/1996) *Ships of the Port of London*. 2 vols English Heritage. Archaeological Reports 3 and 5, London.

Marstrander, S. (1963) *Østfolds Jordbruksristninger*. Skjeberg. 2 vols. Oslo.

Marstrander, S. (1976) Building a hide boat. An archaeological experiment, *IJNA* **5**, 13–22.

Maxwell, S. (1950-1) Discoveries made in 1934 on King

Fergus Isle and elsewhere in Loch Laggan, Inverness-shire, *PSAS* **85**, 160–5 and Plate XVI. Edinburgh.

Merryfield, R. (1983) *London: city of the Romans*. Batsfords.

Middleton, R, (1996) Discovery of the Kilnsea boat. *Hull Natural History Society Newsletter*, 3–6.

Miles, S. (1971–2) Shipwrecks and survival, *Proc. Roy Soc.* (Edin) **73** (B. Biology), 97–103.

Millett, M, and McGrail, S. (1985) The Hasholme logboat, *Ant.* **59**, 117–120.

Millett, M. and McGrail, S. 1987. Archaeology of the Hasholme logboat. *Archaeological Journal* **144**, 96–155.

Milne, G. (1982) Recording timberwork on the London waterfront in McGrail, S. (ed.) (1982), 7–23.

Milne, G. and Hobley, B. (eds) (1981) *Waterfront Archaeology in Britain and Northern Europe*. CBA Research Report 41.

Moll, F. (1926) History of wood preserving in shipbuilding, *MM* **12**, 357–74.

Morcken, R. (1968) Norse nautical units and distance measurements, *MM* **54**, 393–401.

Morcken. R. (1970a) Europas eldste sjømerker? *Sjøfarts-historisk Arbok 1969*, 7–48. Bergen.

Morcken, R. (1970b) Norse day's sailing, *MM* **56**, 347–9.

Morcken, R. (1980) *Langskip, Knarr og Kogge*, Bergen.

Morcken, R. (1983) *Ton og Lest*. Bergen.

Moore, A. (1925) *Last days of mast and sail*. Clarendon Press, Oxford. Reprinted 1970, David and Charles.

Morrison, J. S. (1976) Classical Traditions in Greenhill, B. G. (1976), 155–73.

Morrison, J. (1995) *Age of the Galley*. Conway, London.

Morrison, J. S. and Williams, R. T. (1968) *Greek Oared Ships* 900–322 BC. Cambridge.

Mowat, R. J. C. (1996) *Logboats of Scotland*. Oxbow Monograph 68, Oxford.

Muckelroy, K., Haselgrove, C., and Nash, D. (1978) A pre-Roman coin from Canterbury and the ship represented on it, *PPS* 44, 439–444.

Muller, S. (1886) Votivfund fra Sten-og Bronzealderen, *Aarbøger for Nordisk Oldkyndighed og Historie*. Copenhagen.

Müllner, A. (1892) Ein Schiff im Laibacher Moore, *Argo* **1**, 1–8 Laibach, Austria.

Munro, R. (1882) *Ancient Scottish Lake Dwellings or Crannogs*, Edinburgh.

Munro, R. (1890) *Lake Dwellings of Europe*.

Myrhe, B. (1980) Ny datering av våre eldste båter, *Arkeo* **7**, 27–30.

Nance, R. M. (1911) 15th century Trader, *MM* **1**, 65–7.

Nance, R. M. (1913) Caravels, *MM* **3**, 265–71.

Nance, R. M. (1922) Wicker vessels, *MM* **8**, 199–205.

Nayling, N., Maynard, D. and McGrail, S. (1994) Barland's Farm, Magor, Gwent: a Romano-Celtic boat. *Antiquity* **68**, 596–603.

Nayling, N. and McGrail, S. (forthcoming) *The Boat from Barland's Farm, Gwent*.

Needham, J. (1971) *Science and Civilisation in China* 4 pt. 3. Cambridge.

Needham, S. and Longley, D. (1980) Runnymede bridge, Egham: a late Bronze Age riverside settlement in Barrett, J. and Bradley, R. (eds) *British Later Bronze Age*, 397–436. BAR (Oxford) 83, vol. 2.

Needham, S. and Longley, D. (1981) Runnymede bridge, in Milne and Hobley (1981), 48–50.

Nelson, J. G. (1961) Geography of the Balsa, *American Neptune* **21**, 157–195.

Nicholson, S. (ed.) (1981) *Changing Face of Liverpool*. University of Liverpool.

Nicolyasen, N. (1882) *Viking Ship Discovered at Gokstad in Norway*. Kristiania, Republished 1971. Farnboro.

Nikkilä, E. (1947) En Satakundensisk äsping och dess Eurasiska motsvarigheter, *Folk-Liv* **11**, 33–46.

Nilsson, A. (1943) Primitiva Båtar i Småland, *Fataburen-Nordiske Museets och Skansens Årsbok 1943*, 123–136.

Nishimura, S. (1925) *Ancient Rafts of Japan* (Ancient Ships of Japan, Vol. 10) Toyko.

Nishimura, S. (1931) *Skinboats* (Ancient Ships of Japan, Vols 5, 6, 7, 8) Tokyo.

Nopsca, F. (1925) *Albanien, Bauten, Trackten und Geräte Nordalbaniens*, 145–5.

O'Connell, D. B. (1941) An Irish dugout, *MM* **27**, 81–2 and plate 9a.

Olsen, O. T. (1885) *Fisherman's Seamanship*. Grimsby.

Olsen, O. and Crumlin-Pedersen, O. (1967) Skuldelev Ships (II) *AA* **38**, 73–174 Copenhagen.

Olsson, H. and Sjöberg, J. E. (1971) *Låssby Skaggered Goteborg Stockbåtar Bronsålder/Jarnålder*, Goterborg. Excavation Report published by the Arkeologiska Museum, Goteborg.

O'Meara, J. J. (1978) *Voyage of St. Brendan*. Dolmen. Dublin.

Orme, B. J. and Coles, J. M. (1983) Prehistoric woodworking from the Somerset Levels – 1 Timber in Coles, J. M. (ed.) *Somerset Levels Papers* **9**, 19–43.

Ostler, A. G. (forthcoming) The N. Tyne Trows – a logboat link? *MM* **71**.

Paret, O. (1930) Die einbäume im Federseeried im übrigen Europa, *Prähistorische Zeitschrift* **21**, 76–116.

Parfitt, K. (1993) Dover boat. *Current Archaeology* **133**, 4–8.

Paris, F. E. (1843) *Essai sur la Construction Navale des peuples extra-Européens*. Paris.

Paris, P. (1955) *Esquisse d'une ethnograohie navale des peuples annamites*. Rotterdam.

Peacock, D. P. S. (1978) The Rhine and the problem of Gaulish wine in Britain in Taylor, J.du P. and Cleere, H. (ed.) (1978), 49–51.

Peck, W. (1815) *Topographical account of the Isle of Axholme* **1**.

Petersen, J. (1951) Vikingetidens Redskaper, *Skrifter utgitt av Det Norske Videnskaps-Akademi i Oslo*. Oslo.

Petersen, H. C. (1982) *Instruction in Kayak Building.* Viking Ship Museum, Roskilde.

Petersen, H.C. (1986) *Skin Boats of Greenland.* Viking Ship Museum, Roskilde.

Phillips, C. W. (1941) Some recent finds from the Trent near Nottingham, *Ant. J.* **21**, 133–143.

Phillips-Burt, D. (1971) *History of Seamanship.* George Allen and Unwin.

Phillips-Burt, D. (1979) *Building of Boats.* Stanford Maritime.

Philipsen, J. P. W. (1965) Utrecht ship, *MM* **51**, 35–46.

Philipsen, P. S. (1983) *En Analyse of Danske Stammebåde.* Aarhus.

Pietri, J. B. (1949) *Voiliers d'Indochine,* Saigon.

Piggott, S. (1974) *The Druids.* Harmondsworth.

Piggott, S. and Daniel, G. E. (1951) *Picture book of Ancient British Art,* Cambridge.

Platt, C. (1973) *Medieval Southampton.* R. K. P.

Polanyi, K. (1963) Ports of trade in early societies, *J. Economic History* **23**, 30–45.

Pomey, P. (forthcoming) Mediterranean sewn boats in Antiquity in McGrail, S. and Kentley, E. (ed.) *Sewn Plank Boats.* NMM Greenwich.

Poulson, G. (1840) *History and Antiquities of the Seigniory of Holderness,* vol. 2.

Price, D. J. de S. (1959) Ancient Greek computer, *Scientific American* (June) 60–7.

Prins, A. J. H. (1970) *Origin of Leeboards, MM* **56**, 349–353.

Prins, A. H. J. (forthcoming) *Sewn Boats.* NMM Greenwich.

Prynne, M. (1973) Some general considerations applying to the examination of the remains of old ships, *IJNA* **2**, 227–33.

Purdy, J. (1842) *New Sailing Directory for the English Channel* 9th edn.

Quinn, D. B. and A.M. (eds.) (1973) *Virginian Voyages from Hakluyt.*

Rackham, O., Blair, W. J. and Munby, J. J. (1978) 13th-century roofs and floor of the Blackfriars priory at Gloucester, *Med. Arch.* **22**.

Rausing, G. (1984) *Prehistoric Boats Ships of N.W. Europe,* Från Forntid och Medeltid, 8. University of Lund, Sweden.

Rasmussen, H. (1953) Hasselø-egen, *Kuml,* 15–48.

Rawson, K. J. and Tupper, E. C. (1976–7) *Basic Ship Theory* 2 vols. Longman.

Ray, H. P. and Salles, J-F. (eds) (1996) *Tradition & Archaeology.* Manohar, New Delhi.

Reinders, R. (1979) Medieval ships: recent finds in the Netherlands in McGrail, S. (ed.) (1979), 35–44.

Reinders, R. (1983) Excavation and salvage of a 16th century 'Beurtschip', *IJNA* **12**, 336–8.

Reinders. R. (forthcoming) *Cog finds. from the Ijsselmeer Polders,* Flevobericht No. 248 Lelystad.

Rieth, E. (1978) La constriuction navale chez les Vikings. *Archeologia,* 115: 47–61.

Rieth, E. (1996) *Le Mâitre-Gabarit, la Tablette et le Trebuchet.* C.T.H.S. Paris.

Riley, W. E. (1912) *Ship of the Roman Period discovered on the New County Hall site* (2nd edn) London County Council.

Roberts. O. (1984) Viking sailing performance in McGrail, S. (ed.) (1984a) 123–151.

Roberts, O. (1985) Cots of Rosslare Harbour and Wexford. *MM,* 71: 13–34.

Roberts, O. (1992) Brigg 'raft' re-assessed as a round bilge Bronze Age boat, *IJNA* **21**, 245–258.

Roberts, O. (1995) A response. *IJNA* **24**, 159–160.

Rogers, E. S. (1965) Dugout canoe in Ontario, *American Antiquity* **30**, 454–9.

Rolfsen, P. (1974) *Bånanst på Jaerkysten.* Stavanger Museums Skrifter, 8.

Roop, W. P. (1942) Watercraft in Amazonia, *Revista del Institute de Antropologia de la Universidad National de Tucumán* **2**. (also privately published by author in 1935, Woodbury, New Jersey).

Rose, S. (1982) *Navy of the Lancastrian Kings.* Allen and Unwin.

Rosenberg, G. (1937) *Hjortspring fundet,* Copenhagen.

Rosenquist, A. M. (1979) Undersokelse av Sammensetning og struktur av nagler og spiker fra batgraver fra Vikingtida i Sor-Norge, *Universitets Oslo Årbok,* 1979, 181–8.

Ross, A. (1970) *Everyday Life of the Pagan Celts.* Batsford.

Ross. J. B. and McLaughlin, M. M. (eds) (1977) *Portable Medieval Reader,* 144–9.

Roth, H. L. (1899) *Aborigines of Tasmania* 2nd edn, Halifax.

Rowlands, M. J. (1971) Archaeological interpretation of prehistoric metallurgy. *Word Arch.* **3**, 210–224.

Rudolph, W. (1964) Bootsbau in Brandenburg 75 *Jahre Museum fur Volkskunde,* 117–122. Staaliche Museem zu Berlin.

Rudolph, W. (1966) *Handbuch der Volkstümlichen Boote im östlichen Neider-Deutschland,* Berlin.

Rudolph, W. (1974a) *Inshore Fishing craft of the Southern* Baltic from Holstein to Curonia. Maritime Monograph 14. Greenwich.

Rudolph. W. (1974b) *Boats, Rafts and Ships.* Adlard Coles, London.

Rule, M. and Monaghan, J. (1993) *Gallo-Roman Trading Vessel from Guernsey.* Guernsey Museum Monograph 5.

Ryckaert, M. (1980) Resultaten van het historish-geografisch onderzoek in de Belgische kustvlakte in Verhulst, A. and Gottschalk, M.K.E. (eds) *Transgressies en occupatiegeschiedenis in de Kustgebieden van Nederland en Belgie,* 75–92. Centre belge d'histoire rurale, Gent.

Salaman, R. A. (1967) Tools of the shipwright, *Folk Life* **5**, 19–51.

Salaman, R. A. (1975) *Dictionary of Tools.*

Salemke, G. (1972) Der Einbaum vom Mondsee, Österreich, *Das Logbuch* **8**, 4–8.

Salisbury, W. (ed.) (1958) *Treatise on Shipbuilding* issued as Occasional Publication No. 6 by the Society for Nautical Research.

Salisbury, W. (1965a) Review of 'Nydamskeppen'. *MM* **51**, 278–80.
Salisbury, W. (1965b) Side rudder and steering oars and the Nydam ship, *MM* **51**, 359–61.
Salisbury, W. (1966) Early tonnage measurements in England, *MM* **52**, 41–51.
Salomonsson, B. (1957) Découverte d'une pirogue préhistorique en Scanie, *L'Anthropologie* **61**, 289–94.
Sandahl, B. (1951) *Middle English Sea Terms* 1. Ship's Hull. Studia Anglistica Upsaliensia 8, Uppsala.
Sandahl, B. (1958) *Middle English Sea Terms* 2. Mast, spars and sails. Studia Anglistica Upsaliensia 20. Uppsala.
Sandahl, B. (1982) *Middle English Sea Terms* 3. Standing and running rigging. Studia Anglistica Upsaliensia 42. Uppsala.
Sandison, C. (1954) *Sixareen and her racing descendants*. Lerwick.
Sarsfield, J. P. (1984) Mediterranean whole moulding, *MM*, **70**, 86–88.
Sarsfield, J. P. (1988) Survival of pre-16th century Mediterranean lofting techniques in Bahia, Brazil, in Filguieras, O. L. (ed.) *Local Boats*, 63–86. BAR S438, Oxford.
Sauer, M. (1802) *Account of a Geographical and Astronomical Expedition to the Northern Ports of Russia* ... by Commodore Joseph Billings. Republished 1972 Richmond, Surrey.
Sawyer, P. H. (1984) Ohthere and Viking Age Trade in Lund, N. (ed.), 43–55.
Schaeffer, C. F. A. (1939) Un voilier de l'époque mérovingienne du Nord de la France. *Revue Archéologique* **14**, 181–27.
Schnall, U. (1996) Early shipbuilding & navigation in northern Europe, in Christensen, (1996), 120–128.
Schovsbo, P. O. (1980) Mast and sail in Scandinavia in the Bronze Age, *MM* **66**, 15–16.
Seaby, W. A. (1960) Dug-out boats 368, 369 and 370: 1959, *UJA* **23**, 58–9.
Seaby, W. A. (1973) *Rowing dugout boats from North of Ireland*. (Unpublished ms.)
Serruys, H. (1981) Hun-t'o: tulum, floats and containers in Mongolia and Central Asia, *Bull. School of African and Oriental Studies* **44**, 105–19.
Severin, T. (1978) *Brendan Voyage*. Hutchinson.
Severin, T. (1982) *Sindbad Voyage*. Hutchinson.
Sheppard, T. (1901) Notes on the ancient model of a boat and warrior crew found at Roos in Holderness, *Trans. East Riding Antiquarian Society* **9**, 62–74.
Sheppard, T. (1902) Additional note on the Roos Carr images, *Trans. East Riding Antiquarian Society* **10**, 76–9.
Sheppard, T. (1926) Roman remains in north Lincolnshire, *East Riding Antiquarian Society Transactions* **25**, 170–4.
Sherratt, A. (1980) *Cambridge Encyclopedia of Archaeology* CUP.
Shetelig H. (1903) Fragments of an old boat from Halsnø, *Årbok Bergen* **7**, 8–21.
Shetelig, H. (1971) *Tuneskibet*, Kristiania.
Shetelig, H. (1930) Das Nydamschiff, *AA* **1**, 1–30.
Shetelig, H. and Johannessen, F. (1929) *Kvalsunfundet*. Bergen Museum New Series Volume 2, No. 2.
Sieveking, G. de G. (1954) Recent Archaeological discoveries in Malaya (1952–3), *J. Malayan branch of Royal Asiatic Society* **27**, 224–233.
Simonsen, P. (1953) Smedgraven fra Ytre Elgsnes, *Viking* **17**, Oslo.
Sinor, D. (1961) On water transport in central Eurasia. *Ural-Altaische Jahrbücher* **3**, 156–179.
Sjöberg, J. E. (1970) Hypotetiska båtar – och verkliga, *Göteborgs Arkeologiska museum Årstryck* (1970), 41–6. Göteborg.
Sjovøld, T. (1969) *Oseberg find*. Olso.
Slade, W. J. and Greenhill, B. (1974) *West country Coasting Ketches*. Conway, Greenwich.
Sleeswyk, A. W. (1980) Phoenician joints, *coagmenta punicana*, *IJNA* **9**, 243–4.
Sleeswyk, A. W. (1982) Rudder of the Blussus's boat, *IJNA* **11**, 153–4.
Slettebo, J. (1979) Hårdt Styrbord, *Skalk* 1979/2, 14–15.
Smolarek, P (1972) Szkutnictwo Pomorza Gdańskiego we Wczesnym, Sredniowiec zu in Cieślaka. E. (ed.) *Historia bydownictwa okretowego na Wybrzezu Gdańskim*, 11–68. Gdańsk.
Sølver, C. V. (1946a) Rebaek rudder, *MM* **32**, 115–20.
Sølver, C. V. (1946b) Ladby ship anchor. *Acta Archaeologica*, 17: 117–126.
Speciale, G. C. (1934) Roman anchors found at Nemi, *MM* **17**,309–20.
Stampfuss, R. (1961) Der Einbaum von Gartrop-Buhl, *Bonner Jahrbücher* (Koln) **161**, 300–7.
Startin, W. (1978) Linear pottery culture houses: reconstruction and manpower, *PPS* **44**, 143–59.
Steffy, J. R. (1982) Reconstruction of the 11th-century Serce Liman vessel, *IJNA* **11**, 13–34.
Steffy, J. R. (1984) Structural philosophy and technology-clues from shipwrecks in Coates, J. and McGrail, S. (eds) *Greek trireme of the 5th century* BC. NMM Greenwich.
Steffy, J. R. (1985) Kyrenia ship. *American J. of Archaeology* **89**, 71–101.
Steffy, J. R. (1994) *Wooden Shipbuilding and the Interpretation of Shipwrecks*. College Station, Texas A & M University
Stuart, J (1864/5) Notices of a group of artificial islands in the Loch of Dowalton, Wigtonshire, *PSAS* **6**, 114–78.

Taylor, E. G. R. (1971) *Haven-finding art*. Hollis and Carter.
Taylor, J. du P. and Cleere, H. (ed.) (1978) *Roman Shipping and Trade*. CBA Research Report 24.
Tchernia, A. *et al.* (1978) L'epave romaine de la Madrague de Giens (Var), *Gallia* 34th supplement, Paris.
Thesiger, W. (1978) *Marsh Arabs*. Penguin, Harmondsworth.
Thoen, H. (1978) *De Belgische Kustvlakte in de Romeinse Tijd*. Paleis der Academiën, Brussel.

Thomas, C. (1981) Lord of Goonhilly, *Nat. Trust Studies for 1980*. London.

Thorvildsen, K. (1957) *Ladby-Skibet*, København.

Throckmorton, P. and J. (1973) The Roman wreck at Pantano Longarini, *IJNA* **2**, 243–66.

Tilley A. F. and Fenwick, V. H. (1980) Ours is to reason why: iconography of ancient ships, *IJNA* **9**, 151–4.

Timmermann, G. (1953) Schiffskundliche Betrachtungen zum Hirschsprungboot, *Nordelbingen* **21**, 20–30.

Timmermann, G. (1956) Vom Einbaum zum Wikingerschiff, *Schiff und Hafen* **8**, 133–8.

Tinniswood, J. T. (1949) English galleys, 1272–1377, *MM* **35**, 276–315.

Tooley, M. J. (1990) Sea-level and coastline changes during the last 5000 years, in McGrail, S. (ed.) *Maritime Celts, Frisians & Saxons*, 1–16. CBA Research Report. 71.

Tozer. H. F. (1869) *Researches in the Highlands of Turkey* **1**.

Traeger, P. (1904) Schiffsfahrzeuge in Albanien u Macedonien, *Corr-Blatt d. Deutsch G. F. Anthrop.* **35**, 25–38.

Troels-Smith, J. (1946) Stammebaade fra Aamosen, *Nationalmuseets Arbejdsmark*, 15–23. København.

Tschan, F. J. (ed.) (1959) *History of the Archbishops of Hamburg Bremen.*

Ucelli, G. (1940) *Le Navi di Nemi*. Rome

Unger, R. W. (1980) *Ship in the Medieval Economy 600–1600*. Croom Helm, London.

Unger, R. W. (ed.) (1994) *Cogs, Caravels and Galleons*. Conway, London.

Ullenius, G. (1932–3) Ejstock eller Båtflotte? *Norbotten* 1932–3. Luleå, Sweden.

Upham, N. (1978) *Problem of protecting the underwater hull of the wooden vessel*. Occ. Tech. Paper No. 8. NMM Greenwich.

Upham, N. E. (1983) *Anchors*, Shire.

van der Heide, G. (1974) *Scheepsarcheologie*. Bussum, Strengholt.

Van der Walle, A. L. J. (1969) Some technical analogies between building and other crafts in the use of split wood during the Middle Ages, *Chateau Gaillard* **3**, 152–5.

van Doorninck, F. H. (1976) 4th-century wreck at *Yassi Ada*, *IJNA* **5**, 115–131.

Vebaek, C. L. and Thirslund, S. (1992) *Viking Compass*. Skjern, Denmark.

von der Porten, E. (1978) Mast and sail in Scandinavia in the Bronze Age? *MM* **64**, 284.

van Nouhuys (1928) Dug-outs, *MM* **14**, 274–6.

van Nouhuys, J. W. (1951) The anchor. *MM* 37: 17–47.

van Zeist, W. (1957) De Mesolithische boot van Pesse. Nieuwe Drentse Volksalmanak (1957), 4–11, Assen.

Viires, A. (1969) *Woodworking in Estonia* (trans. by J. Levitan from 1960 Tallinu edition.) published by Smithsonian Institute, Washington.

Vlek, R. 1987 *Medieval Utrecht Boat*. Oxford: BAR. S382

Wakeman, W. F. (1872/3) Curach, *J. Roy. Soc. Ant. of Ireland* 4th series Vol. 2, 74–6.

Wallace, P. F. (1982) Carpentry in Ireland AD 900–1300 – the Wood Quay evidence in McGrail, S. (ed.) (1982), 263–300.

Walton, K. (1974) A Geographer's view of the sea, *Scottish Geographical Magazine* **90**, 4–13.

Waters, D. (1978) *Art of Navigation in England in Elizabethan and Early Stuart Times*. NMM Greenwich.

Watkins, T. (1980) Prehistoric coracle in Fife, *IJNA* **9**, 277–86.

Waugh, F. W. (1919) Canadian Aboriginal Canoes, *Canadian Field Naturalist* **33**, 23–33.

Werner, W. (1973) Einbäume auf Österreichischen seen, *Das Logbuch* **9**, 43–50.

Westerdahl, C. (1980) On oral traditions and place names, *IJNA* **9**, 311–329.

Westerdahl, C. (1982a) Om Barkkanoter i Sverige, *Meddelanden* **5.1**, 24–25. Stockholm.

Westerdahl, C. (1982b) Nytt om Äskekärrsskeppet, *Meddelanden* **5.1**.

Westerdahl, C. (ed.) (1994) *Crossroads in Ancient Shipbuilding*. Oxbow Monograph 40, Oxford.

Wheeler. R. E. M. (1925) *Prehistoric and Roman Wales*. Oxford.

Whitcomb, N. R. (1968) Two prehistoric dugout canoes from the River at Hylton near Sunderland, Co. Durham, *Archaeologica Aeliana* **46**, 297–301.

Wicksteed, O. H. (1978) *Hardanger Faering*. NMM Greenwich Monograph No. 32.

Wilde, W. R. (1857) *A Descriptive Catalogue of the Antiquities of Stone, Earthen and Vegetable materials in the Royal Irish Academy*. Dublin.

Wilde, J. P. (1978) Cross-Channel trade and the textile industry in du Plat Taylor, J. and Cleere, H. (eds) (1978) 79–81.

Willis, D. D. K. (1922) Development from log to clipper, *MM* **8**, 66–72, 133–140, 268–276.

Willis, E. H. (1961) Marine transgression sequences in the English fenlands, *Annals of the New York Academy of Sciences* **95**, 368–76.

Wilson, D. (1851) *Archaeology and Prehistoric Annals of Scotland*.

Wilson, D. M. (1966) Medieval boat from Kentmere, Westmorland, *Med. Arch.* **10**, 81–8.

Wilson, D. M. (1968) Anglo-Saxon carpenters' tools in Claus, M. (ed.) *Studien zur Eurpä]ischen vor-und Frühgeschichte*, 143–50. Neumünster.

Wilson, J. (1972) *Fenland Barges*, Kettering.

Wilson, G. (1994) Computer documentation & analysis of historic ship design. In Westerdahl, (1994), 265–270.

Wood, J. E. (1978) *Sun, Moon and Standing Stones*. O.U.P.

Woodruff, C. E. (1895) *History of the town and port of Fordwich*. Canterbury.

Worcester, G. R. G. (1955) Footboat of Shaohsing, *MM* **41**, 222–6.

Worcester, G. R. G. (1956a) Notes on the Canoes of British Guiana, *MM* **42**, 249–251.

Worcester, G. R. G. (1956b) Origin and observance of the Dragon boat Festival in China, *MM* **42**, 127–37.

Worcester, G. R. G. (1966) *Sail and Sweep in China*. HMSO.

Worthington, E. B. (1933) Primitive craft of the Central African lakes, *MM* **19**, 146–63.

Wright, D. (ed.) (1957) *Beowulf*. Penguin, London.

Wright, E. V. (1976) *The North Ferriby Boats*. NMM Greenwich Monograph No. 23.

Wright, E. V. (1984) Practical experiments in boat stitching in McGrail, S. (ed.) (1984a), 57–84.

Wright, E.V. (1985) N. Ferriby boats – a revised basis for reconstruction, in McGrail, S. and Kentley, E. (eds) *Sewn Plank Boats*: 105–144. BAR Oxford S. 276.

Wright, E. V. (1994) *N. Ferriby boats – a final report*, in Westerdahl, (1994), 29–34.

Wright, E. V. and Switsur, V. R. (1993) Ferriby 5 boat fragments, *Archaeological Journal* **150**, 46–56.

Wright, E. (1990) *Ferriby Boats*. Routledge, London.

Zelenin, D. (1927) *Russische (Ostslavische) Volkskunde*, Berlin.

Anon (1982) Den tredje Näverbåten Lokaliserad, *Meddelanden* **5.4**, 43. Stockholm.

Index of sites: a guide to further reading

This index includes some works which were published after 1985 and which were therefore not in the first edition. See also the Preface to this edition.
Except where otherwise stated, sites are of a raft or boat find, or parts thereof. Where there is no recent definitive work, several references are given.

Index

Note: Italicized page references denote Figures.